T0201568

Geophysical Monograph Series

Including IUGG Volumes
Maurice Ewing Volumes
Mineral Physics Volumes

Geophysical Monograph 106

Faulting and Magmatism at Mid-Ocean Ridges

W. Roger Buck
Paul T. Delaney
Jeffrey A. Karson
Yves Lagabrielle

Editors

American Geophysical Union
Washington, D.C.

Published under the aegis of the AGU Books Board

Library of Congress Cataloging-in-Publication Data

Faulting and magmatism at mid-ocean ridges / W. Roger Buck . . . [et al.].
p. cm. -- (Geophysical monograph ; 106)
Includes bibliographical references.
ISBN 0-87590-089-5
1. Sea-floor spreading (Geology)   2. Faults (Geology)
3. Magmatism.   I. Buck, W. Roger, 1955-   . II. Series.
QE511.7.F38   1998
551.1'36--dc21   98-44135
CIP

ISBN 0-87590-089-5
ISSN 0065-8448

Cover. Perspective views of shaded bathymetric relief for sections of two mid-ocean ridges. The back cover shows an axial valley and flanking abyssal hills along an 80 kilometer long segment of the Southeast Indian Ridge. As shown by the arrow on the inset globe, the view is to the west from 116°E longitude. About 1500 m of relief is shown in the image. The front cover uses the same color scale for depths and shows an axial high along a 160 kilometer long segment of the East Pacific Rise. The overlapping spreading center in the foreground is at about 7°S latitude. The images were made by Bill Haxby using data collected during National Science Foundation-sponsored cruises led by Jim Cochran.

Copyright 1998 by the American Geophysical Union
2000 Florida Avenue, N.W.
Washington, DC 20009

Figures, tables, and short excerpts may be reprinted in scientific books and journals if the source is properly cited.

Authorization to photocopy items for internal or personal use, or the internal or personal use of specific clients, is granted by the American Geophysical Union for libraries and other users registered with the Copyright Clearance Center (CCC) Transactional Reporting Service, provided that the base fee of $1.50 per copy plus $0.35 per page is paid directly to CCC, 222 Rosewood Dr., Danvers, MA 01923. 0065-8448/98/$01.50+0.35.

This consent does not extend to other kinds of copying, such as copying for creating new collective works or for resale. The reproduction of multiple copies and the use of full articles or the use of extracts, including figures and tables, for commercial purposes requires permission from the American Geophysical Union.

Printed in the United States of America.

CONTENTS

PREFACE

Seafloor spreading at mid-ocean ridges was recognized more than 30 years ago as the key which unlocked the plate tectonic revolution. Ridges are not only the locus of the most voluminous magmatic activity on Earth, but they are also the largest and most active extensional tectonic regime on the planet. The abyssal hills of the seafloor, formed at mid-ocean ridges, are the most widespread morphologic features on the planet. Beneath these ridges, oceanic lithosphere forms nearly 60% of the volume of the Earth's rigid outer shell. As impressive as this may be, it is worth recalling that similar volumes have been created repeatedly in the cycle of spreading and subduction of oceanic tracts that characterizes plate tectonics.

This book, a mix of reviews and cutting-edge research, can be regarded as a progress report on the current understanding of faulting and magmatism at mid-ocean ridges. It provides an overview of what we know about the structure and constitution of mid-ocean ridge spreading centers, focusing on the fundamental processes occurring at mid-ocean ridges themselves rather than on the relation of ridges to plate tectonics. We are concerned with the forces that shape the topography of the spreading center, the processes that create the mafic igneous crust, the deformation and metamorphism that modify it, and the relationships among these parameters. Although faulting and magmatism are typical of many lithosphere plate boundaries, they are perhaps nowhere more dramatically and intimately linked than at mid-ocean ridges. Whereas seafloor exposures of oceanic lithosphere and ophiolite complexes afford views of the end-result of these interactions, studies of mid-ocean ridge axes reveal ongoing activity. One of the great advantages of studying magmatic and tectonic processes at ridges is that, unlike analogous continental terranes, they do not have a long and complex geologic history. Furthermore the features seen at ridges are (by geological standards) rapidly evolving works in progress. We can dive on a ridge axis and see a fissure that was not there just a few months before. We can detect earthquakes and see the surface faults that record their displacements. We may soon be able to continually record the movement of faults and magma beneath ridges in real time, thereby "taking the pulse" of spreading centers. Perhaps the most important goal in studying these magmatically and tectonically active areas is to learn about the fundamental principles of faulting and magmatism. For example, if we are to understand the behavior of faults in continental areas, where pre-existing structures and complex erosional processes obscure the tectonic picture, a logical reference point would be the arguably more simple case of faulting at ridges. Ridges are a good place to try to understand dynamic processes and properties of Earth materials because their structure varies so much along different sections of the global ridge system. Ridge topography and structure appear to vary systematically, but not smoothly, as functions of the rate of plate spreading and the magma supply. We might regard various parts of the mid-ocean ridge system as a grand series of ongoing experiments in which the fundamental parameters such as spreading rate, magma supply, and plate boundary geometry vary from place to place.

The volume grew out of a meeting of about 160 scientists and students who focused on many of the issues presented here. The meeting was organized under the auspices of the U.S. Ridge Inter Disciplinary Global Experiments (RIDGE) program of the National Science Foundation, and was held at Lake Tahoe in June 1995. That meeting was the fourth RIDGE Theoretical Institute and, as in past institutes, it combined aspects of a school in covering basic well-known features, as well as a symposium in presenting the latest data and model results.

The editors thank Bob Detrick, Susan Humphris and Anita Norton for advice and help in planning the Lake Tahoe meeting and in organizing this book. We gratefully acknowledge a grant from the Ocean Sciences Division of the National Science Foundation (OCE 9415539 to Bob Detrick) which paid for the meeting. Bill Haxby kindly made the cover images. We also thank the reviewers of these papers, who made many excellent suggestions to improve the final volume.

W. Roger Buck
Lamont-Doherty Earth Observatory, Palisades, NY

Paul T. Delaney
U.S. Geological Survey, Flagstaff, AZ

Jeffrey A. Karson
Duke University, Durham, NC

Yves Lagabrielle
CNRS, Plouzanne, France

Editors

Global Systematics of Mid-Ocean Ridge Morphology

Christopher Small

Lamont-Doherty Earth Observatory, Columbia University, Palisades, NY

Global mid-ocean ridge morphology is characterized by a dichotomy in which slow spreading centers have discontinuous axial valleys and rugged flanking morphology while fast spreading centers have more continuous axial rises and smoother flanking morphology. This study investigates the relationship between axial and flanking morphology and the nature of the dichotomy. Both axial and flanking morphology show similar variations with spreading rate but the transition between axial valley and axial rise morphology suggests that the amplitude of the axial morphology is not the sole determinant of flanking roughness. At slower spreading rates, ridge flank roughness is moderately correlated with axial valley relief, but at intermediate spreading rates significant seafloor deformation can occur without the formation of an axial rise or axial valley. At faster spreading rates, well developed axial rises have smoother flanking morphology and show no correlation between axial relief and flanking roughness. This reflects a fundamental difference in the mechanisms of lithospheric deformation as well as the importance of extrusive volcanism at high spreading rates and thermal environments where an axial magma chamber can be maintained. Axial valley spreading centers exhibit a spreading rate dependence in both amplitude and variability of axial relief, asymmetry, flanking roughness and depth but spreading centers with axial rises show much less variability and no spreading rate dependence for any observable above intermediate (~70 km/Ma) spreading rates. The variability of axial valley environments at a given spreading rate results primarily from intrasegment structure but may also reflect temporal episodicity of magmatic emplacement and extension. The dichotomy, which is also manifest in the subsurface structure of spreading centers, suggests a transition from episodicity and rate dependence at slow spreading rates to a metastable mode of lithospheric accretion at faster spreading rates. The most plausible mechanism for the transition is a stabilization of mid-ocean ridge thermal structure above a critical spreading rate.

INTRODUCTION

Bathymetric measurements at mid-ocean ridges provide three complementary types of information: the actual depth below sea level, the characteristic axial morphology and the smaller scale roughness superimposed on the axis and ridge flanks. The relationships between the different components of morphology are controlled by the processes that form and deform the lithosphere. The tectonic complexity of most ridge flanks precludes simple generalizations relating axial and flanking morphology but global patterns are observed in both types of morphology. The global systematics of these patterns reveal a structure to the system that is not always obvious in studies of individual spreading centers.

Faulting and Magmatism at Mid-Ocean Ridges
Geophysical Monograph 106
Copyright 1998 by the American Geophysical Union

Until recently, our understanding of mid-ocean ridges was based almost entirely on detailed studies of the northern Mid-Atlantic Ridge and the northern East Pacific Rise. These spreading centers conveniently characterize the diverse end members of the global system and offer complementary views of the processes of lithospheric accretion, but to understand how these processes operate globally it is necessary to consider the entire system of mid-ocean ridges. By investigating the relationship between the consistently observed characteristic features of mid-ocean ridges and the variability in the morphology that changes from one location to the another it may be possible to better understand the dynamics of the entire system. The purpose of this study is to quantify the global systematics and to provide constraints on the processes of lithospheric deformation at mid-ocean ridge plate boundaries.

One of the earliest observations of seafloor structure was a dichotomy between the axial morphology of the Mid-Atlantic Ridge and the East Pacific Rise [Heezen, 1960, Menard, 1960]. Further exploration of the global system revealed that slow spreading mid-ocean ridges are generally characterized by deep axial valleys and uplifted flanks while faster spreading mid-ocean ridges have narrow axial rises [e.g. Macdonald, 1982]. It was shown by Macdonald [1986] that the relief of axial valleys tends to diminish with increasing spreading rate and Small and Sandwell [1989] showed that this spreading rate dependence of axial valley structure is contrasted by an insensitivity of axial rise structure to changes in spreading rate. The dichotomy in axial morphology has prompted extensive modeling efforts directed at understanding the spreading rate dependence of ridge axis dynamics (see Poliakov and Buck [this volume] for a review) but less attention has been paid to the actual transition from one morphology to the other. This is largely a result of the fact that most of the intermediate spreading rate ridges that characterize this transition had not been adequately surveyed until recently.

In addition to the dichotomy in axial morphology, it has been similarly observed that the flanking seafloor created at slow spreading centers is generally more rugged than that created at fast spreading centers [Menard, 1967]. Previous analyses of seafloor "roughness" have attempted to quantify this observation but reach no consensus as to whether the spreading rate dependence is discontinuous [Small et al 1989], a power law [Malinverno, 1991], or linear [Hayes and Kane 1991]. The distinction is important because it places a fundamental constraint on the processes that create the morphology of the seafloor and how these processes operate on a global scale. Over long time scales, seafloor roughness is strongly affected by migration of ridge axis offsets and off-axis volcanism but stochastic analyses of multibeam bathymetry [e.g. Goff, 1991; Goff et al, 1997] suggest that some characteristics of abyssal hill morphology vary with spreading rate. The apparent correlation between ridge axis morphology and ridge flank roughness suggests a causal relationship between the two.

Variations in mid-ocean ridge dynamics and structure are generally discussed with respect to spreading rate but "anomalous" spreading centers such as the Reykjanes Ridge and the Australian Antarctic Discordance (AAD) suggest that these variations are not dependent strictly on spreading rate but rather on temperature. Mid-ocean ridges are often parameterized by spreading rate because it kinematically exerts a first order control on both the rate of deformation and the regional thermal structure of the plate boundary. Spreading rate is also one of the few controlling parameters of mid-ocean ridge dynamics that can be accurately determined on a global basis.

This study establishes global systematics for spreading center depth, axial morphology, and near axis topographic roughness using a collection of 228 bathymetric profiles from the global mid-ocean ridge system (Figure 1). Spreading center depths discussed in this study include both zero-age depth and mean profile depth. When referring to axial morphology, profiles on which the zero-age depth is greater than the mean profile depth will be categorized as axial valleys and the term axial rise will be used for profiles on which the zero-age depth is less than the mean profile depth. The finer scale morphology of the ridge axis and flanks, referred to here as roughness, will be described statistically as discussed below.

The justification for using profile data is the widespread availability over the global mid-ocean ridge system which is not yet available from multibeam and sidescan sonar datasets. The trade-off is between local detail and global coverage. A number of studies have used the two dimensional coverage offered by multibeam swath bathymetry data and sidescan sonar imagery to study specific processes such as segmentation evolution [e.g. Carbotte et al, 1991; Gente et al. 1992]. and abyssal hill formation [e.g. Goff and Jordan, 1988; Edwards et al, 1991; Goff, 1991] but profile data can provide a complementary global view of the system. Studies based on profile data alone should however be interpreted in light of more detailed local studies. A global analysis of profile data may establish the first order relationships between axial and flanking morphology to which more detailed local studies may be related.

SEAFLOOR ROUGHNESS AND LITHOSPHERIC DEFORMATION

Mechanisms

In general, the morphology of a mid-ocean ridge flank results from a combination of abyssal hills, volcanic edifices and scars left by migrating offsets of the ridge axis. Both offset propagation [e.g. Hey et al, 1977; Macdonald, 1988] and seamount production [e.g. Batiza, 1982; Epp and Smoot, 1989] are widespread and appear to be influenced by spreading rate, but neither process is as ubiquitous as abyssal hill formation. The creation of

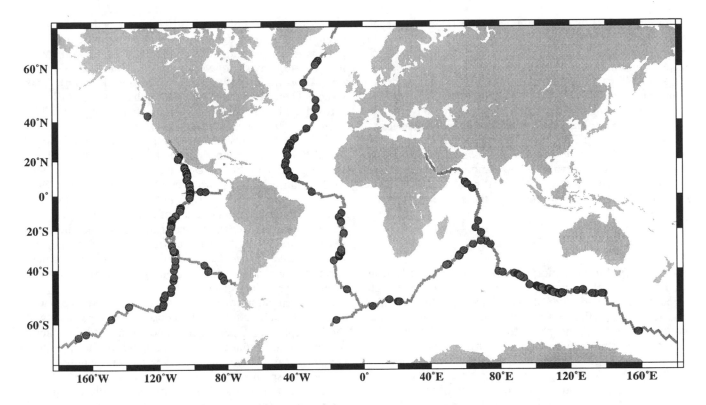

Figure 1. The global mid-ocean ridge system and the geographic distribution of the 228 bathymetric profiles used in this study. All profiles extend at least 40 km on each side of the ridge, cross the ridge within 45° of perpendicular and do not intersect known transforms or fracture zones.

abyssal hills at spreading centers with axial valleys is dominated by faulting [e.g. *Macdonald and Atwater*, 1978; *Macdonald*, 1982; *Shaw and Lin*, 1993] while abyssal hills formed near axial rises are believed to result from a combination of faulting and constructional volcanism [e.g. *Lonsdale*, 1977; *Lewis*, 1979; *Macdonald and Luyendyk*, 1985; *Macdonald et al*, 1996]. It is often assumed that there is a genetic relationship between the ridge flank roughness and the morphology of the spreading center at the time the flanking lithosphere was created [e.g. *Marks and Stock*, 1994; *Sahabi et al*, 1996]. This assumption is impossible to test since current rates of plate motion preclude direct observation of the movement of axial lithosphere to the ridge flank. It is however possible to systematically investigate the relationship between present day axial morphology and near axis roughness in order to understand how axial morphology contributes to lithospheric deformation.

Along axis migration of spreading center offsets also contributes to ridge flank roughness. In past studies, ridge flank roughness has often been assumed to be synonymous with abyssal hill topography but recognition of the importance of small offset propagation [*Macdonald*, 1988; *Carbotte et al.*, 1991; *Grindlay et al.*, 1991] suggests that some fraction of the topographic relief on ridge flanks is

related to scars left by migration of axial discontinuities throughout the history of the ridge segment. Propagation of an offset through ridge flank lithosphere generally produces pseudofaults that are distinct from abyssal hills in both their amplitude and orientation and may represent different deformational mechanisms than those forming abyssal hills at the ridge axis. Detailed ridge flank surveys [e.g. *Gente et al*, 1992; *Tucholke et al*, 1997] reveal that, in addition to the smaller scale abyssal hill topography, there is a larger scale pattern of semi-continuous deeps amid larger (10's of kms) shoals. The continuity of the deeps, and their relation to present day segmentation, suggests that they are created by migrating offsets. Since the majority of ridge axis offsets are shorter than 40 km [*Ludwig and Small*, 1997], it can be inferred that the lithosphere into which the offsets propagate is relatively young but not necessarily within the neovolcanic zone. Thus, the roughness observed on the ridge flanks results from a combination of two distinct types of deformation in addition to some component of constructional volcanism.

Seamount volcanism also contributes to the ridge flank morphology but the variation in roughness discussed here is probably dominated by abyssal hills and propagating offsets. Studies of seamount populations in the Atlantic Ocean indicate that off axis constructional volcanism (seamounts >100 m high) is generally related to hotspots [*Epp and Smoot*, 1989] and that seamounts formed near the axis of slow spreading centers are generally smaller (< 150 m, [*Smith and Cann*, 1992]) than the average abyssal hill height (>140 m, [*Goff et al*, 1995]). Seamounts formed near axial rises tend to be larger than the surrounding abyssal hills but cover only a small percentage of the seafloor. *Scheirer and Macdonald* [1995] find that near axis seamounts on the EPR (8°N - 17°N) have a characteristic height of 240 m but cover only ~6% of the surveyed seafloor. These observations of seamount abundance suggest that constructional volcanic edifices have a negligible effect on the average roughness of the seafloor formed at axial rises.

Controlling Parameters

If near-axis flanking morphology results primarily from the deformational processes at the ridge axis then we might expect the form and relief of the axial morphology to affect the roughness of the seafloor created there. The physical process of moving lithosphere from the axis of a spreading center to the flanks requires that it traverse the axial morphology and suggests that it is deformed in the process. Similarly, the morphology of propagating offsets also appears to be related to the axial morphology of the spreading center [e.g. *Macdonald et al*, 1988; *Gente et al*, 1992]. It stands to reason that the morphology of the ridge flanks may be related to the form and relief of the axial morphology. The degree of correlation between the present day axial morphology and recently created flanking morphology may provide a constraint on the degree of temporal variability of axial morphology and the extent to which ridge crest morphology is a steady state feature of the plate boundary. Temporal variations in the mechanism(s) that create axial topography may result in spatial variations in flanking roughness thereby weakening any global correlation between present day morphology and time averaged roughness; a strong correlation would suggest otherwise.

The roughness of near axis flanking topography should be controlled, in part, by the extensional strength of the lithosphere. Stronger lithosphere will withstand greater stress before undergoing brittle failure [e.g. *Huang and Solomon*, 1988], thereby creating and maintaining greater relief than weaker lithosphere. Extensional strength of the lithosphere is modulated by both temperature and crustal thickness. Higher temperatures at the ridge axis weaken lithosphere by reducing its ductile yield strength [e.g. *Brace and Kohlstedt*, 1980], but also by producing thicker crust [e.g. *Klein and Langmuir*, 1987]. Since oceanic crust is rheologically weaker than oceanic mantle, the net effect of crustal thickening is to reduce the overall strength of the lithosphere. Oceanic lithosphere strengthens monotonically with age as a result of cooling [*McNutt and Menard*, 1982] but the difference in crust and mantle rheologies delays rapid strengthening until the mantle has cooled sufficiently

to undergo brittle deformation. The combined effect of the strengthening with age and the basal stress field on the plate is to concentrate lithospheric deformation in the area near the ridge axis [*Tapponier and Francheteau*, 1978; *Lin and Parmentier*, 1989; *Chen and Morgan*, 1990a,b]. This is consistent with the observation that abyssal hills are formed within several kilometers of the ridge axis [*Le Pichon*, 1969; *Edwards et al*, 1991].

Variations in lithospheric strength also result from along-axis differences in temperature and crustal thickness related to ridge segmentation. Detailed studies of the northern Mid-Atlantic Ridge [*Shaw*, 1992; *Shaw and Lin*, 1993; *Tucholke and Lin*, 1994; *Goff et al*, 1995; *Shaw and Lin*, 1996] highlight the importance of intra-segment variations in lithospheric strength to deformation within the axial valley. Studies on the East Pacific Rise [*Goff et al*, 1991; *Carbotte and Macdonald*, 1994b] also find intra-segment variations in abyssal hill morphology and tectonic fabric at axial rise spreading centers.

Higher temperatures and thicker crust conspire to produce weaker lithosphere but also to reduce the average density of the lithosphere. A persistent reduction in net lithospheric density should reduce the regional depth of the spreading center at scales larger than the flexural wavelengths of the elastic lithosphere. Because the number of seismic measurements of crustal thickness is extremely limited, spreading center depths are sometimes used as a proxy for crustal thickness or thermal structure when no other data are available [e.g. *Klein and Langmuir*, 1987]. The validity of this assumption depends on the compensation of the lithosphere and the importance of dynamic topography at the ridge axis [*Neumann and Forsyth*, 1993]. Seafloor depths increase predictably with age as a result of cooling and thermal contraction [*Parsons and Sclater*, 1977] but flanking and zero-age depths vary considerably on a global basis [*Marty and Cazenave*, 1989]. Since the conductive cooling model does not account for the complexities of ridge axis dynamics, we would not expect thermal subsidence to dominate until the lithosphere has moved some distance away from the spreading center. Some part of the variation in spreading center depth is also expected to be dynamic [e.g. *Sleep and Biehler*, 1970; *Parmentier and Forsyth*, 1985]. It is reasonable to ask whether some part of the variability in regional depth may be related to variations in the thermal and compositional state of the lithosphere and whether these variations are reflected in its deformation. We might therefore expect some global relationship between spreading center depth and near axis roughness.

DETERMINISTIC AND STOCHASTIC SEAFLOOR MORPHOLOGY

Modeling in geodynamics is generally based on deterministic phenomena. This involves measuring a deterministic signal, constructing a physical model to explain it and then finding the parameters of that model which best describe the observations. This methodology has been used to study phenomena that control the evolution of oceanic lithosphere such as the thermal subsidence [e.g. *Parsons and Sclater,* 1977] and elastic flexure [e.g. *Turcotte,* 1979]. Less attention has been devoted to the understanding of the stochastic component [e.g. *Fox and Hayes,* 1985] although some studies [e.g. *Goff and Jordan,* 1988; *Malinverno and Gilbert,* 1989; *Shaw and Smith,* 1990; *Neumann and Forsyth,* 1995] have developed statistical models to quantify abyssal hill morphology. One objective of this study is to quantify the relationship between the "deterministic" and "stochastic" components of spreading center morphology in order to understand the extent to which the dynamics at the ridge axis influence the stochastic morphology preserved on the flanks. The extent to which the time averaged topography of the flanks can be correlated to the present day axial morphology may provide constraints on the temporal variability of the mechanism(s) of lithospheric deformation.

The expressions deterministic and stochastic are not used here in the strict statistical sense but rather to distinguish between the characteristic components of mid-ocean ridge axial morphology and the smaller scale relief that is superimposed on it. The former is assumed to be a direct manifestation of the deterministic continuum process(es) of lithospheric accretion while the latter is a consequence of the stochastic variations in lithospheric response to the continuum processes. The actual form of the deterministic component at a particular location may eventually be predicted by a physical model whereas the stochastic component can only be described and predicted statistically.

In this analysis, a primary objective will be to separate the characteristic features in the axial morphology from the smaller scale tectonic fabric and quantify the relationship between the two. The across axis bathymetry of the seafloor near a spreading center may thus be envisioned as a sum of a regional mean depth, Z_{avg}, a deterministic component, $\mathcal{D}(x)$, characteristic of all spreading centers and a stochastic component, $S(x)$, unique to that particular location. Specifically, the deterministic component is the characteristic axial rise or axial valley relief and the stochastic component is the collection of individual fault scarps and constructional features that have formed at or near the ridge axis and been transported to the flanks. If these components can be separated then each may be analyzed independently and the relationship between them can be investigated.

EMPIRICAL ORTHOGONAL FUNCTION ANALYSIS

This study uses an Empirical Orthogonal Function (EOF) analysis to quantify axial and flanking morphology in the set of bathymetric profiles described below. The theory of application of EOF's to topographic fields is discussed in detail by *Small* [1994] and will be described only briefly here. The objective of this analysis is to find the components of ridge axis morphology which are common to all spreading centers and to separate these components from those which vary from location to location. The EOF analysis allows us to do this because it decomposes the entire collection of profiles into a common set of independent spatial modes (empirical basis functions) and determines the amount of variance associated with each mode. This variance partition establishes the distinction between the different components of the morphology. It is important to point out that the decomposition upon which the EOF analysis is based is not dependent on any *apriori* assumptions about the structure of the dataset. The spatial modes are eigenvectors of the dataset and are determined subject to the constraint that the maximum amount of variance in the original dataset be described with the smallest number of basis functions possible [*Davis,* 1976].

The EOF decomposition allows an individual bathymetric profile to be described as a linear combination of independent spatial modes (empirical basis functions) as

$$Z_{i0}(x) = c_{1i}\mathcal{M}_1(x) + c_{2i}\mathcal{M}_2(x) + ... c_{ri}\mathcal{M}_r(x) + S_i(x)$$

where

$Z_{i0}(x) = Z_i(x) - Z_{avg}$, is a zero-mean bathymetric profile i, a function of distance x.

$\mathcal{M}_1(x), \mathcal{M}_2(x)...\mathcal{M}_r(x)$ are the spatial modes of the entire dataset.

$c_{1i}, c_{2i} ... c_{ri}$ are the coefficients of the modes for the profile i.

$S_i(x)$ is the linear combination of the remaining modes for profile i.

r is the number of significant modes.

With this decomposition, any profile may be partially reconstructed as a linear combination of the spatial modes. If we assume that the sum of the significant modes represents the deterministic component of the system then the remainder is stochastic and the representation above reduces to $Z_0(x) = \mathcal{D}(x) + S(x)$. Eigenvalue decompositions have been used investigate deterministic physical processes in dynamical systems with stochastic variability [e.g. *Davis,* 1976; *Priesendorffer,* 1988; *Vautard and Ghil,* 1989]. but here the analysis is limited to decomposition of the dataset. It is beyond the scope of this study to model the deterministic processes responsible for the formation of ridge axis morphology.

The number of significant modes, r, may be inferred from the spectrum of singular values calculated in the analysis. These singular values are the roots of the eigenvalues of the dataset and give the total percentage of variance that is accounted for by each mode. The distribution of variance given by these singular values provides some basis on which to distinguish between the deterministic and stochastic components of the dataset. In some instances, the dimension of a dynamical system can be estimated by the number of non-zero singular values, each of which corresponds to an eigenfunction of the system [*Vautard and Ghil,* 1989]. The variance related to the stochastic components of the system is distributed over the remaining low-order eigenvectors and the corresponding singular values are smaller but generally non-zero. Because the stochastic component is spatially incoherent from one profile to the next, its variance is distributed over the larger number of low order modes. A clear separation in the size of the singular values allows the components to be distinguished. In this analysis the spectrum of singular values is used to distinguish between the components of the morphology that are consistently present on cross axis profiles from the component which differs from profile to profile.

The representation of ridge axis morphology with EOFs is conceptually similar to the commonly used Fourier representation of geophysical data but has a number of advantages. The spectrum of singular values is analogous to the traditional power spectrum in that both provide estimates of the distribution of variance over a space of basis functions; the difference between the two types of spectra is in the choice of basis functions. A Fourier decomposition represents the signal as a linear combination of analytic sinusoids whereas the EOF decomposition uses empirically determined basis functions. Sinusoids are analytically tractable and offer a natural basis for periodic signals but EOFs are optimum for non-periodic data because they require no assumptions about the form of the basis functions and they are optimized for efficiency rather than analytic tractability. The basis functions used in an EOF analysis are optimum in the sense that they account for as much variance as possible with the smallest number of orthogonal functions. Since periodicity is not built into the basis, no assumptions of statistical stationarity are required in EOF analysis. This is an important factor when dealing with seafloor bathymetry which is generally non-periodic and clearly non-stationary. The EOF decomposition also provides some basis for discriminating between deterministic and stochastic components in the dataset [*Priesendorffer,* 1988; *Vautard and Ghil,* 1989] whereas Fourier analysis of seafloor bathymetry rarely provides an unambiguous spectral separation of these components. The EOF decomposition also allows us to separate the symmetric and antisymmetric parts of the deterministic component of the topography and analyze each separately [*Small,* 1994].

THE DATASET

The analysis presented here is based on a compilation of 228 archival underway profiles (Figure 1) ranging from a PDR record collected in 1959 to Seabeam 2000 centerbeam profiles collected in 1995. Quality control and the logistical requirements of the analysis severely limit the number of usable profiles. In this study, profiles must 1) extend at least 40 km on either side of the ridge axis, 2) cross the axis within 45° of perpendicular, 3) have an average along track sample rate of 1 km or better, and 4) do not intersect known transforms or fracture zones within 40 km of the axis. These requirements were chosen to maximize the number of usable profiles without compromising the resolution of the features that the study seeks to quantify. The basis of this dataset is a set of profiles used in an analysis of axial morphology by *Small* [1994]; the reader is referred to that study for a detailed description of the dataset construction.

This investigation builds upon the analysis of *Small* [1994] but extends the previous study in two ways. First, a considerably larger dataset is used here; the previous study used 156 profiles whereas this study uses 228 profiles. The previous study was limited by a lack of data in many parts of the Southern Ocean - notably the Southeast and Southwest Indian Ridges. Recent surveys of the Southeast Indian Ridge provide continuous coverage of ~2300 km of intermediate spreading rate ridge axis which spans almost the entire global range of axial and flanking morphologies [*Cochran et al,* 1997; *Sempere et al,* 1997]. Also, recent availability of global satellite gravity data [*Sandwell and Smith,* 1997] has allowed the location of many unmapped ridge segments in the Southern Ocean to be determined precisely [*Ludwig and Small,* 1997] so that existing archival profile data may now be used in areas that were poorly understood at the time of the previous study. This has resulted in improved coverage of the Southwest Indian Ridge which is important to constrain the slow end of the spreading rate spectrum. This analysis also includes data from the AAD and Reykjanes Ridge which were not used in the previous study. The second major difference between this and the previous study is that this study focuses specifically on the relationship between axial and flanking morphology whereas the previous study was primarily concerned with variations in axial morphology alone and included only a very limited discussion of the relationship between axial morphology and flanking roughness.

RESULTS AND OBSERVATIONS

Separation of Axial Morphology and Flanking Roughness

The result of the EOF analysis of the 228 profiles is the spectrum of singular values and spatial modes shown in Figure 2. Because the first five singular values are noticeably larger than the others and because the spatial modes corresponding to these five singular values resemble recognizable features of ridge axis morphology, these modes are interpreted as the deterministic component of the dataset. The higher order modes bear no resemblance to ridge axis morphology but rather resemble sinusoids of continuously decreasing wavelength. The first five modes account for 44% of the total variance in the dataset and bear an obvious

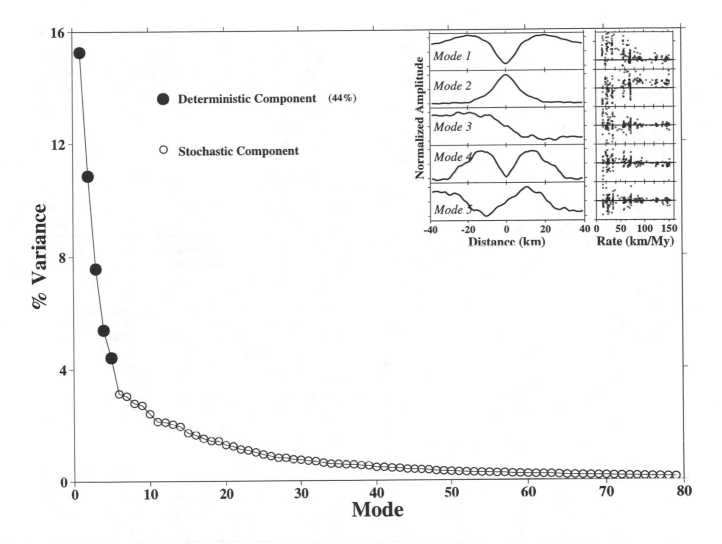

Figure 2. Empirical Orthogonal Function decomposition of the global dataset. The singular values indicate the distribution of variance over the spatial modes given by the decomposition. The spatial modes are common to the entire dataset; the coefficients for each mode correspond to individual profiles. The five largest singular values are interpreted as the deterministic component of the topography and the continuum of smaller singular values represent the stochastic component. The five spatial modes corresponding to the largest singular values and their coefficients for each profile are shown in the insets. These five modes account for 44% of the variance in the dataset.

resemblance to the axial valleys and rises that characterize mid-ocean ridge axial morphology. The component of the topography described by the continuum of higher order modes is considered stochastic and accounts for the remaining variance in the system. Three of the five primary modes are symmetric and two are antisymmetric so it is possible to consider the asymmetry ridge axis morphology separately. The modes estimated from this dataset are effectively identical to those estimated from the subset used by *Small* [1994]. It was shown by *Small* [1994] that the majority of the variance in this system is associated with the high relief of axial valley spreading centers but that the

decomposition provides a robust description of the entire dataset. The deterministic component of the each individual profile may be reconstructed as a linear combination of the first five spatial modes as shown in Figure 3. The stochastic component is a linear combination of the remaining modes and forms a residual profile from which the estimates of roughness are derived.

The deterministic and stochastic components of several example profiles are shown in Figure 4. The deterministic component is superimposed on the original bathymetric profile so it is apparent that the basic shape of the axial topography is present in the deterministic component in each

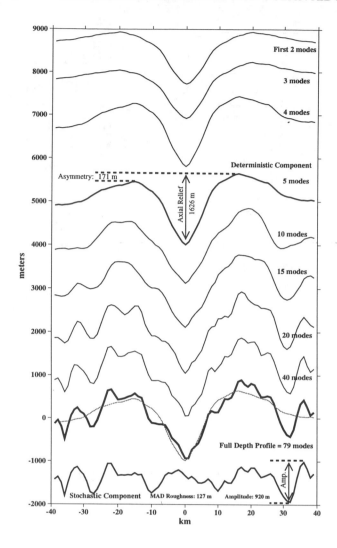

Figure 3. Example of a progressive reconstruction of a single bathymetric profile across the Mid-Atlantic Ridge from its spatial modes. Each plot is a linear combination of an increasing number of the lowest order modes. In the additive model illustrated here, the first five modes are considered the deterministic component of the topography $\mathcal{D}(x)$, the remainder (bottom) forms the stochastic component $S(x)$, and the observed topography is the sum, $Z_0(x) = \mathcal{D}(x) + S(x)$.

case. The residual profiles, shown above each example, illustrate that the variance of the stochastic component of the topography is distributed evenly along the length of the profile and does not result from misfit of the axial morphology. The greatest discrepancy between the bathymetric profile and its deterministic component is seen on some axial rises which are either very narrow or contain a large axial summit caldera. In the case of spreading centers with no discernible axial morphology the deterministic

component shows very little relief and almost all of the variance resides in the stochastic component. Two of the profiles across the Reykjanes Ridge have several hundred meter deep grabens superimposed on a broad (60 km) swell. The polarity of the coefficients for these profiles categorize them as axial valleys because the axial swell they are superimposed upon is so broad.

Axial Relief and Width Estimates

A number of parameters can be estimated from the deterministic and stochastic components of each profile. In this study, axial relief is estimated as the range (maximum - minimum) of the deterministic profile; this is always the difference between the center of the axial valley and the higher of the flanks or the difference between the center of the axial rise and the lower of the flanks. Asymmetry is estimated as the range of the asymmetric component of the deterministic profile (mode 3 + mode 5) and is the difference of the flanking depths. The asymmetry values calculated in this study agree closely with those estimated by *Severinghaus and Macdonald* [1988]. The existence of an axial valley or an axial rise may be determined either visually or by the sign of the sum of the coefficients of the symmetric deterministic modes. Axial valley width is measured as the horizontal distance between the shallowest points (maxima) on each side of the deterministic profile.

Roughness Estimates

The statistical characterization of the stochastic topography discussed here is based on the frequency distribution of the roughness profiles derived from the EOF analysis. The dispersion (often called variance) of these profiles as an indicator of the amount of deformation contributed by the processes responsible for the formation of the stochastic component of the topography.

Median Absolute Deviation . One measure of roughness, the Median Absolute Deviation (MAD), is the median of the absolute differences between each point in the residual profile and the median value of the profile [*Rice,* 1988]. The MAD is analogous to the commonly used Root Mean Square (RMS) but is more robust in the presence of extreme values such as those resulting from off axis seamounts. The roughness values computed using this method are very similar to the RMS heights of abyssal hills calculated by *Goff* [1991]. Since the roughness profile provides separate estimates of the stochastic topography on each side of the ridge, the MAD is estimated for each half profile in the dataset. Estimates such as RMS and MAD are indicators of the average variance of the stochastic profile but they consistently underestimate the actual relief of the features that make up the roughness profile (Figure 3). In order to compare the stochastic topography to the

Axial Valleys 8x V.E.   Axial Rises 16x V.E.

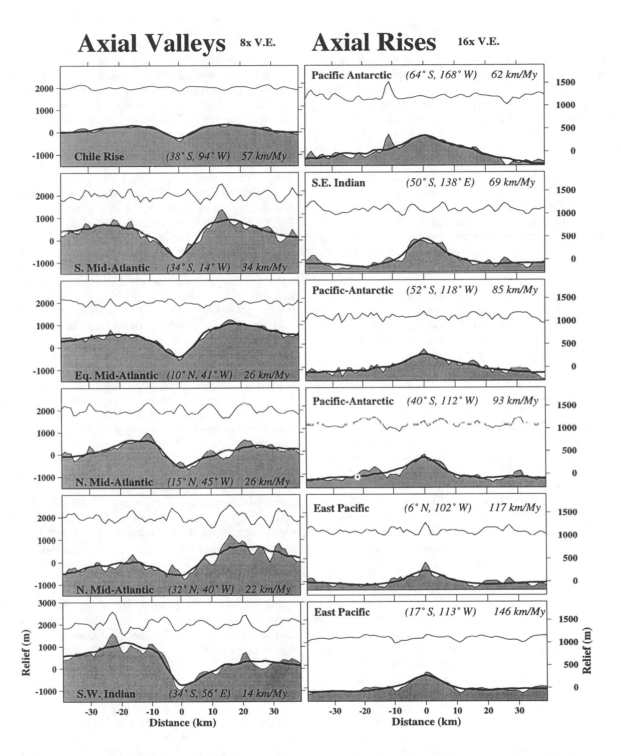

Figure 4. Examples of decomposed profiles for the range of spreading rates and morphologies. The heavy curve is the deterministic component of the topography, $\mathcal{D}(x)$, superimposed on the observed bathymetry $\mathcal{Z}_0(x)$. The stochastic component, $S(x)$, is plotted above each profile. Axial rises and axial valleys are shown at different vertical exaggerations for clarity.

axial topography it is necessary to use an estimate that more accurately reflects the actual relief of the roughness profiles.

Roughness Amplitude. Visual inspection indicates that the extreme (max and min) values on any given stochastic profile are almost always related to one individual feature (Figure 4). These extrema generally appear to be single or nested fault scarps. Although the extrema usually correspond to a single feature, these features are rarely much larger than the other features on the profile. This is reflected by the fact that the frequency distributions of the stochastic profiles are "short tailed". Isolated extrema, such as seamounts, produce long tailed distributions and occur only rarely in this dataset. The extrema may therefore provide an additional estimate of the amplitude of the roughness profile which can be directly compared with the axial relief measurements. It is important however to verify that the extrema are representative of the entire roughness profile.

The difference between the extrema of the stochastic profiles is referred to here as the roughness amplitude. If the stochastic profiles are dominated by short wavelength variance then the amplitude will provide an estimate of the small scale relief generated by the deformation process. The shorter tailed the distribution, the more representative the estimate. A spectral analysis indicates that the majority of the variance in the stochastic profiles occurs at length scales significantly shorter than the half profile length of 40 km. The fact that the stochastic profiles are essentially zero mean (average = 2.2 m) with zero slope (average = 0.2 m/km) supports this observation. Therefore, it is possible to estimate the size of the larger features on the stochastic profile by measuring the difference between the maximum and minimum values of the half profiles. If the profiles are sufficiently short tailed then the extrema are representative of the average relief of the profile. This can be determined using the order statistics of the stochastic profiles. The extrema width is the horizontal distance between the maximum and minimum values of each half profile. Smaller extrema widths support the assertion. The median extrema width for all 456 of the half profiles is 10 km and 83% of the extrema widths are less than 20 km. Only one of the extrema widths longer than 20 km corresponds to a roughness amplitude greater than 1000 m and 75% of the extrema widths longer than 20 km correspond to roughness amplitudes less than 500 m. This suggests that the roughness amplitude of each half profile may provide a reliable estimate of the largest single relief producing feature on each side of the ridge axis. Roughness amplitudes of the half profiles calculated as described above correlate strongly with the MAD roughness estimates (Table 1). Some fraction of the throw of individual faults on the axial valley walls is represented in the relief of the deterministic profile so the roughness amplitudes measured from the stochastic profiles do not represent the actual fault throw. The usefulness of the roughness amplitude is that it provides an

additional parameterization of the amplitude of the flanking morphology that may be compared directly with the axial relief measurements.

The roughness estimates discussed here are not necessarily equivalent to abyssal hill topography. Traces of migrating offsets also contribute to the roughness estimates. Without the full two dimensional coverage provided by multibeam data it is not possible to uniquely determine the origin of the stochastic topography. It is important to note however that the MAD roughness estimates of this study are effectively the same amplitude as RMS abyssal hill heights derived from stochastic analyses [*Goff*, 1991, 1992] suggesting that much of the stochastic variability does result from abyssal hills. Some of the profiles included in this analysis cross seamounts but these are isolated features that have a minimal influence on the MAD operator used to estimate roughness from the stochastic profiles. The 1 km sample spacing used in this study is near the limit of resolution for some Pacific abyssal hills but in most cases is adequate to resolve abyssal hill relief. Stochastic analyses [*Goff*, 1991; *Goff et al*, 1995] find that the mean peak to peak characteristic widths of abyssal hills range from 2 km to 14 km but is usually at least 4 km.

In this study roughness is determined only for regions within 40 km of the ridge axis. The majority of this topography presumably results from faulting related to the deformation and migration of lithosphere. Since profiles crossing the ridge at or near first order offsets were eliminated from the dataset, only the scars of smaller propagating offsets should contribute to the roughness discussed here. Using shorter, near axis profiles also eliminates complications related to sedimentation and thermal subsidence. At greater distances from the ridge axis thermal subsidence is the primary component of long wavelength depth variations [*Parsons and Sclater*, 1977] but the high variability in near axis depths (this analysis and [*Marty and Cazenave*, 1989]) suggests that thermal subsidence does not become a dominant factor until the lithosphere has moved some distance away from the ridge. In light of the arguments given above, it is expected that the majority of the stochastic signal discussed here is related to abyssal hill formation with some minor contribution from offset migration and that the contribution from constructional volcanism is negligible.

Global Variations in Axial Relief, Width, Roughness and Depth

Axial relief and asymmetry are plotted as functions of spreading rate in Figure 5. Spreading rates used in this study were calculated with the Nuvel 1a relative plate motion model [*DeMets et al*, 1994] and are given in km/Ma (rather than mm/yr) to better reflect the spatiotemporal resolution of plate velocity estimates. Axial relief and asymmetry in Figure 5 both diminish with increasing spreading rate for axial valleys and are invariant to spreading rate for

Table 1. Correlations and Trends

Parameter 1	Parameter 2	Correlation Coeff.	Rank Order Coeff	Slope	Intercept
Axial Valley Relief	Spreading Rate	0.67	0.61	17.5	-1786
MAD (rise)	Spreading Rate	-0.14	-0.14	- - - -	- - - - -
MAD (valley)	Spreading Rate	-0.67	-0.77	-1.35	165
Log_{10}(MAD (valley))	Log_{10}(Rate)	-0.68	-0.77	-0.50	2.8
Log_{10}(MAD (All))	Log_{10}(Rate)	-0.74	-0.71	-0.63	2.9
Amp. (rise)	Spreading Rate	-0.13	-0.15	- - - -	- - - - -
Amp. (valley)	Spreading Rate	-0.60	-0.70	-7.70	1000
Log_{10}(Amp (valley))	Log_{10}(Rate)	-0.61	-0.70	-0.47	3.5
Log_{10}(Amp (All))	Log_{10}(Rate)	-0.69	-0.67	-0.58	3.6
MAD (rise)	Axial Relief	0.15	-0.04	- - - -	- - - - -
MAD (valley)	Axial Relief	-0.56	-0.64	-0.05	56
MAD (All)	|Axial Relief|	-0.70	0.64	0.06	31
Amp. (rise)	Axial Relief	0.11	0.01	- - - -	- - - - -
Amp. (valley)	Axial Relief	-0.62	-0.66	-0.30	356
Amp. (All)	|Axial Relief|	0.72	0.64	0.38	224
MAD (rise)	Amp.	0.73	0.76	5.14	81
MAD (valley)	Amp.	0.83	0.85	4.82	158
MAD (All)	Amp.	0.89	0.90	5.25	95

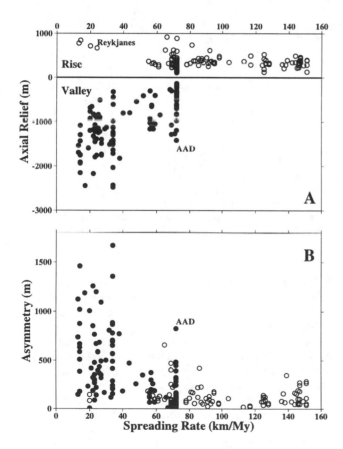

Figure 5. Variation in A) total axial relief and B) asymmetry of axial morphology with spreading rate. Relief and asymmetry of axial valleys diminishes with increasing spreading rate but axial rise relief and asymmetry remains essentially constant over an equivalent range of spreading rates.

axial highs. Both axial relief and asymmetry of axial valleys show considerable scatter and, although the average values diminish with increasing spreading rate, the data show that small (< 1000 m) and symmetric axial valleys are found at all spreading rates between 20 and 75 km/My. The commonly cited dependence of axial valley relief on spreading rate seems less applicable to minimum relief. The presence of the AAD and the relative paucity of mid-ocean ridges with spreading rates between 38 and 60 km/My complicates the interpretation. Axial valley width is relatively uniform (median = 40 km; 50% between 32 and 44 km) and shows no variation with spreading rate.

The variation of both roughness measures with spreading rate (Figure 6) is similar to that of axial relief and asymmetry; a rate dependence for axial valleys with an invariance to spreading rate for profiles characterized by an axial rise. MAD roughness and roughness amplitude are correlated and both show the same variation with spreading rate, thereby supporting the previous assertion that the roughness amplitude is a robust estimate of the average relief of the roughness profile. When plotted in Log_{10} space, the MAD roughness still has considerable scatter and shows a decrease in amplitude with spreading rate for axial valleys but not for axial rises. This observation does not support the continuous power law relationship between roughness and spreading rate proposed by *Malinverno* [1991]

Depth Estimates

Two estimates of spreading center depth are used in this study. Zero-age depth is the single depth value taken at the center of each profile and is identical to the axial depth

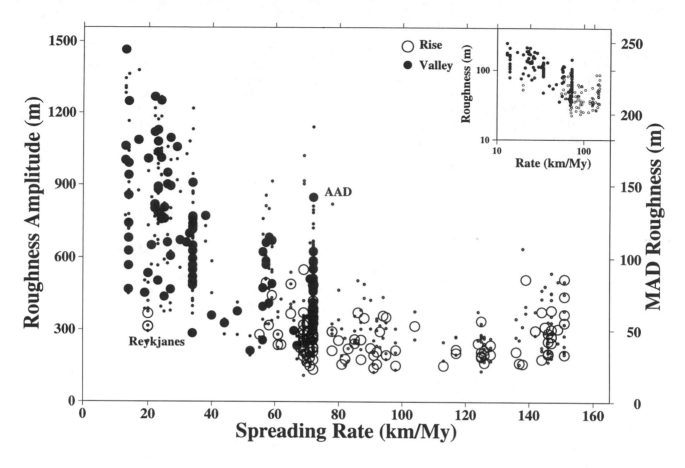

Figure 6. Variation in seafloor roughness with spreading rate. MAD Roughness (large symbols) is the Median Absolute Deviation of the stochastic profile. Roughness amplitude (small symbols) is the maximum relief of the stochastic profile on each side of the ridge axis. Roughness at axial valleys diminishes with increasing spreading rate but minimum values are less dependent on spreading rate. Roughness of axial rise flanking topography is uniformly low and invariant to spreading rate. MAD roughness plotted in Log$_{10}$ space emphasizes the discontinuity between axial valley and axial rise roughness variation and provides little support for a continuous power law model for seafloor roughness.

commonly discussed in studies of mid-ocean ridge segmentation [e.g. *Sempere et al,* 1990]. Profile depth is the arithmetic average of all depth values along the length of a given profile; the median depth along each profile gives nearly the same result. To minimize the influence of the axial morphology, the profile depth can also be estimated as the average depth between distances of 30 and 40 km off axis (Figure 7). Zero-age depth shows a spreading rate variability similar to that of axial relief, asymmetry and roughness. Mean profile depth spans a smaller range and shows greater scatter for profiles with axial valleys than for those with axial rises but otherwise no apparent spreading rate variation.

Correlations

Since this analysis provides estimates of a number of parameters for each profile in the dataset, one way to quantify the relationships between different components of the mor-

phology is to estimate correlations between parameters (Table 1). All correlations are calculated using both the linear correlation coefficient (ρ) and the Spearman rank-order coefficient [*Press et al,* 1992] and generally show close agreement. The correlations provide a convenient way to summarize relationships between the parameters but the results do not generally suggest simple linear relationships between most of the parameters investigated here. In fact, the most significant results of the study may be the lack of correlation and scatter in the relationships between some of the parameters.

It was demonstrated by *Malinverno* [1990] that axial depth and axial valley relief must be correlated since the depth is a component of the relief. It was also shown by *Small* [1994] that the depth of the rift mountains flanking the axial valleys is inversely correlated with axial depth, verifying the common assumption that axial deepening is balanced by flanking uplift. The results of this analysis support the both of these findings and are not shown here.

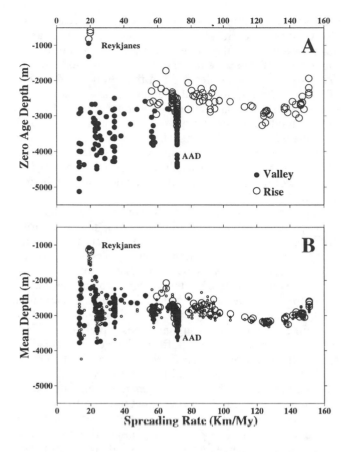

Figure 7. Variation of mid-ocean ridge depths with spreading rate. A) Variation of zero-age depth shows a dichotomy similar to axial relief, asymmetry and roughness. Zero-age depth of axial rises is less variable than that of axial valleys and has no spreading rate dependence. B) Mean spreading center depths within 40 km of the ridge axis have only two thirds the range of zero-age depths and no spreading rate dependence for axial valleys or rises. The greater variability of zero-age depths presumably reflects dynamic topography that is not preserved on the ridge flanks. Mean depths between 30 and 40 km off axis on either ridge flank (small circles) are not influenced by axial topography but show the same pattern. The median spreading center depth for the entire dataset is -2922 m.

The similarity in spreading rate variations of axial relief and roughness suggests a causal relationship. MAD roughness and roughness amplitude are plotted with respect to the axial relief in Figure 8. There are overall correlations ($\rho = 0.7$) between absolute axial relief and both MAD roughness and roughness amplitude but this is a result of generally lower roughness values for axial rises than for axial valleys. Considered separately, MAD roughness and roughness amplitude are both moderately correlated with axial valley relief ($\rho = 0.6$) but not with axial rise relief ($\rho = 0.1$). The same relationship holds true for axial asymmetry. MAD Roughness is also correlated with zero-age depth but this is primarily a result of consistently lower roughness values for axial rises and an overall deepening

trend for axial valleys. No correlation ($\rho = -0.009$) exists between mean profile depth and roughness.

INFERENCES AND IMPLICATIONS

Spreading Center Depth

The variation in zero-age depth with spreading rate is similar to that of axial relief, asymmetry and seafloor roughness but the variation of mean profile depth does not show a similar spreading rate dependence (Figure 7). The global average of profile depths for axial valleys is nearly the same as that of axial rises but the individual profile depths of axial valley spreading centers are more variable, spanning almost twice the 1000 m range of those with axial rises.

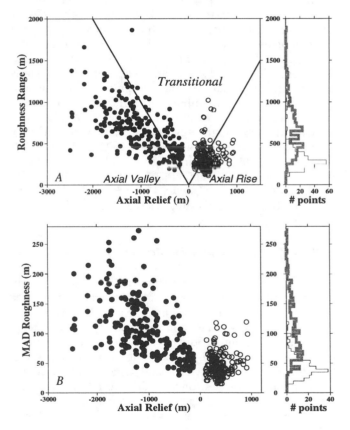

Figure 8. Relationship between roughness and axial relief. Roughness amplitude (A) and MAD roughness (B) are both moderately correlated with axial valley relief but uncorrelated with axial rise relief. Roughness amplitude exceeds axial relief at transitional spreading centers, suggesting that the creation of seafloor roughness does not require significant axial relief. Considerable overlap in axial rise (thin histograms) and axial valley (thick histograms) roughness distributions indicates that the most commonly observed roughness values can be found at either axial rise or axial valley spreading centers, making it difficult to uniquely determine paleo-spreading rate or axial morphology from profile roughness.

Profile depths span only about two thirds the range of zero-age depths which suggests that the remaining third (~1000 m) is dynamically maintained and not preserved once the lithosphere has spread away from the ridge axis. The difference results from dynamic deepening of axial valleys; mean profile depth of axial rise spreading centers is somewhat deeper than zero-age depth but spans essentially the same range. A significant portion of the variation of axial rise profile depths is related to the deepening of the East Pacific Rise in the vicinity of the Galapagos Triple Junction (130 km/My) and shallowing of the robust segment of the Australian-Antarctic Rise adjacent to the Balleny hotspot (65 km/My). The variability of profile depths at any given spreading rate presumably reflects variability in the combined effects of axial dynamics and lithospheric compensation.

A poor correlation (ρ = -0.009) between roughness and profile depth implies that there is no systematic relationship between lithospheric strength and compensation at the scales considered in this study. It is likely that temporal variations in the upwelling patterns of slow spreading ridges result in smaller scale variations in crustal thickness and that their regional influence is limited by the flexural rigidity of the near axis lithosphere. Intrasegment correlations between fault heave and mantle Bouguer anomalies on the Mid-Atlantic Ridge suggest that these variations may occur at spatial scales on the order of ~60 km [Shaw, 1992; Shaw and Lin, 1993] which is significantly shorter than the estimated flexural wavelength of 100's of kilometers for slow spreading ridges [Cochan, 1979; Forsyth, 1992]. The lack of global correlation suggests that this may be the case elsewhere.

Seafloor Roughness and Axial Morphology

A significant result of this study is that seafloor roughness near mid-ocean ridges is only moderately correlated with axial valley relief and is not correlated with axial rise relief. Given the similarity in their variation with spreading rate and the long-standing observation that slow spreading centers produce rough topography it is surprising that the actual correlation between axial valley relief and coincident roughness is not stronger. This correlation is weakened by a symmetric dispersion of values about the main trend, corresponding to large valleys with smooth flanks and small valleys with rough flanks. The implication is that roughness and axial valley morphology may be controlled by the same parameters on a global scale but the relief of the axial valley is not the primary determinant of flanking roughness. Alternatively, back rotation and antithetic faulting [Harrison and Steiltjes, 1977] may tend to reduce the amplitude of flanking topography at spreading centers with large axial valleys. This would explain some of the scatter in Figure 8 but would not explain the frequent occurrence of rough morphology on the flanks of small

axial valleys. Temporal variations in these processes may also weaken the correlation between the instantaneous axial morphology and the time averaged estimate of the roughness. If this were the case however, the limited off-axis extent of the profiles used in this study (< ~2 valley widths) would require that the temporal variations in axial valley relief occur on relatively short time scales. In spite of the moderate correlation of axial valley relief and roughness, the relationship between axial relief and roughness is distinctly different for axial valleys and axial rises (Figure 8). Axial rises show no correlation between the relief of the axial rise and the roughness of the flanking topography.

These results suggest that the variation in roughness with spreading rate is discontinuous and is not necessarily the manifestation of the same mechanism at all spreading centers. The maximum roughness attainable in axial valley environments diminishes with increasing spreading rate and/or thermal structure but the scatter encompasses almost the entire range at any given rate. The latter observation can be explained by intra-segment variability of axial valley spreading centers. The roughness near spreading centers with axial rises shows no significant change with spreading rate (or thermal structure) and is equally variable at all rates. Both the rate dependence and the difference in variability suggest that the mechanism that generates seafloor roughness undergoes a transition coincident with the transition in axial morphology. The Log_{10} plot in Figure 6 emphasizes this discontinuity and does not support the power law dependence on spreading rate proposed by Malinverno [1991]. Neither the power law proposed by Malinverno [1991] nor the best fit power law for the data in this study (Table 1) adequately describes the roughness variation over the entire range of spreading rates because neither power law flattens sufficiently. As a result, both continuous functions over-predict roughness at intermediate spreading rates and under predict roughness at faster spreading rates. Interestingly, the roughness estimates for axial valleys are fit with exponents of approximately 0.5 when roughnesses from axial rise spreading centers are excluded (Table 1). This is consistent with the notion of a $\sqrt{age}$ lithospheric strengthening but the scatter in the roughness estimates (even in Log_{10} space) makes it difficult to draw firm conclusions from the data.

In spite of the discontinuous variation of roughness with spreading rate, it is questionable whether profile roughness can provide accurate estimates of paleo-spreading rates or axial morphologies. The histograms in Figure 8 suggests that very rough seafloor (more than 500 m amplitude) is diagnostic of lower spreading rates but that smoother seafloor (200 - 500 m amplitude) can be generated locally at a wide range of spreading rates and axial morphologies. The considerable scatter limits the inference of paleomorphology to the prediction of an axial valley in the case of anomalously high roughness or an axial rise in the case of anomalously low roughness. The most commonly observed roughness values (> 50% between 200 and 500 m) are found at all spreading rates. This imperfect correlation between axial valley relief and flanking roughness is also apparent in recent regional multibeam surveys on axial valley and transitional spreading centers. Detailed 2D studies of abyssal hill morphology [eg.*Goff et al*, 1997] may be able to discriminate between seafloor formed at axial rises and axial valleys but it seems unlikely that this will be possible with profile data in heavily sedimented older seafloor.

Transitional Spreading Centers

The relationship between axial valley morphology and roughness undergoes a transition at the point where the amplitude of the roughness exceeds the amplitude of the axial morphology. In these cases the spreading center has little or no distinguishable deterministic axial morphology and neither an axial rise nor valley are evident. Examples of this can be seen in multibeam bathymetry surveys of intermediate rate spreading centers such as the Southeast Indian Ridge [*Cochran et al*, 1997; *Sempere et al*, 1997] and the Pacific Antarctic Rise [*Macario*, 1994; *Macario et al.*, 1994]. Recent studies of the Southeast Indian Ridge have shown that a single segment may contain axial morphologies that range from shallow valleys to small rifted rises [*Cochran et al*, 1997; *Ma and Cochran*, 1996; *Sempere et al*, 1997; *Shah and Sempere*, 1998]. Alternatively, single segments at the same spreading rate may have axial morphology that has neither a rise nor a valley in which the pattern of faulting seen on the flanks is also found at the ridge axis in the absence of any characteristic axial morphology (Figure 9). The abyssal hills formed on the Southeast Indian Ridge also have heights, widths and aspect ratios that span the range between those observed at slower and faster rates [*Goff et al*, 1997]. The characteristics of the segmentation of the Southeast Indian Ridge also vary between the extremes observed at faster and slower spreading rates [*Small et al*, 1998; *Ludwig and Small*, 1997].

Incorporating profiles from the Southeast Indian Ridge into the EOF analysis makes it possible to quantitatively compare transitional morphologies to more thoroughly studied axial rise and axial valley spreading centers. Figure 8a shows that the amplitude of the roughness is significantly less than the relief of the axial valley for valleys larger than 1000 m but for valleys smaller than 500 m the amplitude of the roughness frequently exceeds the relief of the valley. In contrast, at well developed axial rise spreading centers the amplitude of the roughness rarely exceeds the relief of the axial rise by a significant margin. Roughness amplitudes greater than 500 m on axial rise profiles result from seamounts or other isolated features. The

majority of the profiles which lie above and between the diagonal lines in Figure 8a correspond to transitional morphologies intermediate between the axial valleys and rises which lie below the diagonals. In the context of the EOF analysis this corresponds to a absence of a characteristic axial morphology and a dominance of the stochastic component. Because the axial relief is estimated by measuring the range of the deterministic component of the topography, this analysis never assigns a exact zero value for axial relief. In some cases, the points which lie above the diagonals in Figure 8 do contain some identifiable axial morphology but the profiles with axial relief estimates less than ~200 m are neither axial valleys nor axial highs and correspond to widely separated extrema on the deterministic profile.

This analysis suggests that the roughness of lithosphere formed at well developed axial rises is lower than that formed at transitional spreading centers without well developed axial morphology. This agrees with the findings of *Goff et al* [1997] in which some abyssal hills on the Southeast Indian Ridge have RMS heights and widths intermediate between those from axial rise and axial valley spreading centers. One possible cause for this further decrease in roughness may be the more frequent occurrence of relief reducing lava flows on magmatically robust axial rise spreading centers [*Macdonald et al*, 1996]. The lack of a steady state magma chamber at intermediate spreading rates may also allow the lithosphere to cool sufficiently to accumulate greater strength and therefore produce larger offset faults than is possible at axial rise spreading centers. The difference in roughness is not large enough however to consistently predict paleo-axial morphology.

The existence of transitional spreading centers with significant topographic roughness but no discernible axial morphology suggests that axial valley relief is not required to generate significant seafloor roughness at intermediate rates (Figure 9). This dataset contains numerous examples where the roughness amplitude exceeds the present axial relief by several hundred meters (Figure 8). This suggests that the roughness is controlled primarily by the extensional strength of the lithosphere and does not rely on deformation related to transport out of a dynamically maintained axial valley. At intermediate spreading rates where an axial rise or valley is not present, the roughness would be a function of the amount of stress that the lithosphere could maintain before faulting and the geometry of the faults produced during failure. Axial morphology also appears to have much greater temporal variability at intermediate spreading rates so the present axial relief may be smaller than that which produced the larger features measured by the roughness amplitude estimate.

In the context of the *Chen and Morgan* [1990*a,b*] model for axial morphology the absence of an axial rise or valley could correspond to a situation in which the axial crust is decoupled from the mantle and is in isostatic equilibrium.

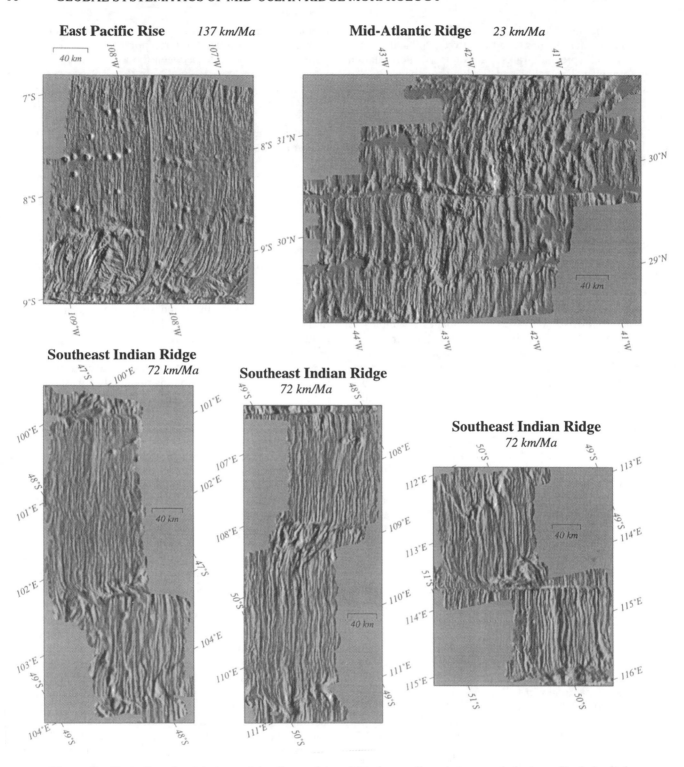

Figure 9. Examples of axial rise, axial valley and transitional spreading center morphologies. Shaded relief images show normalized ridge perpendicular topographic gradients based on multibeam survey data gridded at 500 m resolution. All images are plotted in oblique Mercator projections with the ridge axis vertical. Two dimensional coverage emphasizes the importance of offset propagation to flanking roughness as well as the along axis variability in both axial and flanking morphology. Transitional morphology on the Southeast Indian Ridge shows numerous examples of tectonic fabric formed in the absence of any distinguishable axial morphology. Data were obtained from the RIDGE Multibeam Synthesis (http://www.ldeo.columbia.edu).

Low relief transitional axial morphology could be produced by a temporally variable thermal structure in which the lower crust is weak enough for the system to attain isostatic equilibrium but the mantle is not hot enough to consistently produce an excess of magma and maintain an axial rise.

There exists considerable evidence that the process of abyssal hill formation at axial rises is different from that within axial valleys. Because of the preponderance of steeper inward dipping slopes, normal faulting has long been considered the dominant mode of deformation within axial valleys [*Atwater and Mudie*, 1968; *Searle and Laughton*, 1977; *Macdonald and Atwater*, 1978; *Harrison and Stieltjes*, 1976]. Several authors [e.g. *Dick et al.* 1981; *Karson* 1987; *Brown and Karson*, 1988; *Mutter and Karson*, 1992; *Tucholke and Lin*, 1994; *Goff et al*, 1995] also argue that intrasegment variations in lithospheric strength result in the formation of detachment faults at the ends of axial valley segments which are responsible for much of the lithospheric deformation seen at slow spreading rates. Studies of the intermediate spreading East Pacific Rise [*Normark*, 1976; *Lewis*, 1979], Southeast Indian Ridge [*Sauter et al*, 1991] and Juan de Fuca ridges [*Kappel and Ryan*, 1986] also provide evidence of magmatic and amagmatic episodicity. *Kappel and Ryan* [1986] proposed that abyssal hills on the flanks of the Juan de Fuca Ridge are formed by split axial volcanic ridges which form cyclically and are carried onto the flanks. *Carbotte and Macdonald* [1994] further investigate differences in the morphology of axial rises and propose that those at intermediate spreading centers are volcanic constructs which can be supported by the strength of the lithosphere while the axial highs at fast spreading centers are usually formed on weaker lithosphere and supported isostatically. Stochastic analyses of abyssal hills [*Goff*, 1991; *Goff et al*, 1993] suggest that the size and shape of abyssal hills formed on the East Pacific Rise do not vary significantly with changes in spreading rate but rather with the local segmentation. The results suggest that the dichotomy in axial morphology and the dichotomy in flanking morphology are both controlled by lithospheric strength but that the actual relief of the axial valley is not required to create seafloor roughness.

The Dichotomy in Spreading Center Morphology

The most prominent global characteristic of mid-ocean ridge bathymetric expression is the dichotomy between fast and slow spreading centers. This analysis shows that axial relief, asymmetry, roughness and zero-age depth all have similar variations with spreading rate. For each of these parameters, spreading centers with axial valleys exhibit a variability spanning nearly the full range of each parameter but the maximum attainable value diminishes with increasing spreading rate. This is contrasted by spreading centers characterized by axial rises for which none of these parameters show any significant variation with spreading rate.

This pattern suggests that the processes of lithospheric accretion and deformation are controlled by spreading rate or local mantle temperature until a critical rate or thermal structure is attained. Beyond this critical point these processes apparently cease to vary with increasing spreading rate. This type of threshold behavior suggests that the mechanism of lithospheric accretion is characterized by dynamically stable and unstable states which are thermally controlled and generally modulated by spreading rate.

While spreading rate offers a convenient parameterization, it should not be inferred that the rate of plate separation alone is responsible for the pattern seen in these plots. Profiles from the Reykjanes Ridge are more consistent with those of fast spreading ridges and profiles from the AAD are similar to those from slower spreading centers. These "anomalous" spreading centers are the most commonly cited counterexamples suggesting that differences between fast and slow spreading axial morphology are more directly a result of thermal structure. Extensive modeling efforts [e.g. *Sleep*, 1975; *Parmentier*, 1987; *Phipps Morgan et al*, 1987; *Lin and Parmentier*, 1989; *Chen and Morgan*, 1990a,b; *Phipps Morgan and Chen*, 1993; *Henstock et al*, 1993; *Neumann and Forsyth*, 1993; *Poliakov and Buck, this volume*] have demonstrated the importance of thermal structure in the formation of axial morphology. Models of lithospheric deformation also demonstrate the importance of thermal structure in the formation of abyssal hill morphology [e.g. *Carbotte and Macdonald*, 1994; *Shaw and Lin*, 1996]. In order to demonstrate a thermal origin for the systematic variations observed in all of these parameters it is necessary to address both the dichotomy in spreading rate dependence and the dichotomy in variability.

Spreading Rate Dependence and Variability of Axial Valley Environments

At a single spreading rate, most of the variability in axial valley relief, asymmetry and roughness results from well known variations in the segmentation of slow spreading plate boundaries. As zero-age depth increases approaching first and second order discontinuities [e.g. *Sempere et al*, 1990] axial valley relief and asymmetry also increase and the roughness of flanking topography tends to increase [*Tucholke and Lin*, 1994]. The deepening of the ridge axis is commonly attributed to an intra-segment variation in mantle temperature and crustal thickness [e.g. *Lin et al*, 1990; *Lin and Phipps Morgan*, 1992] and is accompanied by a corresponding increase in axial relief and asymmetry which may be the result of low angle detachment faulting [*Dick et al.* 1981; *Karson* 1987; *Brown and Karson*, 1988; *Mutter and Karson*, 1992; *Tucholke and Lin*, 1994] or possibly the local stress field [*Severinghaus and Macdonald*, 1988]. Intra-segment variations in abyssal hill morphology on the Mid-Atlantic Ridge also correlate with variations in residual gravity anomalies which are inferred to represent variations in crustal thickness [e.g. *Shaw and Lin*, 1993; *Goff et al*, 1995].

The variability related to segmentation is generally believed to result from temporal and spatial variations in the upwelling of low viscosity asthenosphere [e.g. *Whitehead et al*, 1984; *Schouten et al*, 1985]. The control of upwelling on segmentation may be a result of either shallow redistribution of upwelling material [*Bell and Buck*, 1992; *Wang and Cochran*, 1993] or a transition from 3D to 2D upwelling structure [*Lin and Phipps Morgan*, 1992; *Parmentier and Phipps Morgan*; 1990]. Both mechanisms would depend strongly on thermal structure near the ridge axis. Much of the variability seen in axial valley environments is likely to be related to the episodicity of magmatic and amagmatic extension and the time dependence of mantle upwelling at slower spreading rates and/or colder thermal structures [e.g. *Mutter and Karson*, 1992; *Shaw and Lin*, 1993; *Tucholke and Lin*, 1994; *Thatcher and Hill*, 1995]. On a global scale however there appears to be an upper bound on the variability of axial valley environments and this upper bound is presumably related to the thermal structure at the plate boundary.

The results of this study indicate that the average and maximum attainable depth, morphology and roughness of axial valley spreading centers is spreading rate dependent. At any given rate, almost the entire range of any of these parameters is observed. Axial relief and roughness on the Southwest Indian Ridge are not quite as low as those on the faster spreading northern Mid-Atlantic Ridge but the data distribution on the Southwest Indian Ridge is still rather limited. If we assume that the spreading rate dependence of the maxima is thermally controlled then the implication is that it is possible to find almost the entire range of thermal structures at any spreading rate where an axial valley environment can be maintained. Interpreted in the context of the upwelling models discussed above, this would imply that amagmatic extension of strong lithosphere in large, deep axial valleys is thermally controlled by the spreading rate but can be locally overridden by an isolated upwelling of high temperature asthenosphere.

Insensitivity of Axial Rise Environments to Spreading Rate

The variability and spreading rate dependence of slow spreading (10 - 80 km/My) mid-ocean ridges stands in sharp contrast to the relative invariance of axial rise environments to equivalent variations in spreading rate (70-150 km/My). The results of this analysis show no significant variation in axial relief, depth or roughness for axial rise spreading centers. On axial rises, variations in mean depth tend to be regional while variations in axial depth are related to segmentation but neither shows any consistent variation with spreading rate. Detailed analyses of a number of areas on the northern East Pacific Rise indicate that the variability of axial rise spreading centers is controlled by segmentation rather than spreading rate. Cross sectional area of axial rise morphology varies with local segmentation and the presence of an axial magma chamber on the East Pacific Rise but does not show a consistent variation with spreading rate

[*Scheirer and Macdonald*, 1993]. Stochastic analyses of abyssal hill morphology on the East Pacific Rise also show consistent variations in height, width and aspect ratio of abyssal hills with ridge axis segmentation but no appreciable spreading rate dependence [*Goff et al*, 1993].

Some aspects of axial rise spreading centers do vary somewhat with increasing spreading rate but this variability is smaller than the intrasegment variation and considerably smaller than the variation seen in axial valley environments. *Carbotte and Macdonald* [1994b] show that the relative percentage of inward dipping faults diminishes consistently over the entire range of spreading rates. Limited high resolution deep tow surveys also suggest that throw on abyssal hill bounding faults may decrease somewhat over the range of spreading rates (75 - 150 km/My) on the East Pacific Rise [*S. Carbotte, personal communication*]. Seismic studies indicate that depths to the velocity inversion [*Purdy et al*, 1992] and depths to the top of the magma lens [*Phipps Morgan et al*, 1994] decrease somewhat between rates of 108 and 152 km/My but the uncertainties associated with the former span nearly the same range for both areas.

Spreading Rate Dependence of Internal Structure

The morphologic dichotomy between axial valley spreading centers and axial rise spreading centers is also reflected in a number of other parameters (Figure 10). Mantle Bouguer Anomaly (MBA) amplitudes are characterized by large negative values at axial valley spreading centers and relatively constant higher values at axial rise spreading centers. *Lin and Phipps Morgan* [1991] interpret this pattern as the result of a transition between two and three dimensional upwelling structure as predicted by *Parmentier and Phipps Morgan* [1990]. *Wang and Cochran* [1993] however, find evidence for a localized upwelling structure on the fast spreading East Pacific Rise. Mantle Bouguer anomaly gradients, which normalize for differences in segment length, show a similar global dichotomy between axial rise and axial valley environments. Figure 10 shows the compilation of MBA gradients published by *Wang and Cochran* [1995] supplemented with additional data from the Southeast Indian Ridge [*Cochran et al*, 1996], the AAD [*West and Sempere*, 1998], the Mid-Atlantic Ridge [*Detrick et al*, 1996] and the East Pacific Rise [*Magde et al*, 1996]. Wang and Cochran [1994] interpret this pattern as a result of differences in the mechanism of shallow melt migration and redistribution at fast and slow spreading centers.

Mantle Bouguer anomalies are believed to result from both variations in subaxial mantle temperature and variations in crustal thickness. Seismic estimates of crustal thickness (Figure 10) also show a similar pattern of high variability at low spreading rates and relatively constant values at high spreading rates [*Chen*, 1992]. Regardless of whether these variations in crustal thickness result from differences in the nature of the upwelling structure or from differences in the shallow redistribution of melt and partially

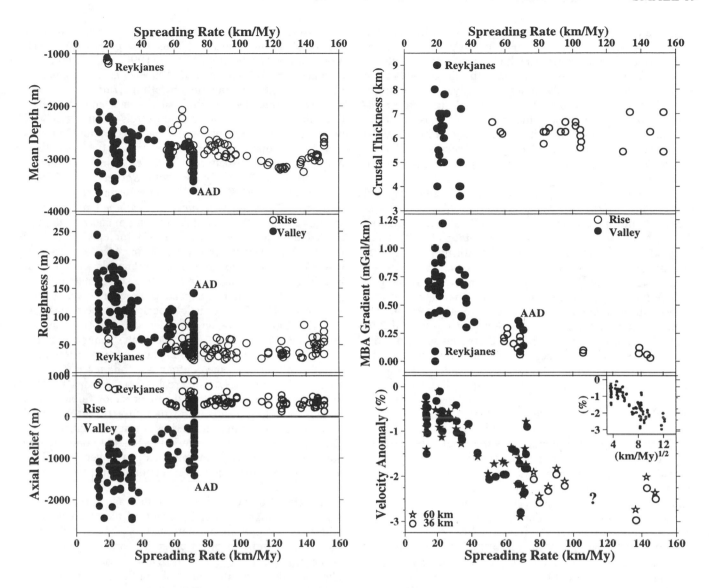

Figure 10. Summary of spreading rate variations of morphologic characteristics estimated in this study and from related studies. Seismically determined crustal thickness [*Chen*, 1990], Mantle Bouguer Anomaly gradient (see text for sources) and seismic velocity anomalies [*Zhang and Tanimoto*, 1993] reveal a dichotomy analogous to that observed in morphology. Variance in crustal thickness and MBA gradients diminishes markedly at higher rates. Velocity anomalies, at depths of 36 and 60 km beneath the ridge axis, show a spreading rate dependence for ridges with axial valleys but no apparent rate dependence for those with axial rises. The inset suggests that the anomalies at the fastest rates are not explained by the lithospheric cooling model. *Zhang and Tanimoto* [1993] did not include data from the EPR between rates of 100 and 140 km/My to avoid hotspot effects.

molten material, the mechanism responsible for these patterns must be strongly temperature dependent and appears to stabilize at intermediate spreading rates.

Global seismic velocity models may provide further support for a stabilization in mid-ocean ridge thermal structure at intermediate spreading rates. Figure 10 shows the results of a global survey of seismic velocity anomalies beneath mid-ocean ridges by *Zhang and Tanimoto* [1993]. These shallow (36 and 60 km) seismic velocity anomalies,

estimated from travel time inversion of shear waves and surface waves, show a pattern similar to that seen in the other data discussed above. Travel time delays (relative to the global average) beneath ridges spreading at rates less than ~80 km/My increase with spreading rate but those beneath ridges spreading at faster rates do not continue to increase significantly. Because *Zhang and Tanimoto* [1993] attempted to avoid areas affected by hotspots, a gap in coverage exists between rates of 100 and 140 km/My.

Nonetheless, the seismic velocity anomalies beneath the fastest spreading ridges are essentially the same as those beneath intermediate rate ridges which suggests that doubling the spreading rate does not increase the magnitude of the shallow velocity anomaly at scales resolved by the inversion. We may expect the velocity anomaly to increase linearly with the square root of the spreading rate as a result of the increase in the average lithospheric age within the fixed region about the ridge sampled by the inversion but the inset in Figure 10 suggests that this is not the case for the fast spreading ridges.

The accuracy of Zhang and Tanimoto's velocity model has been questioned by *Su et al* [1992; 1994] but the discrepancy lies at depths greater than 100 km; at shallow depths the model of *Zhang and Tanimoto* [1993] agrees with other global inversions which use different techniques and different seismic phases [see *Su et al, 1992; 1994*]. While the velocity anomalies shown in Figure 10 suggest that regional thermal structure may not change appreciably above intermediate spreading rates, global inversions are limited in resolution and are subject to geographic differences in sampling by teleseismic earthquakes. More detailed studies will be necessary to determine whether the seismic velocity structure of fast spreading centers is actually this insensitive to changes in spreading rate.

Spreading Rate, Stability and Episodicity

The transition between axial rise and axial valley spreading centers occurs within a relatively narrow range of spreading rates where both axial valleys and axial rises occur as well as transitional morphologies that are neither distinct rises nor valleys. The section of the Southeast Indian Ridge between the Amsterdam/Kerguelen hotspots and the AAD shows little change in spreading rate (~72 ± 1 km/My) but the thermal structure of the mantle is believed to have a significant horizontal gradient along axis between the hotspots and the cold region of the AAD [*Roult et al, 1994; Sempere et al, 1997; West et al, 1997*]. Within this thermal gradient, local perturbations appear to play a major role in local variations in axial and abyssal hill morphology and deformation. The spatial scale of these variations within individual segments suggests that temporal variations between magmatic and amagmatic extension may occur on time scales of less than 10^6 years (Figure 9). The episodicity implied by these spatiotemporal variations does not necessarily require however that the regional thermal structure change significantly at corresponding temporal or spatial scales. The interaction of two or more threshold mechanisms may produce dynamic instability without large changes in the regional thermal structure .

The diversity of morphologies that occur at intermediate spreading rates suggest that the mechanism of lithospheric accretion is easily perturbed and dynamically metastable under the conditions that prevail at intermediate spreading rates. Metastability, marked by vacillation between stable and unstable states, may result from the interaction of various threshold mechanisms such as those proposed for axial morphology [*Chen and Morgan, 1990a,b*], magma lens emplacement [*Henstock et al, 1993; Phipps Morgan and Chen, 1993*], and mantle upwelling [*Parmentier and Phipps Morgan,1990*]. Both spatial and temporal variations in thermal structure would exercise a strong control on any of these mechanisms and may account for the diversity of morphologies observed within the transition. Nonlinear coupling between processes of melt production and migration [e.g. *Sparks and Spiegelman, 1995; Spiegelman and McKenzie, 1987*] and mechanical deformation [e.g. *Poliakov and Buck, this volume*] may induce episodicity and stability transitions in a manner analogous to that envisioned by *Shaw* [1988] for episodic silicic magmatism. Once the spreading rate increases above a critical point the elevated thermal structure would tend to stabilize the system and maintain a relative steady state in comparison to the strong rate dependence seen at lower spreading rates. The characteristic time scale of the metastable behavior appears to be on the order of 10^5 to 10^6 years for most spreading centers. At time scales longer than 10^6 years, changes in large scale plate motion often have significant effects on the configuration and morphology of the spreading center. At time scales shorter than 10^5 years, the lithosphere is still within the inner valley floor at axial valley spreading centers and has only recently moved beyond the edge of most axial rises at fast spreading centers. The characteristic time scale is presumably controlled by the upwelling rate of buoyant asthenosphere which is responsible for the episodic advection of heat to the ridge axis.

The global dichotomy in variability and spreading rate dependence suggests a stabilization of the mechanism of lithospheric accretion and deformation. This stabilization also appears to be manifest in the low order segmentation of the mid-ocean ridge system [*Ludwig and Small, 1997*]. The coincident transition in axial and flanking morphology and the appearance of a steady state magma chamber further support the assertion that the system has attained a stable configuration. If, as previous work suggests, the processes of lithospheric accretion are controlled primarily by the thermal structure of the spreading center then the implication is that the large scale thermal structure itself does not change appreciably once this transition occurs. To some extent, this can be explained by the asymptotic behavior of most mid-ocean ridge thermal models with spreading rate. Changes in the regional thermal structure are more significant for an incremental rate change at slow spreading rates than they are for an equivalent rate change at fast spreading rates. Asymptotic lithospheric strengthening may therefore be sufficient to explain the overall pattern observed in seafloor roughness but the mechanisms responsible for axial morphology, abyssal hill morphology and segmentation evolution are more complex. The implication of this and numerous other studies is that the advective-diffusive processes that dictate the thermal structure of mid-ocean ridge systems have inherent stability regimes that control both the formation and deformation of the lithosphere. More detailed studies will be necessary to understand how the mechanisms of accretion and deformation interact to maintain stability and control the morphogenesis of the spreading center.

SUMMARY

Bathymetric measurements at mid-ocean ridges provide information at three different spatial scales: the mean regional depth, the characteristic axial morphology and the smaller scale roughness superimposed on the axial morphology and adjacent flanks. Empirical Orthogonal Function (EOF) analysis of bathymetric profiles across the global mid-ocean ridge system allows these components to be separated and analyzed at coincident locations. The EOF analysis of the 228 profiles in this study provides a basis for a distinction between the deterministic and stochastic components of mid-ocean ridge topography. These results quantify long standing qualitative observations that both axial and flanking morphology vary with spreading rate and allow the relationship between them to be investigated.

The full range of zero-age depths observed at the ridge axis is not preserved on the adjacent ridge flanks. One third of the variation in axial depth appears to be dynamically maintained while the remaining two thirds are preserved on the flanks. Variation in regional spreading center depths decreases with spreading rate for axial valleys but is invariant to spreading rate for axial rises. Regional depth and roughness are uncorrelated so if regional variations in mean depth are a result of crustal thickness and thermal structure this is evidently not reflected in the lithospheric deformation.

At spreading centers with axial valleys the flanking roughness and axial relief diminish with increasing spreading rate. Axial valley relief is moderately correlated with flanking roughness but at intermediate spreading rates the amplitude of the flanking roughness frequently exceeds the relief of the axial valley suggesting that axial relief is not required to form abyssal hills. Distinctive axial morphologies at intermediate spreading rates may be explained by the presence of a weak lower crust that can isostatically decouple the crust and mantle but does not allow sufficient melting to consistently maintain a well developed axial rise. In contrast, spreading centers with axial rises are more sensitive to local segmentation than to spreading rate. Axial relief and flanking roughness are uncorrelated at axial rise spreading centers. Both axial relief and flanking roughness vary discontinuously with spreading rate.

In spite of the discontinuous variation with spreading rate, seafloor roughness does not provide a unique measure of paleo-spreading rate. The presence of anomalously high roughness may be indicative of an axial valley environment but the occurrence of relatively low roughness at almost all spreading rates suggests that it is not generally possible to uniquely determine paleo-axial morphology from profile roughness. Stochastic modeling of two dimensional abyssal hill morphology certainly provides more information about the tectonic environment but it is not straightforward to apply this type of analysis to the older heavily sedimented seafloor.

The most prominent global characteristic of mid-ocean ridge bathymetric expression is the dichotomy between fast and slow spreading centers. This study demonstrates that axial relief, asymmetry, roughness and zero-age depth, as well as several other non-bathymetric observations, all show similar behavior with respect to spreading rate and, presumably, local thermal structure. For each of these parameters, spreading centers with axial valleys exhibit a variability spanning nearly the full range of each parameter but the maximum attainable value diminishes with increasing spreading rate. This is contrasted by spreading centers characterized by axial rises for which none of these parameters show any significant variation with spreading rate. The East Pacific Rise and Pacific Antarctic Rise make up most of the fast spreading part of the system and show a remarkable invariance with spreading rate over more than half the global range of rates (75 - 151 km/My).

The dichotomy in spreading center structure and morphology suggests a transition from episodicity and rate dependence to a more stable mode of accretion. At transitional spreading centers the morphology exhibits even greater sensitivity to small changes in thermal structure. The implication is that the processes of lithospheric accretion and deformation are controlled by spreading rate or local thermal anomalies until a critical point is attained. Beyond this critical point, these processes cease to vary with increasing spreading rate. The most plausible mechanism for such a transition is a stabilization of mid-ocean ridge thermal structure above a critical spreading rate. This type of threshold behavior suggests that the mechanism of lithospheric accretion is characterized by dynamically stable and unstable states which are thermally controlled but generally modulated by spreading rate. The diversity of morphologies that co-exist within a narrow range of intermediate spreading rates further suggest that the mechanism of lithospheric accretion is easily perturbed and is dynamically metastable.

Acknowledgments. The breadth of topics summarized in this paper makes it virtually impossible to acknowledge all the scientists who have contributed to our collective understanding of the subjects. The citations are certainly not exhaustive. I would like to thank Robert Parker, Russ Davis and Alexei Kaplan for sharing their insight into the nuances of EOF analysis and Suzanne Carbotte and John Goff for their insight into the nuances of abyssal hill formation. Conversations with Roger Buck, John Chen, Jim Cochran, Milene Cormier, Dan McKenzie, Alexei Poliakov, Marc Spiegelman and Xuejin Wang also helped to refine the ideas discussed in this paper. I thank John Goff, Yves Lagabrielle and Dan Scheirer for thorough reviews of the long and tedious manuscript. This research was supported by NSF grant OCE-9302091 and a Lamont-Doherty Postdoctoral Research Fellowship.

REFERENCES

Atwater, T., and J. Mudie, Block faulting on the Gorda Rise, *Science, 159,* 729, 1968.

Batiza, R., Abundances, distribution and sizes of volcanoes in the Pacific Ocean and implications for the origin of non-hotspot volcanoes, *Earth Planet. Sci. Lett., 60,* 195-206, 1982.

Bell, R. E., and W. R. Buck, Crustal control of ridge segmentation inferred from observations of the Reykjanes Ridge, *Nature, 357,* 583-586, 1992.

Brace, W. F., and D. L. Kohlstedt, Limits on lithospheric stress imposed by laboratory experiments, *J. Geophys. Res., 85,* 6248-6252, 1980.

Brown, J. R., and J. A. Karson, Variations in axial processes on the Mid-Atlantic Ridge, *Mar. Geophys. Res., 10,* 109-138, 1988.

Carbotte, S. M., and K. C. Macdonald, The axial topographic high at intermediate and fast spreading ridges, *Earth Planet. Sci. Lett., 128,* 85-97, 1994a.

Carbotte, S. M., and K. C. Macdonald, Comparison of seafloor tectonic fabric at intermediate, fast, and super fast spreading ridges: Influence of spreading rate, plate motions, and ridge segmentation on fault patterns, *J. Geophys. Res., 99,* 13,609-13,631, 1994b.

Carbotte, S. M., S. M. Welch, and K. C. Macdonald, Spreading rates, rift propagation and fracture zone offset histories during the last 5 M.Y. on the Mid-Atlantic Ridge: 25°-27°30'S and 31°-34°30'S, *Mar. Geophys. Res., 13,* 51-80, 1991.

Chen, Y. J., Oceanic crustal thickness versus spreading rate, *Geophys. Res. Lett., 8,* 753-756, 1992.

Chen, Y. J., and W. J. Morgan, Rift valley/no rift valley transition at mid-ocean ridges, *J. Geophys. Res., 95,* 17,571-17,581, 1990a.

Chen, Y. J., and W. J. Morgan, A non-linear rheology model for mid-ocean ridge axis topography, *J. Geophys. Res., 95,* 17,583-17,604, 1990b.

Cochran, J. R., An analysis of isostasy in the worlds oceans, 2, Mid-ocean ridge crests, *J. Geophys. Res., 84,* 4713-4729, 1979.

Cochran, J. R., J. C. Sempere, D. Christie, M. Eberle, L. Geli, J. A. Goff, H. Kimura, Y. Ma, A. Shah, C. Small, B. Sylvander, B. P. West, and W. Zhang, The Southeast Indian Ridge between 88°E and 120°E: Gravity anomalies and crustal accretion at intermediate spreading rates, *J. Geophys. Res., 102,* p.15506-15520, 1997.

Davis, R. E., Predicatability of sea surface termperature and sea level pressure anomalies over the North Pacific Ocean, *J. Phys. Ocean., 6,* 249-266, 1976.

DeMets, C., R. G. Gordon, D. F. Argus, and S. Stein, Effect of recent revisions to the geomagnetic reversal time scale on estimates of current plate motions., *Geophys. Res. Lett., 21,* 2191-2194, 1994.

Detrick, R. S., H. D. Needham, and V. Renard, Gravity anomalies and crustal thickness variations along the Mid-Atlantic Ridge between 33°N and 40°N, *J. Geophys. Res., 100,* 3767-3787, 1995.

Dick, H. J. B., Low angle faulting and steady state emplacement of plutonic rocks at ridge-transform intersections, *EOS Trans. AGU, 62,* 406, 1981.

Dick, H. J. B., H. Schouten, P. S. Meyer, D. G. Gallo, H. Bergh, R. Tyce, P. Patriat, K. T. M. Johnson, J. Snow, and A. Fisher, Tectonic evolution of the Atlantis II fracture zone, *Proc. Ocean Drill. Program Sci. Results, 118,* 359-398, 1991.

Edwards, M. H., D. J. Fornari, A. Malinverno, and W. B. F. Ryan, The regional tectonic fabric of the East Pacific Rise from 12°50'N to 15°10'N, *J. Geophys. Res., 96,* 7995-8017, 1991.

Epp, D., and N. C. Smoot, Distribution of seamounts in the North Atlantic, *Nature, 337,* 254-257, 1989.

Forsyth, D. W., Comment on "A Quantitative Study of the Topography of the Mid-Atlantic Ridge" by A. Malinverno, *J. Geophys. Res., 96,* 2039-2047, 1991.

Forsyth, D. W., Geophysical constraints on mantle flow and melt generation at mid-ocean ridges. in *Mantle flow and melt generation at mid-ocean ridges,* edited by J.Phipps Morgan and D. Blackman, 1-65, American Geophysical Union, 1992.

Fox, C. G., and D. E. Hayes, Quantitative methods for analyzing the roughness of the seafloor, *Reviews of Geophys. and Space Phys., 23,* 1-48, 1985.

Gente, P., G. Ceuleneer, C. Durand, R. Pockalny, C. Deplus, and M. Maia, Propagation rate of segments along the Mid-Atlantic Ridge between 20 degrees and 24 degrees (SEADMA I Cruise), *Eos, 74n,* 43, 569, 1992.

Grindlay, N. R., P. J. Fox, and K. C. Macdonald, Second order ridge axis discontinuities in the South Atlantic: Morphology, Structure, Evolution, *Mar. Geophys. Res., 13,* 21-49, 1991.

Goff, J. A., A global and regional stochastic analysis of near-ridge abyssal hill morphology, *Jour. Geophys. Res., 96,* 21,713-21,737, 1991.

Goff, J. A., J. R. Cochran, Y. Ma, J.-C. Sempere, and A. Shah, Stochastic analysis of seafloor morphology on the flanks of the Southeast Indian Ridge: The influence of ridge morphology on the formation of abyssal hills, *J. Geophys. Res.,102,* p.15521-15534, 1997.

Goff, J. A., and T. H. Jordan, Stochastic modeling of seafloor morphology: Inversion of Sea Beam data for second-order statistics, *J. Geophys. Res., 93,* 13,589-13,608, 1988.

Goff, J. A., A. Malinverno, D. J. Fornari, and J. R. Cochran, Abyssal hill segmentation: Quantitative analysis of the East Pacific Rise flanks 7°S-9°S, *J. Geophys. Res., 98,* 13,851-13,862, 1993.

Goff, J. A., B. E. Tucholke, J. Lin, G. E. Jaroslow, and M. C. Kleinrock, Quantitative analysis of abyssal hills in the Atlantic Ocean: A correlation between inferred crustal thickness and extensional faulting, *J. Geophys. Res., 100,* 22,509-22,522, 1995.

Harrison, C. G. A., and L. Stieltjes, Faulting within the median valley, *Tectonophysics, 38,* 1377-144, 1977.

Hayes, D. E., and K. A. Kane, The dependence of seafloor

roughness on spreading rate, *Geophys. Res. Lett., 18*, 1425-1428, 1991.

Heezen, B. C., The Rift in the Ocean Floor, *Scientific American, 203*, 99-106, 1960.

Henstock, T. J., A. W. Woods, and R. S. White, The accretion of oceanic crust by episodic sill intrusion, *J. Geophys. Res., 98*, 4131-4161, 1993.

Hey, R. N., A new class of pseudofaults and their bearing on plate tectonics: A propagating rift model, *Earth Planet Sci. Lettr., 37*, 321-325, 1977.

Huang, P. Y., and S. C. Solomon, Centroid depths of mid-ocean earthquakes: dependence on spreading rate, *J. Geophys. Res., 93*, 13445-13477, 1988.

Kappel, E. S., and W. B. F. Ryan, Volcanic episodicity and a non-steady state rift valley along northeast Pacific spreading centers: Evidence from SeaMarc I, *J. Geophys. Res., 91*, 13,925-13,940, 1986.

Karson, J. A., et al, Along axis variations in seafloor spreading in the MARK area, *Nature, 328*, 681-685, 1987.

Klein, E. M., and C. H. Langmuir, Global correlations of ocean ridge basalt chemistry with axial depth and crustal thickness, *J. Geophys. Res., 92*, 8089- 8115, 1987.

LePichon, X., Models and structure of the oceanic crust, *Tectonophysics, 7*, 385-401, 1969.

Lewis, B. T. R., Periodicities in volacanism and logitudnal magma flow on the East Pacific Rise at 23°N, *Geophys. Res. Lett., 6*, 753-756, 1979.

Lin, J., and E. M. Parmentier, Mechanisms of lithospheric extension at mid-ocean ridges, *Geophys. Jour. Int., 96*, 1-22, 1989.

Lin, J., and E. M. Parmentier, A finite amplitude necking model of rifting in brittle lithosphere, *J. Geophys. Res., 95*, 4909-4923, 1990.

Lin, J., and J. Phipps Morgan, The spreading rate dependence of three-dimensional mid-ocean ridge gravity structure, *Geophys. Res. Lett., 19*, 13-16, 1992.

Lin, J., G. M. Purdy, H. Shouten, J. C. Sempere, and C. Zervas, Evidence from gravity data for focused magmatic accretion along the Mid-Atlantic Ridge, *Nature, 344*, 627-632, 1990.

Lonsdale, P. F., Structural geomorphology of a fast-spreading rise crest: The East Pacific Rise near 3°25'S, *Mar. Geophys. Res., 3*, 251-293, 1977.

Ludwig, K., and C. Small, Quantifying large scale mid-ocean ridge segmentation, *Eos, 78, 46*, F682, 1997.

Ma, Y., and J. R. Cochran, Transitions in axial morphology along the Southeast Indian Ridge, *J. Geophys. Res., 101*, 15,849-15,866, 1996.

Macario, A., Crustal accretion at intermediate spreading rates: Pacific-Antarctic ridge at 65°S. Ph.D., Columbia University, 1994.

Macario, A., W. F. Haxby, J. A. Goff, W. B. F. Ryan, S. C. Cande, and C. A. Raymond, Flow line variations in abyssal hill morphology for the Pacific-Antarctic Ridge at 65°S, *J. Geophys. Res., 99*, 17,921-17,934, 1994.

Macdonald, K. C., Mid-ocean ridges: Fine scale tectonic volcanic and hydrothermal processes within the plate boundary zone, *Annu. Rev. Earth Planet. Sci., 10*, 155-190, 1982.

Macdonald, K. C., The crest of the Mid-Atlantic Ridge: Models for crustal generation processes and tectonics. in *The Western North Atlantic Region*, edited by P. Vogt and B.

Tucholke, 51-68, Geological Society of America, Boulder, Colo., 1986.

Macdonald, K. C., A new view of the mid-ocean ridge from the behaviour of ridge-axis discontinuities, *Nature, 335,* 217-225, 1988.

Macdonald, K. C., and T. Atwater, Evolution of rifted ocean ridges, *Earth Planet. Sci. Lett., 39*, 319-327, 1978.

Macdonald, K. C., P. J. Fox, R. T. Alexander, R. Pockalny, and P. Gente, Volcanic growth faults and the origin of Pacific abyssal hills, *Nature, 380*, 125-129, 1996.

Macdonald, K. C., R. M. Haymon, S. P. Miller, J. C. Sempere, and P. J. Fox, Deep-Tow and Sea Beam studies of dueling propagating ridges on the East Pacific Rise near 20°40'S, *J. Geophys. Res., 93*, 2875-2898, 1988.

Macdonald, K. C., and B. P. Luyendyk, Investigation of faulting and abyssal hill formation on the flanks of the East Pacific Rise (21°N) using Alvin, *Mar. Geophys. Res., 7*, 515-535, 1985.

Magde, L. S., R. S. Detrick, and the. TERA. group, Crustal and upper mantle contribution to the axial gravity anomaly at the southern East Pacfic Rise, *J. Geophys. Res., 100*, 3747-3766, 1995.

Malinverno, A., A quantitative study of axial topography of the Mid-Atlantic Ridge, *J. Geophys. Res., 95*, 2645-2660, 1990.

Malinverno, A., Inverse square-root dependence of mid-ocean ridge flank roughness on spreading rate, *Nature, 352*, 58-60, 1991.

Malinverno, A., and L. E. Gilbert, A stochastic model for the creation of abyssal hill topography at a slow spreading center, *J. Geophys. Res., 94*, 1665-1675, 1989.

Marks, K. M., and J. M. Stock, Variations in ridge morphology and depth-age relationships on the Pacific-Antarctic ridge, *J. Geophys. Res., 99*, 531-543, 1994.

Marty, J. C., and A. Cazenave, Regional variations in subsidence rate of oceanic plates: a global analysis, *Earth Planet. Sci. Lett., 94*, 301-315, 1989.

McNutt, M., and H. W. Menard, Constraints on yield strength in the oceanic lithopshere derived from observations of flexure, *Geophys. J. R. Astron. Soc., 71*, 363-394, 1982.

Menard, H. W., The East Pacific Rise, *Science, 132*, 1737-1742, 1960.

Menard, H. W., Sea floor spreading, topography and the second layer, *Science, 157*, 923-924, 1967.

Mutter, J. C., and J. A. Karson, Structural processes at slow-spreading ridges, *Science, 257*, 627-634, 1992.

Neumann, G. A., and D. W. Forsyth, The paradox of the axial profile: Isostatic compensation along the axis of the Mid-Atlantic Ridge, *J. Geophys. Res., 98*, 17,891-17,910, 1993.

Neumann, G. A., and D. W. Forsyth, High resolution statistical estimation of seafloor morphology: oblique and orthogonal fabric on the flanks of the Mid-Atlantic Ridge, 34°-35.5°S, *Mar. Geophys. Res., 17*, 221-250, 1995.

Normark, W. R., Delineation of thge main extrusion zone of the East Pacific Rise at latitude 21°N, *Geology, 4*, 681-685, 1976.

Parmentier, E. M., Dynamic topography in rift zones: Implications for lithospheric heating, *Philos. Trans. R. Soc. London. Ser. A, 321*, 23-25, 1987.

Parmentier, E. M., and D. W. Forsyth, Three-dimensional flow beneath a slow spreading ridge axis: A dynamic contribution to the deepening of the median valley toward fracture zones, *J. Geophys. Res., 90*, 678-684, 1985.

Parmentier, E. M., and J. Phipps Morgan, Spreading rate dependence of three-dimensional structure in oceanic spreading centres, *Nature, 348*, 325-328, 1990.

Parsons, B., and J. G. Sclater, Ocean floor bathymetry and heat flow, *J. Geophys. Res., 82*, 803-827, 1977.

Phipps Morgan, J., and Y. J. Chen, The genesis of oceanic crust: magma injection, hydrothermal circulation and crustal flow, *J. Geophys. Res., 98*, 6283-6297, 1993.

Phipps Morgan, J., A. Harding, J. Orcutt, G. Kent, and Y. J. Chen, An observational and theoretical synthesis of magma chamber geometry and crustal genesis along a mid-ocean ridge spreading center. in *Magmatic Systems, edited by M.P. Ryan*, Academic Press, San Diego, CA, 1994.

Phipps Morgan, J., E. M. Parmentier, and J. Lin, Mechanisms for the origin of mid-ocean ridge axial topography: Implications for the thermal and mechanical structure of accreting plate boundaries, *J. Geophys. Res., 92*, 12,823-12,836, 1987.

Poliakov, A.N.B. and W. R. Buck, Mechanics of stretching elastic-plastic-viscous layers: Applications to slow-spreading mid-ocean ridges, *this volume*.

Preisendorfer, R. W., Principal component analysis in meteorology and oceanography. Edited by C.D. Mobley. Amsterdam: Elsevier, 1988.

Press, W. H., B. P. Flannery, S. A. Teukolsky, and W. T. Vetterling, Numerical Recipes. 2 ed. New York: Cambridge University Press, 1992.

Purdy, G. M., L. S. L. Kong, G. L. Christeson, and S. C. Solomon, Relationship between spreading rate and the seismic structure of mid-ocean ridges, *Nature, 355*, 815-817, 1992.

Rice, J., Mathematical Statistics and Data Analysis. Pacific Grove, CA: Wadsworth and Brooks, 1988.

Roult, G., D. Roulland, and J. P. Montagner, Antarctica, II; Upper mantle structure from velocities and anisotropy, *Phys. Earth and Planet.Int., 84*, 33-57, 1994.

Sahabi, M., L. Geli, J. L. Olivet, L. Gilig-Capar, G. Roult, H. Ondreas, P. Beuzart, and D. Aslanian, Moprhological reorganization with the Pacific-Antarctic Discordance, *Earth Planet. Sci. Lett., 137*, 157-173, 1996.

Sauter, D., H. Whitechurch, M. Munschy, and E. Humler, Periodicity in the accretion process on the Southeast Indian Ridge at 27° 40'S, *Tectonophysics, 195*, 47-64, 1991.

Sandwell, D.T., and W.H.F. Smith, Marine gravity anomaly from Geosat and ERS-1 satellite altimetry, *J. Geophys. Res., 102*, 10,039-54, 1997.

Scheirer, D. S., and K. C. Macdonald, The variation in cross-sectional area of the axial ridge along the East Pacific Rise: Evidence for the magmatic budget of a fast-spreading center, *J. Geophys. Res., 98*, 7871-7885, 1993.

Scheirer, D. S., and K. C. Macdonald, Near-axis seamounts on the flanks of the East Pacific Rise, 8°N to 17°N, *J. Geophys. Res., 100*, 2239-2259, 1995.

Schouten, H., K. D. Klitgord, and J. A. Whitehead, Segmentation of mid-ocean ridges, *Nature, 317*, 225-229, 1985.

Searle, R. C., and A. S. Laughton, Sonar studies of the Mid-Atlantic ridge and Kurchatov fracture zone, *J. Geophys. Res., 82*, 5313-5328, 1977.

Sempere, J. C., J. R. Cochran, D. Christie, M. Eberle, L. Geli, J. A. Goff, H. Kimura, Y. Ma, A. Shah, C. Small, B. Sylvander, B. P. West, and W. Zhang, The Southeast Indian Ridge between 88°E and 120°E: Variations in crustal accretion at constant spreading rate, *J. Geophys. Res, 102*, p.14489-15505, 1997.

Sempere, J. C., G. M. Purdy, and H. Schouten, Segmentation of the Mid-Atlantic Ridge between 24°N and 30°40'N, *Nature, 344*, 427-431, 1990.

Severinghaus, J. P., and K. C. Macdonald, High inside corners at ridge-transform intersections, *Mar. Geophys. Res., 9*, 353-367, 1988.

Shah, A., and J.-C. Sempere, Morphology of the transition from an axial high to an axial valley at the Southeast Indian Ridge and the relation to variations in mantle temperature, *J. Geophys. Res.*, In Press, 1998.

Shaw, H. R., Mathematical attractor theory and plutonic-volcanic episodicity. in Modeling Volcanic Processes, edited by C.Y. King and R. Scarpa, Vieweg, Braunschweig, 1988.

Shaw, P. R., Ridge segmentation, faulting and crustal thickness in the Atlantic, *Nature, 358*, 490-493, 1992.

Shaw, P. R., and J. Lin, Causes and consequences of variations in faulting style at the Mid-Atlantic Ridge, *J. Geophys. Res., 98*, 21839-21851, 1993.

Shaw, P. R., and D. K. Smith, Robust description of statistically heterogeneous seafloor topography through its slope distribution, *J. Geophys. Res., 95*, 8705-8722, 1990.

Shaw, W. J., and J. Lin, Models of ocean ridge lithospheric deformation: Dependence on crustal thickness, spreading rate and segmentation, *J. Geophys. Res., 101*, 17,9777-17,933, 1996.

Sleep, N. H., Formation of oceanic crust: Some thermal constraints, *J. Geophys. Res., 80*, 4037-4042, 1975.

Sleep, N. H., and S. Biehler, Topography and tectonics at the intersections of fracture zones with central rifts, *J. Geophys. Res., 80*, 2748-2752, 1970.

Small, C., A global analysis of mid-ocean ridge axial topography, *Geophys. J. Int., 116*, 64-84, 1994.

Small, C., and D. T. Sandwell, An abrupt change in ridge axis gravity with spreading rate, *J. Geophys. Res., 94*, 17,383-17,392, 1989.

Small, C., D. Sandwell, and J. Y. Royer, Discontinuous geoid roughness along the Southeast Indian Ridge, *Eos, 70*, 15, 468, 1989.

Small, C., J. R. Cochran, J. C. Sempéré and D. Christie, The Structure and Segmentation of the Southeast Indian Ridge, *Marine Geology*, In Press, 1998.

Smith, D. K., and J. R. Cann, The role of seamount volcanism in crustal construction at the Mid-Atlantic Ridge (24°-30°N), *J. Geophys. Res., 97*, 1645-1658, 1992.

Sparks, D., and M. Spiegelman, Self-consistent models of heat and mass transfer between upper mantle and oceanic crust, *Eos, 76n*, 594, 1995.

Spiegelman, M., and D. P. MacKenzie, Simple 2-D models for melt extraction at mid-ocean ridges and island arcs, *Earth Planet. Sci. Lett., 83n*, 137-152, 1987.

Su, W., R. L. Woodward, and A. M. Dziewonski, Deep origin

of mid-ocean ridge seismic velocity anomalies, *Nature, 360*, 149-152, 1992.

Su, W., R. L. Woodward, and A. M. Dziewonski, Degree 12 model of shear velocity heterogeneity in the mantle, *J. Geophys. Res., 99*, 6945-6980, 1994.

Tapponnier, P., and J. Francheteau, Necking of the lithosphere and the mechanics of slowly accreting plate boundaries, *J. Geophys. Res., 83*, 3955-3970, 1978.

Thatcher, W., and D. P. Hill, A simple model for the fault-generated morphology of slow spreading mid-oceanic ridges, *J. Geophys. Res., 100*, 561-570, 1995.

Tucholke, B. E., and J. Lin, A geologic model for the structure of ridge segments in slow-spreading ocean crust, *J. Geophys. Res., 99*, 11,937-11,958, 1994.

Tucholke, B., J. Lin, M. Kleinrock, M. Tivey, T. Reed, J. Goff, and G. Jaroslow, Segmentation and crustal structure of the western Mid-Atlantic Ridge flank, 25 degrees 25'-27 degrees 10'N and 0-29 m.y., *J. Geophys. Res., 102*, 10,203-10,223, 1997.

Turcotte, D. L., Flexure, *Advances in Geophysics, 21*, 51-86, 1979.

Vautard, R., and M. Ghil, Singular spectrum analysis in nonlinear dynamics with applications to paleoclimatic time series, *Physica D, 38*, 395-424, 1989.

Wang, X., and J. R. Cochran, Gravity anomalies, isostasy and mantle flow at the East Pacific Rise crest, *J. Geophys. Res., 98*, 19,505-19,531, 1993.

Wang, X., and J. R. Cochran, Along-axis gravity gradients at mid-ocean ridges: Implications for mantle flow and axis morphology, *Geology, 23*, 29-32, 1995.

West, B. P., W. S. D. Wilcock, and J. C. Sempere, Three-dimensional structure of the astenospheric flow benearh the Souteast Indian Ridge, *J. Geophys. Res., 102*, 7783-7802, 1997.

West, B. P., and J. C. Sempere, Gravity anomalies, flexure of axial lithosphere, and along-axis asthenospheric flow beneath the Southeast Indian Ridge, *Earth Planet. Sci. Lett.,* In Press, 1998.

Whitehead, J., H. J. B. Dick, and H. Schouten, A mechanism for magmatic accretion under spreading centres, *Nature, 312*, 146 148, 1984.

Zhang, Y. S., and T. Tanimoto, High resolution global upper mantle structure and plate tectonics, *J. Geophys. Res., 98*, 9793-9823, 1993.

C. Small, Lamont-Doherty Earth Observatory, Palisades, NY 10964, (email, small@ldeo.columbia.edu)

Linkages Between Faulting, Volcanism, Hydrothermal Activity and Segmentation on Fast Spreading Centers

Ken C. Macdonald

Department of Geological Sciences and Marine Sciences Institute, University of California, Santa Barbara, CA

Systematic variations occur along the crests of intermediate- to fast-spreading ridges which suggest linkages between volcanism, faulting, hydrothermal activity and magma supply. Going from mid-segment regions to major (first and second order) discontinuities, the following parameters show decreases: axial cross-sectional area, MgO content of lavas, occurrence of an axial magma chamber (sill or lens), abundance of hydrothermal activity, width (and inferred depth) of fissures; while other parameters increase: axial depth, crustal magnetization, mantle Bouguer anomaly, average lava age, abundance of fissures and near axis scarp height (inferred throw on normal faults). These variations can be attributed to variations along segments in magma supply, hydrothermal heat loss and crustal deformation. Finer scale segments (third and fourth order) do not show as much systematic intra-segment variability, but appear to function as independent crustal accretion units for short time intervals. They appear to be the manifestations of dike intrusion events, and thus represent the most fundamental units of crustal accretion. Faulting dismembers the newly accreted crust, but syntectonic volcanism continues to interact with faulting giving rise to volcanic growth faults and significant damming of lava flows. The resulting crustal architecture is somewhat more complex than previously appreciated.

INTRODUCTION

There is a natural tendency in scientific investigations for increasing specialization; many important advances are made by narrowing one's focus and building on the broad foundation of earlier more general research. However, in studies of Mid-Ocean Ridge (MOR) tectonics, magmatism, volcanism and hydrothermal activity, the greatest excitement is in the linkages between these different fields. For example, geophysicists searched for hydrothermal activity on MORs for many years by towing arrays of thermisters near the seafloor; after all someone looking for hot water should measure the temperature of the water. However, hydrothermal activity was eventually documented primarily by photographing the distribution of exotic vent animals

Faulting and Magmatism at Mid-Ocean Ridges
Geophysical Monograph 106
Copyright 1998 by the American Geophysical Union

[Corliss and others, 1979; Spiess and others, 1980]. Even now, the best indicators we have of the recency of volcanic eruptions and the duration of hydrothermal activity is coming from studying the characteristics of benthic faunal communities. For example, during the first deep sea MOR eruption witnessed from a submersible, divers did not see a slow lumbering cascade of pillow lavas as filmed in "Fire Under the Sea" [Moore, 1975; Tepley and Moore, 1974]. What they saw was completely unexpected: white bacterial matting billowing out of the seafloor, creating a scene much like a mid-winter blizzard in Iceland, covering all of the freshly erupted lava with a thick blanket of bacterial snow [Haymon and others, 1993].

As investigations continue, we are seeing more evidence for important linkages between very diverse kinds of observations, for example: ridge crest axial depth, cross-sectional area of the ridge (a proxy for magmatic budget on fast-spreading ridges), crustal thickness, the geochemistry and inferred eruption temperature of lavas, measurements of crustal magnetization, characteristics of near-axis faulting such as along strike variations if fault throw, the widths and inferred depths of cracks and fissures along the axis, lava ages, presence or absence of a crustal axial magma chamber (or melt lens), intensity of hydrothermal activity, and abundance of hydrothermal vent communities. Marine geophysicists and geochemists often attend the talks of benthic ecologists and vice versa; this would never have happened 20 years ago. So, in spite of recent budgetary traumas, research in Mid-Ocean Ridges is more exciting and interdisciplinary than ever before.

This is by no means a comprehensive review, but a discussion of several topics which are vigorously debated at MOR meetings. I attempt to summarize some of the possible linkages between volcanic, tectonic and hydrothermal processes based on observations primarily of fast- to intermediate-spreading ridges (>60 mm/yr). The greater speed of the MOR "tape recorder" at these spreading rates make some of these possible linkages easier to resolve. In the process of synthesizing some of the outstanding issues, I offer some new speculations which will require new observations to test. For students new to this subject, I suggest reading several of the review papers available on MOR tectonics and segmentation [*Detrick et al.*, 1987; *Detrick et al.*, 1993a; *Langmuir et al.*, 1986; *Lin et al.*, 1990; *Macdonald et al.*, 1988; *Macdonald et al.*, 1993; *Mutter and Karson*, 1992; *Schouten et al.*, 1985; *Sinton and Detrick*, 1992] and for a summary of earlier work [*Macdonald*, 1982; *Menard*, 1986].

LARGE SCALE VARIATIONS IN AXIAL MORPHOLOGY; CORRELATIONS WITH MAGMA SUPPLY AND SEGMENTATION.

The axial depth profile of mid-ocean ridges undulates up and down with a wavelength of tens of km and amplitude of tens to hundreds of meters at fast and intermediate rate ridges [*Lonsdale*, 1989; *Macdonald et al.*, 1984]. This same pattern is observed for slow-spreading ridges as well, but the wavelength of undulation is shorter and the amplitude is larger (Fig. 1). In most cases, ridge axis discontinuities (RADs) occur at local maxima along the axial depth profile, although exceptions to this simple relationship do occur, for example in the Lau back arc basin [*Collier and Sinha*, 1992; *Morton and Sleep*, 1985; *Sinha*, 1995; *Wiedicke and Collier*, 1993]. These discontinuities include transform faults (first order); overlapping spreading centers (OSCs, second order) and higher order (third, fourth order) discontinuities which are increasingly short-lived, mobile and associated with smaller offsets of the ridge (see Table 1 and Fig. 2) [*Macdonald et al.*, 1988; *Macdonald et al.*, 1991].

A much debated hypothesis is that the axial depth profile (Figs. 1, 3) reflects the magma supply along a ridge segment [*Langmuir et al.*, 1986; *Lonsdale*, 1985; *Macdonald and Fox*, 1983; *Macdonald and Fox*, 1988]. According to this idea, the magma supply is enhanced along shallow portions of ridge segments and is relatively starved at segment ends (at discontinuities). In support of this hypothesis is the observation at ridges with an axial high that the cross-sectional area or axial volume varies directly with depth (Figs. 4a,b). Maxima in cross-sectional areas (>2.5 km^2) occur at minima along the axial depth profile (generally not near RADS) and are thought to correlate with regions where magma supply is robust. Conversely, small cross-sectional areas (<1.5 km^2) occur at local depth maxima and are interpreted to reflect minima in the magma supply rate along a given ridge segment. Studies of crustal magnetization show that very highly magnetized zones occur near segment ends which are most easily explained by a locally starved magma supply resulting in the eruption of highly fractionated lavas high in Fe [*Carbotte and Macdonald*, 1992; *Gee and Kent*, 1997; *Sempere*, 1991].

An alternative interpretation of axial depth variations is that discontinuities, especially transform faults, act as significant heat sinks at segment ends; the cold "edge effect" controls the magmatic budget due to enhanced cracking of the crust and hydrothermal circulation [*Francheteau and Ballard*, 1983]. However, this idea does not explain commonly seen axial depth variations in which the shallowest part of a ridge may be near a large transform fault (e.g. Clipperton or Garrett), and small OSC's may be associated with large, long wavelength depth anomalies (e.g. 20°40'S OSC, 9°N OSC) [*Macdonald et al.*, 1984]. This idea is also at odds with thermal models which indicate that only the largest transform faults (none on the East Pacific Rise (EPR)) have a significant thermal edge effect [*Forsyth and Wilson*, 1984].

Multi-channel seismic and gravity data provide important but not conclusive documentation of the axial volume/magma supply/segmentation hypothesis (Figs. 4, 5, 6, 7). A bright reflector, which is phase-reversed in many places, occurs commonly (>60% of ridge length) beneath the axial region of both the northern and southern portions of the fast- and ultra-fast spreading East Pacific Rise (EPR) [*Detrick et al.*, 1987; *Detrick et al.*, 1993a; *Kent et al.*, 1994; *Mutter et al.*, 1995; *Tolstoy et al.*, 1997]. This reflector has been interpreted to be a thin lens of magma residing at the top of a broader axial magma reservoir [*Sinton and Detrick*, 1992; *Vera et al.*, 1990]. The amount of melt is highly variable along strike varying from a lens which is primarily crystal mush to one which is close to 100% melt [*Hussenoeder et al.*, 1996]. Along a 70 km long section of the EPR near 14°S, seismologists observed three 3-4 km long sections at ~20 km intervals which may contain pure melt; in-between the lens is inferred to be filled primarily with crystal mush [*Singh et al.*, 1997]. Interestingly, the scale of along-strike variation of melt to mush is very similar to the scale of 4th order segments. This "axial magma chamber" (AMC) reflector is observed where the ridge is shallow and where the axial high has a broad cross-sectional area. Conversely, it is rare where the ridge is deep and narrow, especially near RADs. In fact the AMC reflector is NEVER present where the cross-sectional area <1 km^2, and is present >90% of ridge length where the

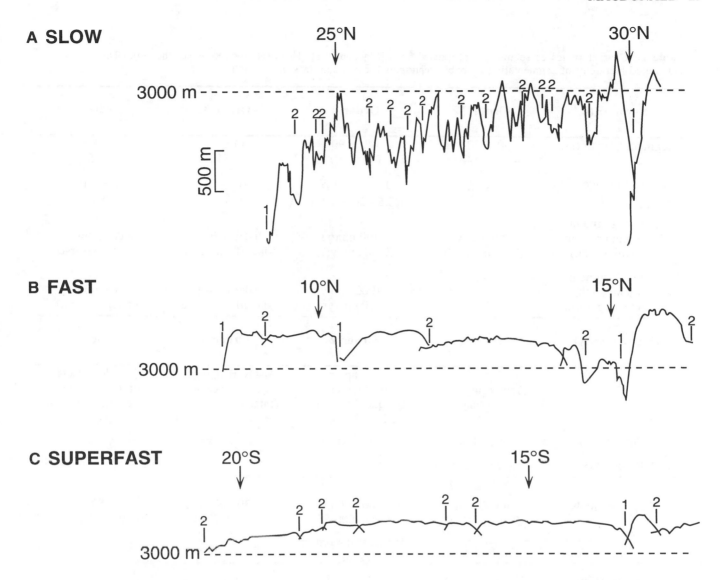

Figure 1. Axial depth profiles for (A) slow-spreading and (B) fast- and (C) ultrafast-spreading ridges (after [*Macdonald et al.*, 1991]; original references are [*Macdonald et al.*, 1992; *Macdonald et al.*, 1988; *Sempere et al.*, 1990]. Discontinuities of orders 1 and 2 typically occur at local depth maxima (discontinuities of orders 3 and 4 not labeled here). The segments at faster spreading rates are longer and have smoother, lower-amplitude axial depth profiles. These depth variations may reflect the pattern of mantle upwelling.

cross-sectional area >3.5 km^2 [*Scheirer and Macdonald*, 1993].

This correlation is particularly notable because it is indirect: the cross-sectional area of the ridge should be linked with the volume of the reservoir of hot rock beneath the ridge which, in turn, might have some correlation with the presence or absence of a magma lens. Melt lens characteristics probably vary over much shorter temporal and spatial scales than properties of the low velocity zone or size of the cross-sectional area [*Hooft et al.*, 1997; *Hussenoeder et al.*, 1996; *Mutter et al.*, 1995]. Thus the seismic measurement

which might best correlate with cross-sectional area is the cross-sectional area of the axial low velocity zone rather than the presence or absence of an AMC reflector [*Hooft et al.*, 1997; *Scheirer and Macdonald*, 1993]. Measurements of the low velocity zone are far too sparse to test this idea at present. Multichannel seismic measurements do indicate a correlation between ridge cross-sectional area and cumulative thickness of layer 2A which is interpreted to be the volcanic extrusive layer [*Carbotte et al.*, 1997]. However, this is only observed where there is significant variation along strike in axial depth and cross-sectional area near 17S.

Table 1. Characteristics of segmentation updated from [Macdonald et al., 1991] (see references therein). This four-tiered hierarchy of segmentation probably represents a continuum in segmentation.

Segments	Order 1	Order 2	Order 3	Order 4
Segment Length (km)	600 ± 300* (400 ± 200)**	140 ± 90 (50 ± 30)	50 ± 30 $(15 \pm 10?)$	14 ± 8 $(7 \pm 5?)$
Segment longevity (yrs)	$> 5 \times 10^6$	$0.5 - 5 \times 10^6$ $(0.5 - 30 \times 10^6)$	$\sim 10^4 - 10^5$ (?)	$\sim 10^2 - 10^4$ (?)
Rate of segment lengthening (long term migration)	0-50 mm/yr (0-30 mm/yr?)	0-100 mm/yr (0-30 mm/yr?)	Indeterminate -no off-axis trace	Indeterminate -no off-axis trace
Rate of segment lengthening (short term propagation)	0-100 mm/yr (?)	0-1000 mm/yr (0-50 mm/yr)	Indeterminate -no off-axis trace	Indeterminate -no off-axis trace
Discontinuities	**Order 1**	**Order 2**	**Order 3**	**Order 4**
Type	Transform, large propagating rifts	Overlapping spreading centers (oblique shear zones, rift valley jogs)	Overlapping spreading centers (Inter-volcano gaps)	Devals, offsets of axial summit caldera (Intra-volcano gaps)
Offset (km)	> 30 km	2-30 km	0.5-2.0 km	< 1 km
Offset age (yrs)***	$> 0.5 \times 10^6$ $(> 2 \times 10^6)$	$< 0.5 \times 10^6$ $(< 2 \times 10^6)$	~ 0	~ 0
Depth anomaly	300-600 m (500-2000 m)	100-300 m (300-1000 m)	30-100 m (50-300 m)	0-50 m (0-100 m?)
Off-axis trace	Fracture zone	V-shaped discordant zone	faint or none	none
High amplitude magnetization?	yes	yes	rarely (?)	no? (?)
Breaks in axial magma chamber?	always	yes, except during OSC linkage? (N.A.)	yes, except during OSC linkage? (N.A.)	rarely
Break in axial low-velocity zone?	yes (N.A.)	no, but reduction in volume (N.A.)	small reduction in volume (N.A.)	small reduction in volume? (N.A.)
Geochemical anomaly?	yes	yes	usually	~50%
Break in high temp venting?	yes	yes	yes (N.A.)	often (N.A.)

*$\pm$ 1σ, **where information differs for slow- versus fast-spreading ridges (< 60 mm/yr), it is placed in (parentheses). N.A. means non-applicable. (?) means not presently known or poorly constrained.
***Offset age refers to the age of the seafloor which is juxtaposed to the spreading axis at a discontinuity.

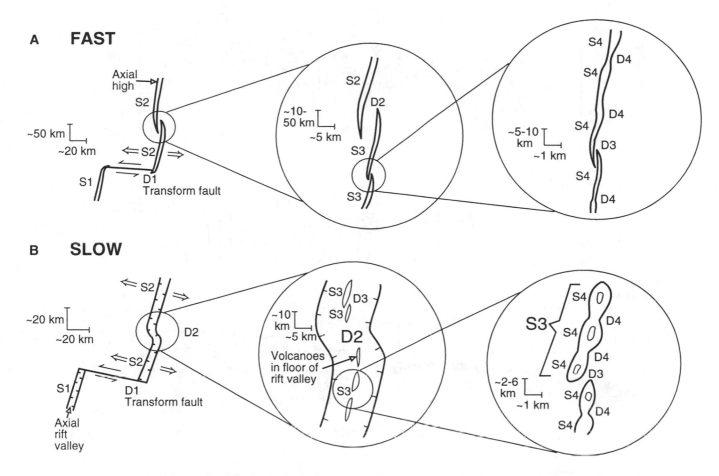

Figure 2. A possible hierarchy of ridge segmentation for (A) fast- and (B) slow-spreading ridges (after [*Macdonald et al.*, 1991]). S1-S4 are ridge segments of order 1-4, and D1-D4 are ridge axis discontinuities of order 1-4. At both fast- and slow-spreading centers, first-order discontinuities are transform faults. Examples of second-order discontinuities are overlapping spreading centers (OSCs) on fast-spreading ridges and oblique shear zones on slow-spreading ridges. Third-order discontinuities are small OSCs on fast-spreading ridges. Fourth-order discontinuities are deviations from axial linearity (devals) resulting in slight bends or lateral offsets of the axis of less than 1 km on fast-spreading ridges [*Langmuir et al.*, 1986]. This four-tiered hierarchy of segmentation is probably a continuum; it has been established, for example, that fourth-order segments and discontinuities can grow to become third-, second-, and even first-order features and vice-versa at both slow- and fast-spreading centers [*Carbotte and Macdonald*, 1992; *Carbotte et al.*, 1991; *Cormier et al.*, 1996; *Grindlay et al.*, 1991; *Perram and Macdonald*, 1990].

Where axial depth and cross-sectional area show little along strike variation, for example near 14°15'S, layer 2A is very uniform in thickness [*Kent et al.*, 1994], and the very small variations measured may indicate the noise level of crustal accretion variability [*Hussenoeder et al.*, 1996]. Variations may also be small because the thickness of the volcanic layer may be self-regulating [*Buck et al.*, 1997]. As layer 2A thickens, the driving pressure for subsequent eruptions is reduced because of the low density of layer 2A volcanics.

There is evidence that major element geochemistry correlates with apparent magmatic budget (axial cross-sectional area). For example, on the EPR at 14°-16°S, where the axial depth profile is exceptionally flat, the cross-sectional area correlates with significant changes in MgO (dredged samples) and depth to the AMC reflector (Fig. 6a). For the EPR 13°-21°S, the correlation coefficient between MgO wt. % and cross-sectional area is R^2=0.61 [*Hooft et al.*, 1997], a respectable correlation for two such seemingly disparate properties.

In addition, the abundance of hydrothermal venting (as measured by light transmission and backscatter in the water column and geochemical tracers) varies directly with the cross-sectional area of the EPR between ~8°-12°N (Fig. 5) [*Baker et al.*, 1994] and between ~14°-19°S (Fig 6b) [*Baker and Urabe*, 1996]. It has been suggested that hydrothermal venting does not correlate with cross-sectional area and

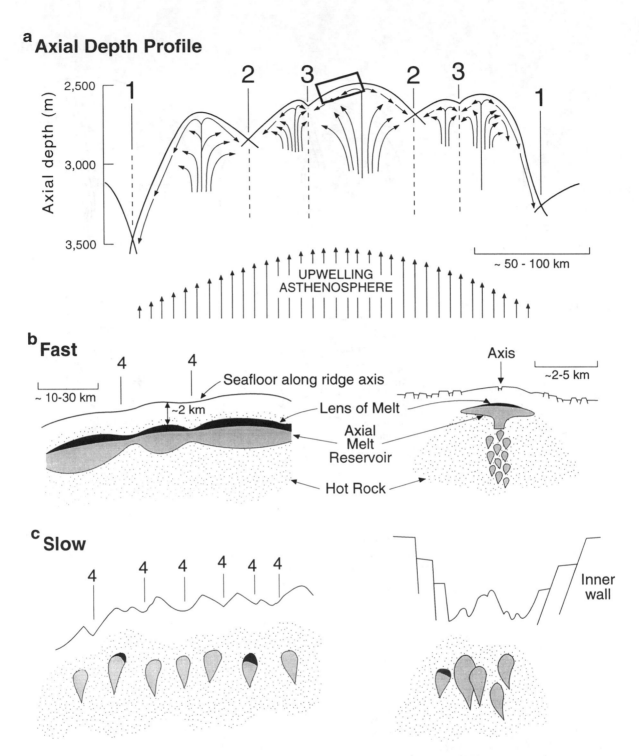

a Axial Depth Profile

b Fast

c Slow

Figure 3. Schematic diagram of how ridge segmentation may be related to mantle upwelling (a), and the distribution of magma supply (b and c) (after [*Macdonald et al.*, 1991] and adapted from [*Langmuir*, 1987; *Macdonald et al.*, 1988]). In (a), the depth scale applies only to the axial depth profile; numbers denote discontinuities and segments of orders 1 to 3. Decompression partial melting in upwelling asthenosphere occurs at depths 30-60 km beneath the ridge. As the melt ascends through a more slowly rising solid residuum, it is partitioned at different levels to feed segments of orders 1-3. Mantle upwelling is hypothesized to be "sheetlike" in the sense that melt is upwelling along the entire length of the ridge; but the supply of melt is thought to be enhanced beneath shallow parts of the ridge away from major discontinuities. The rectangle is an enlargement to show fine-scale segmentation for (b) a fast-spreading example, and (c), a slow-spreading example. In (b) and (c) along-strike cross-sections showing hypothesized partitioning of the magma supply relative to fourth-order discontinuities (4's) and segments are shown on the left. Across-strike cross-sections for fast- and slow-spreading ridges are shown on the right.

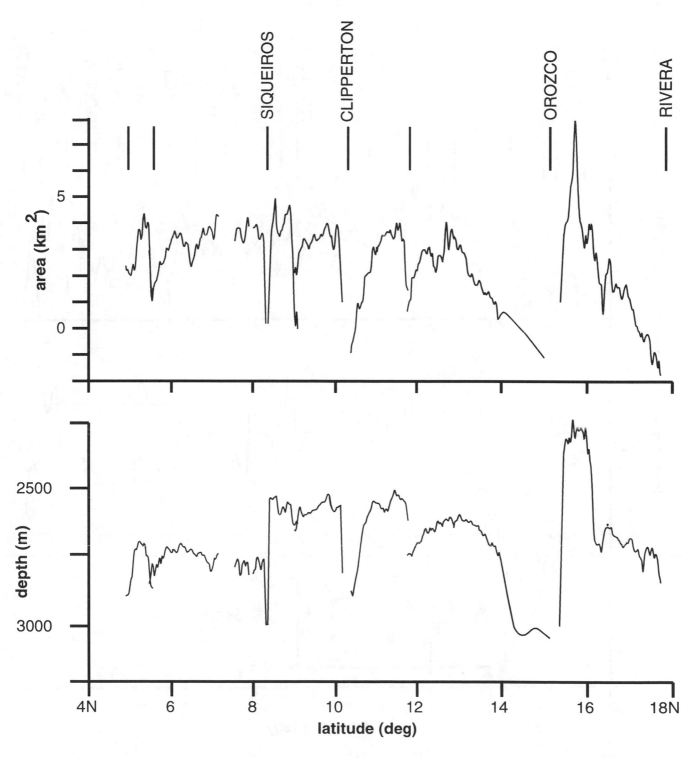

Figure 4. Profiles of the along-axis cross-sectional area and depth variation of the a) northern and b) southern EPR (updated from [*Scheirer and Macdonald*, 1993]). The locations of first- and second-order discontinuities are denoted by vertical bars (first order discontinuities are named); each occurs at a local minimum of the ridge area profile, and a local maximum in ridge axis depth. Two-way travel time to the axial magma chamber reflector is plotted beneath depth for the southern EPR for ~13°-21°S [*Detrick et al.*, 1993a].

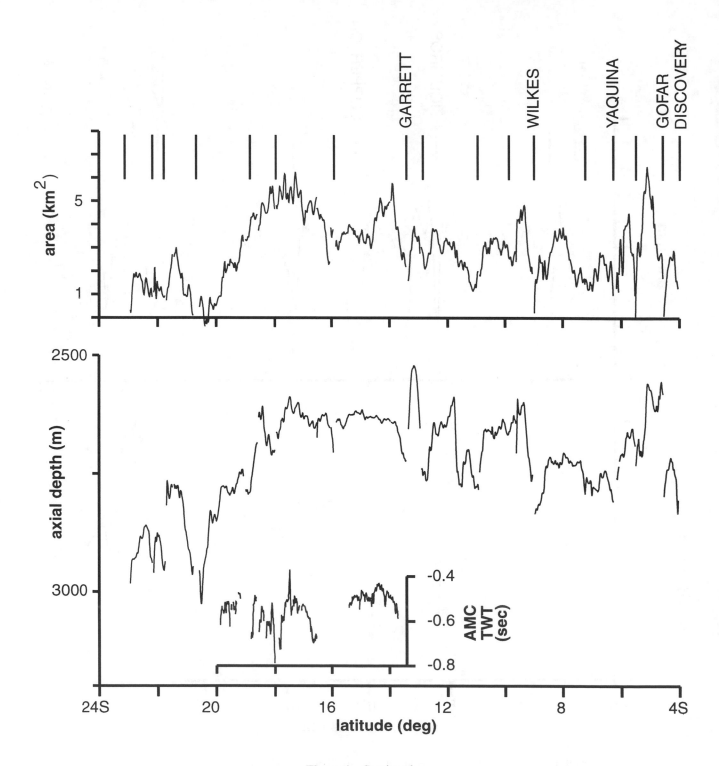

Figure 4. Continued.

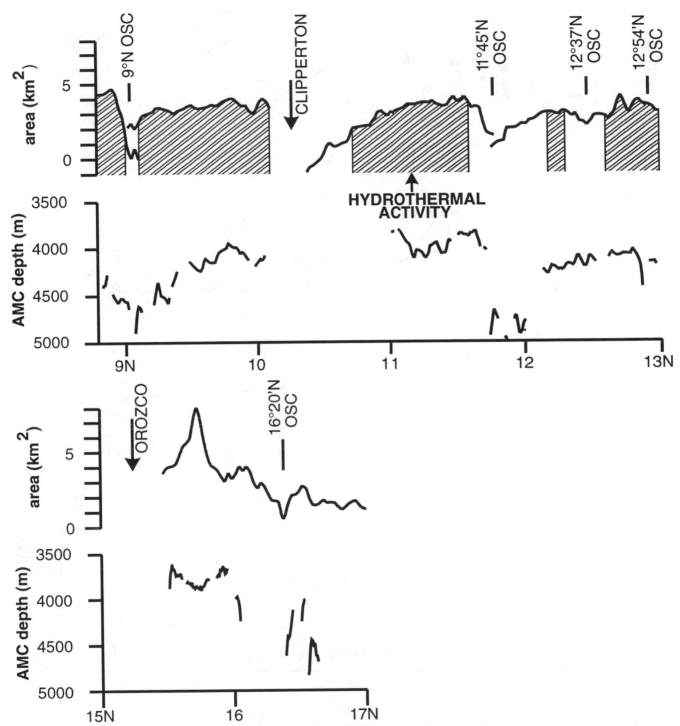

Figure 5. Enlarged view of the cross-sectional area of the northern East Pacific Rise where multi-channel seismic data is available for ~9°-13°N [*Detrick et al.*, 1987] and ~15°-17°N [*Carbotte et al.*, 1996; *Carbotte et al.*, 1997], and where hydrothermal measurements are available for ~9°-13°N [*Baker et al.*, 1994; *Ballard*, 1988; *Charlou et al.*, 1991; *Haymon et al.*, 1991; *Sempere and Macdonald*, 1986a]. The presence of hydrothermal activity is indicated by the diagonally hatched region beneath the graph of cross-sectional area; coverage is very sparse for 11°45'-12°37'N. Hydrothermal activity and an axial magma chamber reflector are more likely to be found where the cross-sectional area is large. However, the depth to the AMC reflector shows little correlation with area.

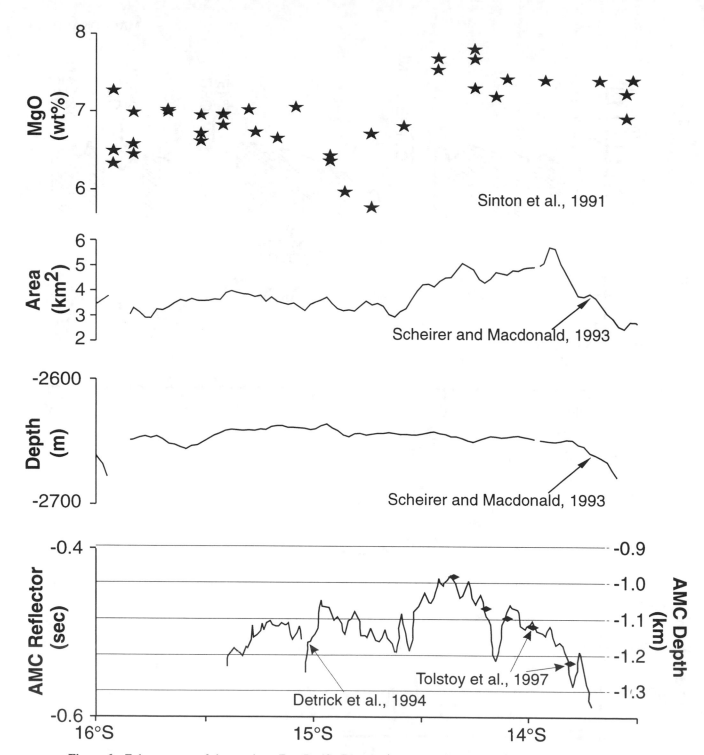

Figure 6. Enlargements of the southern East Pacific Rise for two particularly interesting areas. 6(a) EPR 13.5°-16°S, comparison of dredge sample MgO content, ridge cross-sectional area, axial depth and AMC reflector depth (in seconds of two-way travel time for the solid line from [*Detrick et al.*, 1993a]; diamonds show depth in km where velocity corrected [*Tolstoy et al.*, 1997]). Changes in MgO content, ridge area and AMC reflector depth occur near 14.5°S even though there is no change in axial depth. The long-wavelength deepening of the AMC toward the north is associated with proximity to the Garrett transform fault [*Tolstoy et al.*, 1997].
6(b) EPR ~14°-19°S Depth to AMC reflector in two-way travel time [*Detrick et al.*, 1993a], cross-sectional area of the ridge [*Scheirer and Macdonald*, 1993], and nephel intensity in bottom waters above the ridge crest (an indicator of hydrothermal plumes) [*Baker and Urabe*, 1996]. The correlation is intriguing given the very different crustal properties represented by each measurement.

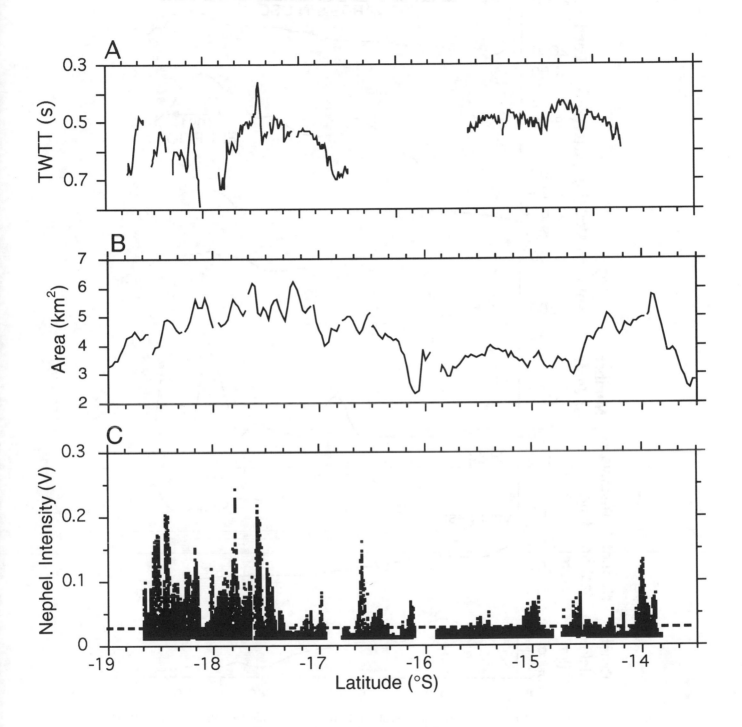

Figure 6. Continued.

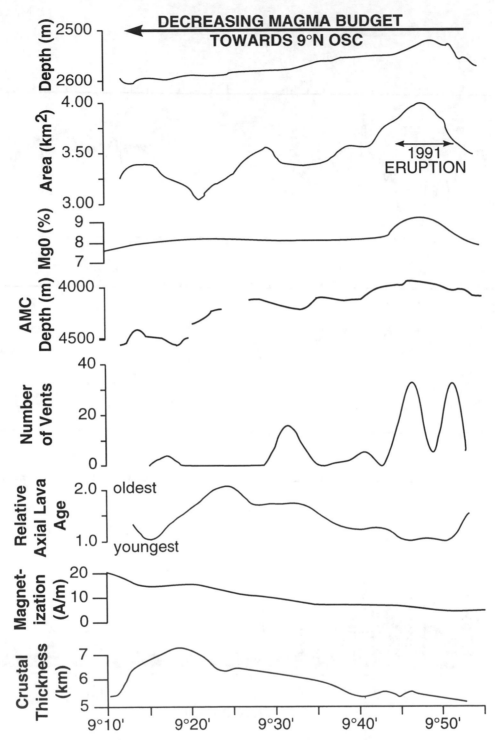

Figure 7. Along-axis and near-axis variation in geological properties for the East Pacific Rise 9°10'-9°55'N. The region of proposed mantle upwelling is centered near 9°40'-50'N. From top to bottom: axial depth [*Macdonald et al.*, 1984]; axial cross-sectional area [*Scheirer and Macdonald*, 1993]; MgO wt. % (heavily smoothed from [*Langmuir et al.*, 1986]); depth to axial magma chamber reflector ([*Detrick et al.*, 1987], updated interpretation has reflector following seafloor depth more closely [*Harding et al.*, 1993]); number of hydrothermal vents binned at two minute intervals of latitude [*Haymon et al.*, 1991]; relative axial lava age [*Haymon et al.*, 1991]; axial magnetization [*Carbotte and Macdonald*, 1992]; near axis crustal thickness [*Barth and Mutter*, 1996]. Other properties which vary systematically along strike toward 9°40'-50'N are: fewer fissures [*Wright et al.*, 1995a]; fissures which are wider and deeper [*Wright et al.*, 1995b]; smaller throw on inward- and outward-facing fault scarps [*Alexander and Macdonald*, 1996].

other long term indicators of magma budget along the southern EPR [*Hooft et al.*, 1997]. Certainly the correlation for short wavelength variations along strike is poor, but at a wavelength of ~100 km, the correlation is good (Fig. 6b). The correlation is also quite good along the northern EPR at this wavelength (Figs. 5, 7). It is not often that one sees a correlation between two such different kinds of measurements. It is all the more remarkable considering that the measurements of hydrothermal activity are sensitive to changes on a time scale of days to months [*Macdonald et al.*, 1980; *Von Damm*, 1990], while the cross-sectional area probably reflects a time scale of change measured in tens of thousands of years [*Carbotte and Macdonald*, 1994a; *Hooft et al.*, 1997; *Scheirer and Macdonald*, 1993].

On the Mid-Atlantic Ridge (MAR), seismic and gravity data indicate that the oceanic crust thins significantly near many transform faults, even those with a small offset [*Cormier et al.*, 1984; *Detrick et al.*, 1993b; *Tolstoy et al.*, 1993]. This is thought to be the result of highly focused mantle upwelling near mid-segment regions, with very little along axis flow of magma away from the upwelling region. Focused upwelling is inferred from "bulls-eye"-shaped residual gravity anomalies [*Kuo and Forsyth*, 1988; *Lin et al.*, 1990] and by crustal thickness variations documented by seismic refraction and microearthquake studies [*Kong et al.*, 1992; *Tolstoy et al.*, 1993; *Wolfe et al.*, 1995]. At slow-spreading centers, melt probably resides in small, isolated and very short-lived pockets beneath the median valley floor (Fig. 3c) [*Smith and Cann*, 1992] and beneath elongate axial volcanic ridges [*Sinha et al.*, 1997]. An alternative view is that the observed along strike variations in topography and crustal thickness can be accounted for by along strike variations in mechanical thinning of the crust by faulting [*Mutter and Karson*, 1992], although there is no conflict between these models, and both focused upwelling and mechanical thinning may occur along each segment [*Tucholke and Lin*, 1994].

One might expect the same to hold at fast spreading centers, i.e., crustal thinning adjacent to OSC's. This does not appear to be the case at 9°N on the EPR, where seismic data suggest a thickening of the crust toward the OSC and a widening of the AMC reflector [*Barth and Mutter*, 1996; *Kent et al.*, 1992]. There is no indication of crustal thinning near the Clipperton transform fault either [*Barth et al.*, 1994; *Begnaud et al.*, 1997; *Van Avendonk et al.*, in press]. And yet, as one approaches the 9°N OSC from the north, the axial depth plunges, the axial cross-sectional area decreases (Fig. 7), the AMC reflector deepens (even in the recalculated reflector [*Harding et al.*, 1993]), average lava age increases [*Haymon et al.*, 1991], MgO in dredged basalts decreases [*Langmuir et al.*, 1986]; hydrothermal activity decreases dramatically [*Haymon et al.*, 1991; *Sempere and Macdonald*, 1986a], crustal magnetization increases significantly (suggesting eruption of more fractionated basalts in a region of decreased magma supply) [*Carbotte and Macdonald*, 1992; *Sempere*, 1991; *Sempere et al.*, 1984], crustal fracturing and inferred depth of fracturing increases (indicating a greater ratio of extensional strain to magma supply) [*Sempere and Macdonald*, 1986a; *Wright et al.*, 1995a; *Wright et al.*, 1995b] and the throw of off-axis normal faults increases (suggesting thicker lithosphere and greater strain) [*Alexander and Macdonald*, 1996]. How can these parameters all correlate so well, indicating a decrease in the magmatic budget and an increase in amagmatic extension, yet the seismic data suggest crustal thickening off-axis from the OSC and a wider magma lens near the OSC?

One possibility is that mantle upwelling and the axial magmatic budget is enhanced away from RADs even at fast spreading centers, but that subaxial flow of magma "downhill" away from the injection region redistributes magma (Fig. 8) [*Cormier et al.*, 1995; *Wang and Cochran*, 1995; *Haymon et al.*, 1991]. This along strike flow and redistribution of magma may be unique to spreading centers with an axial high such as the EPR or Reykjanes where the axial region is sufficiently hot at shallow depths to facilitate subaxial flow [*Bell and Buck*, 1992]. It is well-documented in Iceland and other volcanic areas analogous to MORs that magma can flow in subsurface chambers and dikes for distances of many tens of km away from the source region before erupting [*Ewart et al.*, 1990; *Sigurdsson and Sparks*, 1978]. In this way, thicker crust may occur away from the mid-segment injection points, proximal to discontinuities such as the OSC at 9°N [*Barth and Mutter*, 1996; *Wang et al.*, 1996]. Field evidence in the Oman ophiolite supports this idea. In regions of inferred diapiric upwelling of upper mantle, the crust is thin; away from these upwelling regions the crust is somewhat thicker [*Nicolas et al.*, 1996].

Based on these observations from the EPR and Oman, a modified version of the originally proposed "magma supply" model is proposed (Fig. 9). This model (modified considerably from [*Hooft et al.*, 1997]) explains the intriguing correlation between approximately a dozen structural, geochemical and geophysical variables within a first or second order segment. It also addresses the initially puzzling observation that crust is sometimes thinner in the mid segment region where upwelling is supposedly enhanced. Intuitively, one might expect crust to be thickest over the region where upwelling is enhanced as observed on the MAR. However, along-axis redistribution of melt may be the controlling factor on fast-spreading ridges where the subaxial melt region may be well-connected for tens of km. In this model, temporal variations in along-axis melt connectivity may result in thicker crust near mid-segment when connectivity is low (most often a slow-spreading ridges), and thicker crust closer to the segment ends when connectivity is high (most often, but not always the case at fast-spreading ridges).

When applying this modified magma supply model, it is important to consider the rapidly evolving geometry of OSCs and the resulting temporal variations in melt distribution. The well-studied OSC near 9°N is a good example.

NICOLAS ET AL.: VARIABLE CRUSTAL THICKNESS IN THE OMAN OPHIOLITE

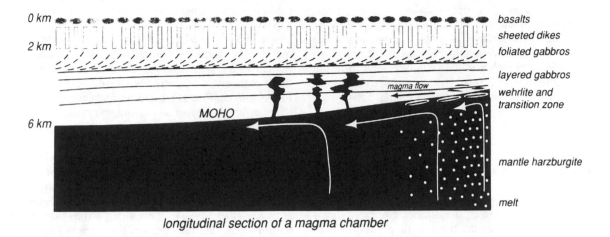

longitudinal section of a magma chamber

Figure 8. Schematic along axis cross-section extending from mid-segment (right) to a first- or second-order discontinuity based on field studies in the Oman ophiolite (after [*Nicolas et al.*, 1996]). Contrary to some earlier interpretations, this presents a possible explanation for crustal thickening near the end of a segment, even though mantle upwelling (vertical portions of arrows) may be concentrated near the mid-segment region to the right. If the axial magma chamber is continuous, melt may flow "downhill", away from the region of upwelling, producing locally thicker crust (primarily in layer 3 gabbros) near one or both segment ends.

At most OSCs, the offset measured at the OSC is ~20% less than the total net offset of the neighboring spreading segments measured away from the OSC due to a curving of the segment tips toward each other caused by shear stresses [*Pollard and Aydin*, 1984; *Sempere and Macdonald*, 1986b]. However, at the 9°N OSC, the net offset of the neighboring ridge segments is significantly less than the ~8 km offset at the OSC (i.e., the neighboring segments are approximately colinear) [*Macdonald et al.*, 1984]. How can this be? The OSC is in a rapidly evolving stage [*Macdonald et al.*, 1987] in which the west limb of the OSC is being sheared and rafted off onto the Pacific plate and thus curves far off to the west (see Fig. 11 in [*Carbotte and Macdonald*, 1992]). When near zero offset occurs at a discontinuity on fast-spreading ridges, this is an optimal time for magma to pool and the AMC to widen near the OSC. A recent 3D seismic survey in this area should be able to test this possibility.

The degree of focusing of mantle upwelling as a function of spreading rate is an important and controversial unknown [*Forsyth*, 1992; *Magde and Sparks*, 1997; *Magde et al.*, 1997]. It has been suggested that upwelling is sheetlike at fast- and intermediate-rate spreading centers, unlike the highly focused upwelling at slow-spreading centers inferred from gravity, bathymetry and theoretical models[*Lin and Phipps Morgan*, 1992; *Parmentier and Phipps Morgan*, 1990; *Tucholke et al.*, 1997]. Perhaps the best qualitative indicators of the degree of focusing of upwelling are along axis variations in cross-sectional area for fast spreading centers, and along-axis variations in either axial depth (Fig. 1) or "mantle Bouguer anomaly" for all spreading centers

[*Wang and Cochran*, 1995]. If this is true, then mantle upwelling is most focused at slow-spreading centers characterized by a deep rift valley, and the degree of focusing steadily decreases as a function of spreading rate (just as along axis variations in depth decrease (Fig. 1)). But focusing of mantle upwelling does not vanish entirely, not even at the fastest spreading rates [*Cormier et al.*, 1995; *Scheirer et al.*, in review].

FINE SCALE VARIATIONS IN RIDGE MORPHOLOGY WITHIN THE AXIAL NEOVOLCANIC ZONE

The axial neovolcanic zone occurs on or near the axis of the axial high on fast spreading centers, or within the floor of the rift valley on slow spreading centers (Fig. 3 b, c). Studies of the widths of the polarity transitions of magnetic anomalies, including in situ measurements from ALVIN, document that ~90% of the volcanism which creates the extrusive layer of oceanic crust occurs in a region 1-5 km wide at most spreading centers [*Klitgord et al.*, 1975; *Macdonald et al.*, 1983; *Sempere et al.*, 1988]. Direct qualitative estimates of lava age at spreading centers using submersibles and ROVs tend to confirm this [*Choukroune et al.*, 1984; *CYAMEX*, 1981; *Embley et al.*, 1991; *Haymon et al.*, 1991; *Haymon and others*, 1993] as well as recent high resolution seismic measurements which show that layer 2A achieves its full thickness within 1-3 km of the rise axis [*Carbotte et al.*, 1997; *Christeson et al.*, 1994; *Harding et al.*, 1993; *Kent et al.*, 1994; *Vera and Diebold*,

MAGMA SUPPLY MODEL
For Fast-Spreading Ridges

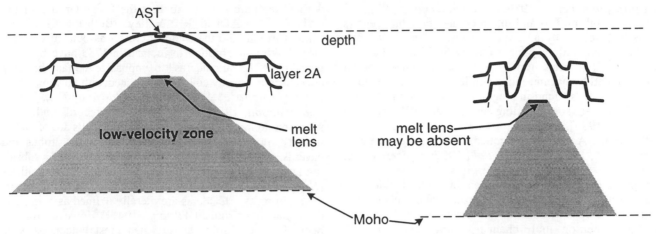

A. High Magma Budget

-shallow water depth

-broad cross-sectional area

-wide low-velocity zone

      -high MgO content

      -high temperature

      -low density

-melt lens ubiquitous, depth and width variable*

-thinner crust (melt flows "downhill" away from mantle upwelling zone)*

-axial summit trough present unless in eruptive phase*

-hydrothermal venting abundant*

-less crustal fissuring*

-younger average lava age*

-lower crustal magnetization*

-smaller throw on flanking fault scarps

B. Low Magma Budget

-deeper water depth

-small cross-sectional area

-narrow low velocity zone (except where 2 LVZs coalesce at an OSC)*

      -lower MgO content

      -lower temperature

      -higher density

-melt lens less common, highly variable width, depth where present, disrupted at RADs*

-thicker crust, but highly variable*

-axial summit trough rare

-hydrothermal venting rare*

-more crustal fissuring*

-older average lava age*

-higher crustal magnetization*

-larger throw on flanking fault scarps

Figure 9. Modified magma supply model for fast-spreading ridges. I started with the model of [*Hooft et al.*, 1997] and modified it significantly to incorporate recent interpretations from the EPR and Oman (main changes marked by *). (A) represents a segment with a robust magmatic budget, (B) represents a segment with a diminished magma budget. (A) may also apply to mid-segment areas overlying regions of enhanced mantle upwelling; conversely, (B) may apply to segment ends overlying regions of diminished mantle upwelling. (AST is the axial summit trough).

1994]. However, there are significant exceptions, including small volume off-axis volcanic constructions [*Alexander and Macdonald*, 1996; *Perfit et al.*, 1994; *White et al.*, in review] and voluminous off-axis floods of basaltic sheet flows [*Macdonald et al.*, 1989; *Macdonald et al.*, 1997].

The axial high on fast and intermediate spreading centers is usually bisected by an axial summit trough ~10-200m deep that is found along approximately 60-70% of the axis. At intermediate spreading rates, an axial rift valley (~500m deep) bounded by normal faults similar to that on the slow-spreading MAR is found in magma-starved regions, for example, the Galapagos spreading center near 86°W [*Klitgord and Mudie*, 1974], the EPR near 23°N [*Macdonald et al.*, 1979], and the Australia-Antarctica Discordance zone area [*Palmer et al.*, 1993; *Sempere et al.*, 1997; *Sempere et al.*, 1991].

Along the axial high of fast-spreading ridges, side-scan sonar records show that there is an excellent correlation between the presence of an axial summit trough and an AMC reflector as seen on multi-channel seismic records (>90% of ridge length) [*Macdonald and Fox*, 1988]. Neither axial summit troughs nor AMCs occur where the ridge has a very small cross-sectional area (<1.5 km^2).

In rare cases, an axial summit trough is *not* observed where the cross-sectional area is large (>3.5 km^2). In these locations, volcanic activity is occurring at present or within the last decade. For example, on the EPR near 9°45-52'N, a volcanic eruption documented from the submersible, ALVIN was associated with a single major dike intrusion [*Haymon and others*, 1993; *Wright et al.*, 1995a], similar to the 1993 eruption on the Juan de Fuca Ridge [*Dziak et al.*, 1995]. Side scan sonar records showed that an axial trough was missing from 9°52'N to 10° 02'N and in subsequent dives, it was found that dike intrusion had propagated into this area producing very recent lava flows (vintage 1992) and hydrothermal activity complete with bacterial "snow-storms" [*Macdonald et al.*, 1992; *Rubin et al.*, 1994]. A similar situation has been thoroughly documented at 17°25'-30'S on the EPR where the axial cross-sectional area is large but the axial summit trough is partly filled [*Auzende*, 1996; *Haymon et. al.*, 1997]. Our interpretation is that the axial summit trough has been flooded with lava so recently that magma withdrawal and summit collapse is just occurring now [*Macdonald et al.*, 1993]. Thus, the presence of an axial summit trough along the axial high of a fast-spreading ridge is a good indicator of the presence of a subaxial lens of partial melt (AMC); where an axial summit trough is not present but the cross-sectional area is large, this is a good indicator of very recent or current volcanic eruptions; where an axial summit trough is not present and the cross-sectional area is small, this is a good indicator of the absence of a magma lens (AMC).

There is considerable controversy over what to call the axial summit trough because of disagreement over its origin. For years it has been called a graben, or "axial summit graben" [*Lonsdale*, 1977; *Macdonald and Fox*, 1988]. However, even these authors emphasized that the axial summit trough has much more in common with collapse structures along volcanic rift zones than with traditionally defined tectonic grabens such as the Rhine Graben or the graben of the African Rift Valley. Both the EPR axial summit trough and the chain of craters associated with the large Lakigigar eruption in Iceland in 1783 have been referred to as "linear calderas" to emphasize the fact that they are caused in part by collapse associated with volcanic eruptions [*Lonsdale*, 1977; *Macdonald and Fox*, 1988; *Thorarinsson*, 1969]. Side-scan sonar records and direct submersible investigations show that the walls are often extremely sinuous (Fig. 10), contain hollow lava tubes and caverns, and are characterized by cooling ledges of a subsiding lava lake, much like the "bathtub rings" on lava pillars [*Haymon et al.*, 1991; *Haymon and others*, 1993; *Fornari et. al.*, in press]. Calderas are literally defined as being caldron-shaped, or round [*Williams*, 1941]. However, this aspect of the definition arose when most volcanoes were known to be approximately round, and well before anyone knew about seafloor spreading and volcanic ridges associated with spreading >100 km long but <10 km wide. To emphasize the volcano-tectonic collapse aspects of axial summit trough origin, some have chosen to refer to the trough as an "axial summit caldera" [*Haymon et al.*, 1991]. The use of the term caldera is controversial in this context and perhaps inappropriate [*Fornari et. al.*, in press; *Chadwick and Embley*, in press], but it has served the purpose of focusing attention on the importance of volcanic processes and collapse associated with magma withdrawal, as opposed to use of the term graben which emphasizes the role of extension and tectonic subsidence associated with graben formation. Even if the trough is primarily a volcanic collapse feature, it may begin as a small graben which is caused by dike injection [*Chadwick*, 1997; *Chadwick and Embley*, in press; *Pollard et al.*, 1983].

The scale of axial summit collapse and underlying causes of this collapse is also controversial [*Fornari et al.*, in press; *Chadwick*, 1997; *Haymon et al.*, 1991]. There is general agreement that many places along the trough on a scale of tens to hundreds of meters are fossil lava lakes whose roofs have collapsed onto the floor of the trough due to magma withdrawal. However, there is disagreement as to whether some of the larger scale collapse structures (tens of km along strike) are caused by collapse of the axial region over an AMC following eruption. The excellent correlation between the occurrence of an axial summit trough and an AMC lends some support to this argument [*Macdonald and Fox*, 1988]. In addition, it is found that the dimensions of the axial summit trough correlate with the depth to the AMC; where the AMC is shallower the trough is deeper [*Lagabrielle and Cormier*, in review]. The thickness of the AMC magma lens or sill is approximately 20-100m [*Hussenoeder et al.*, 1996] which is similar to the

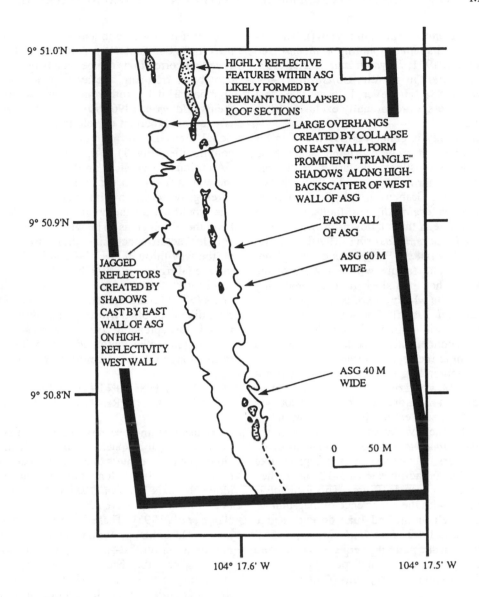

9° 51.0'N

HIGHLY REFLECTIVE
FEATURES WITHIN ASG
LIKELY FORMED BY
REMNANT UNCOLLAPSED
ROOF SECTIONS

B

LARGE OVERHANGS
CREATED BY COLLAPSE
ON EAST WALL FORM
PROMINENT "TRIANGLE"
SHADOWS ALONG HIGH-
BACKSCATTER OF WEST
WALL OF ASG

9° 50.9'N

EAST WALL
OF ASG

ASG 60 M
WIDE

JAGGED
REFLECTORS
CREATED BY
SHADOWS
CAST BY EAST
WALL OF ASG
ON HIGH-
REFLECTIVITY
WEST WALL

ASG 40 M
WIDE

9° 50.8'N

0   50 M

104° 17.6' W   104° 17.5' W

Figure 10. Interpretation of ARGO 100kHz side scan sonar record showing details of the axial summit trough wall on the EPR near 10°N [*Haymon et al.*, 1991]. Heavy lines show limits of survey area, light lines show trough walls. The trough walls are far too sinuous with too many re-entrants to be the plan view traces of normal faults, instead they are attributed to volcano-tectonic collapse.

depth range for the axial summit trough. Thus, multiple episodes of magma withdrawal from the AMC could easily account for collapses of the appropriate magnitude.

I suspect that all three positions presented here are correct at various phases of an eruptive cycle. The occurrence of small grabens caused by the focusing of extensional deformation above individual dike intrusions is well documented in Iceland, Afar and Hawaii, and is common at the beginning of an eruption [*Gudmundsson*, 1983; *Gudmundsson*, 1990; *Pollard et al.*, 1983; *Rubin*, 1990; *Rubin*, 1992]. In a magmatically robust environment like the EPR, it is

likely that these dike-induced graben are buried almost immediately by lava if the dike reaches the surface and feeds an eruption [*Chadwick and Embley*, in press]. Local collapses caused by withdrawal of support of lava lake roofs as lava floods the ridge axis is well documented [*Fornari et al.*, in press; *Haymon and others*, 1993]. Subsequent structural or caldera-like collapse of the axial summit trough floor above a waning magma reservoir is indicated by circumstantial evidence: correlation of the existence of an AMC with the existence of an axial summit trough [*Macdonald and Fox*, 1988], and the inverse correlation of AMC depth with trough size [*Lagabrielle and Cormier*, in review]). The origin of axial summit troughs remains an important unsolved problem [*Haymon et al.*, 1991; *Haymon and others*, 1993; *Lagabrielle and Cormier*, in review; *Lonsdale*, 1977; *Macdonald and Fox*, 1988]. However, I do not think that any of the proposed processes are mutually exclusive; they may occur in sequence.

As far as the terminology is concerned, Williams in his classic paper on "Calderas and Their Origins" said it best, "few terms in the volcanologist's vocabulary have suffered more vicissitudes than the term caldera." To that, I think I would add "graben" too. "Grabens are physically defined by two conjugate and *converging* dip-slip faults, so that, in profile, they contain between them a down-thrown, wedge-shaped fault block" ([*Price and Cosgrove*, 1990]; a similar definition is given in [*Holmes*, 1965]). As discussed below, it is likely that crustal failure within 1-2 km of the rise axis is tensional (rather than shear failure), producing vertical offsets along faults which do *not* converge at depth [*Carbotte and Macdonald*, 1994a]. For now, it may be best to stay with the very generic term, axial summit trough.

Very little is known about eruption frequency. It has been estimated based on some indirect observations that at any given place on a fast-spreading ridge eruptions occur approximately every 50-100 yrs [*Lonsdale*, 1977], and that on slow-spreading ridges it is more like every 10,000 yrs [*Smith and Cann*, 1992]; suggesting, in a very speculative way, that eruption frequency varies inversely with the spreading rate squared [*Macdonald*, 1982]. On intermediate- to fast-spreading centers, if one assumes a typical dike width of ~50 cm. and a spreading rate of 5-10 cm/yr, then an eruption could occur ~every 5-10 yrs. This estimate is in reasonable agreement with the occurrence of megaplumes and eruptions on the well-monitored Juan de Fuca Ridge [*Baker and Hammond*, 1992; *Fox and Cowen*, 1996]. However, observations in sheeted dike sequences in Iceland and ophiolites indicate that only a small percentage of the dikes reach the surface to produce eruptions [*Gudmundsson*, 1990].

On fast-spreading centers, the axial summit trough is so narrow (30-1000 m) and well-defined in most places that tiny offsets and discontinuities of the rise axis can be detected (Table 1, Fig. 2). This finest scale of segmentation (fourth order segments and discontinuities) probably corresponds to individual fissure eruption events similar to the Krafla eruptions in Iceland or the Kilauea east rift zone eruptions in Hawaii. Given a magma chamber depth of 1-2 km, an average dike ascent rate of ~0.1km/hr and an average lengthening rate of ~1 km/hr [*Dvorak*, 1995], typical diking events would give rise to segments 10-20 km long. This agrees with observations of fourth order segmentation [*Macdonald et al.*, 1992] and the scale of the recent diking event on the Juan de Fuca Ridge and in other volcanic rift zones [*Dvorak*, 1995; *Dziak et al.*, 1995]. The duration of such segments is thought to be very short, ~100-1000 yrs.

(too brief in any case to leave even the smallest detectable trace off-axis, Table 1). Yet even at this very fine scale, excellent correlations can be seen between average lava age, density of fissuring, the average widths of fissures, and abundance of hydrothermal vents within individual segments [*Haymon*, 1996; *Haymon et al.*, 1991]. In fact there is even an excellent correlation between ridge cross-sectional area and the abundance of benthic hydrothermal communities (Fig. 7) [*Wright et al.*, 1995a].

A curious observation on the EPR along 9°-10° N is that the widest fissures occur in the youngest lava fields. If fissures grow in width with time and increasing extension, one would expect the opposite; the widest fissures should be in the oldest areas. The widest fissures are ~5 m. Using simple fracture mechanics, these fissures probably extend all the way through layer 2A and into the sheeted dike sequence [*Wright et al.*, 1995b]. These have been interpreted as eruptive fissures, and this is where high temperature vents (>300°C) are concentrated. In contrast to the magma rich, dike-controlled hydrothermal systems which are common on fast-spreading centers, magma-starved hydrothermal systems tend to be controlled more by the penetration of sea water along faults near the ridge axis [*Haymon*, 1996].

FAULTING ON INTERMEDIATE- TO FAST-SPREADING RIDGES

Extension at mid-ocean ridges causes fissuring and normal faulting. The lithosphere is sufficiently thick and strong on slow-spreading centers to support shear failure on the axis, so normal faulting along dipping fault planes can occur along the axis [*Gudmundsson*, 1992; *Kong et al.*, 1992; *Macdonald and Luyendyk*, 1977; *Toomey et al.*, 1985; *Wolfe et al.*, 1995]. These faults produce grabens as defined above. In contrast, normal faulting on dipping planes is not common on fast-spreading centers within +/- 2 km of the axis, probably because the lithosphere is too thin for shear failure to occur [*Carbotte and Macdonald*, 1994a; *Edwards et al.*, 1991]. In addition, near-axis faulting may be impeded at fast-spreading centers because the upper crust is decoupled from extensional stresses by the AMC or by a ductile lower crust [*Chen and Morgan*, 1990; *Goff*, 1991; *Shaw and Lin*, 1996].

Fault strikes tend to be perpendicular to the least compressive stress; thus they also tend to be perpendicular to the spreading direction. While there is some "noise" in the fault trends, most of this noise can be accounted for by perturbations to the least compressive stress direction due to shearing in the vicinity of active or fossil ridge axis discontinuities [*Alexander and Macdonald*, 1996]. Once this is accounted for, we find that fault trends faithfully record changes in the direction of opening to within +/- 3° and can be used to study plate motion changes on a finer scale than that provided by seafloor magnetic anomalies [*Carbotte and Macdonald*, 1994b; *Cormier et al.*, 1996; *Perram and*

Macdonald, 1990] although there is some disagreement over the exact resolution [*Edwards et al.*, 1991]. Because the azimuth distributions of faults are skewed (i.e., non-Gaussian), modes are better than averages in determining spreading directions [*Cormier et al.*, 1996].

(Note: In the following discussion, "inward dipping" faults dip toward the spreading axis; "outward dipping" faults dip away from the axis, and they are inferred in most cases from bathymetric measurements of inward and outward facing scarps).

Studies of the cumulative throw of normal faults, seismicity, and fault spacing suggest that most faulting occurs within +/- 10 km of the axis independent of spreading rate [*Bicknell et al.*, 1987; *Macdonald*, 1982; *Macdonald and Luyendyk*, 1977; *Macdonald and Luyendyk*, 1985; *Solomon et al.*, 1988]. Recently however, more complete studies along the EPR 9°-10°N indicate that normal faults remain active out to ~40 km off-axis in lithosphere ~ 0.7 Ma (Figs. 11, 12) [*Alexander and Macdonald*, 1996; *Crowder and Macdonald*, 1997; *Lee*, 1995; *Macdonald et al.*, 1996; *Wilcock et al.*, 1992]. On the Galapagos Spreading Center, there is some evidence of faulting in even older crust [*Kent et al.*, 1996].

There is a spreading rate dependence for the occurrence of inward and outward dipping faults. Most faults dip toward the axis on slow-spreading centers (~80%), but there is a monotonic increase in the occurrence of outward dipping faults with spreading rate (Fig. 13) [*Carbotte and Macdonald*, 1990]. Inward and outward facing faults are approximately equally abundant at very fast spreading rates. This can be explained by the smaller mean normal stress across a fault plane which dips toward the axis, cutting through thin lithosphere, vs a fault plane which cuts through a much thicker section of lithosphere dipping away from the axis (Fig. 14). Given reasonable thermal models, the difference in the thickness of the lithosphere cut by planes dipping toward vs. away from the axis (and the mean normal stress across those planes) decreases significantly with spreading rate, making outward dipping faults more likely at fast-spreading rates.

At all spreading rates, important along-strike variations in faulting occur within major (first- and second-order) spreading segments. Fault throws (inferred from scarp heights) decrease in the mid-segment regions away from discontinuities (Fig. 15a) [*Alexander and Macdonald*, 1996; *Carbotte and Macdonald*, 1994b; *Goff*, 1991; *Goff et al.*, 1993; *Macario et al.*, 1994; *Shaw*, 1992; *Shaw and Lin*, 1993]. Shaw [*Shaw*, 1992] has proposed that this is caused by a combination of thicker crust, thinner lithosphere, greater magma supply and less amagmatic extension away from RADS in the mid-segment region (Fig. 15b). Another possible explanation for along strike variations in fault throw is along-strike variations in the degree of coupling between the mantle and crust [*Shaw and Lin*, 1996]. A ductile lower crust will tend to decouple the upper crust from

extensional stresses in the mantle, and the existence of a ductile lower crust will depend on spreading rate, the supply of magma to the ridge and proximity to major discontinuities.

Estimates of crustal strain due to normal faulting vary from 10-20% on the slow-spreading MAR [*Macdonald and Luyendyk*, 1977; *Shaw and Lin*, 1993; *Solomon et al.*, 1988] to ~3-5% on the fast-spreading EPR [*Carbotte and Macdonald*, 1994b; *Cowie et al.*, 1993]. In contrast to slow-spreading ridges, faulting at fast-spreading ridges rarely produces earthquakes of magnitude > 5. Cowie et al [*Cowie et al.*, 1993] note that the level of seismicity measured at fast-spreading ridges can only account for a very

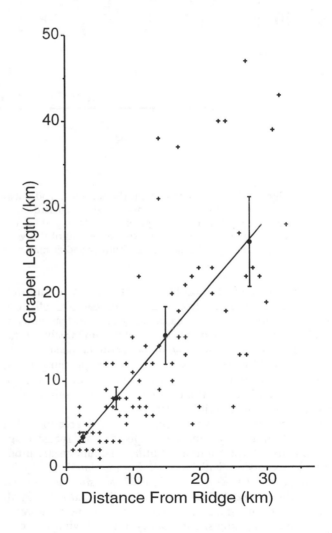

Figure 11. Graben length versus distance from the ridge axis on the EPR 9°-10°N after [*Macdonald et al.*, 1996]. Dots show average graben length in bins 0-5 km, 5-10 km, 10-20 km and 20-35 km off-axis; error bars are +/- 2 standard errors; line is linear regression. Faults bounding the grabens lengthen by propagation and linkage at a rate of ~60 mm/yr.

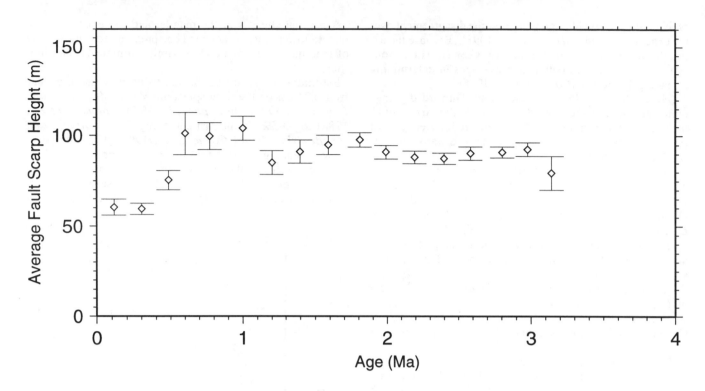

Figure 12. Fault scarp height (both inward- and outward-facing) vs. crustal age near EPR 8°-10°N [*Crowder and Macdonald*, 1997]. Error bars are +/- 2 standard error. Continued lengthening and linkage of faults is inferred to occur at least to a crustal age of ~0.7 Ma or ~60 km off-axis. Large average scarp heights in the first bin are due to scarps which have not yet experienced all of the volcanic burial and reduction in scarp height which eventually occurs during volcanic growth faulting and damming of lava flows by scarps.

small percentage of the observed strain due to faulting, whereas fault strain at slow ridges is comparable to the observed seismic moment release. They suggest that faults in fast-spreading environments accumulate slip largely by stable sliding (aseismically) due to the warm temperatures and associated thin brittle layer. At slower spreading rates, faults will extend beyond a frictional stability transition into a field where fault slip occurs unstably (seismically) because of a thicker brittle layer.

An important unsolved problem is the degree to which abyssal hills reflect alternating episodes of robust magmatic activity with minimal faulting, and amagmatic lithospheric extension with strain accommodated mostly by faulting (Fig. 16) [*Karson et al.*, 1987; *Karson and Winters*, 1992; *Mutter and Karson*, 1992; *Thatcher and Hill*, 1995]. Significant amagmatic extension seems to be peculiar to slow-spreading ridges, particularly at the "high inside corners" of ridge-transform intersections [*Severinghaus and Macdonald*, 1988] where significant detachment faults have unroofed lower crust and upper mantle [*Cann et al.*, 1997; *Karson and Dick*, 1983; *Karson and Elthon*, 1987; *Tucholke et al.*, 1997]. There is no indication of this type of extreme amagmatic extension on fast-spreading ridges [*Carbotte and Macdonald*, 1994b].

VOLCANIC GROWTH FAULTS AND ABYSSAL HILLS IN THE PACIFIC

Ever since the discovery of abyssal hills on the floor of the Pacific Ocean nearly 50 years ago, their origin has been the source of vigorous debate [*Dietz et al.*, 1954]. In the Pacific, they are typically 10-30 km long, 2-5 km wide and 50-300 m high and cover virtually all of the ocean floor except where they are buried beneath sediment. The proponents of a horst and graben structure for the hills [*Bicknell et al.*, 1987; *Lonsdale*, 1977] are persuaded by deep tow high resolution bathymetric records showing many of the hills to be bounded by very steep escarpments facing toward and away from the spreading axis. However, multi-beam bathymetric records (lower resolution but superior coverage) seem to support an opposing school of thought; that volcanoes are erupted along the spreading axis and split in two to create the rolling relief of the abyssal hills [*Kappel and Ryan*, 1986; *Lewis*, 1979]. Other hypotheses posit that the abyssal hills are whole, intact volcanoes rafted away from the spreading axis, or back-tilted fault blocks (Fig. 17).

Based on observations made from the submersible ALVIN on the flanks of the EPR near 9°-10°N, the outward facing slopes of the hills are neither simple outward dipping

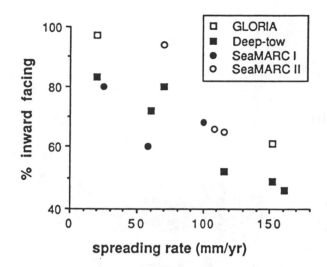

Figure 13. Spreading rate vs. percent of scarps (interpreted to be normal faults) that are inward-facing; after [*Carbotte and Macdonald*, 1990]. An increase in the percentage of inward-facing scarps occurs at slower spreading rates.

normal faults, as would be predicted by the horst/graben model, nor are they entirely volcanic-constructional, as would be predicted by the split volcano model. Instead, the outward facing slopes are "volcanic growth faults" (Figs. 18a, b). Outward-facing scarps produced by episodes of normal faulting are buried near the axis by syntectonic lava flows originating along the axial high [*Macdonald et al.*, 1996]. Repeated episodes of dip-slip faulting and volcanic burial result in structures resembling growth faults, except that the faults are episodically buried by lava flows rather than being continuously buried by sediment deposition. In contrast, inward dipping faults act as tectonic dams to lava flows [*Edwards*, 1991; *Hurst*, 1994]. Based on these observations, the abyssal hills are horsts and the intervening troughs are grabens with the important modification to the horst/graben model that the outward facing slopes are created by volcanic growth faulting rather than traditional normal faulting. Thus, volcanism is a more important aspect of the morphology of abyssal hills than previously appreciated in the horst/graben model. The integrated result of these processes is an asymmetric abyssal hill bounded by a steeply dipping normal fault on the side facing the spreading axis, and bounded by a volcanic growth fault on the opposing side.

Recently collected multichannel seismic data from the very fast-spreading southern EPR show that the base of layer 2A closely follows the shape of the seafloor above it (Fig. 19) [*Carbotte et al.*, 1997]. This could only be the case if the relief of the abyssal hills is primarily attributable to faulting; if they were volcanic constructions rather than volcanically draped horsts, the seafloor and base of layer 2A would have completely different shapes. Seismic reflection

records delineating the base of layer 2A also indicate that significant fault burial occurs both by volcanic draping of outward-facing scarps and damming of lava flows by inward-facing scarps. In one example near 17°30' S, a 100 m high scarp is associated with a 300 m offset in layer 2A indicating 200 m of burial by lava flows (see Fig. 5 in [*Carbotte et al.*, 1997]). If this degree of burial is common, then estimates of crustal extension on fast-spreading ridges (~3-5%) may underestimate extensional strain by a factor of two. A strain of ~10% is probably more accurate. If so, strain on fast-spreading ridges may be almost as large as strain on slow-spreading ridges.

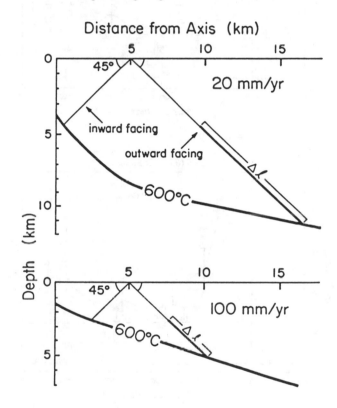

Figure 14. Possible explanation for the increased percentage of inward dipping faults at slower spreading rates after [*Carbotte and Macdonald*, 1990]. Using the 600° C isotherm as an approximation to the base of the brittle layer, it can be seen that outward dipping faults must extend deeper to penetrate the brittle layer than inward dipping faults at a given distance from the ridge, and this difference is much greater at slow spreading rates (100 mm/yr, top) than fast spreading rates (20 mm/yr, bottom). In this example, a dip of 45° is assumed, but the same argument holds for any dip other than 90°. Greater mean stress on a fault plane which extends to greater depth will require greater shear stress to initiate failure, hence predominance of inward-facing faults at slow spreading centers. At faster spreading rates, the difference in mean stress across inward- vs. outward-dipping faults decreases and approximately equal numbers of inward- and outward-facing scarps are observed.

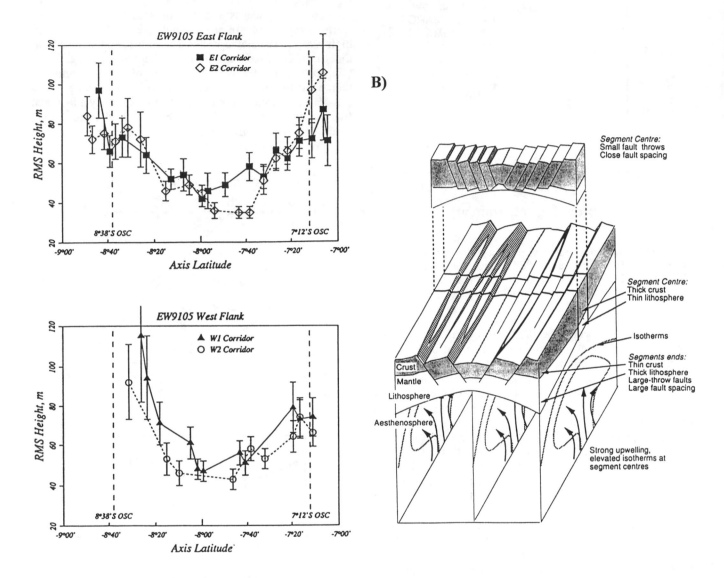

Figure 15. Along axis variations in fault scarps. a) RMS scarp height vs. latitude for the fast-spreading EPR ~7°-8°S after [*Goff et al.*, 1993]. Note that for both the east (top) and west (bottom) flanks of the ridge, scarps are highest near the ends of the segment, in this case a second order segment bounded by two OSCs. Similar along axis variations are observed at all spreading rates.

b) A geologic interpretation for along-axis variations in scarp height, and more closely spaced scarps near mid-segment on the MAR after [*Shaw*, 1992]. Cross-section through segment center (top) shows more closely-spaced, smaller-throw faults than at the segment ends (bottom). Focused mantle upwelling near the segment center causes this region to be hotter, the lithosphere will be thinner while increased melt supply creates a thicker crust. Near the segment ends, the lithosphere will be thicker and magma supply is less creating thinner crust. Along axis variations in scarp height and spacing reflect these along axis variations in lithospheric thickness. Amagmatic extension across the larger faults near segments ends may also thin the crust. Along-axis variations in coupling between crust and mantle may also play an important role [*Shaw and Lin*, 1996].

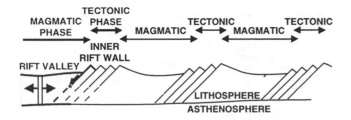

Figure 16. Alternating episodes of increased tectonic (faulting) activity and magmatic activity may be reflected in the abyssal hills of the Atlantic although this idea is still controversial after [*Thatcher and Hill*, 1995].

Abyssal hills and intervening grabens are observed to lengthen off-axis (an observation consistent with faulting but hard to explain if the hills are volcanoes or split volcanoes (Fig. 11)). The rate of lengthening of faults which bound these hills is approximately 60 mm/yr. for the first ~0.7 My (Macdonald in prep.). This lengthening occurs both by along-strike propagation of individual faults as well as by linkage of approximately colinear faults (Fig. 20) [*Alexander and Macdonald*, 1996].

This process describes the origin of most abyssal hills at fast-spreading centers characterized by an axial high, although the degree of volcanic draping of outward dipping faults may vary significantly with local magmatic budget. It may also vary along strike with proximity to discontinuities. At slow- to intermediate-rate spreading centers and adjacent to ridge axis discontinuities, other processes are likely to be important. Near ridge axis discontinuities on fast-spreading centers, the lithosphere appears to be sufficiently thick to support axial volcanic constructions [*Carbotte and Macdonald*, 1994a]. Subsequent propagation and decapitation [*Macdonald et al.*, 1987] of the end of the spreading segment may result in the rafting off of whole volcanoes as abyssal hills (Fig. 17c). At intermediate-rate spreading centers, abyssal hill structure may vary with the local magmatic budget. Where the budget is starved and the axis is characterized by a rift valley, abyssal hills are generally back-tilted fault blocks [*Klitgord and Mudie*, 1974]. Where the magmatic budget is robust and an axial high is present, the axial lithosphere is episodically thick enough to support a volcanic construction which may then be split in two along the spreading axis, resulting in split volcano abyssal hills [*Carbotte and Macdonald*, 1994a; *Kappel and Ryan*, 1986] (Fig. 17d). At slow-spreading centers characterized by an axial rift valley, back-tilted fault blocks and half-grabens may be the dominant origin of abyssal hills (Fig. 17a), although there is continued controversy over the role of high- versus low-angle faults, listric faulting versus planar faulting, and the possible role of punctuated episodes of volcanism versus amagmatic extension [*Gente et al.*, 1995; *Karson et al.*, 1987; *Mutter and Karson*, 1992; *Pockalny et al.*, 1988].

FUTURE WORK

There are a number of outstanding questions. For example, what controls mid-ocean ridge segmentation? Is it controlled by the locations of focused mantle upwelling (mantle "diapirs"), or is it controlled entirely by lithospheric processes such as along-strike variations in amagmatic extension, crust-mantle coupling and hydrothermal heat loss? Is the supply of melt to the ridge two-dimensional (sheetlike) or three-dimensional (concentrated in the mid-segment regions), and does this vary with spreading rate? Do longer segments extract melt from deeper regions than short segments?

There are a number of issues concerning the magmatic and volcanic budgets of spreading centers under different circumstances. How does the axial magma reservoir vary with spreading rate and along-strike? Is it always a narrow, thin magma sill as it appears to be at present beneath the EPR, or are there times when it is much larger? How do the dimensions of the low velocity zone (indicative of the larger, time-integrated magma reservoir) vary along-strike and with spreading rate? Does the magmatic budget and its numerous indicators (such as cross-sectional area, MgO, etc.) vary significantly with time and if so, over what time scales? How frequently do eruptions occur and what are their volumes? How wide and how variable is the zone of dike intrusion relative to average distances that lavas flow from fissure eruptions?

There are also issues concerning deformation of oceanic crust. How does crustal deformation vary with spreading rate? Does extensional strain decrease at faster spreading rates or are our estimates inaccurate due to volcanic growth faulting (volcanic burial)? Do faults on fast spreading centers dip ~50° as on the Mid-Atlantic Ridge (based on focal mechanisms), or are they close to vertical? Is detachment faulting an important process only at the inside-corners of slow-spreading ridges, or does it occur on fast-spreading centers as well during hiatuses in magma supply? Why is most faulting on fast-spreading centers apparently aseismic?

The list is much longer than this and we still have a great deal to learn from high-resolution seagoing investigations of mid-ocean ridges.

Acknowledgments. I want to thank Suzanne Carbotte, Dan Fornari, Lindsay Parson, Jeff Karson and Rachel Haymon for reviewing and assisting with ideas presented in this paper. Thank you also to my students and a long list of colleagues, including most of the authors in this volume, for many stimulating discussions and debates on these subjects, especially my long term partner in ridge crest studies, Jeff Fox. Thank you to Antoinette Padgett and Philip Sharfstein for help with the illustrations and word processing. Lastly, I want to thank the U. S. National Science foundation (NSF grants OCE94-15632 and 94-16996) as well as the Office of Naval Research (ONR grant N00014-94-10678) for their support, and I hope that the Office of Naval Research renews its interest in the deep ocean in the future.

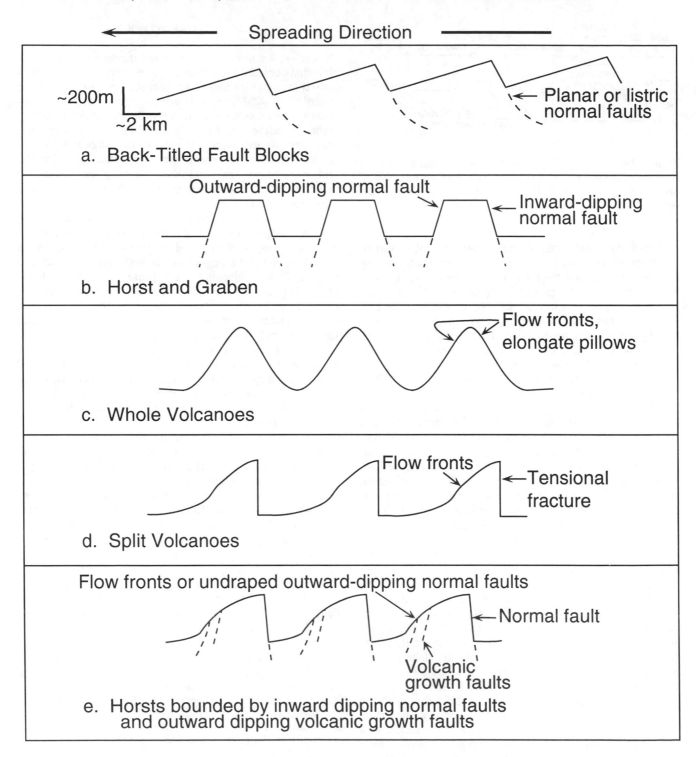

Figure 17. Five models for the development of abyssal hills after [*Macdonald et al.*, 1996]. (a) back-tilted fault blocks (episodic inward-dipping normal faulting off-axis), (b) horst and graben (episodic inward- and outward-dipping faulting off-axis), (c) whole volcanoes (episodic volcanism on-axis), (d) split volcanoes (episodic volcanism and splitting on-axis), (e) horsts bounded by inward-dipping normal faults and outward-dipping volcanic growth faults (episodic faulting off-axis and episodic volcanism on or near-axis).

A)

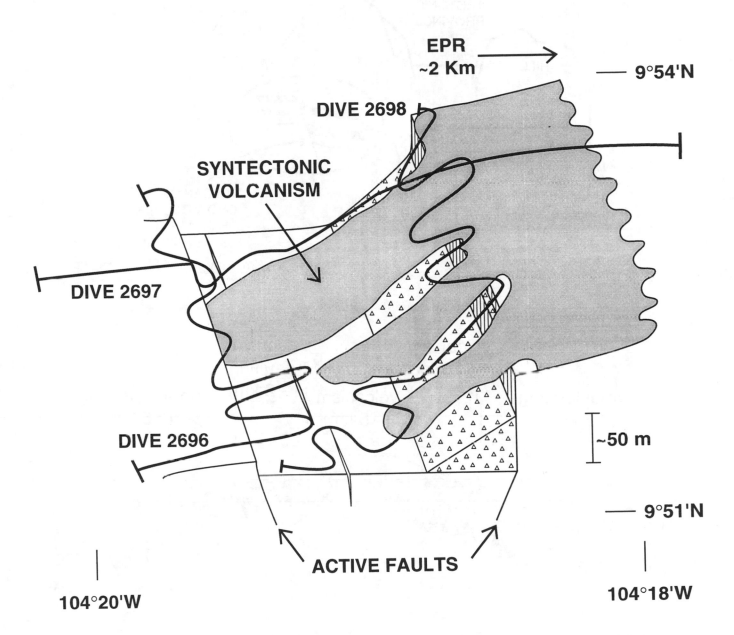

EPR
~2 Km

9°54'N

DIVE 2698

SYNTECTONIC
VOLCANISM

DIVE 2697

DIVE 2696

~50 m

9°51'N

ACTIVE FAULTS

104°20'W

104°18'W

Figure 18. Volcanic growth faults after [*Macdonald et al.*, 1996]. (a) lavas cascading over outward facing scarps based on dives in the 9°54'N near-axis graben.
(b) cross-sectional depiction of the development of volcanic growth faults. Volcanic growth faults are common on fast-spreading centers and explain some of the differences between inward- and outward-facing scarps as well as the morphology and origin of most abyssal hills on fast-spreading centers.

B)

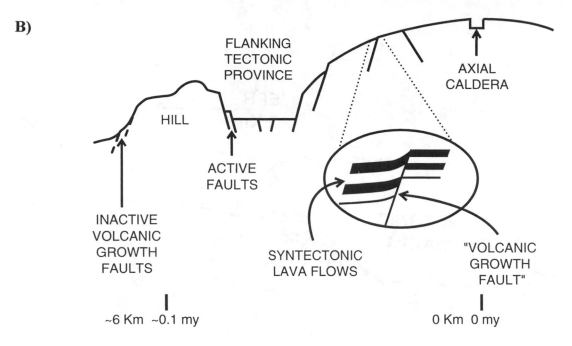

Figure 18. Continued.

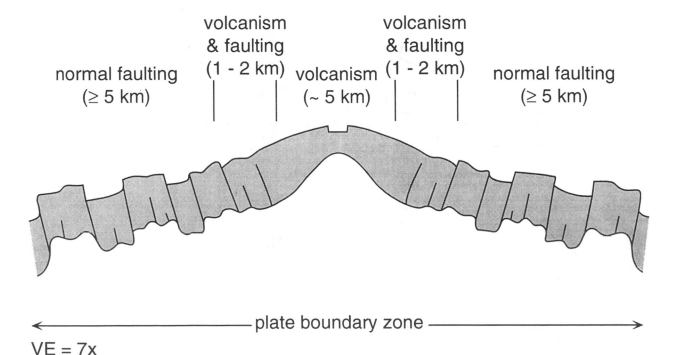

Figure 19. Schematic cross-section of plate boundary zone on the EPR near 17°S based on multi-channel seismic reflection records after [*Carbotte et al.*, 1997]. The volcanic layer accumulates within a region ~5 km wide centered about the ridge axis. Normal faults begin to develop at the edges of this zone but may be partially buried by lava flows. Large-scale normal faulting beyond the neovolcanic zone gives rise to fault bounded abyssal hill topography on the ridge flanks. Both buried fault scarps (mostly outward-facing) and inward facing fault scarps (less vulnerable to complete volcanic burial) account for undulations in the base of layer 2A. These undulations would not exist if the abyssal hills were built by constructional volcanism.

A)

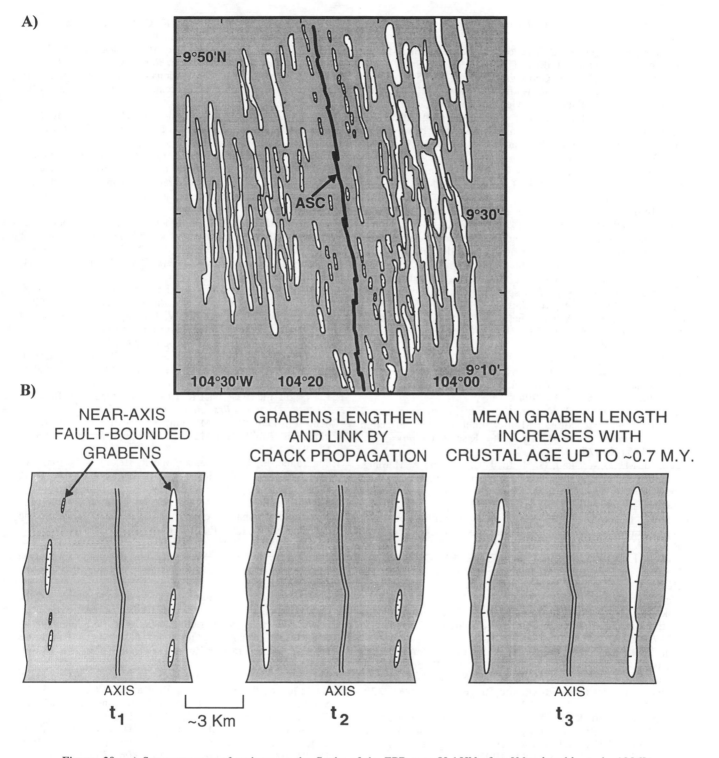

B)

Figure 20. a) Summary map of grabens on the flanks of the EPR near 9°-10°N after [*Macdonald et al.*, 1996] based on Sea Beam and Sea MARC II data; ticks point downslope. Note that the grabens tend to get longer with distance off-axis.
b) proposed time sequence of along-strike propagation and linkage of near-axis grabens which define the edges of abyssal hills.

REFERENCES

Alexander, R. T., and K. C. Macdonald, Sea Beam, Sea MARC II, and ALVIN-based studies of Faulting on the East Pacific Rise 9°20'-9°50'N, Mar. Geophys. Res., 18, 557-587, 1996.

Auzende, J. M., V. Ballu, R. Batiza, D. Bideau, J.L. Charlou, M.H. Cormier, Y. Fouquer, P. Geistdoerfor, Y. Lagabrielle, J. Sinton, P. Spadea, Recent tectonic, magmatic and hydrothermal activity on the East Pacific Rise between 17° and 19°S: Submersible observations, J. Geophys. Res., 101, 17,995-18,010, 1996.

Baker, E. T., R. A. Feely, M. J. Mottl, F. J. Sansone, C. G. Wheat, J. A. Resing, and J. E. Lupton, Hydrothermal plumes along the East Pacific Rise, 8°40' to 11°50'N: Plume distribution and relationship to the apparent magmatic budget, Earth Planet. Sci. Lett., 128, 1-17, 1994.

Baker, E. T., and S. R. Hammond, Hydrothermal venting and the apparent magmatic budget of the Juan de Fuca Ridge, J. Geophys. Res., 97, 3443-3456, 1992.

Baker, E. T., and T. Urabe, The distribution of hydrothermal venting along the superfast-spreading East Pacific Rise, 13°30'-18°40'S, J. Geophys. Res., 101, 8685-8695, 1996.

Ballard, R. D. e. a., ARGO Studies of hydrothermal activity along the EPR 10°40'-11°50'N, Canadian Mineralogist, 26, 467-486, 1988.

Barth, G. A., K. A. Kastens, and E. M. Klein, The origin of bathymetric highs at ridge-transform intersections- a multi-disciplinary case study at the clipperton fracture zone., Mar. Geophys. Res, 16, 1-50, 1994.

Barth, G. A., and J. C. Mutter, Variability in oceanic crustal thickness and structure: Multichannel seismic reflection results from the northern East Pacific Rise, J. Geophys. Res., 101, 17951-17975, 1996.

Begnaud, M., J. McClain, G. Barth, J. Orcutt, and A. Harding, Structure of the eastern Clipperton ridge-transform intersection, East Pacific Rise, from three-dimensional seismic tomography, EOS, 78, 675, 1997.

Bell, R. E., and W. R. Buck, Crustal control of ridge segmentation inferred from observations of the Reykjanes Ridge, Nature, 357, 583-586, 1992.

Bicknell, J. D., P. J. Fox, J.-C. Sempere, and K. C. Macdonald, Tectonics of a fast spreading center: A Deep-Tow and Sea Beam survey of the East Pacific Rise at 19°30'S, Mar. Geophys. Res., 9, 25-45, 1987.

Buck, W. R., S. M. Carbotte, and C. Mutter, Controls on Extrusion at Mid-Ocean Ridges, Geology, 25, 935-939, 1997.

Cann, J. R., D. K. Blackman, D. K. Smith, E. McAllister, B. Janssen, S. Mello, E. Avgerinos, A. R. Pascoe, and J. Escartin, Corrugated slip surfaces formed at ridge-transform intersections on the Mid-Atlantic Ridge, Nature, 385, 329-332, 1997.

Carbotte, S., and K. C. Macdonald, East Pacific Rise 8°-10°30'N: Evolution of ridge segments and discontinuities from SeaMARC II and three-dimensional magnetic studies, J. Geophys. Res., 97, 6959-6982, 1992.

Carbotte, S., C. Mutter, J. Mutter, M. Spiegelman, G. Correa, M. McNutt, V. Baht, and R. Cruz-Orozco, Influence of spreading rate and magma supply on crustal magma bodies: Results from a recent seismic study of the shallow East Pacific Rise north of the Orozco Transform Fault, RIDGE Events, 7, 1-4, 1996.

Carbotte, S., S. M. Welch, and K. C. Macdonald, Spreading rates, rift propagation, and fracture zone offset histories during the past 5 my on the Mid-Atlantic Ridge; 25°-27°30'S and 31°-34°30'S, Mar. Geophys. Res., 13, 51-80, 1991.

Carbotte, S. M., and K. C. Macdonald, Causes of variation in fault-facing direction on the ocean floor, Geology, 18, 749-752, 1990.

Carbotte, S. M., and K. C. Macdonald, The axial topographic high at intermediate and fast spreading ridges, Earth Planet. Sci. Lett., 128, 85-97, 1994a.

Carbotte, S. M., and K. C. Macdonald, Comparison of seafloor tectonic fabric at intermediate, fast and super fast spreading ridges: influence of spreading rate, plate motions, and ridge segmentation on fault patterns, J. Geophys. Res., 99, 13609-13631, 1994b.

Carbotte, S. M., J. C. Mutter, and L. Xu, Contribution of volcanism and tectonism to axial and flank morphology of the southern East Pacific Rise, 17°10'-17°40'S, from a study of layer 2A geometry, J. Geophys. Res., 102, 10,165-10184, 1997.

Chadwick, W. W. J., Is the "axial summit caldera" on the East Pacific Rise really a "graben"?, Ridge Events, 8, 8-11, 1997.

Chadwick, W. W. J., and R. W. Embley, Graben formation associated with recent dike intrusions and volcanic eruptions on the mid-ocean ridge., J. Geophys. Res., in press.

Charlou, J. L., H. Bougault, P. Appriou, P. Jean-Baptiste, J. Etoubleau, and A. Birolleau, Water column anomalies associated with hydrothermal activity between 11° 40' and 13°N on the East Pacific Rise: Discrepancies Between Tracers., Deep Sea Res., 38, 569-596, 1991.

Chen, Y., and W. J. Morgan, Rift valley/no rift valley transition at mid-ocean ridges, J. Geophys. Res., 95, 17,571-17,581, 1990.

Choukroune, P., J. Francheteau, and R. Hekinian, Tectonics of the East Pacific Rise near 12°50'N: A submersible study, Earth Planet. Sci. Lett., 68, 115-127, 1984.

Christeson, G. L., G. M. Purdy, and G. J. Fryer, Seismic constraints on shallow crustal emplacement processes at the fast spreading East Pacific Rise, J. Geophys. Res., 99, 17,957-17,973, 1994.

Collier, J. S., and M. C. Sinha, Seismic mapping of a magma chamber beneath the Valu Fa Ridge, Lau Basin, J. Geophys. Res., 97, 14,031-14,053, 1992.

Corliss, J., and a. t. others, Submarine thermal springs on the Galapagos rift, Science, 203, 1073-1082, 1979.

Cormier, M.-H., R. S. Detrick, and G. M. Purdy, Anomalously thin crust in oceanic fracture zones: New seismic constraints from the Kane Fracture Zone, J. Geophys. Res., 89, 10249-10266, 1984.

Cormier, M.-H., D. S. Scheirer, and K. C. Macdonald, Evolution of the East Pacific Rise at 16°-19°S since 5 Ma: Splitting and rapid migration of axial discontinuities, Marine Geophysical Researches, 18, 53-84, 1996.

Cormier, M. H., K. C. Macdonald, and D. S. Wilson, A three-dimensional gravity analysis of the East Pacific Rise from 18°-21°30'S., J. Geophys. Res, 100, 8063-8082, 1995.

Cowie, P. A., C. H. Scholz, M. Edwards, and A. Malinverno, Fault Strain and Seismic Coupling on Mid-Ocean Ridges, J. Geophys. Res., 98, 17911-17920, 1993.

Crowder, L. K., and K. C. Macdonald, Implications for the Width of the Zone of Active Faulting on the East Pacific Rise: How Wide Is It?, EOS, 78, 647, 1997.

CYAMEX, First manned submersible dives on the East Pacific Rise at 21°N (Project RITA): General results, *Mar. Geophys. Res., 4,* 345-379, 1981.

Detrick, R. S., P. Buhl, E. Vera, J. Mutter, J. Orcutt, J. Madsen, and T. Brocher, Multi-channel seismic imaging of a crustal magma chamber along the East Pacific Rise, *Nature, 326,* 35-41, 1987.

Detrick, R. S., A. J. Harding, G. M. Kent, J. A. Orcutt, J. C. Mutter, and P. Buhl, Seismic structure of the Southern East Pacific Rise, *Science, 259,* 499-503, 1993a.

Detrick, R. S., R. S. White, and G. M. Purdy, Crustal structure of North Atlantic Fracture Zones, *Reviews of Geophysics, 31,* 439-458, 1993b.

Dietz, R. S., H. W. Menard, and E. L. H. E. o. t. M.-P. Expedition, Echograms of the Mid-Pacific Expedition, *Deep-Sea Res., 1,* 258-272, 1954.

Dvorak, J. J., Volcano geodesy: Results of 100 years of Surveillance, *Cahiers du Centre Europeen de Geodynamique et de Seismologie, 8,* 1-19, 1995.

Dziak, R. P., C. G. Fox, and A. E. Schreiner, The June-July 1993 seismo-acoustic event at CoAxial segment, Juan de Fuca Ridge: Evidence for a lateral dike injection, *Geophys. Res. Lett., 22,* 135-138, 1995.

Edwards, M. E., D. J. Fornari, A. Malinverno, W. B. F. Ryan, and J. Madsen, The regional tectonic fabric of the East Pacific Rise from 12°50'N to 15°10'N, *J. Geophys. Res., 96,* 7995-8017, 1991.

Edwards, M. H., *The morphotectonic fabric of the East Pacific Rise: Implications for fault generation and crustal accretion,* Ph.D., Columbia, p. 1991.

Embley, R. W., W. Chadwick, M. R. Perfit, and E. T. Baker, Geology of the northern Cleft segment, Juan de Fuca Ridge: Recent lava flows, sea-floor spreading, and the formation of megaplumes, *Geology, 19,* 771-775, 1991.

Ewart, J. A., B. Voight, and A. Bjornsson, Dynamics of Krafla caldera, north Iceland: 1975-1985, in *Magma Transport and Storage,* edited by M. P. Ryan, p. 225-276, J. Wiley & Sons, 1990.

Fornari, D. J., R. M. Haymon, M. R. Perfit, T. K. P. Gregg, and M. Edwards, Axial summit caldera of the East Pacific Rise 9°-10°N: Geologic Characteristics and Evolution of the Axial Zone on Fast-Spreading Mid-Ocean Ridges, *J. Geophys. Res.,* in press.

Forsyth, D. F., and B. Wilson, Three-dimensional temperature structure of a ridge-transform-ridge system, *Earth Planet. Sci. Lett., 70,* 355-362, 1984.

Forsyth, D. W., *Geophysical Constraints on Mantle Flow and Melt Generation Beneath Mid-Ocean Ridges,* Geophysical Monograph 71: Mantle Flow and Melt Generation at Mid-Ocean Ridges, p. American Geophysical Union, Washington, D.C., 1992.

Fox, C., and J. Cowen, Northern Gorda Ridge: Event detection and response 1996, *RIDGE Events, 7,* 1-4, 1996.

Francheteau, J., and R. D. Ballard, The East Pacific Rise near 21°N, 13°N and 20°S: Inferences for along-strike variability of axial processes of the mid-ocean ridge, *Earth Planet. Sci. Lett., 64,* 93-116, 1983.

Gee, J., and D. V. Kent, Magnetization of axial lavas from the southern East Pacific Rise (14°-23°S): Geochemical controls on magnetic properties, *J. Geophys. Res., 102,* 24, 873-24,886, 1997.

Gente, P., R. A. Pockalny, C. Durand, C. Deplus, M. Maia, G.

Ceuleneer, C. Mevel, M. Cannat, and C. Laverne, Characteristics and evolution of the segmentation of the Mid-Atlantic Ridge between 20°N and 24°N during the last 10 million years, *Earth Planet. Sci. Lett., 129,* 55-71, 1995.

Goff, J. A., A global and regional analysis of near-ridge abyssal hill morphology, *J. Geophys. Res., 96,* 21713-21737, 1991.

Goff, J. A., A. Malinverno, D. J. Fornari, and J. R. Cochran, Abyssal hill segmentation: Quantitative analysis of the East Pacific Rise Flanks 7°S-9°S, *J. Geophys. Res., 98,* 13851-13862, 1993.

Grindlay, N. R., P. J. Fox, and K. C. Macdonald, Second-order ridge-axis discontinuities in the South Atlantic: Morphology, structure, evolution and significance, *Mar. Geophys. Res., 13,* 21-49, 1991.

Gudmundsson, A., Form and dimensions of dikes in eastern Iceland, *Tectonophysics, 95,* 295-307, 1983.

Gudmundsson, A., Emplacement of dikes, sills, and crustal magma chambers at divergent plate boundaries, *Tectonophysics, 176,* 257-275, 1990.

Gudmundsson, A., Formation and growth of normal faults at the divergent plate boundary in Iceland, *Terra Nova, 4,* 464-471, 1992.

Harding, A. J., G. M. Kent, and J. A. Orcutt, A multichannel seismic investigation of upper crustal structure at 9°N on the East Pacific Rise: Implications for crustal accretion, *J. Geophys. Res., 98,* 13925-13944, 1993.

Haymon, R., The response of ridge-crest hydrothermal sytems to segmented, epispdic magma supply, in *Tectonic, Magmatic, Hydrothermal and Biological Segmentation of Mid-Ocean Ridges,* vol. 118, edited by C. J. MacLeod, P. A. Tyler and C. L. Walker, p. 157-168, Geologcial Society Special Publication, 1996.

Haymon, R. M., D. J. Fornari, M. H. Edwards, S. Carbotte, D. Wright, and K. C. Macdonald, Hydrothermal vent distribution along the East Pacific Rise Crest (9°09'-54'N) and its relationship to magmatic and tectonic processes on fast-spreading mid-occan ridges, *Earth Planet. Sci. Lett., 104,* 513-534, 1991.

Haymon, R. M., and a. l. others, Volcanic eruption of the mid-ocean ridge along the East Pacific Rise crest at 9°45'-52'N: Direct submersible observations of seafloor phenomena associated with an eruption event in April, 1991, *Earth Planet. Sci. Lett., 119,* 85-101, 1993.

Haymon, R. M., K. C. Macdonald, S. Baron, L. Crowder, J. Hobson, P. Sharfstein, S. White, B. Bezy, E. Birk, F. Terra, D. Scheirer, D. Wright, L. Magde, C. V. Dover, S. Sudarikov, and G. Levai, Distribution of Fine-Scale Hydrothermal, Volcanic and Tectonic Features Along the EPR Crest, 17°15'-18°30'S: Results of Near-Bottom Acoustic and Optical Surveys, *EOS, 78,* 705, 1997.

Holmes, A., *Principles of Physical Geology,* p. 1288, Thomas Nelson & Sons, Ltd., London, 1965.

Hooft, E. E. E., R. S. Detrick, and G. Kent, Seismic structure and indicators of magma budget along the southern East Pacific Rise, *J. Geophys. Res., 102,* 27,319-27,340, 1997.

Hurst, S. D. a. J. A. K., Paleomagnetism of tilted dikes in fast spread oceanic crust exposed in the Hess Deep Rift: Implications for spreading and rift propagation, *Tectonics, 13,* 789-802, 1994.

Hussenoeder, S. A., J. A. Collins, G. M. Kent, R. S. Detrick and T. Group, Seismic analysis of the axial magma chamber reflector along the southern East Pacific Rise from conventional reflection profiling, *J. Geophys. Res., 101,* 22087-22105, 1996.

Kappel, E. S., and W. B. F. Ryan, Volcanic episodicity and a non-steady state rift valley along Northeast Pacific spreading centers: Evidence from Sea MARC I, *J. Geophys. Res., 91,* 13925-13940, 1986.

Karson, J. A., and H. J. B. Dick, Tectonics of ridge-transform intersections at the Kane Fracture Zone, *Marine Geophys. Res., 6,* 51-98, 1983.

Karson, J. A., and D. A. Elthon, Evidence for variations in magma production along oceanic spreading centers: A critical appraisal, *Geology, 15,* 127-131, 1987.

Karson, J. A., G. Thompson, S. E. Humphris, J. M. Edmond, W. B. Bryan, J. R. Brown, A. T. Winters, R. A. Pockalny, J. F. Casey, A. C. Campbel, G. Kinkhammer, M. R. Palmer, R. J. Kinzler, and M. M. Sulanowska, Along-axis variation in seafloor spreading in the MARK area, *Nature, 328,* 681-685, 1987.

Karson, J. A., and A. T. Winters, Along-axis variations in tectonic extension and accomodation zones in the MARK Area, Mid-Atalntic Ridge 23°N latitude, in *Ophiolites and their Modern Oceanic Analogues,* vol. Special Publication No. 60, edited by L. M. Parson, B. J. Murton and P. Browning, p. 107-116, Geological Society of America, Boulder, 1992.

Kent, G., A. J. Harding, and J. A. Orcutt, Distribution of magma beneath the East Pacific Rise near the 9°03'N overlapping spreading center from forward modeling of CDP data, *J. Geophys. Res., 98,* 13971-13996, 1992.

Kent, G. M., A. J. Harding, J. A. Orcutt, R. S. Detrick, J. C. Mutter, and P. Buhl, Uniform accretion of oceanic crust south of the Garrett Transform at 14°15'S on the East Pacific Rise, *J. Geophys. Res., 100,* 9097-9116, 1994.

Kent, G. M., S. A. Swift, R. S. Detrick, J. A. Collins, and R. A. Stephen, Evidence for active normal faulting on 5.9 Ma old crust near Hole 504B on the southern flank of the Costa Rica Rift, *Geology, 24,* 83-86, 1996.

Klitgord, K. D., S. P. Huestis, J. D. Mudie, and R. L. Parker, An analysis of near-bottom magnetic anomalies: Seafloor spreading and the magnetized layer, *Geophys. J. Roy. Astr. Soc., 43,* 387-424, 1975.

Klitgord, K. D., and J. D. Mudie, The Galapagos spreading center: A near-bottom geophysical survey, *Geophys. J. Roy. Astr. Soc., 38,* 563-586, 1974.

Kong, L. S., S. C. Solomon, and G. M. Purdy, Microearthquake characteristics of a mid-ocean ridge along-axis high, *J. Geophys. Res., 97,* 1659-1685, 1992.

Kuo, B. Y., and D. W. Forsyth, Gravity anomalies of the ridge transform intersection system in the South Atlantic between 31 and 34.5°S: Upwelling centers and variations in crustal thickness, *Mar. Geophys. Res., 10,* 205-232, 1988.

Lagabrielle, Y., and M.-H. Cormier, Fast spreading ridge crest morphology controlled by magma reservoir characteristics: the origin of the large summit trough alonv the East Pacific Rise at 17°56'-18°35'S, *Earth Planet. Sci. Lett.,* in review.

Langmuir, C. H., A magma chamber observed?, *Nature, 322,* 422-429, 1987.

Langmuir, C. H., J. F. Bender, and R. Batiza, Petrological and tectonic segmentation of the East Pacific Rise, 5°30'N-14°30'N, *Nature, 322,* 422-429, 1986.

Lee, S.-M. a. S. C. S., Constraints from Sea Beam bathymetry on the development of normal faults on the East Pacific Rise, *Geophys. Res. Lett., 22,* 3135-3138, 1995.

Lewis, B. T. R., Periodicities in volcanism and longitudinal magma flow on the East Pacific Rise at 23°N, *Geophys. Res. Lett., 6,* 753-756, 1979.

Lin, J., and J. Phipps Morgan, The spreading rate dependence of three-dimensional mid-ocean ridge gravity structure, *Geophys. Res. Lett., 19,* 13-16, 1992.

Lin, J., G. M. Purdy, H. Schouten, J.-C. Sempere, and C. Zervas, Evidence from gravity data for focused magmatic accretion along the Mid-Atlantic Ridge, *Nature, 344,* 627-632, 1990.

Lonsdale, P., Structural geomorphology of a fast-spreading rise crest: The East Pacific Rise near 3° 25' S, *Mar. Geophys. Res., 3,* 251-293, 1977.

Lonsdale, P., Linear volcanos along the Pacific-Cocos plate boundary, 9°N to the Galapagos triple junction, *Tectonophysics, 116,* 255-279, 1985.

Lonsdale, P., Segmentation of the Pacific-Nazca spreading center, 1°N-20°S, *J. Geophys. Res., 94,* 12197-12226, 1989.

Macario, A., W. F. Haxby, J. A. Goff, W. B. F. Ryan, S. C. Cande, and C. A. Raymond, Flow line variations in abyssal hill morphology for the Pacific-Antarctic Ridge at 65°S, *J. Geophys. Res., 99,* 17,921-17,934, 1994.

Macdonald, K., J.-C. Sempere, and P. J. Fox, East Pacific Rise from Siqueiros to Orozco Fracture Zones: Along-strike continuity of axial neovolcanic zone and structure and evolution of overlapping spreading centers, *J. Geophys. Res., 89,* 6049-6069, 1984.

Macdonald, K. C., Mid-ocean ridges: Fine scale tectonic, volcanic and hydrothermal processes within the plate boundary zone, *Ann. Rev. Earth Planet. Sci., 10,* 155-190, 1982.

Macdonald, K. C., E. T. Baker, R. Haymon, M. D. Lilley, M. R. Perfit, D. J. Fornari, and K. V. Damm, Time-series water column measurements and submersible observations at EPR 9°30'N-10°02'N from 4/91 to 3/92: Evidence for changing hydrothermal activity, *EOS Trans. AGU, 73,* 530, 1992.

Macdonald, K. C., K. Becker, F. N. Spiess, and R. D. Ballard, Hydrothermal heat flux of the "black smoker" vents on the East Pacific Rise, *Earth Planet. Sci. Lett., 48,* 1-7, 1980.

Macdonald, K. C., and P. J. Fox, Overlapping spreading centers: New accretion geometry on the East Pacific Rise, *Nature, 302,* 55-58, 1983.

Macdonald, K. C., and P. J. Fox, The axial summit graben and cross-sectional shape of the East Pacific Rise as indicators of axial magma chambers and recent volcanic eruptions, *Earth Planet. Sci. Lett., 88,* 119-131, 1988.

Macdonald, K. C., P. J. Fox, R. T. Alexander, R. Pockalny, and P. Gente, Volcanic growth faults and the origin of Pacific abyssal hills, *Nature, 380,* 125-129, 1996.

Macdonald, K. C., P. J. Fox, S. Carbotte, M. Eisen, S. Miller, L. Perram, D. Scheirer, S. Tighe, and C. Weiland, The East Pacific Rise and its flanks, 8°-18°N: History of segmentation, propagation and spreading direction based on SeaMARC II and Sea Beam studies, *Mar. Geophys. Res., 14,* 299-344, 1992.

Macdonald, K. C., P. J. Fox, L. J. Perram, M. F. Eisen, R. M. Haymon, S. P. Miller, S. M. Carbotte, M.-H. Cormier, and A. N. Shor, A new view of the mid-ocean ridge from the behaviour of ridge-axis discontinuities, *Nature, 335,* 217-225, 1988.

Macdonald, K. C., R. Haymon, and A. Shor, A 220 km2 recently erupted lava field on the East Pacific Rise near lat 8°S, *Geology, 17,* 212-216, 1989.

Macdonald, K. C., R. M. Haymon, S. M. White, and P. J. Sharfstein, Evidence for Extensive Lava Flows on the Southern East Pacific Rise, *EOS, 78,* 705, 1997.

Macdonald, K. C., K. Kastens, F. N. Spiess, and S. P. Miller, Deep tow studies of the Tamayo Transform Fault, *Mar. Geophys. Res., 4,* 37-70, 1979.

Macdonald, K. C., and B. P. Luyendyk, Deep-tow studies of the structure of the Mid-Atlantic Ridge crest near lat 37° N, *Geol. Soc. America Bull., 88,* 621-636, 1977.

Macdonald, K. C., and B. P. Luyendyk, Investigation of faulting and abyssal hill formation on the flanks of the East Pacific Rise (21°N) using ALVIN, *Mar. Geophys. Res., 7,* 515-535, 1985.

Macdonald, K. C., S. P. Miller, B. P. Luyendyk, T. M. Atwater, and L. Shure, Investigation of a Vine-Matthews magnetic lineation from a submersible: The source and character of marine magnetic anomalies, *J. Geophys. Res., 88,* 3403-3418, 1983.

Macdonald, K. C., Scheirer, S. Carbotte, and P. J. Fox, It's only topography, *GSA Today, 3,* 1, 24-25, 29-30, 34-35, 1993.

Macdonald, K. C., D. S. Scheirer, and S. M. Carbotte, Mid-ocean ridges: Discontinuities, segments and giant cracks, *Science, 253,* 986-994, 1991.

Macdonald, K. C., J.-C. Sempere, P. J. Fox, and R. Tyce, Tectonic evolution of ridge axis discontinuities by the meeting, linking, or self-decapitation of neighboring ridge segments, *Geology, 15,* 993-997, 1987.

Magde, L. S., and D. W. Sparks, Three-dimensional mantle upwelling, melt generation, and melt migration beneath segmented slow-spreading ridges, *J. Geophys. Res., 102,* 20,571-20,583, 1997.

Magde, L. S., D. W. Sparks, and R. S. Detrick, The relationship between buoyant mantle flow, melt migration, and gravity bull's eyes at the Mid-Atlantic Ridge between 33°N and 35°N, *Earth Planet. Sci. Lett., 148,* 59-67, 1997.

Menard, H. W., *The Ocean of Truth,* p. 353, Princeton University Press, Princeton, N.J., 1986.

Moore, J. G. 6.-2., Mechanism of formation of pillow lava, *American Scientist, 63,* 269-277, 1975.

Morton, J. L., and N. H. Sleep, Seismic reflections from a Lau Basin magma chamber, in *Geology and Offshore Resources of Pacific Island Arcs - Tonga Region,* Earth Science Series, vol. 2, edited by D. W. Scholl and T. L. Vallier, p. 441-453, Circum-Pacific Council for Energy and Mineral Resources, Houston, 1985.

Mutter, J. C., S. M. Carbotte, W. Su, L. Xu, P. Buhl, R. S. Detrick, G. Kent, J. Orcutt, and A. Harding, Seismic Images of active magma systems beneath the East Pacific Rise 17° to 17°35'S, *Science, 268,* 391-395, 1995.

Mutter, J. C., and J. A. Karson, Structural processes at slow-spreading ridges, *Science, 257,* 627-634, 1992.

Nicolas, A., F. Boudier, and B. Ildefonse, Variable crustal thickness in the Oman ophiolite: Implication for oceanic crust, *J. Geophys. Res., 101,* 17,941-17,950, 1996.

Palmer, J., J.-C. Sempere, D. M. Christie, and J. Phipps Morgan, Morphology and tectonics of the Australian-Antarctic Discordance between 123°E and 128°E, *Mar. Geophys. Res., 15,* 121-152, 1993.

Parmentier, E. M., and J. Phipps Morgan, Spreading rate dependence of three-dimensional structure in oceanic spreading centers, *Nature, 348,* 325-328, 1990.

Perfit, M. R., D. J. Fornari, M. C. Smith, J. F. Bender, C. H. Langmuir, and R. M. Haymon, Small-scale spatial and temporal variations in mid-ocean ridge crest magmatic processes, *Geology, 22,* 375-379, 1994.

Perram, L. J., and K. C. Macdonald, A one-million-year history of the 11°45'N East Pacific Rise discontinuity, *J. Geophys. Res., 95,* 21,363-21,381, 1990.

Pockalny, R. A., R. S. Detrick, and P. J. Fox, Morphology and tectonics of the Kane Transform from Sea Beam bathymetry data, *J. Geophys. Res., 93,* 3179-3193, 1988.

Pollard, D. D., and A. Aydin, Propagation and linkage of oceanic ridge segments, *J. Geophys. Res., 89,* 10,017-10,028, 1984.

Pollard, D. D., P. T. Delaney, W. A. Duffield, E. T. Endo, and A. T. Okamura, Surface deformation in volcanic rift zones, *Tectonophysics, 94,* 541-584, 1983.

Price, N. J., and J. W. Cosgrove, *Analysis of Geological Structures,* p. 502, Cambridge University Press, Cambridge, 1990.

Rubin, A. M., A comparison of rift-zone tectonics in Iceland and Hawaii, *Bull. Volcanol., 52,* 302-319, 1990.

Rubin, A. M., Dike-induced faulting and graben subsidence in volcanic rift zone, *J. Geophys. Res., 97,* 1839-1858, 1992.

Rubin, K. H., J. MacDougall, and M. Perfit, 210Po-210Pb dating of recent volcanic eruptions on the seafloor., *Nature, 368,* 841-844, 1994.

Scheirer, D. S., D. W. Forsyth, M.-H. Cormier, and K. C. Macdonald, Shipboard Geophysical Indications of Asymmetry and Melt Production Beneath the East Pacific Rise near the MELT Experiment, *Science,* in review.

Scheirer, D. S., and K. C. Macdonald, Variation in cross-sectional area of the axial ridge along the East Pacific Rise: Evidence for the magamatic budget of a fast-spreading center, *J. Geophys. Res., 98,* 7871-7885, 1993.

Schouten, H., K. D. Klitgord, and J. A. Whitehead, Segmentation of mid-ocean ridges, *Nature, 317,* 225-229, 1985.

Sempere, J.-C., High magnetization zones near spreading center discontinuities, *Earth Planet. Sci. Lett., 107,* 389-405, 1991.

Sempere, J.-C., J. R. Cochran, and S. S. Team, The Southeast Indian Ocean Ridge between 88° and 118°E: Variations in crustal accretion at constant spreading rate., *J. Geophys Res., 102,* 15489-15507, 1997.

Sempere, J.-C., and K. C. Macdonald, Deep-Tow studies of the overlapping spreading centers at 9°03'N on the East Pacific Rise, *Tectonics, 5,* 881-900, 1986a.

Sempere, J.-C., and K. C. Macdonald, Overlapping spreading centers: Implications from crack growth simulation by the displacement discontinuity method, *Tectonics, 5,* 151-163, 1986b.

Sempere, J.-C., K. C. Macdonald, and S. P. Miller, Overlapping spreading centers: 3-D inversion of the magnetic field at 9°03'N on the East Pacific Rise, *Geophys. J. Roy. Astr. Soc., 79,* 799-811, 1984.

Sempere, J.-C., K. C. Macdonald, S. P. MIller, and L. Shure, Detailed study of the Brunhes/Matuyama reversal boundary on the East Pacific Rise at 19°30'S:Implications for crustal emplacement processes at an ultrafast spreading center, *Mar. Geophys. Res., 9,* 1-25, 1988.

Sempere, J.-C., J. Palmer, D. M. Christie, J. Phipps Morgan, and A. N. Shor, Australian-Antarctic discordance, *Geology, 19,* 429-432, 1991.

Sempere, J.-C., G. M. Purdy, and H. Schouten, Segmentation of the Mid-Atlantic Ridge between 24°N and 30°40'N, *Nature, 344,* 427-431, 1990.

Severinghaus, J. P., and K. C. Macdonald, High inside corners at ridge-transform intersections, *Mar. Geophys. Res., 9,* 353-367, 1988.

Shaw, P. R., Ridge segmentation, faulting and crustal thickness in the Atlantic Ocean, *Nature, 358,* 490-493, 1992.

Shaw, P. R., and J. Lin, Causes and consequences of variations in faulting style at the Mid-Atlantic Ridge, *J. Geophys. Res., 98,* 21839-21851, 1993.

Shaw, W. J., and J. Lin, Models of ocean ridge lithosphere deformation: Dependence on crustal thickness, spreading rate and segmentation, *J. Geophys. Res., 101,* 17,977-17,993, 1996.

Sigurdsson, H., and S. R. J. Sparks, Lateral magma flow within rifted Icelandic crust, *Nature, 274,* 126-130, 1978.

Singh, S. C., J. Collier, G. Kent, and A. Harding, Along-axis variations in crustal magma properties at the southern East Pacific Rise : Melt to mush, *EOS, 78,* 670, 1997.

Sinha, M. C., Segmentation and rift propagation at the Valu Fa Ridge, Lau Basin: Evidence from gravity data, *J. Geophys. Res., 100,* 15,025-15,043, 1995.

Sinha, M. C., D. A. Navin, L M MacGregor, S. Constable, C. Peirce, A. White, and M. A. Inglis, Evidence for accumulated melt beneath the slow-spreaing Mid-Atlantic Ridge, *Phil. Trans. Roy. Soc., A, 355,* 233-253, 1997.

Sinton, J. M., and R. S. Detrick, Mid-ocean ridge magma chambers, *J. Geophys. Res., 97,* 197-216, 1992.

Smith, D. K., and J. R. Cann, The role of seamount volcanism in crustal construction at the Mid-Atlantic Ridge (24°30'N), *J. Geophys. Res., 97,* 1645-1658, 1992.

Solomon, S. C., P. Y. Huang, and L. Meinke, The seismic moment budget of slowly spreading ridges, *Nature, 334,* 58-61, 1988.

Spiess, R. N., and a. 2. others, East Pacific Rise: Hot springs and geophysical experiments, *Science, 207,* 1421-1433, 1980.

Tepley, L., and J. G. Moore, *Fire under the sea--the origin of pillow lava,* Moonlight Productions, Mountain View, California., 1974.

Thatcher, W., and D. P. Hill, A simple model for fault-generated morphology for slow-spreading mid-oceanic ridges, *J. Geophys. Res., 95,* 561-570, 1995.

Thorarinsson, S., The Lakagigar eruption of 1783, *Bull. Volcan., 33,* 910-927, 1969.

Tolstoy, M., A. Harding, and J. Orcutt, Crustal thickness on the Mid-Atlantic Ridge: bulls-eye gravity anomalies and focused accretion, *Science, 262,* 726-729, 1993.

Tolstoy, M., A. J. Harding, J. A. Orcutt, and T. Group, Deepening of the axial magma chamber on the southern East Pacific Rise toward the Garrett Fracture Zone, *J. Geophys. Res., 102,* 3097-3108, 1997.

Toomey, D. R., S. C. Solomon, G. M. Purdy, and M. H. Murray, Microearthquakes beneath the median valley of the Mid-Atlantic Ridge near 23°N: Hypocenters and focal mechanisms, *J. Geophys. Res., 90,* 5443-5458, 1985.

Tucholke, B. E., and J. Lin, A geological model for the structure of ridge segments in slow spreading ocean crust, *J. Geophys. Res., 99,* 11937-11958, 1994.

Tucholke, B. E., J. Lin, M. C. Kleinrock, M. A. Tivey, T. B. Reed, J. Goff, and G. E. Jaroslow, Segmentation and crustal structure of the western Mid-Atlantic Ridge flank, 25°25'-27°10'N and 0-29 m.y., *J. Geophys. Res., 102,* 10,203-10,223, 1997.

Van Avendonk, H., A. Harding, and J. Orcutt, A 2-D tompgraphjic study of the Clipperton transform fault, *J. Geophys. Res.,* in press.

Vera, E. E., and J. B. Diebold, Seismic imaging of oceanic layer 2A between 9°30'N and 10°N on the East Pacific Rise from two-ship wide-aperture profiles, *J. Geophys. Res., 99,* 3031-3041, 1994.

Vera, E. E., J. C. Mutter, P. Buhl, J. A. Orcutt, A. J. Harding, M. E. Kappus, R. S. Detrick, and T. M. Brocher, The structure of 0- to 0.2-m.y.-old oceanic crust at 9°N on the East Pacific Rise from expanded spread profiles, *J. Geophys. Res., 95,* 15529-15556, 1990.

Von Damm, K. L., Seafloor hydrothermal activity: Black smoker chemistry and chimneys, *Ann. Rev. Earth Planet. Sci., 18,* 173-204, 1990.

Wang, X., and J. Cochran, Along-axis gravity gradients at mid-ocean ridges: Implications for mantle flow and axial morphology, *Geology, 23,* 29-32, 1995.

Wang, X., J. R. Cochran, and G. A. Barth, Gravity anomalies, crustal thickness and the pattern of mantle flow at the fast spreading East Pacific Rise, 9-10 degrees N: Evidence for three-dimensional upwelling, *J. Geophys. Res, 101,* 17927-17940, 1996.

White, S. M., K. C. Macdonald, D. S. Scheirer, and M.-H. Cormier, Distribution of isolated volcanoes on the flanks of the East Pacific Rise, 15.3°-20°S, *J. Geophys. Res.,* in review.

Wiedicke, M., and J. Collier, Morphology of the Valu Fa spreading ridge in the southern Lau Basin., *J. Geophys. Res, 98,* 11769 - 11782, 1993.

Wilcock, W. S. D., G. M. Purdy, S. C. Solomon, D. D. L., and D. R. Toomey, Microearthquakes on and near the East Pacific Rise, 9°-10°N, *Geophys. Res. Lett., 19,* 2131-2134, 1992.

Williams, H., Calderas and Their Origin, *Bulletin of the Department of Geological Sciences, U.C. Berkeley, 25,* 239-346, 1941.

Wolfe, C. J., G. M. Purdy, D. R. Toomey, and S. C. Solomon, Microearthquake characteristics and crustal velocity structure at 29°N on the Mid-Atlantic Ridge: The architecture of a slow spreading segment, *J. Geophys. Res., 100,* 24,449-24,472, 1995.

Wright, D. J., R. M. Haymon, and D. J. Fornari, Crustal fissuring and its relationship to magmatic and hydrothermal processes on the East Pacific Rise crest (9°12' to 54'N), *J. Geophys. Res., 100,* 6097-6120, 1995a.

Wright, D. J., R. M. Haymon, and K. C. Macdonald, Breaking new ground: Estimates of crack depth along the axial zone of the East Pacific Rise (9°12'-54'N), *Earth Planet. Sci. lett., 134,* 441-457, 1995b.

K. C. Macdonald, University of California Department of Geological Sciences, Santa Barbara, CA 93106

Magmatism at Mid-Ocean Ridges: Constraints from Volcanological and Geochemical Investigations

Michael R. Perfit

Department of Geology, University of Florida, Gainesville

William W. Chadwick, Jr.

Oregon State University, Hatfield Marine Science Center, Newport

The morphological, structural, and volcanic characteristics of the neovolcanic zone at mid-ocean ridges (MOR) vary strongly with spreading rate. At fast-spreading ridges, the neovolcanic zone is narrow, has low-relief both across and along strike, is dominated by the products of fluid, fissure-fed eruptions, and exhibits morphologic and magmatic continuity along axis. At slow-spreading ridges, the neovolcanic zone is wider, has greater relief, is characterized by many discrete point-source constructs, and exhibits less morphologic and magmatic continuity along axis compared to fast-spreading ridges. Intermediate-spreading ridges typically have characteristics that vary in time and space between these two extremes. Lava flow morphology also varies markedly with spreading rate - sheet flows are dominant on fast-spreading ridges whereas pillow lavas are dominant at slow-spreading ridges. The morphological differences primarily reflect a difference in extrusion rates, and indicate that dikes are intruded at higher magma pressure at fast-spreading ridges. Even though volcanism appears to be concentrated within the neovolcanic zone, off-axis eruptions add significant volumes to the crust. Off-axis volcanism may be fed by the distal sections of magma lenses or, in the case of long-lived, near-axis seamounts, from magma sources that are independent of sub-axial magma bodies. The timing, locations, and volumes of volcanic events on the MOR are still largely unknown, but the documentation of recent eruptions have provided new insights and the first quantitative information regarding active volcanic processes on the ridge-crest. Documentation of historical eruptions has been realized by some good luck and detailed surveying of the neovolcanic zone along the southern Juan de Fuca Ridge (JdFR) and northern East Pacific Rise (EPR), but the most recent eruptions have been detected in real-time by listening with hydrophones for acoustic T-waves that are generated by small earthquakes during shallow crustal intrusive/extrusive events. Narrow grabens have formed adjacent to some of the new lava flows,

Faulting and Magmatism at Mid-Ocean Ridges
Geophysical Monograph 106
Copyright 1998 by the American Geophysical Union

where dikes have intruded near the surface. Similar dike-induced graben faulting has also been documented on rift zones of subaerial volcanoes.

Fine-scale mapping and sampling of a few neovolcanic zones and their adjacent crestal terrains, coupled with geochemical investigations and U-series radiometric dating, have provided critical information regarding the time and spatial scales of MOR magmatism. These more accurate and precise sampling and dating techniques have allowed us to better quantify rates and volumes of magmatic events and to evaluate if changes in mid-ocean ridge basalt (MORB) chemistry are temporally or spatially related (or both). ^{210}Po-^{210}Pb systematics have been successfully used to date and confirm young eruptions (age < 2 yr). New techniques for dating young MORB by mass spectrometric measurement of ^{238}U-^{230}Th, ^{230}Th-^{226}Ra, and ^{235}U-^{231}Pa disequilibria have been successful, but show that samples from neovolcanic zones yield ages (on the order of a few ka) that must be considered "crustal residence ages" rather than true ages of eruption. Along the 9°-10°N segment of the EPR, U-Ra dates show a regional trend consistent with axial variations in topography, axial magma chamber depth and extent of magmatic fractionation which allow constraints to be placed on crystallization rates and construction of the oceanic crust. The identification of anomalously young lavas up to 4 km off-axis on the northern EPR using U-series disequilibria data, also indicates a significant amount of magmatic activity occurs off-axis and that this volcanism can result in the observed thickening of seismic layer 2A (the layer of the oceanic crust that is assumed to be composed of extrusive lavas based on seismic wave velocities and seismic reflection profiles). The chemical diversity and non-systematic distribution of lava types and ages observed on a small-scale (<600 m) across the axis of the EPR may reflect rapid changes in magma chemistry that occur during crystallization and replenishment in small magma lenses coupled with the effects of frequent low-volume eruptions both within and outside of the axial summit trough.

Although significant differences in sources and melting parameters have been shown to control the compositions of MORB on a regional/global basis, local chemical variations appear primarily to be controlled by fractional crystallization and magma mixing at shallow levels (0.5 - 15 km). In conjunction with detailed seafloor mapping, compositional data and quantitative models have been used to examine the volumes, rates of eruption and ranges of geochemical variation within several individual lava flows recently erupted on the MOR. Although all the flows show some degree of chemical variability, sheet flows on the northern EPR and JdFR exhibit greater homogeneity than pillow flows from the JdFR and northern Mid-Atlantic Ridge. The available data from recent eruptions are consistent with a model in which fast-spreading ridges have frequent, relatively homogeneous, small-volume eruptions whereas slow-spreading ridges have infrequent, more heterogeneous, larger eruptions. The fundamental influence on the variables that change with spreading rate along the MOR system seems to be whether or not a steady-state magma reservoir can be sustained at a given location. Where magma supply is continuous and robust, volcanic output dominates over tectonic process; where the supply is intermittent, tectonism may dominate. The critical spreading rate between steady-state and non-steady-state magma reservoirs is approximately 50 mm/yr.

Table 1.   Characteristics of the Neovolcanic Zone at Different Spreading Rates.

Rate of spreading	Range of spreading rates (mm/yr, full rate)	Typical width of neovolcanic zone (meters)	Speculative recurrence interval between eruptions (yrs)	Estimated volume of individual eruptions ($\times 10^6$ m^3)
fast	80-160	100-200	5	1-5
intermediate	40-80	200-2000	50	5-50
slow	10-40	2000-12000	5000	50-1000

1. INTRODUCTION

The mid-ocean ridge (MOR) is the largest, most continuous volcanic system on Earth, yet we have directly observed, mapped or sampled less than one percent of its total surface area. Nevertheless, it is clear that there are major differences in the morphology, structure, and temporal and spatial scales of magmatism on ridges with differing spreading rates. Geophysical, volcanological and geochemical results have helped to explain some of these differences in terms of the competing interaction between volcanic and tectonic processes. Quantitative information on the temporal variability of ridge crest phenomena on geologically short time scales (days to thousands of years) has only become available within the last decade. During the past ten years, detailed investigations at a few sites on the MOR have provided important insights into our understanding of the timing, locations, and volumes of recent magmatic events, as well as their associated hydrothermal and biological impacts. In addition, short-lived U-series isotopes in young lavas are now being used to provide temporal constraints on magmatic process. This review paper considers the current state of knowledge of magmatism and related tectonism on the MOR crest, with an emphasis on the results from documented eruption sites on the East Pacific Rise (EPR) and Juan de Fuca Ridge (JdFR), and how they can be used to infer the relationships between magmatic and tectonic processes at divergent plate boundaries.

2. MAGMATISM AT MID-OCEAN RIDGES

2.1. The Neovolcanic Zone

The "neovolcanic zone" at a MOR is the region along the plate boundary within which volcanic eruptions and high-temperature hydrothermal activity are concentrated. It is the region where magmatism is focused at the ridge crest. The part of the ridge crest that encompasses the neovolcanic zone has variously been described as: the axial valley, rift valley, inner valley floor, median valley, elongate summit depression, axial summit graben, and axial summit caldera. Differences in description are largely a result of the various morphological expressions of the axis of the ridge crest. The continuous spectrum of axial morphologies that are observed is a function of two competing processes that vary

in time and space: volcanic activity which results in the construction of relatively smooth features on the seafloor, and faulting/rifting which results in the creation of rough, linear features at many scales. The width of the neovolcanic zone, its structure, and the style of volcanism within it, varies considerably with spreading rate (Table 1). In the following sections we compare and contrast the neovolcanic zone at fast-, intermediate-, and slow-spreading ridge environments, based on results from selected, well-studied sites.

2.1.1. Fast-spreading ridges. Perhaps the most striking characteristic of the neovolcanic zone at fast-spreading ridges (80-160 mm/yr full rate) is that it is very narrow (generally <250 m and in many places less than 50 m), indicating that most intrusions are tightly focused beneath the ridge. This focusing of ridge magmatism is apparently the direct consequence of the fast rate of plate spreading, greater magma supply, a shallower and more steady-state magma reservoir, and more frequent intrusive events, all of which promote volcanism within a very narrow zone. There is also a relatively high continuity of morphology and magma supply along strike, as compared with slower spreading ridges [*Detrick, et al.*, 1987; *Macdonald, et al.*, 1988; *Detrick, et al.*, 1993]. Magma bodies beneath fast-spreading ridges are thought to consist of a thin lens of partial melt overlying a larger zone of crystal mush [*Sinton and Detrick*, 1992].

The EPR is the fastest-spreading MOR on Earth; the segments that are spreading most rapidly are found on the southern EPR, south of the equator. The cross-sectional shape of the ridge axis on the EPR, which varies from a narrow ridge to a broad crestal plateau 2-5 km wide (Figure 1), and the along-axis depth profile (Figure 2), both are thought to reflect the underlying long-term local supply of magma [*Lonsdale*, 1977b; *Macdonald*, 1982; *Macdonald, et al.*, 1988; *Lonsdale*, 1989; *Scheirer and Macdonald*, 1993; *Hooft, et al.*, 1997]. On the other hand, the depth and width of the magma lens are believed to change over shorter time scales, related to intrusion or eruption events [*Mutter, et al.*, 1995; *Hooft, et al.*, 1997]. There is no axial valley on fast-spreading ridges (Plate 1), but along most of the crest of the EPR, there is a narrow, linear depression or trough, which is typically 5-40 m deep, and 40-250 m wide [*Lonsdale*, 1977b; *Renard, et al.*, 1985; *Gente, et al.*, 1986; *Macdonald and Fox*, 1988; *Haymon, et al.*, 1991]. In some locations, the trough is even larger, up to 100 m deep and 2 km wide, but in many of these cases a narrower trough is nested within the larger one [*Lagabrielle, et al.*,

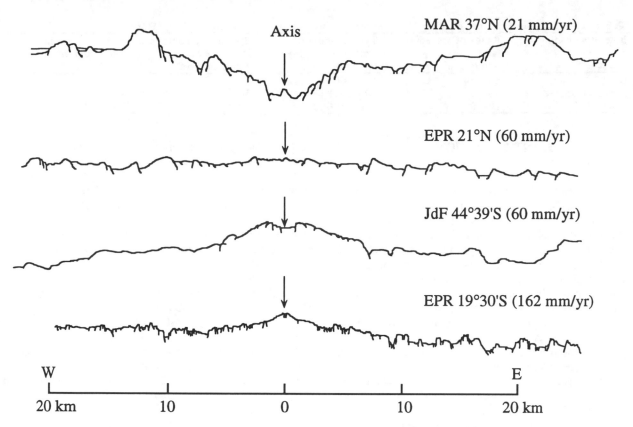

Figure 1. Cross-axis bathymetric profiles of selected mid-ocean ridges with different spreading rates. Profiles across fast-spreading (Southern East Pacific Rise) and slow-spreading (Northern Mid-Atlantic Ridge) ridges show the morphologic contrast between an axial high and a rift valley whereas intermediate spreading rate ridges (Juan de Fuca Ridge) have transitional features. Profiles are modified from Macdonald [1986].

1996]. In some places along the ridge, the depression has been called an "axial summit caldera" (ASC) or "axial summit collapse trough" [*Haymon, et al.*, 1991; *Fornari, et al.*, in press-a] because the trough appears to be more morphologically related to volcanism and collapse than to faulting. In other areas, it has been called an "axial summit graben" because of the clear association with paired, inward-facing faulted walls. For the sake of continuity, we use the more generic term axial summit trough (AST) below with the understanding that the constant interplay between volcanism and tectonism determines the morphologic character of the axis at any given time.

Detailed photographic, submersible and sonar investigations of the axis of a few sections of the EPR have provided us with a picture of the AST and adjacent crestal plateau of fast-spreading centers [e.g. *Ballard, et al.*, 1981; *Francheteau and Ballard*, 1983; *Ballard, et al.*, 1984; *Hekinian, et al.*, 1985; *Renard, et al.*, 1985; *Gente, et al.*, 1986; *McConachy, et al.*, 1986; *Haymon, et al.*, 1993; *Auzende, et al.*, 1996; *Fornari, et al.*, in press-a]. In general, the neovolcanic zone on fast spreading ridges is

characterized by one of three different morphological styles. The first is essentially unfaulted, although it may be extensively fissured, and dominated by broad mounds of lobate lavas and lava ponds (e.g. 9°51'-10°15'N and ~17°30'S). The second has a narrow trough (<200 m) that is characterized by abundant young sheet flows, lava ponds and drain-back features (e.g. 9°40'-51'N, ~11°N, ~13°N, and 18°30'S). The third morphologic style is characterized by a wider (>200 m) trough with numerous faults and a variety of volcanic features and lava types of various ages (e.g. 9°24'-37'N and 21°30'S). Some have interpreted these morphologies as tectonomagmatic stages in an evolutionary progression [*Renard, et al.*, 1985; *Gente, et al.*, 1986; *Fornari, et al.*, in press-a].

Most of the axial volcanism on the EPR occurs within the axial summit trough, where the trough is present. The AST is the structural consequence of the frequent, repeated, narrowly focused dike intrusions on axis, although several mechanisms have been proposed for its formation, including surficial collapse and dike-induced graben subsidence [*Chadwick and Embley*, in press; *Fornari, et al.*, in press-a]. Sonar data and visual observations made during

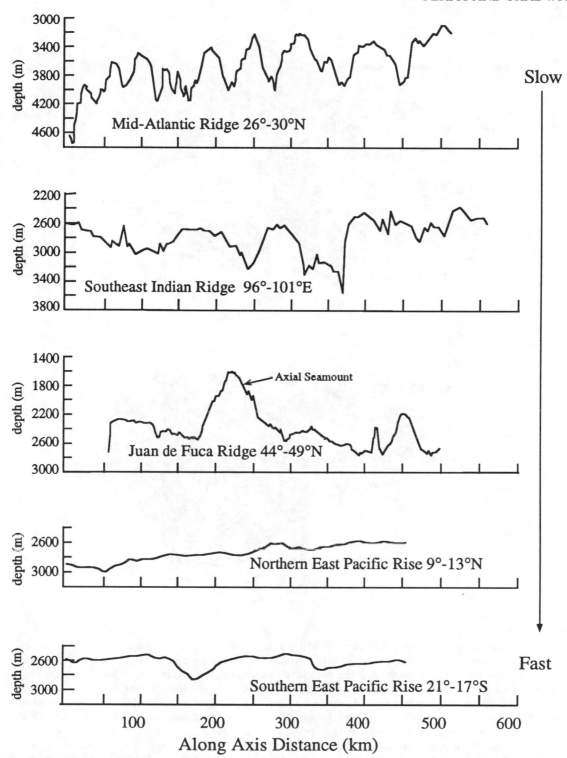

Figure 2. Along-axis, minimum depth bathymetric profiles of mid-ocean ridges with different spreading rates. The magnitude and wavelength of depth variations as well as the absolute depths change systematically with spreading rate. Note that the depth ranges differ for each ridge and the large excursion in the center of the Juan de Fuca Ridge profile is Axial Seamount, a hot spot volcano on the ridge axis (modified from Fornari and Embley [1995] and references therein). Depth profile along the Southeast Indian Ridge illustrates the range of axial morphologies observed along this intermediate spreading ridge (from Cochran et al. [1997]).

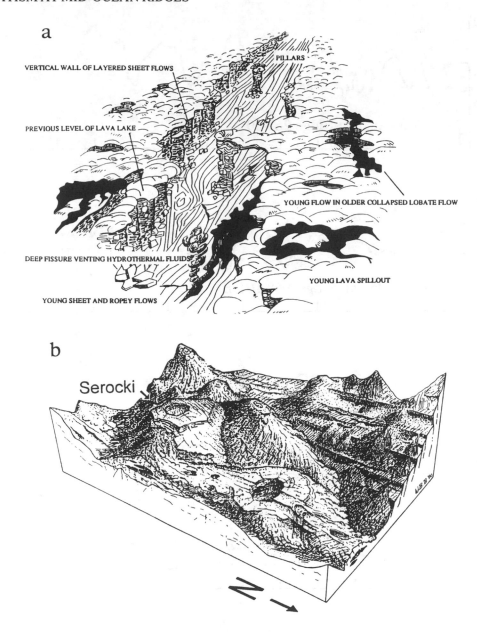

Figure 3. Volcanic features that characterize the axis of some fast vs. slow spreading ridges. (a) Composite sketch of the axial summit trough (AST) along the East Pacific Rise at 9°45'N showing some of the characteristic volcanic features. The irregular boundaries of the AST are due to drainback and collapse of lava that filled the trough. Pillars commonly support archways along uncollapsed sections of the top of flows and hydrothermal activity is concentrated along the margins of the trough. Sheet flows dominate the floor of the AST while lobate flows are more common on the flanks of the trough. Width of the view is approximately 100 m (b) Conceptual view of Serocki volcano and its surroundings on the Mid-Atlantic ridge at 22°55'N (after Bryan et al. [1994]). Serocki may not be a true volcano but rather a megatumulus or inflated lava delta, which drained and collapsed to form the crater. The summit plateau is about 800m in diameter, rises 60 m above the surrounding seafloor, and is comprised of bulbous and elongate pillow lavas. Some sheet flows are present at the rim of the crater. To the north and east of Serocki (lower right corner in figure), there is another megatumulus with similar volcanic features. The entire volcanic edifice is cut by narrow (<1m) post eruptive fissures. Long side of figure is approximately 5.5 km and vertical exaggeration is about 10:1.

ARGO surveys and *Alvin* dives in the 9°17'-9°50'N region of the EPR, show that in plan-view, the boundaries of the AST are irregular and cuspate due to the presence of numerous embayments in the wall created by extensive collapse of lava (Figure 3, 5d) [*Haymon, et al.*, 1993; *Fornari, et al.*, in press-a]. Along the margins of the AST, lava pillars commonly support archways of uncollapsed roof sections of ponded flows that have drained away. The walls of the AST often consist of irregular stair-steps with one or two narrow (1-3 m wide) benches formed by still-stands of lava that represent eruptions which partially filled the trough. Observations indicate that the AST initiates as a series of elongate and irregularly shaped collapse features forming over drained out lava ponds and channels over zones of primary fissuring and diking [*Fornari, et al.*, in press-a]. There are essentially no seamounts or point-source constructs within the neovolcanic zone at fast-spreading ridges (Plate 1b) [*Scheirer and Macdonald*, 1995; *White, et al.*, 1997]. The lavas erupted within the AST are dominantly sheet flows [*Francheteau and Ballard*, 1983; *Bonatti and Harrison*, 1988; *Haymon, et al.*, 1993; *Fornari, et al.*, in press-a], and they vary from remarkably flat and thin (< 4 cm) to ropy and jumbled varieties with chaotically folded and deformed surfaces that can be 10's of centimeters thick.

At present, there is some debate regarding the mechanism for forming narrow AST's on fast spreading ridges. Fornari et al. [in press-a] propose that narrow, AST's on fast-spreading MOR crests principally form through near-surface (<50 m depth) collapse of volcanic flows above primary fissures and feeder dike networks, caused by withdrawal of lava back into the shallow plumbing system, or out onto the ridge flanks through subsurface tube and channel systems. They suggest that ~1-10 m wide eruptive fissures and open cracks form above intruding dikes but do not appear to create opposing normal faults because the crust is too thin and weak to sustain shear failure [*Carbotte and Macdonald*, 1994]. In contrast, Chadwick and Embley [in press] make the case that ~10-100 m wide and 5-15 m deep grabens at recent eruption sites on the JdFR formed by stress perturbations above intruding dikes, and speculate that dike-induced grabens could also underlie the AST on the EPR.

Adjacent to the AST on the crestal plateau of the EPR, lobate lava flows predominate and comprise an extensive collapse zone ~50-200 m-wide that marks the region most often affected by lavas that have flooded and overflowed the AST. An extensive network of subseafloor tubes and channels has developed in this relatively flat area because of the repeated flooding by fluid lavas and their subsequent downslope drainout. The open network creates large-scale permeability within the upper few hundred meters of seafloor at fast-spreading MOR's which promotes the distribution of lavas across the crestal plateau (typically ~100-300 m from the AST wall) and also has great potential for distributing hydrothermal fluids [*Fornari and Embley*, 1995]. Multiple pulses of filling and drainback impart a pseudostratigraphy in the shallow crust adjacent to the AST, because of this high permeability. Hooft et al. [1996; 1997] have proposed that the width of the intrusion zone and lava flows erupting on axis and flowing up to several km off axis may cause the observed thickening of seismic layer 2A, but bottom observations on the northern EPR suggest that most lavas erupted from the AST extend no farther than a few hundred meters from it. Instead, eruptions originating outside of the summit trough (within ~4 km of the axis) apparently form small volume, discontinuous pillow ridges (<10 m high) along ridge parallel fissures and faults [*Perfit, et al.*, 1994], which contribute to layer 2A.

2.1.2. Intermediate-spreading ridges. The neovolcanic zone at intermediate rate spreading ridges (40-80 mm/yr) has morphological characteristics that are transitional between those at fast and slow ridges (Figures 1 and 2). However, individual ridge segments on an intermediate ridge can be significantly different in character from one another; some closer to one end member or the other. The morphology of intermediate ridges is variable, but generally consists of a small axial valley, 1-5 km wide, with bounding faults 50-1000 m high. The neovolcanic zone is located within the floor of that valley. Well studied intermediate ridges include the Juan de Fuca and Gorda Ridges, the Galapagos Ridge, the EPR at 21°N, and the Southeast Indian Ridge.

The zone of focused accretion on intermediate rate ridges tends to be wider than at fast ridges but narrower than at slow ridges. The eruptive products from recent eruptions lie within a zone that is about 1 km wide [*van Andel and Ballard*, 1979; *Ballard, et al.*, 1981; *Luyendyk and Macdonald*, 1985; *Embley and Chadwick*, 1994; *Embley, et al.*, 1995a]. Individual accretion events, even closely spaced in time, can be separated by 100's of meters across axis [*Chadwick, et al.*, 1995]. The frequency of events is moderate, but probably variable and cyclic, more frequent during a magmatically robust phase and less frequent between such phases. For example, evidence for three separate eruptions during the last 15 years has been found on the CoAxial segment of the JdFR [*Chadwick, et al.*, 1995; *Embley, et al.*, 1995a], but it is unlikely that this rate of events is the long-term average. Other segments on the JdFR, like Endeavour, have not had eruptions for at least hundreds and possibly as much as thousands of years [*Karsten, et al.*, 1990].

The extrusive products of accretion events are also varied and intermediate in morphologic style between those found at fast and slow ridges. Both sheet flow and pillow lava eruptions are common. For example, the sheet flows that have been mapped at both the southern and northern ends of the Cleft segment have many similarities to those observed on the EPR. The lobate sheet flows at south Cleft cover the entire floor of the axial valley and erupted out of a narrow "cleft" or trough that bisects it for 10 km along strike [*Normark, et al.*, 1986; *Normark, et al.*, 1987]. Here the trough is 10-30 m deep and 30-50 m wide, is the site of hydrothermal venting, and exhibits many of the features of the AST on the EPR [*Normark, et al.*, 1986; *Smith, et al.*, 1994]. The "young sheet flow" mapped at north Cleft is 4.2 km^2 in area, is probably only decades old, and was erupted from a linear fissure system that hosts hydrothermal activity [*Embley, et al.*, 1991; *Embley and Chadwick*, 1994]. The young sheet flow has a lobate upper surface but is extensively collapsed, revealing lava pillars and ropy, lineated, and jumbled sheet flows in its interior. In contrast, the eruption of long, narrow ridges of pillow lava (5-50 m high, 100-500 m wide, and 1-5 km long) is also common on the Cleft segment, and these are more similar to constructional features observed on slow spreading ridges. For example, the "new pillow mounds" that erupted at north Cleft in the mid-1980's created a line of steep-sided, hummocky flows [*Chadwick and Embley*, 1994]. Unlike sheet flows, these pillow mounds are thick stacks of individual pillow lobes and have no collapse features, but they are porous and can channel hydrothermal fluids though their interiors during cooling. The recent CoAxial and Gorda eruptions on the JdFR also produced pillow flows of similar morphology [*Embley, et al.*, 1995b; *Embley, et al.*, 1996; *Chadwick, et al.*, in press]. Both the pillow mounds and sheet flows on intermediate ridges are produced during short-lived fissure eruptions fed by dikes.

While the balance between intrusion and extrusion at MOR axes, in general, has been difficult to volumetrically constrain, it appears that a significant amount of crustal construction occurs by intrusion at all spreading rates. However, in certain geographic and tectonic environments that are close enough to continental margins to be inundated with terrigenous sediments, intrusive volcanism is clearly dominant. Examples of such sites include Middle Valley on the JdFR [*Davis and Villinger*, 1992; *Goodfellow and Franklin*, 1993], Escanaba Trough on the Gorda Ridge [*Morton, et al.*, 1994; *Zierenberg, et al.*, 1994], and the Guaymas Basin in the Gulf of California [*Lonsdale and Lawver*, 1980; *Lonsdale and Becker*, 1985]. At these sites, MOR magmatism tends to produce sill intrusions into the thick sediments overlying the ridge axis at hydrostatic levels where the intrusive forces are balanced by the overburden of water and sediment, and consequently, extrusions onto the seafloor are rare. The intrusion of sills uplifts the sediments above them, producing faults and pathways for rising hydrothermal fluids and sulfide mineralization [*Lonsdale and Lawver*, 1980; *Einsele*, 1985; *Morton, et al.*, 1987; *Turner, et al.*, 1993; *Zierenberg, et al.*, 1993].

Morphologic, petrologic and structural studies of the JdFR, suggest that it evolves through cycles of accretion related to magmatic output [*Kappel and Ryan*, 1986; *Kappel and Normark*, 1987; *Smith, et al.*, 1994]. During the first cycle, which is dominated by magmatism and volcanic construction, the ridge axis has morphologic characteristics like fast spreading ridges, whereas in

subsequent cycles when volcanism wanes, collapse of the summit region and tectonic stretching dominate, resulting in an axis with a more rugged, faulted appearance more similar to slow spreading ridges. This model suggests the development of intermediate ridges is more episodic than steady-state in nature.

The variability of ridge crest morphology at intermediate rates is especially well displayed on the Southeast Indian Ridge, where recent surveys have documented abrupt morphologic transitions from axial high to axial valley at nearly constant spreading rates [*Sempéré, et al.*, 1996; *Sempéré, et al.*, 1997]. The morphologic transitions correlate with changes in mantle Bouguer gravity anomalies, suggesting that they are caused by variations in crustal thickness and/or mantle temperature [*Cochran, et al.*, 1997]. A similar conclusion was reached to explain the large contrast in morphology between the southern JdFR (which has an axial high) and the northern Gorda ridge (which has an axial valley), even though they spread at the same rate (55-60 mm/yr) [*Hooft and Detrick*, 1995]. Likewise, the Galapagos Ridge has an axial high where its magma supply is locally enhanced by the Galapagos hotspot [*Canales, et al.*, 1997].

The key influence on axial morphology at intermediate ridges may be the presence or absence of a quasi-steady-state magma chamber on a given ridge segment, which would strongly influence the local magma supply, the strength of the lithosphere, and its mechanical evolution [*Sempéré, et al.*, 1996; *Cochran, et al.*, 1997; *Sempéré, et al.*, 1997]. Modeling by Phipps Morgan and Chen [1993a; 1993b] showed that fast spreading ridges can support a steady-state magma lens at shallow depth beneath the ridge axis, but slow spreading ridges cannot (Figure 4). Ridges spreading at intermediate rates, however, are in a delicate balance in which they may or may not have a steady-state magma lens, depending on the local crustal thickness and magma supply, and this balance may change with time. In other words, at a given time one segment on an intermediate ridge may have a magma lens and the next segment may not. This is consistent with the evidence for episodic and spatially discontinuous magma supply to intermediate ridges, and helps explain the wide diversity of morphology, apparent magma budgets, and hydrothermalism observed along intermediate ridges [*van Andel and Ballard*, 1979; *Ballard, et al.*, 1982; *Kappel and Ryan*, 1986; *Karsten, et al.*, 1986; *Baker and Hammond*, 1992; *Embley and Chadwick*, 1994; *Cochran, et al.*, 1997].

2.1.3. Slow-spreading ridges. Slow spreading ridges (10-40 mm/yr) have large axial valleys (8-20 km wide and 1-2 km deep), and the neovolcanic zone can extend across the entire axial valley floor (5-12 km wide). The deeper axial valleys reflect the thicker mechanical lithosphere and cooler crust at slow ridges. The neovolcanic zone is more difficult to define on slow spreading ridges because magmatism there is relatively unfocused, both across and along axis, and recent volcanic events have been difficult to

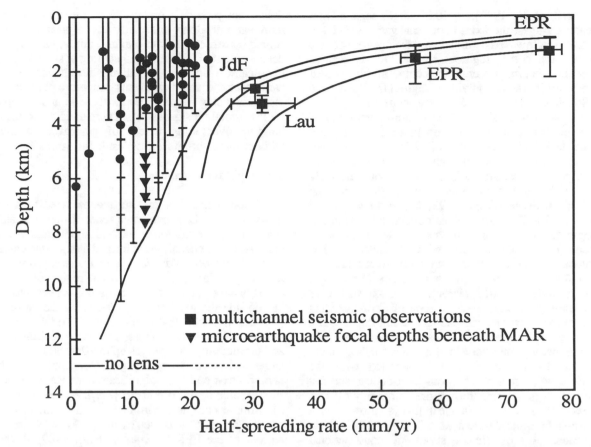

Figure 4. Depth to the top of the magma lens or depth of brittle lithosphere as a function of spreading rate. Filled squares are depth of magma lens determined from multichannel seismic data. Solid dots and associated bars are axial earthquake centroid depths and inferred total rupture depth. Two lower curves are results of numerical experiments by Phipps Morgan and Chen [1993a] that vary the hydrothermal cooling parameters in calculating the thermal structure of ridges. The model results suggest a steady-state magma lens only exists at half-spreading rates greater than about 20-30 mm/yr. Upper curve is the modeled 750°C isotherm which shows good agreement with the depth of the seismically observed brittle-ductile transition on slow spreading ridges (modified from Phipps Morgan and Chen [1993a]).

identify (or are extremely rare). Commonly, a discontinuous axial volcanic ridge (AVR) that can be up to 1-5 km wide, several hundred meters high, and tens of km wide, lies at the center of the inner valley floor (Plate 1a) [*Ballard and van Andel*, 1977; *Crane and Ballard*, 1981; *Brown and Karson*, 1988; *Sempéré, et al.*, 1990; *Gente, et al.*, 1991; *Smith and Cann*, 1993]. However, there is often a wide variation in volcanic and tectonic character along strike, both within and between ridge segments [*Karson, et al.*, 1987; *Brown and Karson*, 1988; *Grindlay, et al.*, 1992]. In a few locations there is evidence for volcanic eruptions along faults both within and along the edges of the valley floor [*Zonenshain, et al.*, 1989; *Karson and Rona*, 1990; *Bryan, et al.*, 1994].

Even among slow-spreading ridges there is a variety of axial morphologies, again probably related to differences in magma supply. Fox et al. [1991] mapped the southern Mid-Atlantic Ridge (MAR) from 31°S-34°30'S and found

the morphology of segments varied from wide, deep valleys to narrower, shallower ones (similar to the rifted axial highs of intermediate ridges), all at the same spreading rate (3.5 cm/yr full-rate). Similarly, Sempéré et al. [1993] described two contrasting segment morphologies on the northern MAR between 24°00'N-30°40'N - one with wide, deep valleys with U-shaped profiles, and the other with narrower valleys that are hourglass-shaped in map view and have V-shaped profiles. In both these areas, the narrower, hourglass-shaped segments have circular, negative mantle Bouguer anomalies corresponding to their axial highs, which have been interpreted to indicate focused mantle upwelling and/or along-axis variations in crustal thickness [*Kuo and Forsyth*, 1988; *Lin, et al.*, 1990; *Tolstoy, et al.*, 1993; *Fujimoto, et al.*, 1996]. Therefore, segments with morphologies that are closer to the intermediate-spreading end of the spectrum (the narrower, hourglass-shaped valleys) are interpreted to be more magmatically robust [*Fox, et al.*,

1991; *Sempéré, et al.*, 1993]. Within single segments on slow-spreading ridges, the axial depth can vary by 1-2 km between the shallow axial highs at the center of the segments and the deeper segment ends. This is seen as further evidence that magma emplacement is highly focused at the depth minimum of individual segments [*Macdonald, et al.*, 1988; *Fox, et al.*, 1991; *Sempéré, et al.*, 1993].

Lava morphology on slow ridges is dominantly pillow lava, which tends to construct hummocks (< 50 m high, < 500 m diameter), hummocky ridges (1-2 km long), or small circular seamounts (10s-100s m high and 100s-1000s m in diameter) that often coalesce to form AVR's along the inner valley floor (Plate 1a) [*Smith and Cann*, 1990; *Smith and Cann*, 1992; *Smith and Cann*, 1993; *Bryan, et al.*, 1994; *Head, et al.*, 1996]. Small seamounts have variable morphologic forms (e.g. smooth textured to hummocky, flat topped to domed, with and without craters). The various types may be isolated, form individual chains, or pile up to form larger scale volcanic ridges [*Smith and Cann*, 1993; *Smith, et al.*, 1995a]. The prevalence of small seamounts in the neovolcanic zone of slow-spreading ridges is in marked contrast with fast-spreading ridges where virtually no seamounts or large constructional edifices are found in the neovolcanic zone and at intermediate ridges where seamounts are only rarely associated with the neovolcanic zone (Figure 5). At both intermediate and fast spreading centers, seamounts form more commonly off-axis than on-axis (Plate 1b). It is clear that shallow crustal structure must be quite different at different spreading rate ridges, because of the different styles of constructional volcanism and especially the different scales and spatial continuity of volcanic constructs [*Fornari, et al.*, 1987; *Smith and Cann*, 1990; *Scheirer and Macdonald*, 1995].

Near-bottom observations show that the inner valley floor is more faulted and fissured than at faster spreading rate ridges [*Ballard and van Andel*, 1977; *Kong, et al.*, 1988; *Mutter and Karson*, 1992]. Similarly, large teleseismic earthquakes (indicating major faulting events) are much more common on the MAR than on faster spreading ridges [*Sykes*, 1967; *Sykes*, 1970; *Solomon, et al.*, 1988; *Bergman and Solomon*, 1990]. These observations and the large axial valleys reflect the dominance of tectonism over volcanism at slow ridges. In fact, the nature and scale of ridge segmentation on slow ridges may be primarily due to deformation processes rather than magmatic processes, which control segmentation on fast-spreading ridges [*Mutter and Karson*, 1992].

Slow-spreading ridges probably do not maintain steady-state magma reservoirs in the shallow crust [*Detrick, et al.*, 1990; *Phipps Morgan and Chen*, 1993b]. Sinton and Detrick [1992] suggest that a low magma supply intermittently feeds an irregular mush zone beneath the axial valley which occasionally erupts, but otherwise crystallizes to form the lower oceanic crust. Therefore, individual magmatic pulses tend to be limited in extent and longevity, and thus are relatively independent in space and

time from preceding ones. Since the magma supply is more variable along strike, it tends to promote more point source constructs. Lower magma supply, thicker crust, and deeper extent of hydrothermal cooling at slow ridges all lead to volcanic events being relatively infrequent. On the other hand, tectonic faulting events are more frequent than at faster spreading ridges, because there is more opportunity for horizontal tensional stresses perpendicular to the ridge to build-up over long periods of gradual plate divergence without interruption from volcanic intrusions.

2.2. Off-Axis Volcanism

Although most of the magma delivered to a MOR is focused within the neovolcanic zone, off-axis volcanism and near-axis seamount formation add significant volumes of material to the oceanic crust formed along ridge crests.

2.2.1. Eruptions outside the neovolcanic zone. Detailed sampling of lavas from the crestal region of the EPR at ~9°31'N have shown that some lavas were erupted up to 4 km outside the axial summit trough on the crestal plateau [*Goldstein, et al.*, 1994b; *Perfit, et al.*, 1994]. This recent off-axis volcanism is expressed as young-looking lava fields and prominent pillow ridges up to 20 m high which have a larger and more diverse range of compositions than the lavas on-axis and show significant chemical variations over small spatial scales (<600 m) [*Perfit, et al.*, 1994] (Figure 6). Similar complexity in the distribution of compositions off-axis has been documented in the 11°30'N and 12-13°N regions of the EPR [*Hekinian, et al.*, 1989; *Reynolds, et al.*, 1992]. Evidence for off-axis eruptions includes the fact that some lavas on the crestal plateau have very little sediment cover compared to the surrounding volcanic terrain, much younger appearances (e.g. fresher glass, little Mn coating) and younger U-series disequilibrium ages than would be predicted by their distance from the axis and the spreading rate [*Goldstein, et al.*, 1994b]. The observational and chemical data suggest that these off-axis eruptions are fed from a different magma plumbing system than the ones at the axis (or the distal edges of subaxial magma bodies), and that small, distinct, and spatially restricted magma bodies periodically erupt lavas on the EPR crestal plateau [*Goldstein, et al.*, 1994b; *Perfit, et al.*, 1994]. Similarly, there is some limited evidence for off-axis lavas being younger than predicted on the Endeavour segment of the JdFR [*Goldstein, et al.*, 1992], suggesting that this process may also take place on intermediate rate ridges.

These results provide an explanation for the observed thickening of seismic layer 2A (interpreted as extrusive) within 1-4 km of the spreading axis on the EPR [*Harding, et al.*, 1993; *Christeson, et al.*, 1994; *Vera and Diebold*, 1994]. Carbotte et al. [1997] noted that where the magma supply is higher on the southern EPR, the layer 2A accumulation zone is wider. They proposed a model, consistent with layer 2A geometry, in which a zone of volcanic accumulation (~5 km wide, centered on axis) is

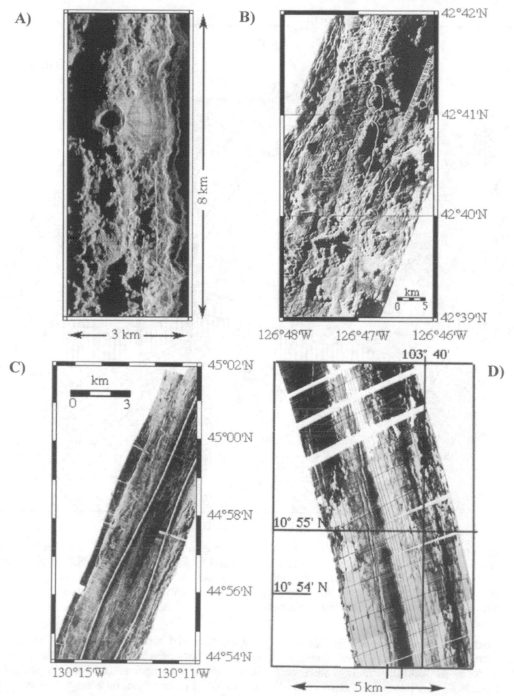

Figure 5.   A comparison of sidescan sonar imagery showing the morphologic character of volcanic constructs at different spreading rates (white is high reflectivity, black is acoustic shadow). Slow-spreading ridges tend to create "lumpy" terrain by erupting pillow lava from point-source vents. Fast-spreading ridges tend to create "smooth" terrain by erupting sheet and lobate flows from fissure vents. Intermediate ridges can have a morphology similar to either the slow or fast end of the spectrum, depending on the site. (a) northern MAR (slow) showing an axial volcanic ridge comprised of hummocks, hummocky ridges and seamounts. The cratered seamount is 220 m high (from Smith and Cann [1993]). (b) northern Gorda Ridge (intermediate) showing hummocky pillow mounds with the 1996 lava flow outlined in white (from Chadwick et al. [in press]), (c) northern Cleft segment, JdFR (intermediate) with smooth seafloor covered by sheet and lobate flows (from Embley and Chadwick [1994]), (d) northern EPR (fast) composed of low relief sheet flows around the narrow axial summit trough (AST) (width of AST near bottom of image is indicated by two short lines) and lobate flows on the flanks of the axis (from Ryan and Fox, unpublished data).

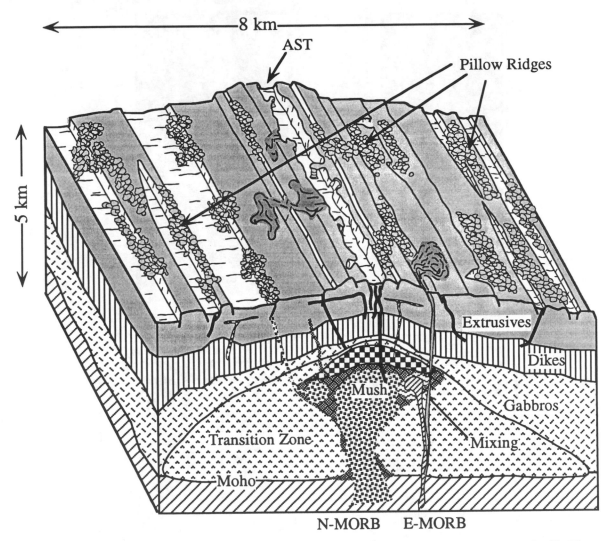

Figure 6. Diagrammatic three-dimensional representation of oceanic crust formed along the northern East Pacific Rise showing relation between on-axis eruptions of predominantly sheet and lobate flows associated with the axial summit trough (AST) and off-axis volcanism of predominantly pillow lavas associated with fissures and faults (modified from Perfit et al. [1994]). All axial basalts are N-MORB whereas a significant proportion of off-axis pillow mounds contain T- and E-MORB. See Figure 21 and text for further discussion.

flanked by zones of active faulting, and that in between there is an overlap zone where both processes compete. This concept of a zone of faulting and syntectonic volcanism slightly off-axis is consistent with observations of abyssal hills near the EPR from recent submersible dives by Macdonald et al. [1996]. In their model, abyssal hills are bounded on both sides by normal faults, but on the side facing away from the ridge the faults are draped by lava flows that are mainly intact elongate pillows. These "volcanic growth faults" form in a "flanking tectonic province" 2-6 km from the ridge axis, and are repeatedly covered by syntectonic lava flows. The lavas are interpreted to have been erupted either from the ridge axis itself or from a zone 2-3 km off-axis, consistent with the geochemical studies discussed above. This model for abyssal hills is specific to fast-spreading ridges, and other processes may be important at intermediate and slow ridges [Macdonald, et al., 1996]. For example, on a ridge that is magma-starved, abyssal hill faults may rarely get covered with lava.

Hooft et al. [1996] developed a stochastic model for the emplacement of dikes and lava flows in a narrow zone at the axis of a fast spreading ridge that can explain the seismic structure of the upper oceanic crust and the off-axis (1-3 km) thickening of layer 2A as well as the crustal structure observed in oceanic drill holes. Their model relies on a bimodal distribution of lava flows: frequent short flows confined to the AST and occasional voluminous flows that erupt outside of the AST or spill out and pond at considerable distance to build up layer 2A. This model implies that much of the upper part of the extrusive section will be composed of flows that have been transported a long distance (a few km) off-axis and that this can explain some of the anomalously young ages obtained by dating off-axis flows on the northern EPR. Although this model is consistent with upper crustal structure, detailed submersible, sidescan and photographic surveys of the northern EPR have yet to identify any large flows extending more than a few 100 m from the AST [*Haymon, et al.*, 1993; *Fornari, et al.*, in press-a].

Only a few examples of extremely large lava flows (or lava fields) have been found on or near some sections of the MOR. Basalt flows up to 60 km long that may have originated from the Dellwood Knolls and the JdFR in the northeast Pacific were described by Davis [1982]. Macdonald et al. [1989] documented a young lava field at 8°S on the EPR with an area of 220 km^2 and a volume of 15 km^3, extending from an apparent source 4 km east of the ridge axis up to 18 km off-axis. Subsequent petrological sampling has shown that this flow field has a diverse chemistry, and so it is almost certainly the product of multiple eruptions over some period of time [*Hall and Sinton*, 1996]. More recent mapping with sidescan sonar has revealed more of these large (10-100 km^2) lava fields along the southern EPR, generally within 30 km of the axis [*Macdonald, et al.*, 1997]. A recent survey with GLORIA-B sidescan sonar of the EPR, south of the Easter microplate near 29°S [*Hey, et al.*, 1995], revealed two other fields of apparently huge, near-axis lava flows. These flows are up to 40 km in length and appear to originate from the tips of two large-offset propagating rifts [*Johnson*, 1996], but are, as yet, unsampled. These types of flows represent tremendous volumes of lava added to the oceanic crust, but it is unknown how often these kinds of events may occur, if they are the result of single or multiple eruptions, and what combination of regional tectonic and/or magmatic processes provoke such large-scale outpourings of lava near spreading axes.

2.2.2. Off-axis seamounts. Another aspect of infrequent but volumetrically significant off-axis volcanism is the formation of near-axis seamounts. These consist of small (typically <500 m high) singular circular volcanoes, or slightly larger (< 1 km high) cones in chains, that appear to have developed above point sources near, but off the ridge axis. (Here we consider only non-hot spot seamounts).

On fast spreading ridges, seamounts do not form on the ridge crest at all, but apparently form within a zone 5-15 km from the axis [*Shen, et al.*, 1993; *Scheirer and Macdonald*, 1995] (Figure 7a and Plate 1b). Nevertheless, there are greater numbers of near-axis seamounts adjacent to parts of the ridge that have a larger cross-sectional area, suggesting that more magma is available for off-axis volcanism where the axial supply is larger [*Scheirer, et al.*, 1996; *White, et al.*, 1997]. On intermediate and slow spreading ridges increasing numbers of seamounts are found

in the neovolcanic zone with decreasing spreading rate [*Barr*, 1974; *Davis and Karsten*, 1986; *Smith and Cann*, 1993; *Smith, et al.*, 1993]. This dependence on spreading rate as to whether or not seamounts are excluded from the neovolcanic zone is related to the relative continuity of magma supply and the focusing of melt beneath the ridge axis. At fast spreading ridges, seamounts do not form on axis because most of the eruptible melt generated in the mantle directly beneath the ridge is focused upward into the magma lens and associated crystal mush zone. Consequently, these seamounts can only form by melt diapirs that occasionally segregate and rise outside this zone of focused upwelling [*Batiza and Vanko*, 1983; *Fornari, et al.*, 1988; *Allan, et al.*, 1989; *Batiza, et al.*, 1989; *Leybourne and Van Wagoner*, 1991; *Shen, et al.*, 1995]. The relatively mafic and chemically (i.e. incompatible element) depleted nature of lavas from many near-axis seamounts at intermediate and fast spreading ridges indicate that the seamount magmas are not usually stored in crustal magma chambers, have separate plumbing systems from the ridge, and must be derived, in part, from sources more depleted than typical mid-ocean ridge basalt mantle sources. On the other hand, trace element evidence from about 50 near-ridge seamounts in the eastern equatorial Pacific (5°-15°N) indicate that some seamounts have erupted both enriched and depleted basalts and that enriched samples exist as close as 4 km from the EPR axis [*Niu and Batiza*, 1997]. These data suggest seamount melts are derived from heterogeneous mantle sources consisting of a very depleted component and one that is more enriched in incompatible elements than average ocean island basalts [*Niu and Batiza*, 1997]. With decreasing spreading rate, the zone of upwelling beneath the ridge axis is less focused and the lateral continuity of the magma lens breaks down, allowing the possibility of individual melt diapirs to rise directly beneath the ridge to form seamounts on axis. The model of Smith and Cann [1990; 1992] suggests the multitude of small seamounts observed on the axial valley floor of the MAR results from eruptions from many individual magma bodies that reside in the crust.

Hammond [1997] has noted a distinct morphologic pattern on many off-axis seamounts on the JdFR in which their craters or calderas tend to be offset from the center of the seamount in the direction toward the ridge (Figure 7b). These observations are consistent with a model of formation in which early, rapid, voluminous extrusion is followed by a long waning stage of growth, culminating in collapse. The duration of seamount growth is an average of 50,000 years, long enough for plate motion to move the seamount off its relatively stationary sub-plate source before collapse [*Hammond*, 1997]. The Lamont seamounts on the EPR also show this offset geometry of collapse craters [*Fornari, et al.*, 1984; *Barone and Ryan*, 1990]. It remains unclear what deep-seated processes are involved in generating these chains of volcanoes and why they erupt from point sources over long periods of time to form

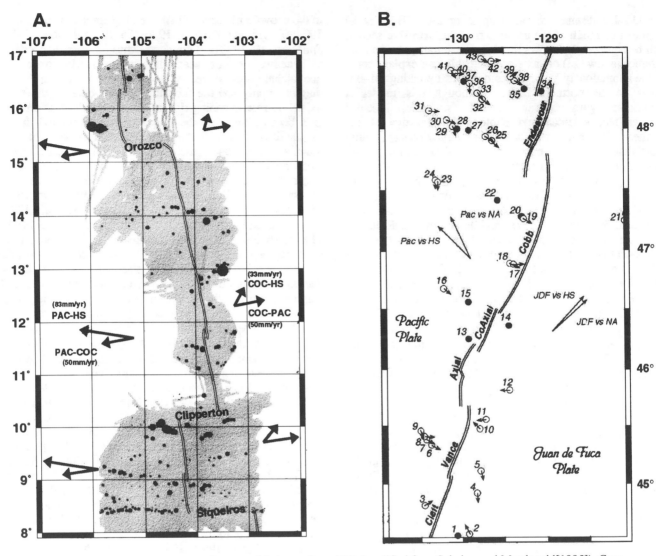

Figure 7. Near-axis seamounts adjacent to (a) the northern EPR (modified from Scheirer and Macdonald[1995]). Gray shaded areas have complete bathymetric coverage and seamounts >200 high are shown as filled circles. (b) The JdFR showing seamounts with (open circles) and without (solid circles) offset calderas or craters (modified from Hammond [1997]). Arrows show azimuth from center of volcano to center of collapse structure in crater.

distinct voluminous cones. Geometrical and geochemical arguments suggest small thermal anomalies cause episodic melting of depleted but heterogeneous mantle sources up to 50 km away from ridge axes and rapid transport of melt via dikes to central vents [*Sleep*, 1988; *Scott and Stevenson*, 1989].

Off-axis volcanism is best documented on intermediate and fast-spreading ridges where the neovolcanic zone is relatively narrow. However, at least one section of the MAR (~26°S) has numerous off-axis seamounts on the ridge flanks and, similar to the EPR and JdFR, they consist of depleted lavas more mafic than recovered in the axial

valley [*Batiza, et al.*, 1989]. Perhaps the lack of many examples of off-axis volcanism at slow-spreading ridges, is because we have defined the neovolcanic zone to be so wide along these ridges (including the entire axial valley floor). The concept of "off-axis" volcanism may only be meaningful where the "axis" is narrow, well-defined, and is the site of most eruptions, from which off-axis activity can be contrasted. In any case, off-axis volcanism is clearly a volumetrically significant addition to the young ocean crust in many areas, but it is likely to be a much less frequent occurrence than eruptions on-axis within the neovolcanic zone.

A) Mid-Atlantic Ridge

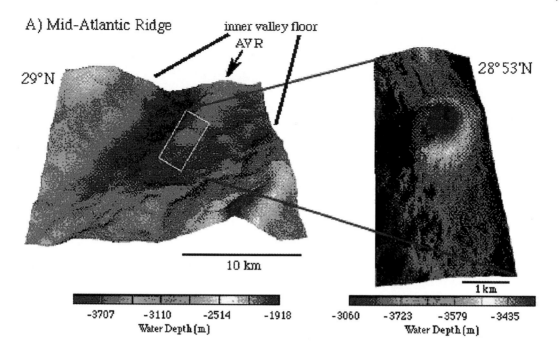

B) East Pacific Rise

Plate 1. Comparison of the morphology slow and fast spreading ridge crests. (a) SeaBeam bathymetry of a section of the MAR showing the deep inner valley floor and flanking crestal mountains. Abundant semicircular volcanoes form in the broad axial valley and commonly join to form axial volcanic ridges (AVR). Inset shows an example of a flat-topped seamount built by a long-lived eruption from a central vent. (from Smith and Cann [1993]). (b) SeaBeam bathymetry of the southern EPR showing the very narrow rise crest with no axial valley and singular off-axis volcanoes that are not directly related to axial magmatism (from Macdonald, [1993]).

Plate 2. Photographs of the four main lava morphologies discussed in the text: pillow, lobate, sheet, and jumbled (photographs from Fox et al. [1988]).

2.3. Morphological Variations in Submarine Lava Flows

A wide variety of lava flow morphologies (Plate 2) have been described from near-bottom observations on the MOR [*Ballard, et al.*, 1975; *Ballard and Moore*, 1977; *Lonsdale*, 1977a; *Ballard, et al.*, 1979; *Crane and Ballard*, 1981; *Francheteau and Ballard*, 1983; *Renard, et al.*, 1985; *Bonatti and Harrison*, 1988; *Fox, et al.*, 1988; *Embley, et al.*, 1990; *Embley and Chadwick*, 1994]. One fundamental conclusion from these observations is that the dominant lava flow morphology on a ridge is clearly a function of spreading rate; slow spreading ridges almost exclusively produce pillow lavas, fast spreading ridges produce mostly sheet flows, and intermediate ridges produce both types of morphologies, although pillowed forms predominate (Figure 8).

The key differences between various flow morphologies are the size of individual flow units, their inter-connectivity, and the rate of crust formation during emplacement. At one end of the spectrum, pillow lavas extrude as individual spherical or cylindrical tubes of lava which cool and crust over on all sides, generally preventing coalescence with neighboring pillows [*Moore*, 1975]. Pillow flows are produced by the piling up of individual pillow lava lobes. As it grows, the newest pillows are erupted from the top of the stack and flow outward a limited distance before freezing, a process which tends to produce steep-sided mounds or ridges [*Chadwick and Embley*, 1994]. At the other end of the spectrum are morphologies formed when lava spreads out as a continuous sheet. In the middle of the spectrum are lobate flows, which advance one lobe at a time, but rapidly enough that the lobes may quickly coalesce back together into one inter-connected flow (much like "hummocky", "tube-fed", or "toey" pahoehoe on land [*Wentworth and Macdonald*, 1953; *Swanson*, 1973]). In this way, lobate flows develop an upper crust that preserves the original lobate surface morphology, but which overlies a molten flow interior that is sheet-like after individual lobes have coalesced. Lobate flows commonly spread out and then inflate upwards (analogous to terrestrial inflated

A.

B.

C.

D.

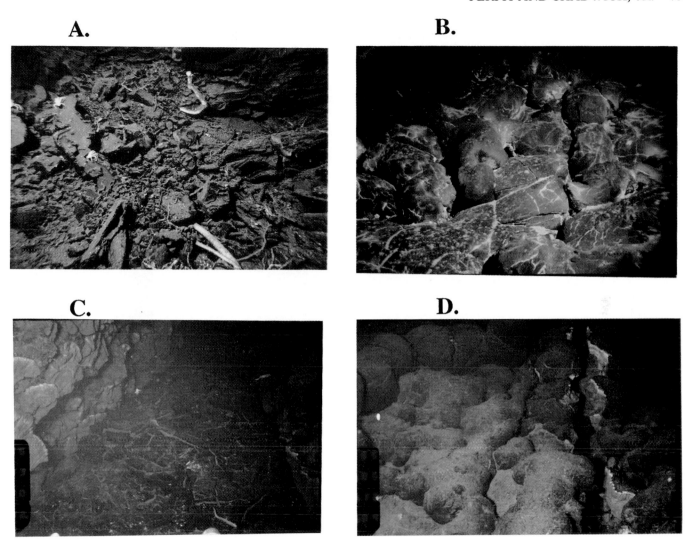

Plate 3. Deep sea photographs of recently active portions of the mid-ocean ridge axis taken from *Alvin*. (a)"Tubeworm Barbecue" site on the EPR at 9°50.6'N where a hydrothermal vent community was overrun by a recently erupted, jumbled sheet flow in March-April 1991. The lack of black glassy surfaces on the flow is due to a thin coating of sulfide-rich "ash" that settled on the surface of the flow and dead animals immediately after the eruption. This flow was dated by Po-Pb techniques and chemically analyzed to evaluate homogeneity (see text for discussion and Figures 16 and 18). The photo was taken on May 27, 1991. Field of view is approximately 1 m across. (b) Photograph taken during *Alvin* dive 2792 in July 1994 of young pillow lavas that form the north end of the 1993 CoAxial Flow near 46°28.5'N on the JdFR (see Figure 11c). Yellowish stains on pillows are remnants of precipitates formed during low-temperature hydrothermal venting the previous year. (c) Photograph taken with *Alvin*'s bow camera (dive 2497) in 1992 looking north along the narrow AST around 9°52'N on the EPR showing young, glassy lava filling a 2-3 m wide fissure that began forming in 1991. This flow has a significantly more fractionated composition than the BBQ flow, less than 1 km to the south and Po-Pb dates indicate this flow erupted in 1992 [*Rubin, et al.,* 1994]. Field of view in foreground is approximately 3 m. Skylights in the surface of the flow suggest some drainout has occurred. (d) Photo also taken on Alvin dive 2497 north of (c) showing narrow (~0.25m) fissure in lobate and pillow lavas near 9°53'N that is part of the AST/eruptive fissure system. In some places, young lava could be seen within the fissure.

East Pacific Rise ~ 9° 31'N

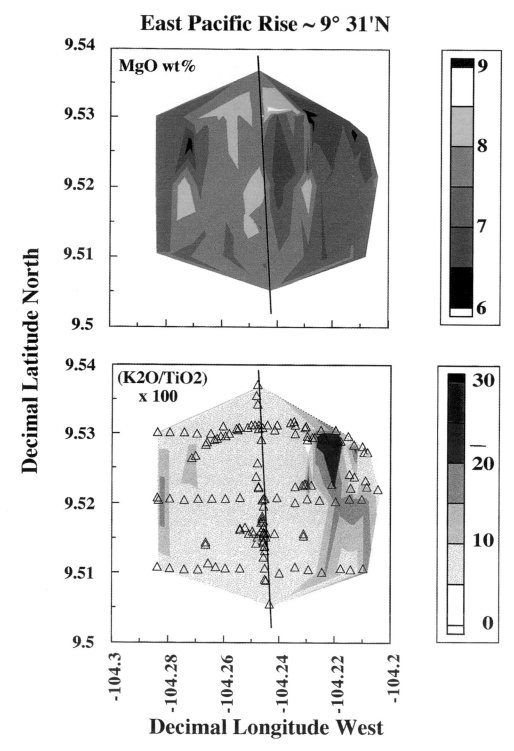

Plate 4. Contour map of chemical variations of lavas on the East Pacific Rise from 9°30.3'N (9.505°) to 9°32.1'N (9.535°). Location of samples recovered by rock core and *Alvin* are shown in the lower diagram as open triangles. Lavas within the AST comprise a relatively homogenous population of mafic N-MORB whereas lavas outside the AST on the crestal plateau exhibit chemical variation over small (<600 m) spatial scales [*Perfit, et al.*, 1994]. Note the asymmetrical distribution of lava composition, particularly the concentration of fractionated T- and E-MORB 3-4 km to the east of the axis.

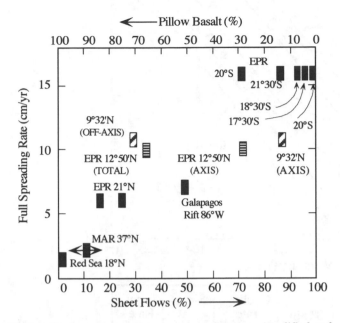

Figure 8. Lava morphology vs. spreading rate (modified and expanded from Bonatti and Harrison [1988]). Values only take areal distributions into account. Additional data added from the EPR ~9°30'-32'N (shown as 9°32'N Off-axis and Axis pair). Two values (Axis and Total) are also shown for the EPR ~ 12°50'. Diagram shows the dominance of sheet flows at fast-spreading centers and pillow lavas at slow-spreading centers. Although data are limited, it appears that pillows are more abundant than sheet flows in off-axis areas on some fast-spreading ridges; a feature that may be related to the low effusion rate of off-axis volcanism.

pahoehoe sheet flows [Hon, et al., 1994]) producing relatively flat upper crusts that may later collapse if the molten interior of the flow subsides. Such flows that become ponded and later collapse are often referred to as "lava lakes", and commonly display "lava pillars" in their hollow interiors [Ballard, et al., 1979; Francheteau, et al., 1979]. Gregg and Chadwick [1996] recently proposed a model in which lava pillars form in inflated lobate flows, beginning initially as gaps between thin flow lobes where water that is trapped beneath the flow is heated and channeled upward, promoting rapid crust growth around the gaps. The gaps between lobes grow upwards into pillars during lava flow inflation, when continued input of lava from the vent into the ponded flow causes its upper surface crust to rise uniformly.

Another common flow morphology is sheet flows. These flows are typically laterally extensive with low surface relief, but may also be erupted within confined depressions, like the AST on parts of the EPR. Sheet flows initially form as broad sheets of lava with an upper crust over a molten interior, but without the lobate upper surface. Instead, the upper crust can display a variety of textures from smooth and very flat, lineated, ropy (with

lava coils or whorls), to jumbled and hackly (Plate 2 and Plate 3a). All of these textures may exist on single flow units depending on the extent to which the lava sheets are affected by underlying roughness or constraining boundaries (e.g. faults, talus, other lava flows). Jumbled and hackly sheet flows are analogous to a'a' lava texture on subaerial volcanoes which forms by the physical disruption of surface crust during lava flow movement, particularly when the viscosity, rate of shear, or the eruption rate increase [Peterson and Tilling, 1980; Rowland and Walker, 1990]. Sheet flows are often thin (<6 cm), consisting mostly of glass, but can be up to a few meters in thickness with thin glass surfaces and more crystalline interiors. Thin sheet flow crusts have been recovered with glassy lava "drips", cuspate septa, or pipe vesicles on their undersides suggesting the interaction of water/vapor with lava during solidification. The surface crust insulates the molten interior of the flow and allows it to stay fluid longer, similar to tube-fed flows on land [Gregg, et al., 1996b].

Collapse areas in submarine lava flows occur in lobate and sheet flows when the molten interior of a flow begins to subside under a thin upper crust that cannot remain intact without support from below. The remnants of these upper crusts are commonly only a few centimeters thick, indicating that the crust was growing in contact with the molten lava for only a matter of minutes or tens of minutes before lava subsidence began [Griffiths and Fink, 1992; Gregg and Fink, 1995]. Thus collapse areas reflect rapidly changing conditions during a submarine eruption. The change from lava inflation to lava subsidence, which causes collapse, may occur due to lava "drainback", or "drain-out", or both. During lava drainback [Takahashi and Griggs, 1987] cessation of effusion at the vent can lead to already-erupted lava receding back into the vent. Lava "drain-out", on the other hand, refers to lava subsidence that occurs when lateral spreading and a continued increase in flow area outpaces the effusion rate.

Recent laboratory model experiments have greatly increased our understanding of the significance of submarine flow morphologies [Fink and Griffiths, 1990; Griffiths and Fink, 1992; Gregg and Fink, 1995]. These experiments involve the extrusion of molten wax into a cold liquid under various conditions. As the wax cools and forms a surface crust it mimics the behavior of submarine lava. These experiments clearly show that submarine flow morphology is a function of extrusion rate [Griffiths and Fink, 1992] and the angle of the slope over which lava is flowing [Gregg and Fink, 1995]. Pillow lavas form at the slowest extrusions rates, lobate flows at intermediate rates, and sheet flows at the highest extrusion rates (Figure 9). Since the viscosities of most MOR lavas are similar, it is essentially a battle between the rate of lava extrusion and the rate of lava crust growth that controls submarine lava morphology. Quantification of these relationships allows the duration of individual eruptions to be estimated based on the morphology and volume of the flow [Gregg and Fink,

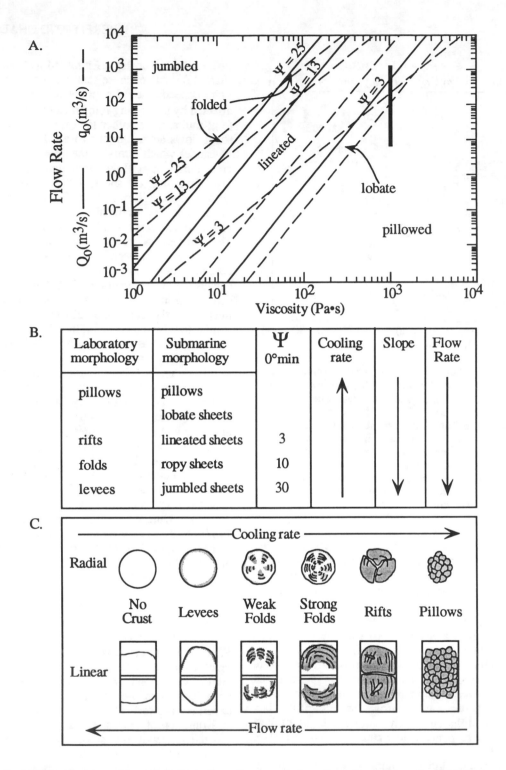

Figure 9. (a) Effusion rate and lava viscosity vs. lava flow morphology for submarine basalts on a 10° slope. Dashed lines separate fields assuming a linear source and solid lines for a point source (from Gregg and Fink [1995]). These results from laboratory simulations show that extrusion rate is probably the dominant influence on submarine lava morphology. (b) Correlation between laboratory-derived y values and laboratory and submarine flow morphologies. y is the ratio of time required for solid crust to form on the surface of a flow to a characteristic time scale for horizontal advection and includes the effects of viscosity, heat capacity, thermal diffusivity, and density of the flow and its environment. (c) Schematic diagrams showing characteristic sequence of morphologic types of flows developed for increasing cooling rates and/or decreasing eruption rates, for radial (upper) and linear (lower) geometry in lab experiments. Solidifying crust is indicated by shading (from Fink and Griffiths [1992] and Gregg and Fink [1995]).

1995]. For example, Gregg et al. [1996b] estimate that the 1991 eruption at 9°50'N on the EPR (Plate 3a) probably filled the AST in less than 2 hours whereas the pillow lavas on the CoAxial segment (Plate 3b) may have erupted rapidly at first and continued to form for approximately 10 days.

Putting these two fundamental observations together - that flow morphology relates to extrusion rate, and that different flow morphologies are dominant at different spreading rates - leads to the conclusion that fast spreading ridges typically have high-effusion rate eruptions, whereas eruptions at slow spreading ridges have lower-effusion rates. This suggests that dikes may be intruded at higher magma pressures at fast spreading ridges than at slow spreading ridges and is consistent with the concept that magmatism dominates over tectonism at fast ridges. At intermediate spreading rate ridges like the JdFR, where the ridge may go through cycles dominated by tectonism or magmatism, the eruption style may vary with time [Smith, et al., 1994].

If dike intrusions occur more frequently on fast ridges, tensional stresses perpendicular to the ridge axis (which gradually build due to divergent plate motion) will be kept at relatively low levels, because each intrusion superimposes new compressive stresses on the system. In this environment, each new dike must have a relatively high internal magma pressure, since the ambient stress is not strongly tensional. In contrast, at slow ridges, dikes can intrude at lower magma pressures because the ridges are magma starved. Less frequent magmatic events means that plate motion is more likely to raise the ambient tensional stress perpendicular to the ridge until either faulting or intrusion occurs. This notion is also consistent with evidence from the formation of dike-induced grabens, discussed below. The greater proportion of pillow flows off-axis at fast spreading ridges may be due to less frequent eruptions there, resulting in relatively higher local tensional stresses, allowing dikes to be intruded with lower internal magma pressure than those on-axis. Also the off-axis lavas tend to be erupted with higher viscosities (due to lower temperature and higher crystallinity) which favors pillow formation.

3. RECENT ERUPTIVE EVENTS AND ASSOCIATED STRUCTURES

3.1. Detection and Observation of Volcanic Events on the Mid-Ocean Ridge

One of the most fundamental unknowns in MOR science is the frequency of magmatic events - how often do dikes intrude, lavas erupt, and hydrothermal and biological systems get perturbed? Before 1990, no historical eruptive events had been documented on the global MOR system, despite the fact that approximately 75% of the Earth's magmatic flux occurs there [Crisp, 1984]. Many "young-looking" lava flows had been found, but there was no way

of knowing "how young" they really were. However, since 1990, eight historical eruptions have been documented - 5 on the JdFR, 2 on the northern EPR, and 1 on the Gorda Ridge (this is still probably a vast under-sampling, but is a dramatic improvement in detection capability!). What contributed to this sudden change? It was primarily the pay-off from multi-year water-column and geologic studies in selected areas, the sudden availability of previously classified U.S. Navy monitoring technology, and some very good luck.

The luck came first. The first documented MOR eruption was found on the Cleft segment of the JdFR. The first piece of evidence was fortuitously finding some unusually large and shallow hydrothermal plumes (called "megaplumes" or "event plumes") over the area in 1986 and 1987 [Baker, et al., 1987; Baker, et al., 1989]. These unusual plumes had apparently formed from the sudden release of a large volume of hydrothermal fluid that had been previously stored in the crust. One way to expel such a burst of fluid is by cracking open the crust during an intrusion or eruption [Baker, et al., 1989; Cann and Strens, 1989; Lowell and Germanovich, 1995]. Subsequent geologic mapping in the area identified extensive hydrothermal venting on the seafloor along a fissure system, a young sheet flow, and a string of equally young pillow mounds (Figure 10) [Embley, et al., 1991; Chadwick and Embley, 1994; Embley and Chadwick, 1994]. Discrepancies were noted between microbathymetry along camera tow tracks over the pillow mounds and the pre-existing multibeam bathymetry. The area was resurveyed with multibeam sonar and a quantitative comparison technique was developed which showed that significant depth changes had occurred exactly where the pillow mounds were mapped [Chadwick, et al., 1991; Fox, et al., 1992]. Making comparisons with other SeaBeam data effectively bracketed the eruption of the pillow mounds between 1983 and 1987 - the same time interval during which the event plumes were found. Repeated sidescan sonar surveys over the area showed that the young sheet flow was already in place in 1982, before the pillow mounds were erupted [Embley and Chadwick, 1994], but its very young appearance and rapid changes in hydrothermal venting along its eruptive fissure strongly suggested that it was only a few years older than the pillow mounds. The pillow mound eruption was interpreted as the product of a lateral dike intrusion that extended at least 17 km, and reached the surface intermittently along its length [Embley and Chadwick, 1994]. Geochemical analyses of samples from the sheet flow and pillow mounds are consistent with recent northward dike propagation but also suggest the event was associated with a recent influx of magma during a subaxial recharge event [Smith, et al., 1994]. The eruptive features produced were mostly thick (up to 45 m), steep-sided mounds of pillow lavas, with some smooth-surfaced lobate flows near vent areas, and their total volume was estimated at 55×10^6 m^3 [Chadwick and Embley, 1994; Chadwick, et al., in press] (Table 2).

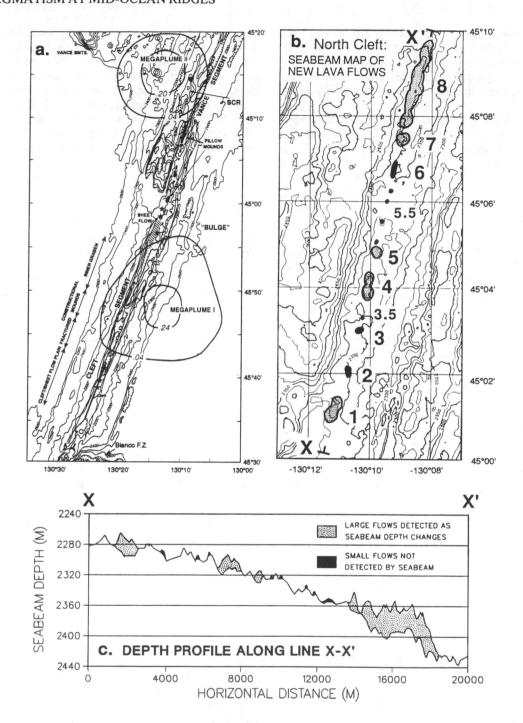

Figure 10. Mid-1980's eruption at the northern Cleft segment, JdFR. (a) The location of the newly erupted pillow mounds was midway between the two event plumes (or "megaplumes") found in 1986 and 1987 (modified from Embley and Chadwick [1994]). (b) Map of new pillow mounds erupted between 1983-87; larger flows (stippled) were detected as depth changes by repeated SeaBeam surveys, smaller flows (solid) were mapped by camera tows. (c) Depth profile along line of lava flows, between X and X' in figure (b) (from Chadwick and Embley [1994]).

Table 2. Estimated Volumes from Individual Eruptive Units on Different Spreading Rate Ridges.

Eruption site	Spreading rate	Estimated volume ($\times 10^6$ m³)	Source of volumes or dimensions
Mercury, FAMOUS segment, MAR	slow	800*	Crane and Ballard, 1981
Venus, FAMOUS segment, MAR	slow	600*	Crane and Ballard, 1981
Pluto, FAMOUS segment, MAR	slow	240*	Crane and Ballard, 1981
Uranus, FAMOUS segment, MAR	slow	360*	Crane and Ballard, 1981
Mars, FAMOUS segment, MAR	slow	380*	Crane and Ballard, 1981
Saturn, FAMOUS segment, MAR	slow	65*	Crane and Ballard, 1981
Serocki, MARK segment, MAR	slow	60*	Bryan et al., 1994
East MARK cones, MARK segment, MAR	slow	40-300*	Bryan et al., 1994
Small seamounts, 24-30°N, MAR	slow	40-80*	Smith and Cann, 1990
New pillow mounds, North Cleft, JdFR	intermediate	55†	Chadwick et al., in press
Young sheet flow, North Cleft, JdFR	intermediate	10.5**	Embley and Chadwick, 1994; this study
1993 flow, FLOW site, CoAxial, JdFR	intermediate	8.7†	Chadwick et al., in press
1982-91 flow, FLOW site, CoAxial, JdFR	intermediate	7.8†	Chadwick et al., in press
1981-91 flow, FLOC site, CoAxial, JdFR	intermediate	10**	Chadwick et al., in press
1996 flow, northern Gorda	intermediate	18†	Chadwick et al., in press
1991-92 BBQ site, 9°50'N, EPR	fast	1.0**	Gregg et al., 1997; this study
ODP flow, 9°30'N, EPR	fast	0.03**	Fornari et al., in press-a

* Volumes calculated using the formula for an ellipsoidal segment and the published length, width, and height dimensions of individual volcanic constructs.

† Volumes calculated using SeaBeam depth differences (which are minimums since they cannot detect the thinner parts of flows) and then adding an estimate for the volume of each flow "missing" from the Seabeam anomalies. The "missing" flow volume is estimated by multiplying the area of "missing" flow (the difference between the total flow area and the Seabeam anomaly area) by a constant thickness of 10 m (a value just below the Seabeam detection threshold).

** Volumes calculated based on mapped flow area and an estimated average flow thickness. New estimate for area of the Cleft young sheet flow is 4.2×10^6 m². For the Cleft and BBQ sheet flows, we assumed a post-emplacement average flow thickness of 2.5 m and 3m respectively.

The events at Cleft showed three important things concerning event detection on the MOR: 1) that event plumes above the seafloor can be a manifestation of eruptive activity on the bottom; 2) that repeat surveys by multibeam sonar can be used to detect new seafloor eruptions and constrain their ages and volumes, if the flows are large enough, and 3) that the geochemistry of basalts can be used to identify individual flow units and to ascertain possible genetic associations between temporally related eruptions.

The second stroke of luck was returning to a carefully mapped part of the northern EPR with the *Alvin* submersible just after an eruption in April 1991. In this case, the changes from the year before were so dramatic and the hydrothermal discharge and bacterial bloom were so intense, that it was clear an eruption had just happened [*Haymon, et al.*, 1991; *Haymon, et al.*, 1993]. Fresh new sheet flows (the "tubeworm BBQ" site) were found in the AST that had clearly just recently overridden biological communities (Plate 3a) [*Haymon, et al.*, 1993]. Geochemical dating (Po-Pb) of the basalt later confirmed that the lavas were only months old when observed for the first time [*Rubin, et al.*, 1994]. The volume erupted was relatively small (1.0×10^6 m³) (Table 2) and the eruption was probably brief in duration (~2 hours) [*Gregg, et al.*, 1996b]. Follow-up cruises to this area since 1991 have documented two smaller eruptions, that probably occurred in 1992 (Plate 3c) [*Rubin, et al.*, 1994, Perfit, unpublished data]. In addition, changes in vent-fluid temperatures measured between 1994 and 1995 and biological observations made at the hydrothermal vents in the 9°49'-51'N region, have been interpreted as the result of an additional small dike intrusion not associated with any obvious lava extrusion [*Fornari, et al.*, in press-b].

Observational data suggest that the primary eruptive fissure for the tubeworm BBQ eruption was nearly continuous along strike over a distance of ~8.5 km within the AST, had small en echelon offsets (~3-5 m), and varied in width between 1-4 m [*Haymon, et al.*, 1993; *Wright, et al.*, 1995; *Gregg, et al.*, 1996b]. The narrowness of the AST and eruptive fissures and lack of structural features associated with the 1992 eruption around 9°52'N are shown in Plates 3c and 3d. In this region, the AST/fissure system is generally continuous (except for small en echelon offsets of a few m) for over 3 km. Along this entire length, outpourings of lava and drainback have only been documented in a few places. Eruptive fissures in the floor of the AST in the 9°49'-51'N region are identified based on relationships between the lava flows and the fissure edge, including down-warping of original flow surfaces towards the fissure and the extensive drainback and collapse features associated the fissure system, similar to those associated with drainback into fissures in Hawaii [*Green and Short*, 1971; *Takahashi and Griggs*, 1987]. Detailed surveys and observations have found no large (>10-m-high), centralized volcanic constructs in the AST along the length of the eruptive fissure. The AST around 9°50'N ranges in depth between 5-8 m. Lava pillar tops are also 5-8 m high and spill-outs rarely extend more than ~200 m beyond the lip of the trough, suggesting that the 1991 flow was largely contained within the AST and its maximum thickness was about 8 m [*Gregg, et al.*, 1996b].

Because of the way lava pillars form, features observed on their interior and exterior walls have been used to reveal information about the waxing and waning stages of eruption. Information about the dynamics and compositional variability of the 1991 EPR eruption is preserved in the glassy selvages of pillars recovered from the AST [*Gregg, et al.*, 1996b]. Major element compositions of glass comprising pillar walls and forming selvages along the outsides of the pillars show little compositional variation. Most elements are within the 2-sigma limits determined from analyses of nearly two dozen samples of the 1991 lava flow suggesting the lava pillars most likely formed during a single eruptive episode, and that the episode is not sufficiently long or voluminous for lava composition to change during the event. Pillar walls tend to be ~4-7 cm thick, suggesting that they required about the same amount of time to grow (~2 hours) as did the lobate crusts on the surface of the ponded 1991 flow. Slight variations in thickness along the length of pillars suggest the bases were in contact with molten lava for a greater amount of time than the pillar tops; possibly 3-4 hours longer, constraining the time required for the lava level to rise and fall [*Gregg, et al.*, 1996a; *Gregg, et al.*, 1996b].

The most important new advance for the detection of MOR eruptions has been the availability of the U.S. Navy SOSUS (SOund SUrveillance System) hydrophones in the NE Pacific and the development of sophisticated processing of these data for locating earthquakes [*Fox, et al.*, 1994]. Detecting eruptions on the MOR in real time had been impossible before SOSUS became available because they generally are too far away from land-based seismometers, which can only detect distant earthquakes down to about magnitude 4. However, earthquake swarms associated with volcanic rifting events are typically below magnitude 3 [*Brandsdottir and Einarsson*, 1979; *Einarsson and Brandsdottir*, 1980]. The only way to detect such events remotely is with hydrophones, because some of an oceanic earthquake's energy is converted into acoustic noise (T-waves), and sound travels much more efficiently through water than P- and S- waves travel through the solid earth. The development of the SOSUS system for earthquake detection suddenly made monitoring of the micro-seismicity associated with submarine volcanic events in the NE Pacific a reality, and opened up the possibility of rapid scientific response efforts to such events.

In fact, just days after the system began operating in real time in June 1993, it detected its first volcanic event on the CoAxial segment of the JdFR [*Fox, et al.*, 1995]. The distribution and character of the T-wave events were interpreted as evidence for the lateral intrusion of a dike along the segment for up to 60 km [*Dziak, et al.*, 1995; *Schreiner, et al.*, 1995]. Within a few weeks, a ship equipped with CTD's and the ROPOS remotely operated vehicle (ROV) confirmed the existence of event plumes at the site and a new pillow lava flow on the seafloor that was still hot and venting warm water (Figure 11) [*Baker, et al.*, 1995; *Embley, et al.*, 1995b]. This event conclusively linked the formation of event plumes to seafloor eruptive activity. Extensive areas of diffuse hydrothermal venting were mapped on the new flow (Plate 3b) and along fissure systems that marked the path of the dike intrusion, including sites with large bacterial blooms, similar to those reported for the 1991 EPR eruption [*Haymon, et al.*, 1993; *Embley, et al.*, 1995b; *Holden, et al.*, 1995; *Juniper, et al.*, 1995]. Repeat multibeam mapping over the CoAxial segment found depth changes up to 30 m over the new 1993 flow, but also found two other areas of depth change between 1981 and 1991 surveys (Figure 11), which subsequent camera tows, sidescan sonar surveys, and dives by submersible and ROV have shown also correspond to very young lava flows [*Chadwick, et al.*, 1995; *Embley, et al.*, 1995a]. Thus the CoAxial segment apparently has had 3 separate eruptive events since 1981, and may be in an unusually active phase. However, chemical analyses of the lavas suggest the 1993 flow was not genetically related to the nearby older flows [*Smith, et al.*, 1995b]. Each of these lava flows has an estimated volume of about 7-11 x $10^6 m^3$ (Table 2), and consists of single or multiple ridges of pillow lava (Plate 3b) that become more lobate in morphology near their eruptive vents.

Another T-wave swarm was detected by SOSUS on the Blanco Fracture Zone in January 1994, but the exact nature of this event is still unclear - it appeared to be volcanic but may have been an intrusion rather than an eruption [*Dziak, et al.*, 1996]. More recently, in February 1996, the

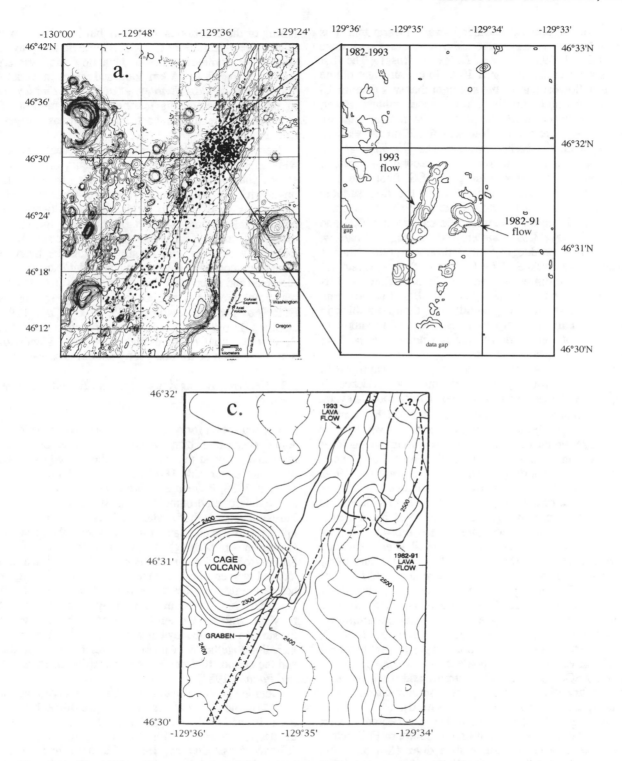

Figure 11. The 1993 eruption on the CoAxial segment, JdFR. (a) Regional map of earthquake epicenters during the CoAxial earthquake swarm - the eruption was located at the center of the densest cluster (from Dziak et al.[1995]). (b) Map showing areas of significant depth change - including the 1993 flow and a slightly older flow - between SeaBeam surveys in 1982 and 1993 (a third survey in 1991 constrains the age of the older flow). The two eruptions are closely spaced in time, but are separated by 700 m across axis, showing that the neovolcanic zone is at least that wide on the JdFR (from Chadwick et al. [1995]). (c) Map of 1993 and 1982-91 lava flows based on initial submersible and ROV observations (from Embley et al. [1995b]).

SOSUS system detected another T-wave swarm that was clearly volcanic in character on the northern Gorda ridge [*Fox and Dziak*, 1996; *Fox and Dziak*, in press]. The first event-response cruise in March 1996 found an event plume near the shallowest part of the segment that was roughly 10 km in diameter and 1 km thick in the water column [*Baker, et al.*, 1996; *Baker, et al.*, in press]. On a later response cruise in April, a new lava flow was found on the seafloor directly beneath the earlier event plume [*Embley, et al.*, 1996; *Chadwick, et al.*, in press]. A second event plume found in April was seeded with a RAFOS float, and then was later re-located and re-sampled when the float surfaced 10 km off-axis in July [*Lupton, et al.*, 1996]. Subsequent cruises have visited this site with sidescan sonar, CTD's and ROV's. The lava flow that erupted is long and narrow, again clearly the product of a brief fissure eruption where a dike reached the surface. The flow is 400 m wide and 2.6 km long (similar dimensions to the CoAxial flow), and its thickness varies markedly along strike. Based on SeaBeam depth changes, the flow is generally less than 20 m thick in the southern half but is up to 75 m thick in the northern half, and the total eruptive volume is estimated to be 18 x 10^6 m^3 [*Chadwick, et al.*, in press].

There are a few more instances in which a MOR volcanic event probably occurred, but the evidence is incomplete or the timing is less well constrained. Several possible intrusive events have been detected by seafloor instruments with bottom pressure recorders which have been deployed in the caldera of Axial seamount on the JdFR to look for magmatic inflation or deflation of the volcano's summit. An apparent deflation event was recorded in 1988, coincident with sudden changes from a temperature sensor and a time-lapse camera, suggesting that a dike intruded into the nearby rift zone and perturbed the local hydrothermal system [*Fox*, 1990]. Since then, these types of seafloor instruments have been deployed at Axial every year, and have shown other possible events [*Fox*, 1993], and many brief T-wave swarms have occurred there since 1991 [*Dziak and Fox*, 1997]. Photography and sampling of very young looking basalt from Axial caldera and its north rift zone suggest that it has recently experienced several eruptions, probably within the last few decades [*Embley, et al.*, 1990, and unpublished data]. Just before this paper went to press, another T-wave swarm and apparent intrusion/eruption was detected by SOSUS on Axial's summit and south rift zone in late January 1998 [C. Fox, pers. communication, 1998; http://newport.pmel.noaa.gov/axial98.html].

Another example of an area that has experienced a recent eruption is the 17.5°S segment on the southern EPR where observations from recent submersible dives [*Renard, et al.*, 1985; *Auzende and Sinton*, 1994; *Auzende, et al.*, 1996; *Embley, et al.*, 1997; *Embley, et al.*, in press], vent fluid sampling [*Butterfield, et al.*, 1997; *Lupton, et al.*, 1997], and hydrothermal plume mapping [*Urabe, et al.*, 1995; *Baker and Urabe*, 1996; *Feely, et al.*, 1996] strongly suggest very recent volcanic and hydrothermal activity. The

timing of the eruption is uncertain, but it probably occurred within a few years prior to 1993. Geologic mapping in the area is not complete, but the lava flow that was erupted appears to be up to 25 km long and varies in width from 400 m up to 1.5 km [*Embley, et al.*, 1997; *Embley, et al.*, in press]. Whether this is one flow or multiple flows erupted over a relatively short time period has not yet been determined.

Despite having documented these recent events and additional evidence for others, the frequency of eruptions on slow, intermediate, and fast MORs is still largely unknown. It will take many more years of real-time monitoring (and luck) to begin to piece together a full picture. Real-time SOSUS monitoring for earthquake detection currently only exists in the NE Pacific. However, since May 1996 an array of six autonomous hydrophones have been moored in the eastern equatorial Pacific which are recovered and redeployed every 6 months (the data are non-real-time) in order to monitor seismicity along most of the EPR [*Fox and Matsumoto*, 1995; *Fox, et al.*, 1997, see also http://www.pmel.noaa.gov/vents/oceanseis.html]. Several potential eruptive sites have been identified, but none have been confirmed yet by bottom observations.

3.2. Graben Faulting Associated with Recent Eruptive Events

Grabens often form above intruding dikes in volcanic rift zones because a dike creates two zones of maximum tensional strain at the surface, parallel to the dike axis but offset to either side [*Pollard, et al.*, 1983; *Mastin and Pollard*, 1988; *Rubin and Pollard*, 1988; *Rubin*, 1992]. In the tectonic environment of a spreading ridge, a dike intrusion releases horizontal tensional stress that has built up by divergent plate motion since the last rifting (intrusion or faulting) event [*Björnsson, et al.*, 1979; *Sigurdsson*, 1980; *Tryggvason*, 1984]. When a dike-induced graben forms above a dike, the faulting occurs before the dike reaches the surface [*Pollard, et al.*, 1983]. Once an eruption begins, lava may then fill or partially bury the graben. Graben faults tend to be near-vertical at the surface because they begin as simple tension cracks and later link together with other fractures between the dike tip and the surface to form a zone of distributed shear [*Mastin and Pollard*, 1988].

Detailed geologic mapping of the recent eruptive sites on the Cleft and CoAxial segments of the JdFR has revealed the existence of distinct grabens, 10-100 m wide and 5-15 m deep, associated with the new flows (Figure 12) [*Chadwick and Embley*, 1994; *Chadwick and Embley*, in press]. These grabens are interpreted to have formed directly over the dike that fed the eruptions as it was intruding [*Chadwick and Embley*, in press]. In most places, the AST on the EPR has similar dimensions to the grabens observed on the JdFR, 5-40 m deep and 40-250 m wide [*Lonsdale*, 1977b; *Hekinian, et al.*, 1983; *Ballard, et al.*, 1984;

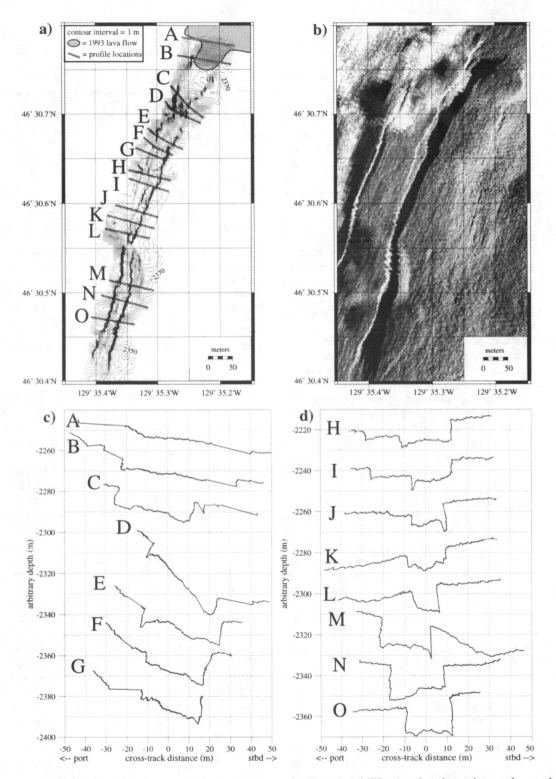

Figure 12. Mesotech and sidescan sonar data from the CoAxial Flow site, JdFR, showing the graben at the southern end of the 1993 lava flow (from Chadwick and Embley [in press]). (a) Mesotech sonar bathymetry (1-m contours). Lettered lines (A-O) show locations of individual Mesotech depth files in (c) and (d). Light gray stipple shows southern end of 1993 lava flow. (b) AMS-60 sidescan sonar image (insonification from SE; white is high reflectivity, black is acoustic shadow). Lower slope of Cage Volcano is in the upper left. (c and d) Individual Mesotech depth

Choukroune, et al., 1984; Renard, et al., 1985; Gente, et al., 1986; McConachy, et al., 1986; Macdonald and Fox, 1988; Haymon, et al., 1991; Auzende, et al., 1996; Fornari, et al., in press-a], and therefore may also be related to dike-induced graben subsidence [*Chadwick and Embley, in press*]. This interpretation is debatable because there is little direct evidence for structural control of the AST at ~9°50'N on the EPR [*Fornari, et al., in press-a*]. Nevertheless, if we assume for a moment that AST's (both "graben" and "calderas") on the EPR are dike-induced subsidence features similar to the grabens documented near recent eruptive sites on the JdFR, there are some interesting structural and mechanical implications.

Working from this assumption, it is noteworthy that the dike-induced grabens on the JdFR and the EPR are narrower and deeper than grabens that formed during well-documented dike intrusions on slow-spreading rifts in Iceland and Afar, which were 1-6 km wide and 0.5-1.5 m deep [*Abdallah, et al., 1979; Björnsson, et al., 1979; Ruegg, et al., 1979; Sigurdsson, 1980*]. Two results from mechanical modeling are helpful in understanding the significance of this difference in the scales of dike-induced faulting. First, as a dike intrudes from a deep to a shallow level in the crust, the distance between the two zones of maximum tensional strain at the surface (where graben faults are most likely to form) decreases as the dike intrudes upwards (Figure 13). Secondly, the magnitude of that dike-induced tensional strain increases as the dike intrudes upwards [*Pollard, et al., 1983; Mastin and Pollard, 1988*]. Together, these results suggest that a narrow graben (as observed on the JdFR and EPR) would form when a dike was very shallow and when the stress perturbation from the dike was very high. Another key point is that a graben will form over a dike as soon as the sum of the dike-induced stress plus the ambient stress, reaches the critical point for brittle failure. The ambient horizontal tensional stress perpendicular to a spreading ridge is mainly a function of how long it has been since the last rifting event - it will be low immediately after an event and will gradually increase with time due to divergent plate motion. Therefore, the width of a dike-induced graben is a qualitative measure of the ambient stress state [*Rubin, 1992; Chadwick and Embley, in press*]. In other words, a wide graben implies that only a small stress perturbation was necessary from a deep dike to initiate faulting; it had been a long time since the last event and the ambient tensional stress was already near failure from plate motion - characteristic of a slow-spreading ridge. On the other hand, a narrow graben implies that a high stress perturbation was required from a shallow dike to cause faulting; it had only been a short time since the last event and the ambient tensional stress was still low and not "primed" for failure - characteristic of a fast-spreading ridge.

Thus the narrow dike-induced grabens observed on the JdFR and EPR imply that typically a relatively low level of ambient horizontal tensional stress exists perpendicular to the ridges (or it may even be compressive). In general, the

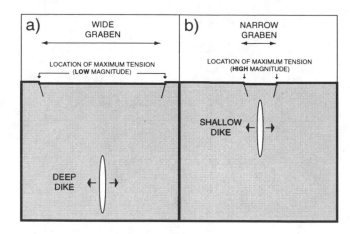

Figure 13. Diagrammatic ridge axis cross section showing the effect of dike depth on graben width, the magnitude of dike-induced stress perturbation, and the location of zones of maximum tension at the surface where fissure/graben formation is most likely (from Chadwick and Embley [in press]). An individual dike starts deep (a) and migrates upward (b), but faulting will only occur if and when the ambient stress plus the dike-induced stress perturbation exceeds the tensional strength of the rock.

character of intermediate rate ridges is somewhere in between those of slow and fast ridges. The grabens associated with recent eruptions at the Cleft and CoAxial segments of the JdFR may be at the narrow end of the spectrum (and similar to those on the EPR) because both these segments had experienced earlier eruptions just years before the most recent events, which would have previously "reset" their ambient stress states. On the EPR, the high magma supply and the high frequency of dike intrusions means that continuous plate spreading rarely gets enough time to significantly increase the horizontal tensional stress. This implies that for dikes to intrude in a fast spreading environment, they must have relatively high internal magma pressure (since the ambient horizontal tensional stress is typically low) [*Chadwick and Embley, in press*]. This would naturally lead to high effusion rate eruptions, which is consistent with the distribution of lava morphologies discussed earlier (mostly sheet flows at fast ridges, mostly pillows at slow ridges), since flow morphology is mainly a function of effusion rate. On fast spreading ridges (and sometimes on intermediate ones), magmatism apparently dominates over tectonism in regulating the local stress state. At slow spreading ridges tectonism dominates.

4. ANALYTICAL DATING TECHNIQUES FOR YOUNG LAVAS

Newly developed dating techniques in conjunction with more accurate sampling methods have given us another tool with which to confirm recent eruptions on the MOR and to

Table 3. Analytical techniques used for "dating" MORB

Method	Half-life(yr)	"Age" resolution[1]	Age Limit
Paleomagnetics			
Distance vs. spreading rate	10 - 30 ka	NA	
High NRM		only < ~10 yr *	~10 yr
Observation			
Visual recognition		only < ~10 yr	< ~10 yr ?
Repeat bathymetric surveys	< year - #	#	
Radiometric			
$^{40}Ar-^{39}Ar$	1.25×10^9	20-50 ka	6.25×10^6 ka
$U-^{230}Th$	75,380	10 -30 ka	380 ka
$U-^{231}Pa$	32,760	2 -10 ka	230 ka
$Th-^{226}Ra$	1,600	0.1 - 1.0 ka	11 ka
$Ra-^{210}Pb$	22.3	~ 20 yr	112 yr
$Pb-^{210}Po$	0.4	1 - 2 mo.	2 yr

[1] Age resolution is an estimate of the time intervals that could potentially be resolved (depending on the methods/elements used and analytical precision) in determining the age of the seafloor or a seafloor basalt. For radiogenic isotopes, maximum ages are generally up to 5 half-lives, but when exesses are large, maximum ages may be up to 7 half-lives. Ar-Ar incremental heating methods can measure ages in the range 0.2 -1.0 Ma with 10% (1s) uncertainty but few samples < 350 ka have been analyzed (Duncan and Hogan, 1994).
* The high natural remnant magnetism measured in some MORB only appears to last for less than 10 years.
# Depends on interval of time between repeat surveys.

assess the temporal dimension of magmatic processes on geologically short time scales (days to thousands of years). Critical to our understanding of the time scales and spatial distribution of volcanism at ridge crests is our ability to determine the age of individual flow units on the seafloor. A number of techniques for sampling and dating mid-ocean ridge basalt (MORB) have allowed us to examine the spatial and temporal variability of ocean ridge volcanism with various degrees of resolution (Table 3). Below some of these methods are reviewed and their benefits and shortcomings are discussed.

4.1. Sampling and Qualitative Dating Methods

Many studies of the temporal and spatial variability of MOR magmatism have relied on sampling by dredge from a surface ship [e.g. *Bryan, et al.*, 1981; *Thompson, et al.*, 1985; *Langmuir, et al.*, 1986; *Hekinian and Walker*, 1987; *Hekinian, et al.*, 1989; *Batiza and Niu*, 1992]. Although determination of dredge positions has improved in recent years due to the combined use of GPS navigation, multibeam mapping, bottom photography, and in some instances, transponder navigated dredging, a significant

amount of uncertainty in the location of sample recovery remains. Rock dredging and dating based solely on location of recovery and spreading rate (determined by paleomagnetics) do not provide the precision or resolution necessary to evaluate the relatively rapid geochemical and volcanological changes that occur on ridge crests nor do they allow us to date "young" events associated with the neovolcanic zone. Additionally, age determinations based on the calculated age of the seafloor assume the surficial lavas were erupted at the ridge axis, as opposed to off-axis, or have not flowed far from the axis, and are therefore contemporaneous with the underlying crust. Clearly, recent observational evidence has shown that is not always true. Nor can one assume that all lavas recovered from the neovolcanic zone are "zero" age because of the periodicity of volcanism along axis that has been documented on many ridge crests. This is particularly true at slow-spreading ridges where volcanism is relatively infrequent and widely distributed in the axial valley.

Rock coring from surface ships has allowed us to do more rapid and site-specific sampling with ~200-400 m target accuracy. This technique has been used to study the spatial variability in EPR and JdFR basalt geochemistry

[*Reynolds, et al.,* 1992; *Perfit, et al.,* 1994; *Smith, et al.,* 1995b] of relatively young seafloor (< ~100 ka because thick sediment cover in older terrain inhibits sample recovery). Small sample size (often < 1 gm, but typically 1-10 gm recovered) is adequate for major and trace element analysis (by microprobe, laser ablation, or ICP-MS) but not always enough for U-series dating, and limits estimates of age based on sample appearance. Sampling with manned submersibles or ROV's coupled with detailed mapping both remotely (e.g. sidescan, repeat multibeam surveys) and with *in situ* observation, currently provides the best method of determining the spatial and temporal relationships between lavas and eruptive events [*Smith, et al.,* 1996].

In the absence of more quantitative methods, estimates of relative ages of MORB are commonly made on the basis of their visual appearance. The "freshest" samples with the least sediment cover and glassiest surfaces are considered the youngest while those with increasing sediment cover and Mn coatings and decreasing glassy luster are considered increasingly older. Although this method of relative dating is useful within limited areas, it cannot be applied globally because of differing rates of pelagic sedimentation, seafloor weathering/alteration, and hydrothermal sediment and Mn accumulation. Age estimates from lava appearance must be made cautiously, because differences in apparent ages can be due solely to differences in surface texture. For example, very young (<5 years old) striated pillows tend to appear older than smooth-surfaced lobate or sheet flows of the same age because surface glass on the pillows is disrupted and their rough surface texture can trap sediment more readily. However, with time (>100 years) flat lobate or sheet flows generally accumulate sediment more rapidly than high-standing pillows. Studies of known eruption sites have noted that new lava flows do not stay pristine for long (~1 year) and that previous "uncalibrated" attempts to estimate ages based on lava appearance were probably over-estimates by about an order of magnitude [*Chadwick and Embley,* 1994].

4.2. Radiometric Dating of MORB

Accurate isotopic dating of MORB can be used to establish a chronological framework for geochemical/petrological data and to quantify the spatial and temporal characteristics of erupted basalts and crustal magmatic reservoirs. K-Ar and ^{40}Ar-^{39}Ar dating of MORB are restricted to basalts that are older than ~40 ka and the techniques have had only limited success because of the low potassium content of typical MORB, excess Ar in many samples, and K-uptake in glasses as they age on the seafloor. Some recent success in dating MORB in the age range of 0.2 to 1.0 Ma with ~10% uncertainty has been obtained by using a ^{40}Ar-^{39}Ar incremental heating method [*Duncan and Logan,* 1994].

Newly developed methods for measuring U-series disequilibria in MORB glasses show great promise for determining what are known as "crustal residence ages" (eruption + magma storage ages) and the potential exists for dating crystallization ages if equilibrium mineral separates can be analyzed. Measurements of ^{230}Th-^{226}Ra [*Rubin and Macdougall,* 1988; *Rubin and Macdougall,* 1990; *Volpe and Goldstein,* 1993], ^{235}U-^{231}Pa and ^{238}U-^{230}Th [*Goldstein, et al.,* 1992; *Goldstein, et al.,* 1993] disequilibria in young MORB can be used to establish a chronology for magmatic processes at ocean ridges on a 0.1-350 ka time scale (Table 3). ^{210}Po-^{210}Pb systematics can also be used to date very young eruptions (<2 yr) with an age resolution of <60 days [*Rubin, et al.,* 1994]. U-series disequilibria have also been used to investigate source characteristics, melting or upwelling rates, melt fractions, and magmatic differentiation in oceanic environments [*Gill and Condomines,* 1992; *Goldstein, et al.,* 1992; *Goldstein, et al.,* 1993; *Volpe and Goldstein,* 1993; *Lundstrom, et al.,* 1995; *Sims, et al.,* 1995].

The isotopes of the U- and Th- decay series are ideal for examining the nature and time-scales of magmatic processes during MORB petrogenesis and for dating young off-axis MORB, for several reasons: 1) the daughter isotopes ^{230}Th (t1/2 = 75,200 yr.), ^{231}Pa (t1/2 = 32,420 yr.) and ^{226}Ra (t1/2 = 1,620 yr.) of the ^{238}U decay series have half-lives which bracket the time scales over which melting and melt extraction are thought to occur, 2) the large disequilibrium ratios observed in oceanic basalts, and a limited number of experimental measurements, suggest there are significant (although not great) differences in mineral/melt partition coefficients between U-Th, Th-Ra and U-Pa, and 3) steady-state requirements constrain the abundances and activities of a decay series prior to chemical fractionation (e.g. melting); chemical fractionation perturbs this steady-state, therefore measurement of disequilibria among the isotopes of the U- and Th-decay series provide unique and direct constraints on the timing and extent of chemical fractionation occurring as a result of magma generation and delivery to the crust and seafloor. Details of the geochemistry and systematics of U-series isotopes are beyond the scope of this paper but can be found elsewhere [*Condomines, et al.,* 1988; *Gill and Condomines,* 1992; *Gill, et al.,* 1992; *Macdougall,* 1995]. Here, we provide an overview of their uses and limitations in dating young volcanism at ridges.

In the U-series, ages are calculated based on the differences between the measured values (or activities) and their known values at equilibrium (activity ratios equal to one). Secular equilibrium is attained after about five half-lives of the longest lived nuclide. Consequently, it is assumed that all of the nuclides in the U-series decay chain are in secular equilibrium in the suboceanic mantle and thus at the onset of melting of any MORB source. Large disequilibria (activity excesses of ^{230}Th, ^{226}Ra, and ^{231}Pa relative to their parents) have been measured in many young MORB from a number of ridge segments. The extent of the observed excesses of daughter isotopes can be due to differences in amounts and styles of melting, and/or source lithology in the mantle [*Goldstein, et al.*, 1992; *Goldstein, et al.*, 1993; *Volpe and Goldstein*, 1993; *Lundstrom, et al.*, 1995].   U-series disequilibria for young samples vary systematically for specific ridge segments and vary as a function of Th/U ratio of the samples (enriched MORB have higher Th/U).   Linear trends of measured excesses in very young samples on isochron diagrams (Figure 14) are interpreted as a reflection of mixing of enriched and depleted mantle melts [see *Lundstrom, et al.*, 1995].

Along individual ridge segments, where mantle sources are believed to be homogeneous, differences between measured excesses in young (so-called "zero-age") basalts and samples of unknown age can be used to constrain the ages of MORB. Mass spectrometric measurement of ^{238}U-^{230}Th, ^{230}Th-^{226}Ra, and ^{235}U-^{231}Pa disequilibria have been successfully applied to "dating" MORB from the JdFR and EPR [*Goldstein, et al.*, 1989; *Rubin and Macdougall*, 1990; *Goldstein, et al.*, 1992; *Goldstein, et al.*, 1993; *Volpe and Goldstein*, 1993; *Goldstein, et al.*, 1994b] However, the "ages" that are determined are not actual dates of eruption but rather must be considered model ages or "crustal residence ages". There are a number of assumptions and complicating factors that must be accounted for in interpreting these ages. The most important assumption is that the suite of samples being dated must have been derived from the same or very similar sources and that melt transit times to the crust are nearly constant. It also assumes that magma resides in the crust for significant periods of time (i.e. in a crustal magma chamber) but it could well be that magma storage also occurs in the shallow mantle or at the crust mantle boundary. For these reasons, the calculated "crustal residence age" should be interpreted in the broadest sense. Figure 15 diagramatically shows some of the processes that may effect the measured disequilibria and transit/storage times that must be taken into account.

U-series closed system model ages are calculated based on the following equations:

$$T = -\frac{1}{l_D} \ln \frac{(D/P) - 1}{(D/P)_o - 1}$$

where D represents the daughter isotopes (e.g. ^{230}Th, ^{231}Pa, ^{226}Ra) generated from P, the respective parent isotopes (e.g. ^{238}U, ^{235}U, ^{230}Th). (D/P) is the measured activity ratio, and $(D/P)_o$ is the initial activity ratio upon eruption, and l_D is the decay constant for the daughter isotope in question. The ^{230}Th age equation is of slightly different form in that the daughter and parent isotopes are normalized to ^{232}Th which is considered a stable isotope relative to the time scales in question.

Model ages are calculated by, 1) estimating an initial activity ratio based on data for the youngest erupted basalt and by assuming constancy of this initial activity ratio over time, or 2) estimating an initial activity ratio based on model assumptions for primary melts derived from the mantle. For example, in calculating ages of samples from

the 9-10°N region of the EPR, a basalt known to have erupted in 1991 was used to determine the initial ^{226}Ra/^{230}Th and ^{231}Pa/^{235}U activity ratios. The decreased radioactive excesses in older samples provide closed system "whole-rock" model ages.

Model ages actually represent age differences between samples since U-Th-Ra-Pa fractionation during partial melting in the mantle. Consequently, the model age T is given by

$$T = T_t + T_{mc} + T_e$$

where T_t is the transit time for magma from mantle to the upper parts of the lithosphere, T_{mc} is the magma residence time within crustal (or subcrustal) magma bodies, and T_e is the time since eruption on the seafloor. It is generally assumed that transit times are small or constant for samples from similar locations and sources (e.g. similar incompatible element (Th/U) and radiogenic isotope ratios) [*Rubin and Macdougall*, 1990; *Goldstein, et al.*, 1992]. Consequently, the model age can be replaced by the crustal residence age T_{cr}, where

$$T \sim T_{cr} = T_{mc} + T_e$$

The crustal residence age provides an upper limit on the eruption age and/or the magma residence age:

$$T_e < T_{cr} \text{ and } T_{mc} < T_{cr}.$$

For off-axis basalts where $T_{mc} \ll T_e$ is the usual situation, then

$$T_e \sim T_{cr}$$

and the crustal residence age is approximately equal to the eruption age [*Goldstein, et al.*, 1994b]. For axial basalts, if $T_e \ll T_{cr}$, then

$$T_{mc} \sim T_{cr}$$

and the crustal residence age is approximately equal to the residence time of melts within the crust and upper mantle.

Clearly, U-series dating can provide us with temporal constraints that were not previously available. However, one must be aware of the assumptions made to calculate the ages and must be careful in their interpretation. Young-looking samples from neovolcanic zones yield ages of a few ka, which must be considered crustal residence ages rather than eruption ages. One must also keep in mind that source heterogeneity, differing melt ascent rates from the mantle and, to a lesser extent, chemical fractionation in subaxial magma bodies can significantly influence the calculated ages. In particular, it seems unlikely that subaxial magma bodies are closed systems and therefore the effects of open system behavior must also be considered in

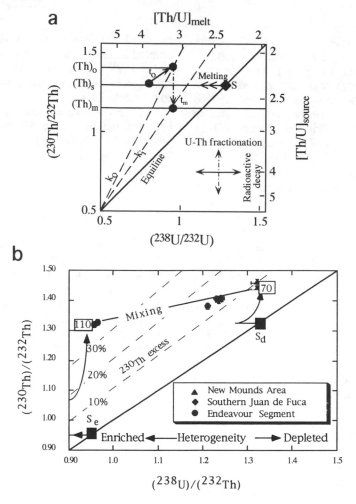

Figure 14. (a) Schematic (^{230}Th)-(^{238}U) isochron diagram that shows major aspects of U-series decay schemes and dating technique discussed in text (after Gill and Condomines [1992]). Disequilibrium occurs by fractionation of ^{230}Th from ^{238}U during partial melting. Magma (solid circles) is produced by melting of homogeneous source such as S which is initially in radioactive equilibrium (on the equiline where $(^{230}\text{Th})/(^{238}\text{U}) = 1$). Assuming $D^{Th} < D^{U}$, then $(^{230}\text{Th})/(^{238}\text{U})$ will be >1.0 (a ^{230}Th-excess) in the melt and the melt will have higher Th/U. Dashed lines "k" represent lines of constant $(^{230}\text{Th})/(^{238}\text{U})$ ratios that are generally shown as percent ^{230}Th-excesses (see diagram below). For example, a disequilibrium $(^{230}\text{Th})/(^{238}\text{U}) = 1.4$ is equivalent to a 40% ^{230}Th-excess. Small extents of melting accentuate the Th/U fractionation leading to increased ^{230}Th-excesses. Increasing degrees of melting decreases the ^{230}Th enrichment and the Th/U ratios in the melt of S decrease (upper axis) bringing the composition closer to the equiline. At the end of melting (time t_0) the $(^{230}\text{Th})/(^{232}\text{Th})$ ratio in the melt is $(\text{Th})_0$ which exceeds $(\text{Th})_s$ if t_0 is long. After the melt ceases to be in chemical equilibrium with its source, Th/U of the melt remains constant but the $(^{230}\text{Th})/(^{232}\text{Th})$ ratio decreases with time (t_i) as a result of unsupported radioactive decay. Recently erupted samples are assumed to have values at or near t_0 if melt transfer times are negligible but will have values between t_0 and t_m as time since eruption (or magma storage) increases. In order to calculate ages, the values at t_0 must be known and present day values at t_m are measured. (b) Dependence of $(^{230}\text{Th})/(^{232}\text{Th})$ on $(^{238}\text{U})/(^{232}\text{Th})$. $(^{230}\text{Th})/(^{238}\text{U})$ from the Juan de Fuca Ridge (modified from Lundstrom et al. [1995]). Using the highest observed disequilibria (youngest samples) minimizes the effect of post eruption decay. Dashed lines show mixing trends between melts from depleted and enriched sources. The linear trend between two trace elements normalized to a third is most easily explained by mixing. Open squares represent equilibrium porous flow melt models at 110 km and 70 km from enriched (S_e) or depleted (S_d) sources. The black square represents the position of the original equilibrium state for all models. ^{230}Th excess is produced by the melting process, as a function of the ratio of two elements (^{232}Th and ^{238}U) which are stable on the time scale of the melting process. The evolutionary paths during melting of the enriched (S_e) and corresponding depleted (S_d) melts are shown. Note that the enriched melt path initially trends off of the diagram to the left. The linear trend of the young MORB from the JdFR between the calculated end members suggests mixing of melts from different sources and depths.

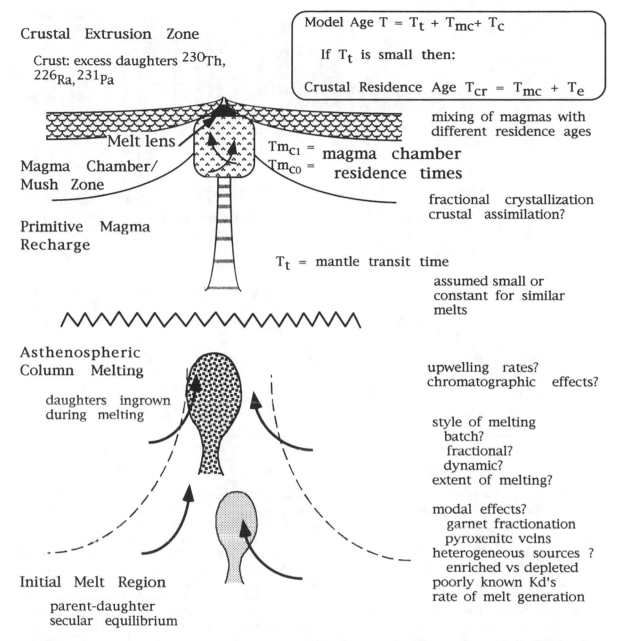

Model Age $T = T_t + T_{mc} + T_c$

If T_t is small then:

Crustal Residence Age $T_{cr} = T_{mc} + T_e$

Crustal Extrusion Zone

Crust: excess daughters ^{230}Th, ^{226}Ra, ^{231}Pa

Melt lens

Magma Chamber/ Mush Zone

$Tm_{c1} =$ magma chamber
$Tm_{c0} =$ residence times

mixing of magmas with different residence ages

fractional crystallization crustal assimilation?

Primitive Magma Recharge

$T_t =$ mantle transit time

assumed small or constant for similar melts

Asthenospheric Column Melting

daughters ingrown during melting

upwelling rates?
chromatographic effects?

style of melting
   batch?
   fractional?
   dynamic?
extent of melting?

modal effects?
   garnet fractionation
   pyroxenite veins
heterogeneous sources ?
   enriched vs depleted
poorly known Kd's
rate of melt generation

Initial Melt Region

parent-daughter secular equilibrium

Figure 15. Cartoon showing the processes that potentially can effect measured U-series disequilibria in MORB. Also shown are the various components of time that comprise the calculated U-series model ages. See text for discussion.

calculating residence times of melts in crustal magma chambers. Condomines [1994] discusses the effects that open system, steady-state magma evolution coupled with fractional crystallization have on U-series modeling.

The Po-Pb system has recently been used to date the time of eruption of very recent flows to within a few months [*Rubin, et al.*, 1994; *Garcia, et al.*, in press; *Rubin, et al.*, in press]. ^{210}Po is another member of the ^{238}U decay chain with a half-life of 138.4 days. Po is a

volatile element that is observed to be essentially completely (> 95%) degassed in newly erupted subaerial lavas [*Bennett, et al.*, 1982; *Gill, et al.*, 1985] and 75% to 100% degassed from basalts erupted in >2 km water depths such as on the East Pacific Rise [*Rubin, et al.*, 1994]. Po degassing during eruption starts a "clock" within the lava that results in a measurable radioactive disequilibrium between ^{210}Po and grandparental ^{210}Pb that persists for 2-2.5 years,

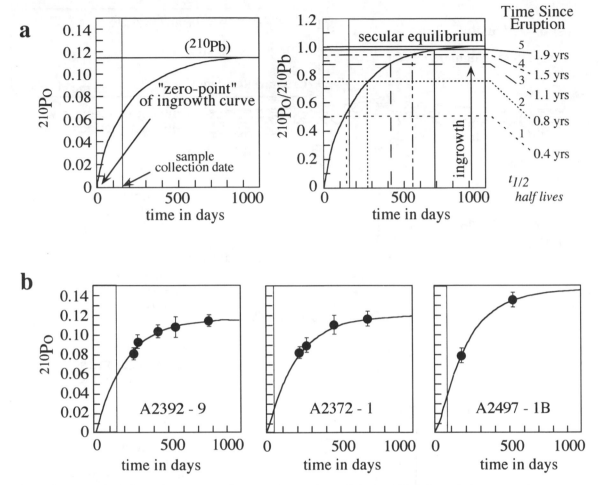

Figure 16. Po-Pb dating of recent eruptions on the northern EPR (modified from Rubin et al. [1994]. (a). Diagrammatic plot of ^{210}Po ingrowth curves showing Po-Pb dating systematics. Po degasses from lavas upon eruption but grows back into secular equilibrium with ^{210}Pb (horizontal line). Multiple measurements of ^{210}Po are made at regular time intervals and the data are regressed to an equation with the functional form of the ingrowth curve. The timescale is fixed by the decay constant and the known intervals between analyses. The lines on the upper right-hand plot show how the data would change if a measurement was taken on each half-life after eruption. Time t is set to 0 at ^{210}Po = 0 to determine the eruption time if degassing is complete upon eruption; this corresponds to the maximum age of the lava flow. The eruption "window" is fixed between t and the date the sample was recovered. (b). Representative ^{210}Po ingrowth curves for samples from the northern EPR ~ 9°51'N (BBQ site) and a younger eruption site at ~9°53'N (see Plate 3). Note that the quality of the "date" is highest where more measurements have been made throughout the grow-in period, and especially early within it.

$$(^{210}Po)t = (^{210}Po)t=0 \ [e^{-lt}] + (^{210}Pb) \ [1-e^{-lt}]$$

(where t is time since eruption; lt is the decay constant, 0.005007 day^{-1}; and parentheses indicate activities). Ingrowth regressions (Figure 16) are determined by finding the best value of time (t) that fits the data to the ingrowth function, where t is the time in days between eruption (unknown) and the date of the first ^{210}Po analysis in the time series (known). Since little or no ^{210}Pb degasses upon eruption, (^{210}Po/^{210}Pb) of a freshly erupted lava is governed by the loss of ^{210}Po only and because freshly erupted lavas have little or no ingrown ^{210}Po, the ^{210}Po-^{210}Pb clock is set upon eruption [*Rubin, et al.*, 1994; *Garcia, et al.*, in press]. An age is determined by repeatedly analyzing (by alpha-counting spectrometry) the activity of ^{210}Po in a lava over a few of its half lives and then fitting the resulting data to an exponential ingrowth curve for the ^{210}Po-^{210}Pb pair (Figure 16). Best-fit ingrowth curves are determined based on ^{210}Po activities and the known time elapsed between sequential measurements. The maximum age is the time corresponding to $(^{210}Po) = 0$ on the ingrowth curve [*Rubin, et al.*, 1994]. Because time since eruption diminishes the magnitude of radioactive disequilibrium, the longer it has been between eruption and the first ^{210}Po analysis, the poorer the age resolution on a given sample. For this reason, analyses should begin as soon as practical following sample collection (a few weeks). This technique is most accurate if many measurements are taken over the time of decay and if complete degassing has occurred. Assuming complete degassing, the maximum age of a lava is determined from the value "t" (time of eruption) where the ingrowth function intercepts the origin. If degassing is incomplete (as is the case for submarine lavas erupted under high hydrostatic pressure), lower values of "t" are obtained and the age of the lava is over-estimated. Determining the true age in these cases requires estimates of the extent of Po degassing during the eruption. This uncertainty results in an estimated eruption time "window" extending back approximately 2 months from the maximum calculated age. Thus eruption dates are fixed most precisely for samples collected shortly after eruption, or for which initial ^{210}Po can be estimated independently [K. Rubin, pers. communication, 1996].

The ^{210}Po dating method is particularly good for confirming recent eruptive events to within an eruption window that is less than ~60 days. It is not particularly useful for determining ages of lavas more than a few years old. Currently the method suffers from requiring relatively large sample size (grams), requiring long periods of measurement of single samples, and the uncertainties of Po degassing in lavas at ridge depths.

4.3. Age Constraints from Dating

Only a few U-series disequilibria studies reported to date have addressed the temporal aspects of MOR volcanism in a limited area. ^{238}U-^{230}Th and ^{235}U-^{231}Pa disequilibria studies of axial and off-axis basalts at ~9°30'N on the EPR have been used to investigate the spatial extent of young volcanism and temporal crustal accretion at this fast spreading ridge axis [*Goldstein, et al.*, 1993; *Goldstein, et al.*, 1994b]. Many of the crustal residence ages calculated from the U-series data for off-axis basalts along this ridge segment were younger than those expected based on distance from the AST and spreading rate (Figure 17). The anomalously young ages and the great variability in lava chemistry of the basalts on the crestal plateau compared to those in the AST indicate that a significant amount of volcanism may occur 0.5-2 km outside the AST and that some volcanism may occur as far as 4 km off-axis [*Goldstein, et al.*, 1994b; *Perfit, et al.*, 1994]. Recently acquired U-series data from samples recovered off-axis from the 9°50'N area indicate a few of the more evolved MORB are up to 70 ka younger than expected from their position based on spreading rate [*Sims, et al.*, 1997b]. Some of the off-axis volcanism in these two sections of the northern

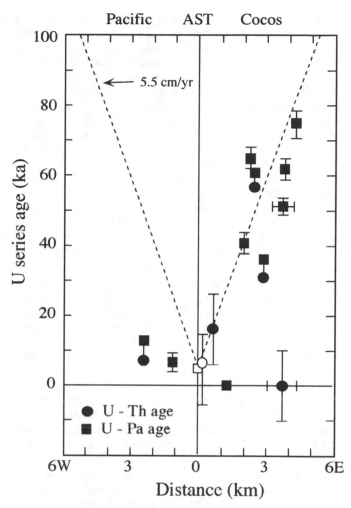

Figure 17. U-series age (ka) versus distance off-axis for MORB from the 9°31'N area of the EPR (after Goldstein et al. [1994b]). Dashed lines indicate age of crust based on paleomagnetically determined spreading rate. Error bars refer to the uncertainty in sample location recovered by dredge (vertical axes) and estimated 2 sigma uncertainties in ^{230}Th and ^{231}Pa ages (horizontal axes). Although many samples from the axis (AST) have been measured, only the average and range of data are plotted (open symbols) for clarity. Data points for samples with both ^{230}Th and ^{231}Pa ages are connected by tie lines and error bars are omitted. Anonymously young ages for many samples outside of the AST indicate a off-axis volcanism is common and suggest eruptions can occur within a zone ~4-8 km wide.

EPR may be related to the initial formation of abyssal hills discussed by Macdonald et al. [1996].

Normal mid-ocean ridge basalts (N-MORB; those with typical depleted incompatible element characteristics) from within the AST in the ~9°30'-10°N section of the EPR, have significantly different ^{226}Ra/^{230}Th ratios. The observed variations among surface axial samples can be explained by differences in magma residence ages rather than significant differences in ages of eruption [*Goldstein, et al.*, 1994a; *Sims, et al.*, 1997a]. On this time scale, ^{226}Ra ages are much less sensitive to small source variations or shallow-level fractionation than ^{231}Pa, and so ^{226}Ra ages should generally be more accurate than ^{231}Pa ages. However, effects of plagioclase crystallization on ^{226}Ra ages may be significant, and so ^{226}Ra age differences are likely an upper limit. Mixing of different magma batches from different parts of the crust-upper mantle system, which is supported by seismic and geochemical data [*Batiza and Niu*, 1992; *Perfit, et al.*, 1994; *Toomey, et al.*, 1994], also may complicate the dating systematics (Figure 15).

Young basalts from the AST in 9°30'-10°N section of the EPR show a regional trend in ^{226}Ra/^{230}Th that are consistent with axial variations in topography, axial magma chamber (AMC) depth and extent of magmatic fractionation [*Goldstein, et al.*, 1994a; *Sims, et al.*, 1997a]. Highest ^{226}Ra/^{230}Th activity ratios (2.8) are found in relatively mafic lavas (Mg# = 64) at the 1991 eruption site at 9°50'N, whereas lower ^{226}Ra/^{230}Th activity ratios (2.0-2.2) are characteristic of more fractionated samples (Mg# = 58-62) from the older-looking, more southerly parts of this ridge segment. The N-MORB samples appear to have been derived from similar (although not identical) source compositions and extents of melting based on isotope and trace element systematics [*Langmuir, et al.*, 1986; *Perfit, et al.*, 1991; *Batiza and Niu*, 1992], hence, mantle transfer times should also be similar. "Absolute" magma residence ages for MORB lavas can be estimated by extrapolation of the ^{226}Ra/^{230}Th data to primitive, unfractionated basalt (Mg# = 71) ratios. Based on calculated initial activity ratios, melt residence times for the axial lavas range from 0.3-2.0 ka for ^{226}Ra [*Goldstein, et al.*, 1994a; *Sims, et al.*, 1997a]. These relatively short "residence times" for these lavas is corroborated by measured ^{231}Pa excesses which are large, but independent of the samples' Mg#.

The ^{226}Ra residence times are significantly smaller than those based on thermal modeling (~10 ka) of a large, hydrothermally cooled AMC [*Lister*, 1983]. Assuming these residence ages are correct, we can use the age data together with liquid line of descent calculations to better constrain rates of crystallization and cooling. For example, assuming lavas in the AST in the 9°30'-10°N segment of the northern EPR are genetically related, the average crystallization and cooling rates calculated from the ^{226}Ra closed system residence times are approximately 30%/ka and 40°C/ka, respectively [*Goldstein, et al.*, 1994a]. These preliminary calculations exemplify the types of information that can be obtained using U-series dating. At this time, the use of U-series isotopes to place temporal constraints on MOR magmatic processes seems very promising but much more careful work and well controlled studies must be completed before the technique can routinely be used.

^{210}Po-^{210}Pb systematics do not provide direct information regarding magmatic processes but do allow us to determine eruption dates of very young flows (< few years) within relatively narrow "eruption windows" [*Rubin, et al.*, 1994; *Garcia, et al.*, in press; *Rubin, et al.*, in press]. Po-Pb eruption age determinations of MORB were first applied to lavas recovered in 1991 and 1992 from ~9°50'N on the EPR. The results confirmed that the young-looking samples recovered from the BBQ site and associated with vigorous hydrothermal activity in the AST, probably were erupted just before the *Alvin* dive series [*Rubin, et al.*, 1994]. More importantly, the ^{210}Po data indicate that this small section of the EPR experienced several eruptions between the spring of 1991 and early 1992 (Figure 18). The latest eruption occurred ~5 km north of the BBQ site from a fissure believed to have formed between 1991 and the dive series in spring 1992 (Plate 3c). The distinctly more evolved chemical composition of the 1992 flow suggests either very rapid along-axis differentiation or, more likely, eruption from another magma body that underlies this robust portion of the EPR. As more young flows are identified and sampled on other segments of the MOR, such temporal information will allow us to assess the frequency and duration of MOR eruptions (and their association with hydrothermal systems), volumes of material erupted during magmatic cycles, and short-term variations in MORB geochemistry.

5. COMPOSITIONAL VARIATIONS OF MID-OCEAN RIDGE BASALTS

The existing MORB database shows the great diversity of lava compositions recovered along MOR crests. Certainly, part of the signal inherent in MORB chemistry is that which is superimposed by shallow lithospheric processes which act to modify "primary" liquids delivered to the melt lens [e.g. *Sinton and Detrick*, 1992] prior to their intrusion into the shallow crust or eruption onto the seafloor. Until recently, however, we have been unable to ascertain the extent to which the geochemical diversity of the lavas reflects variability in source composition or differences in magmatic and volcanic phenomena, and whether these variations are temporally or spatially controlled, or both. Most ridge and MORB studies have not been able to address the fine-scale distribution of lavas and their spatial and temporal relationships to tectonism (e.g. rifting; diking) and hydrothermal activity (vent development, fluid chemistry) because of widely spaced sampling, the inability to date lava flows and a lack of integration with other data sets.

During the past 10 years a large number of rock samples have been collected along the strike of intermediate-, fast- and superfast- spreading MOR crests by dredging, submersible and rock coring [e.g. *Thompson, et al.*, 1985; *Langmuir, et al.*, 1986; *Hekinian, et al.*, 1989; *Thompson, et al.*, 1989; *Perfit, et al.*, 1991; *Sinton, et al.*, 1991; *Batiza and Niu*, 1992; *Reynolds, et al.*, 1992; *Perfit, et al.*, 1994; *Smith, et al.*, 1994]. While global geochemical and

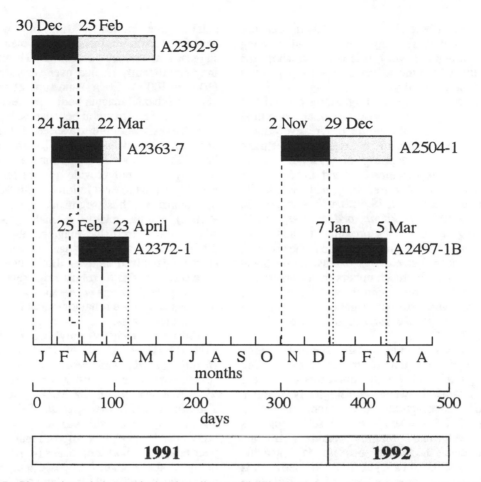

Figure 18. Po-Pb eruption windows calculated as discussed in Figure 16 for samples from the northern EPR (after Rubin et al. [1994]. Black bars are best estimates of eruption windows based on 100% and 75% degassing of initial Po. Collection date places limits on the youngest eruption date of a sample (stippled area). The data indicate there were two eruptive episodes separated by ~7 months; a conclusion which is supported by geochemical and observational evidence.

petrological databases have allowed us to determine some of the fundamental magmatic processes operating at ridge axes and to estimate mantle source characteristics [e.g. *Schilling, et al.*, 1982; *Schilling, et al.*, 1983; *Hart*, 1984; *Melson and O'Hearn*, 1986; *Klein and Langmuir*, 1987; *Langmuir*, 1988; *Langmuir, et al.*, 1992; *Humler, et al.*, 1993; *Shen and Forsyth*, 1995], local databases, particularly when combined with temporal information, have provided us with the opportunity to investigate the details of magmatic evolution, plumbing, eruption and crustal growth in both space and time.

Sampling along the axis and on the adjacent crestal plateau of some segments of the northern EPR and JdFR has been dense enough to allow geologic maps to be constructed on the basis of basalt geochemistry [*Reynolds, et al.*, 1992; *Perfit, et al.*, 1994; *Smith, et al.*, 1995b]. A few fine-scale and systematic petrologic studies have recently been completed at slow-spreading ridges [e.g.

Humphris, et al., 1990; *Frey, et al.*, 1993; *Bryan, et al.*, 1994; *Niu and Batiza*, 1994] but morphologic and tectonic complexities along slow-spreading ridges make interpretations difficult. Below, we discuss what is currently known about the spatial variability of MORB on the EPR and JdFR ridges where extensive sampling has been completed. We also present a comparison of compositional data from well-studied individual flow units on slow, intermediate and fast spreading ridges.

5.1. Segment-Scale and Local Variability

Relatively closely spaced sampling (~1-2 km) of lavas along the axis of the MOR has shown that even very small morphological offsets in the ridge crest such as overlapping spreading centers (OSCs) and small offsets or changes in the azimuth of a ridge axis (deviations from axial linearity or "devals") can mark boundaries in magma supply systems [e.g. *Langmuir, et al.*, 1986; *Sinton, et al.*, 1991]. Because most sampling has been done along axis, and dating young MORB is not commonplace, very little is known about the chemical variability with time at ridge crests or the scale and magnitude of the variability.

Early attempts at detailed sampling along the axis of ridge crests suggested primitive lavas occur at the central high points of ridge segments, while progressively more evolved lavas erupt away from these centers and that these variations may be due to tapping of a zoned magma chamber or to temporal variations related to tapping a chamber that was continually evolving and periodically recharged [*Bryan and Thompson*, 1977; *Bryan, et al.*, 1979; *Thompson, et al.*, 1985; *Hekinian and Walker*, 1987]. Fine-scale sampling of ridge axes proximal to transforms and/or associated with propagating rift tips showed that these areas were characterized by extensive fractional crystallization in crustal magma chambers as a consequence of the relatively cooler thermal regimes and the magmatic processes associated with rift propagation [*Christie and Sinton*, 1981; *Clague, et al.*, 1981; *Perfit, et al.*, 1983; *Sinton, et al.*, 1983].

Petrological studies have addressed the continuity of melt beneath the axis and the extent of transport and mixing along axis. Langmuir et al. [1986] found that OSC's and devals were boundaries between petrologically distinct, independently supplied magmatic units along the EPR. The diversity of MORB on small spatial scales suggests that long-lived, well-mixed magma chambers are the exception rather than the norm beneath fast-spreading MOR. Although seismic data suggest the AMC reflector is continuous along axis in some sections of the EPR [e.g. *Kent, et al.*, 1993], petrologic data suggest that magma is supplied from several centers of injection along axis and these may coalesce to form continuous but incompletely mixed magma bodies over significant ridge lengths [*Sinton and Detrick*, 1992]. Alternatively, magma may spend only a short time in a subaxial chamber: the observed compositional variations might be temporal in nature and reflect varied inputs from the mantle. The discussion above emphasizes that more precise spatial and temporal control is needed in our investigations of the MOR.

Other studies have explored the temporal variability of MORB by sampling across-axis. Closely spaced sampling of the EPR from 12°N to 12°30'N out to ~200 ka show spatial systematics in MORB geochemistry that reflect temporal variation in magmatism and ridge segmentation over 10 ka time intervals [*Reynolds, et al.*, 1992]. Reynolds et al. [1992] also showed that T-MORB are randomly distributed and cover only small areas of the seafloor which may result from their: 1) formation during the earliest stages of a magmatic cycle, 2) eruption off-axis away from an axial magma chamber, or 3) eruption at ridge crest discontinuities. In all three situations, small volume T-MORB retain their geochemical characteristics by avoiding mixing with the more voluminous N-MORB

axial magma chamber. Although Reynolds et al. [1992] have few constraints on the age of their samples, their data suggest that the composition of the axial magma system may significantly change over very short periods of time (100's to 1000's of years) and support geophysical data that suggest subaxial magma bodies are relatively thin and may be separated by structural boundaries.

Cross-axis sampling traverses of three segments of the northern EPR out to ~800 ka [*Batiza, et al.*, 1996] do not have the temporal resolution of other recent EPR investigations, but provide a more time-averaged view of MOR magmatic evolution. MORB recovered from magmatically robust segments of the EPR (11°20'N and 9°30'N) exhibit relatively small geochemical variations whereas lavas from the magmatically "starved" section at 10°30'N have a great range in composition, including extremely evolved samples. Batiza et al. [1996] suggest the differences are due to robust ridge segments having steady-state magma chambers versus starved segments characterized by non-steady-state behavior (variable temperatures) on 200-500 ka time scales.

Until very recently, only two ridge segments on the EPR (11°-13°N and ~21°N) [*Hekinian and Walker*, 1987; *Hekinian, et al.*, 1989; *Reynolds, et al.*, 1992] three on the JdFR [*Karsten, et al.*, 1990; *Smith, et al.*, 1994; *Smith, et al.*, 1996] and one on the MAR (~36°-37° N) [*Bryan and Thompson*, 1977; *Bryan, et al.*, 1979; *Frey, et al.*, 1993] had been systematically sampled along and across-axis in order to understand the compositional variability of MORB over relatively short time scales (< 10 -100 ka). Although many of these studies proposed temporal models for the magmatic evolution of the ridge crest, most were hampered by a paucity of seafloor observational data and little control on the age of samples so the time-scales of the models are poorly constrained. Nevertheless, some important general conclusions can be made from these fine-scale studies: 1) Surface lavas on MOR crests are quite chemically heterogeneous on small spatial scales. 2) Crystal fractionation at crustal levels primarily controls the compositional trends observed in MORB suites from each area. However, some of the variations, which are most clearly seen in incompatible elements abundances and ratios, must be a consequence of processes such as mixing of magmas that came from slightly different parental melts (due to mantle heterogeneities), variable extents of melting of similar MORB sources which produce melts that vary from extremely depleted to moderately enriched (E-MORB) and transitional varieties in between (T-MORB), or *in situ* fractionation processes that occur at the crystallizing boundaries of magma chambers or lenses. 3) Ridge segment boundaries or small offsets in ridge axes (OSC's, devals) typically are reflected in petrologic boundaries on the seafloor. Greater proportions of evolved N-MORB and E-MORB found at these offsets reflect low magma budgets, cooler crustal temperatures, and/or the edge of a magma lens. 4) Near-axis seamounts, particularly on the JdFR and

EPR, are commonly comprised of lavas that are more mafic and chemically diverse than the adjacent ridge crest. There is also strong evidence that many of the seamount basalts were derived from mantle sources more depleted than those that generate normal MORB. 5) MORB chemistry of individual ridge segments is, in general, controlled by the relative balance between tectonic and magmatic activity which in turn may determine whether a steady-state magma chamber exists and for how long. Models portraying the tectonomagmatic evolution of ridges with different spreading rates are remarkably similar. A few representative models of the temporal evolution of ridge axes are shown in Figure 19. Ultimately, the tectonomagmatic evolution is controlled by temporal variations in input of melt from the mantle. Global correlations of abyssal peridotite and MORB geochemical data suggest that the extent of mantle melting beneath normal ridge segments increases with increasing spreading rate and that both ridge morphology and composition may be related to spreading rate [*Niu and Hekinian*, 1997].

The discovery and subsequent investigations of documented recent eruptions along the Cleft and CoAxial segments of the JdFR and on the EPR at ~9°50'N have greatly aided our understanding of volcanic and magmatic processes at moderate to fast spreading MORs. Determination of the extent and volumes of the erupted lava have been facilitated by changes in seafloor morphology and by the youthful appearance of the flows (see section 3 above). In each recently active area, observational and geophysical data suggest that the basalt flows resulted from discrete and relatively short-lived (less than ~ 10 days) eruptive episodes. Geochemical/petrologic data have been used in conjunction with the mapping efforts to define the extent of these eruptions but have also provided the opportunity to examine the ranges of geochemical variation in single flows [*Smith, et al.*, 1996].

Multiple field programs (AdVenture I - VI) during 1991-1996 in the 9°17'-10°N region of the EPR have provided important observational and geochemical data related to the volcanic and magmatic development of a fast-spreading ridge crest. Detailed mapping and sampling with *Alvin* have concentrated on the section of the AST that experienced an eruption during 1991 and early 1992 between 9°45' and 9°52'N and in a somewhat older portion of the EPR around 9°29'-32'N [*Haymon, et al.*, 1993; *Perfit, et al.*, 1994] where ODP Leg 142 drilled bare rock in the AST [*Harp*, 1995]. In addition to more than a dozen *Alvin* traverses up to ~5 km outside the AST, sampling with a rock corer every 300-500m in approximately 8 x 8 km grids across the EPR crestal terrain has provided a more complete picture of the magmatic history over the past ~80 - 100 ka [*Perfit, et al.*, 1994; *Perfit, et al.*, 1995].

In the 9°31'N region the terrain on the crestal plateau up to at least 2.5 km away from the AST (and as far as 4 km in one instance) is covered by relatively unsedimented lobate and pillow lava flows some of which are younger

than the age of the crust predicted by the distance from the axis (Figure 17). Elongate pillow mounds (<20m high) principally focused along ridge parallel (NNW trending) fissures are in stark contrast to the dominant sheet and lobate flows that comprise the terrain in and around the AST (Figures 6 and 20). There is no observational evidence for voluminous sheet or lobate flows extending 2-3 km outward from the AST as suggested by Hooft et al. [1996].

The most recent magmatic events associated with the present AST erupted relatively homogeneous and mafic N-MORB whereas the off-axis lavas are more geochemically diverse and generally more evolved (Figures 20 and 21). All of the samples from within the AST in the 9°31'N area are moderately evolved N-MORB (avg. MgO = 7.70 ± 0.24 wt%) with compositions similar to that of the thick, ponded flow drilled on ODP Leg 142. In contrast, the total range in lava compositions outside the axial summit caldera in this area is nearly as great as that previously reported for the entire East Pacific Rise axis between the 9°03'N overlapping spreading center and the Clipperton transform at ~10°15'N. Lavas outside the AST on the crestal plateau exhibit distinct, asymmetric patterns of chemical variation over small (<600 m) spatial scales (Figure 20 and Plate 4). For example, on the east side of the axis these variations define areas <1 km wide composed of distinctly mafic lavas (MgO = 8.1-8.5 wt%), which are, in turn, flanked by ~1-km-wide swaths of seafloor where evolved enriched-MORB lavas ($K_2O > 0.30$ wt%; $[K_2O/TiO_2]$ x 100 > 15) were recovered. On the west side of the axis, a completely different pattern of chemical types is present [*Perfit, et al.*, 1994] (Plate 4). T-MORB and E-MORB have not been recovered in or proximal to the AST and in fact, have not been recovered in the AST anywhere the 9°17'-10°N section of the EPR. The locations of T- and E-MORB appear to be related to prominent off-axis fault scarps or fissures from which pillows were erupted (Figure 21) and may be related to the initial formation of abyssal hills [*Macdonald, et al.*, 1996].

The complex spatial variability in lava compositions observed in the 9°31'N area is due to differences in magma chemistry generated as a result of rapid (< 5 ka) replenishment and frequent low-volume eruptions coupled with frequent off-axis volcanism. The geochemical data in conjunction with seismic data also suggest that magma bodies feeding the ridge axis in this area are relatively small, extending <5-10 km along axis and only a few km wide. The diversity of MORB observed on small spatial scales suggests that long-lived (> 10 ka), well-mixed magma chambers are the exception rather than the norm beneath fast-spreading MOR. Conceptually, this contradicts the conclusions of Batiza et al. [*Batiza and Niu*, 1992], but the more closely spaced, recent sampling allows much finer temporal resolution. On this segment of the EPR, relatively mafic volcanism appears to be related to magmatic episodes associated with diking focused in the

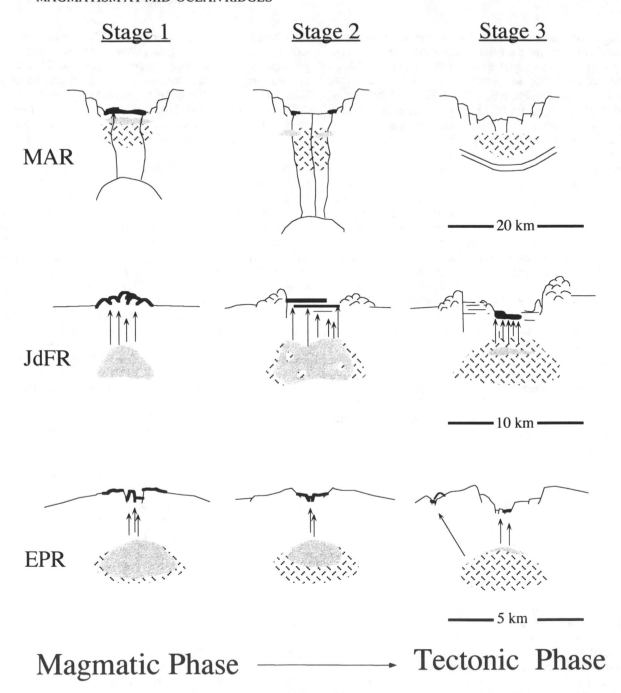

Figure 19. Tectonomagmatic models showing the generalized temporal evolution of specific slow, intermediate- and fast-spreading ridges (modified from [*Frey, et al.*, 1993; *Smith, et al.*, 1994; *Fornari, et al.*, in press-a]). In each case, the ridge is believed to progress from a robust magmatic stage to one that is largely tectonic but on very different time scales (see Figure 24). In the AMAR region of the MAR and Cleft segment of the JdFR, the magmatic stage is associated with the voluminous eruption of mafic basalts. Subsequent evolution is associated with rift failure (MAR), low magmatic output and cooling of the subaxial magma lens (gray pattern) which leads to the eruption of more evolved lava types that sometimes contain xenocrysts from the mush zone (hatched pattern). In the third stage, tectonism may be associated with amagmatic extension (MAR), or low volume eruptions and the development of axial valleys on the JdFR and northern EPR. Note differences in horizontal scale for each ridge.

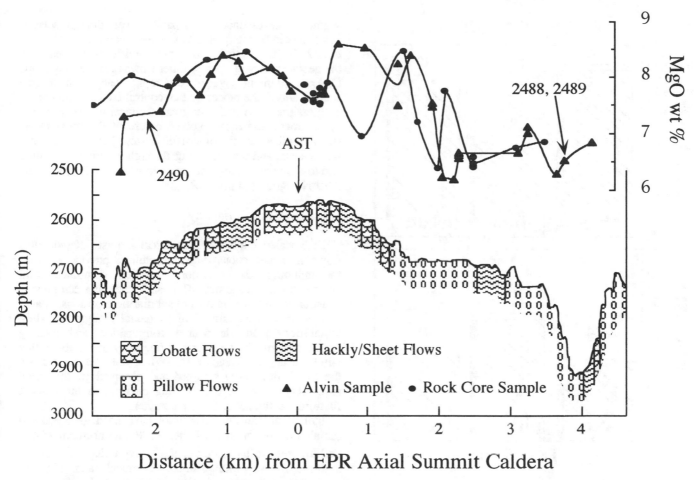

Figure 20. Diagrammatic representation of the bathymetry and surficial lava types across the axis of the EPR at ~9°32'N based on *Alvin* dives 2488, 2489 and 2490. A map of this area is shown in Plate 4, where a thick flow was drilled in the AST at 9°30.85'N during ODP Leg 142. Lava morphology is dominated by sheet and lobate flows in and near the AST whereas lobate and pillow flows are more abundant off-axis. Many of the pillow flows are associated with ridge-parallel fissures and are not continuous along strike. Other pillow flows appear to be related to major faults (e.g. ~ 3-4 km east of the AST) that may be associated with the formation of abyssal hills [*Macdonald, et al.*, 1996]. The chemistry of the flows exhibits significant variability on small spatial scales as indicated above the cross-section by the large variations in MgO (wt%) [*Perfit, et al.*, 1991; *Perfit, et al.*, 1994]. Samples recovered by *Alvin* (filled triangles) are well-located and from the specific lava morphologies shown. Rock cores (filled dots), however, are less well located (~ 200 m accuracy) and therefore do not exactly correspond to the *Alvin* samples. Regardless, the small scale variability off-axis is apparent, particularly when compared to the homogeneity of flows in and proximal to the AST.

AST whereas some off-axis flows may be fed by dikes from the edges of a magma lens or smaller and cooler off-axis magma bodies (Figure 6) [*Perfit, et al.*, 1994]. It is of interest to note that the most mafic lavas recovered on the EPR between 9°17' and 10°N are the 1991 BBQ basalts that overlie the shallowest reflections of the AMC. Although the BBQ flow was relatively small, the eruptive event was marked by vigorous and sustained hydrothermal activity in the AST, suggesting this segment of the ridge may be undergoing a period of magma replenishment from the mantle.

Three recent (<20 a) eruptive events have been recognized and mapped along the CoAxial Segment of the JdFR since 1993 (Figure 11) [*Chadwick, et al.*, 1995; *Embley, et al.*, 1995a]. The recent eruptions are believed to result from lateral dike injections, similar to that documented along the northern Cleft segment of the JdFR [*Embley and Chadwick*, 1994]. The three lava flows, the 1993 CoAxial flow (~46°31'N) and two others erupted between 1981 and 1991, one just east of the 1993 flow and the other just west of the Floc site (~46°18'N), have been extensively sampled by *Alvin* and rock core [*Smith, et al.*,

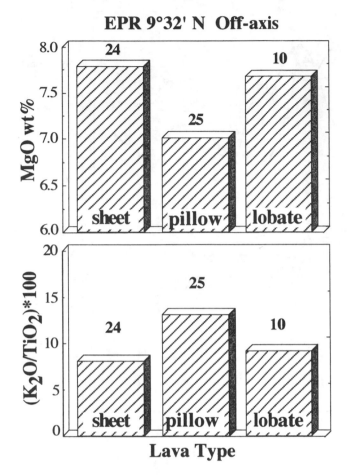

Figure 21. Comparison of average composition of different lava flow types recovered by *Alvin* outside of the AST at ~9°32'N on the EPR crestal plateau (see Figures 20 and Plate 4). Pillow lavas (sampled from off-axis flow ridges and fault scarps) commonly are more evolved (lower MgO) and enriched (higher K/Ti) than sheet and lobate flows. Numbers on top of bars represent the number of samples in each group.

1995b]. The individual flows have rather limited and distinct chemical compositions (Figure 22; Table 4) compared to the wide range of MORB compositions (e.g. MgO values from 9.5 to 6.5 wt. %) observed along the entire CoAxial Segment. Lavas from the 1993 and 1981-1991 flows are more evolved and more crystalline (up to 30% microphenocrysts) than older nearby lavas but have the same incompatible element depleted and non-radiogenic isotopic characteristics as other MORB from the segment. The three young flows from this segment represent the most depleted MORB recovered from the entire JdFR and they are geochemically distinct from slightly enriched MORB from Axial Seamount and its North Rift Zone which overlaps on the west with the southern end of the CoAxial segment. The geochemical and geological data clearly show that the magmatic systems of these two segments are distinct and separate, even though seismic evidence can be interpreted to suggest a connection [*Fox, et al.,* 1995]. In addition, subtle but significant differences in the geochemistry of slightly older flows near the axis of the CoAxial segment suggest very rapid changes in magma sources and/or the processes that controlled the evolution of each magma. Overall, the great chemical diversity within the segment and more evolved nature of the most recent flows implies that the magmatic system has not been in a steady-state, and may be going through a period of cooling and low magma supply, opposite to what is inferred for the 9-10°N segment of the EPR.

5.2. Flow-Scale Variability

Fine-scale sampling and chemical analyses of individual flows or constructional volcanic edifices provide a means for mapping the seafloor. When combined with observational and age data, flow-scale mapping can provide constraints on the dynamics of submarine eruptions, spatial and temporal evolution of subaxial magma bodies, variability in mantle source composition and melting parameters, and the number and frequency of eruptions that build the upper oceanic crust. While only a few areas on the MOR have been sampled and observed in the detail required, we can use the existing database to make some first-order estimates of these parameters.

Individual flows or flow units have been identified and sampled on the northern EPR, JdFR and northern MAR. Drilling results have shown that flow units, comprising multiple flows with similar chemical and physical properties, are common in the oceanic crust. It is unknown if these flow units represent one or more eruptive episodes that occurred over relatively short time periods. The young flows on the northern EPR and CoAxial segment have been constrained to have formed in less than a few days to weeks, but it is less clear how much time is involved in constructing larger volcanic features such as the seamounts in the FAMOUS and AMAR sections of the MAR (36°30'N-37°N) (Table 2). Macdonald [1986] suggested that eruptions required to build these edifices may take place over 1-100 years. The relatively detailed sampling by submersible and geochemical studies that have been done in the FAMOUS/AMAR axial valley show that there are systematic temporal/spatial distributions of lava types related to the tectonomagmatic evolution of the ridge, but the local diversity of compositions is indicative of multiple magmatic histories rather than a single (or a few) liquid line of descent [*Bryan and Moore,* 1976; *Stakes, et al.,* 1984; *Frey, et al.,* 1993; *le Roex, et al.,* 1996]. With the exception of Serocki Volcano, fine-scale sampling of individual seamounts or large volcanic features on the MAR that may represent single eruptive units has not yet been accomplished.

Large lava fields have been observed on the super-fast spreading southern EPR but only one young-looking flow

Table 4.   Major element analyses of individual lava flows from mid-ocean ridges

Flow	BBQ Flow[1]		ODP Flow[2]		S. Cleft Sheet Flow[3]		N. Cleft Sheet Flow[4]		93 CoAxial Flow[5]		Serocki Volcano[6]		Average Anal. Precision
Location	EPR 9°50'N		EPR 9°30'N		S. JdFR 44°40'N		S. JdFR 44°56'N		JdFR 46°31'N		N. MAR 22°55'N		
#	21	2s	23	2s	15	2s	20	2s	17	2s	22	2s	2s
SiO2 (wt%)	49.88	0.40	49.65	0.54	50.24	0.38	50.04	0.70	50.48	0.67	50.03	0.60	0.25
TiO2	1.29	0.07	1.67	0.06	1.84	0.06	1.53	0.08	1.69	0.06	1.67	0.25	0.044
Al2O3	15.52	0.23	14.51	0.72	13.94	0.21	14.78	0.86	13.63	0.35	15.87	0.73	0.14
FeO^T	9.31	0.20	10.71	0.41	12.10	0.28	10.52	0.64	12.73	0.34	9.92	0.97	0.20
MnO	0.17	0.02	0.20	0..02	0.22	0.07	0.20	0.00	0.24	0.03	nd	nd	0.024
MgO	8.55	0.33	7.58	0.34	6.73	0.37	7.58	0.16	6.71	0.23	7.51	0.61	0.13
CaO	12.26	0.44	11.77	0.54	11.28	0.17	11.85	0.18	11.41	0.24	11.21	0.32	0.16
Na2O	2.56	0.12	2.74	0.12	2.65	0.18	2.47	0.10	2.61	0.14	2.93	0.14	0.090
K2O	0.09	0.03	0.14	0.04	0.15	0.03	0.12	0.02	0.12	0.03	0.14	0.05	0.017
P2O5	0.12	0.02	0.16	0.02	nd	nd	0.14	0.02	0.14	0.02	0.17	0.05	0.019

Notes: number = number of individual lavas analyzed, 2s = 2 sigma, all analyses by electron microprobe except ODP flow by ICP-AES and N. Cleft Sheet Flow by XRF. nd= not determined

References: 1. Gregg et al., 1996; 2. Bostrom and Bach, 1995; 3. Perfit, unpublished data;
                     4. Embley et al., 1991; 5. M.C. Smith, unpublished data; 6. Humphris, et al., 1990

has been investigated in any detail. Macdonald et al. [1989] suggested that an approximately 220 km^2 "flow" erupted from cones and fissures 2.5 km off-axis at around 8°S where the ridge appears to be magmatically robust. SeaMARC II sonar images of the area do not have the resolution to show whether the flow is a single unit but recent sampling from five sites [Hall and Sinton, 1996] has shown that it is very heterogeneous and unlikely to have been erupted during a single event or even from the same magma chamber.

Below we discuss the compositional variability of well documented flow units from slow, moderate and fast spreading centers and their relation to regional geochemical trends. The geologic setting and ages of some of these flows are discussed above. The composition of each flow and the standard deviation of each major oxide are presented in Table 4. With the exception of the ODP flow, all of the samples listed in Table 4 were collected with Alvin, and based on seafloor observations and the physical characteristics of the samples, they are believed to have come from individual eruptive units or flows. Each flow was sampled at numerous localities (~10-20 sites) so that the compositions are representative of the areal extent of the flow (although not necessarily the full thickness). Samples from the ODP flow are fine-grained basalts drilled from a 3 m thick ponded flow in the AST ~ 9°30'N of the EPR [Harp, 1995; Bostrom and Bach, 1995]. Although most of the samples discussed here are nearly aphyric, analysis of glass from the outer rinds of the flows avoids the possibility of compositional variability due to the presence of excess or xenocrystic phenocrysts that sometimes occurs when more crystalline "whole rocks" are analyzed. In this regard, it is interesting to note that the crystalline ODP flow appears to be nearly as homogeneous as the glasses from the other locals.

The flows are relatively homogeneous compared to the range of lava compositions from their respective ridge segments (Figure 22). However, major element chemical variability is greater than expected based on the precision of microprobe analytic techniques (Table 4). This is also the case for many trace elements (e.g. Sr, Y, Zr) in the glasses where precision is at the 5% level or better [Embley, et al., 1991; Smith, et al., 1994; Gregg, et al., 1996b, Perfit, unpublished data]. Relative to well-studied Hawaiian flows, the 1991 EPR BBQ flow and two sheet flows on the Cleft segment of the JdFR are significantly more homogeneous than lavas along Kilauea's East Rift Zone [Wolfe, et al., 1987] but similar to a 1984 Mauna Loa eruption [Rhodes, 1983]. The data show that for most major elements, the BBQ, south Cleft, and north Cleft sheet/lobate flows are more homogeneous than the more voluminous, pillowed flows from the 1993 eruption at CoAxial and Serocki Volcano. The CoAxial flow formed rapidly above a propagating dike [Embley, et al., 1995b] whereas Serocki volcano probably formed as a lava delta by the accumulation of numerous pillow and lesser sheet flows [Humphris, et al., 1990] but there are no constraints on the period of time over which Serocki formed. The data presented here indicate sheet/lobate flows are more

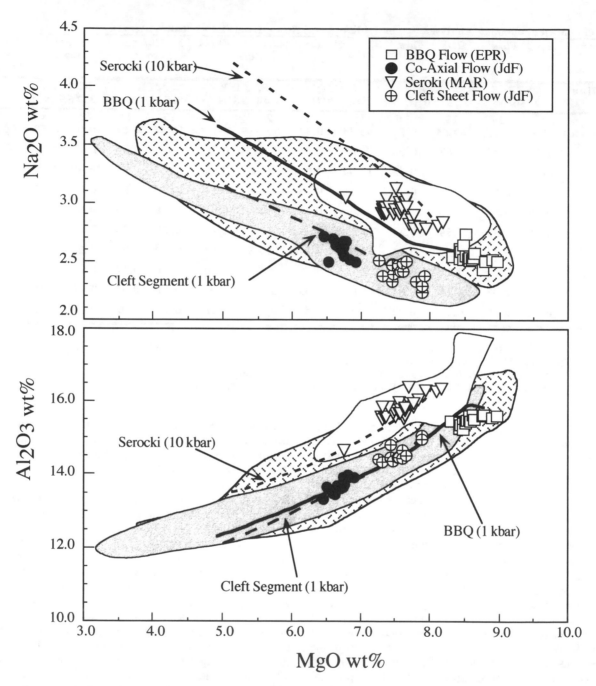

Figure 22. Major element variation diagrams showing compositional range of individual MOR flows. Lines show calculated liquid lines of descent [*Weaver and Langmuir*, 1990] at 1 and 10 kb for different parental magmas. Fields are: JdFR (shaded); NEPR (pattern); MARK (white). See text for discussion.

homogeneous than pillowed flows suggesting a relationship between eruption style (rate) and extent of chemical variability. Not enough single flow units from different ridges have been studied to relate intraflow chemical variability to spreading rate.

Humphris et al. [1990] note the significant variations in major and trace element compositions of Serocki volcanics and suggest they are due to crystal fractionation rather than variable mantle processes or sources. Smith et al. [1995b] suggest spatially systematic chemical variations in the

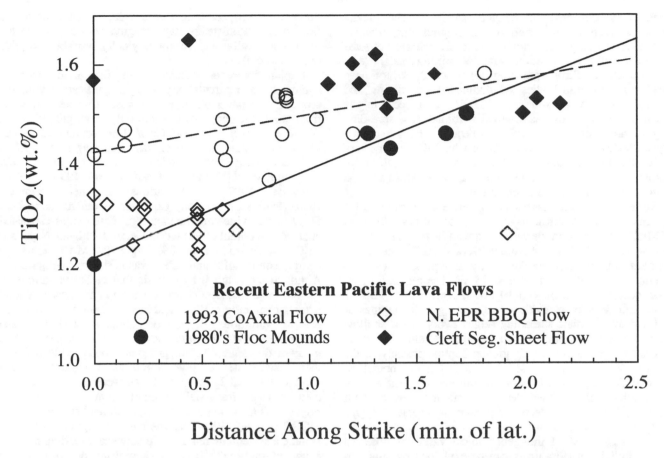

Figure 23. Chemical variations in individual recent lava flows versus distance along strike of the Juan de Fuca Ridge and East Pacific Rise. TiO_2 (wt%) exhibits good correlation with distance along strike in pillowed flows whose interpreted emplacement was via lateral dike injection [CoAxial Flow (dashed line) and Floc (solid line) sites, both from left to right]. The observed increase in TiO_2 corresponds with other elemental variations and may reflect shallow-level fractional crystallization during dike emplacement. No significant latitudinal variation is measured for sheet/lobate flows whose emplacement may have been via a more centralized eruptive vent and subsequent surficial flow (BBQ and Cleft sheet flows).

CoAxial flow and Floc Site flow, farther south, result from shallow level crystal fractionation (Figure 23). Although, the BBQ and Cleft sheet flows also exhibit elemental variations compatible with shallow-level fractionation (Figure 22), neither show any systematic spatial differences (Figure 23). Calculated liquid lines of decent (LLD) for presumed parental magmas from Serocki volcano, the BBQ flow and the northern Cleft sheet flow are shown in Figure 22. A point to be made here is that the compositional variations in single flow units follow the same general trends as the lavas from their respective segments and those paths are predicted from the LLD calculations. The trends of increasing Na_2O and decreasing Al_2O_3 with decreasing MgO are typical of many MORB suites. The chemical trends of the liquids (and most other major element trends within a suite of lavas) are primarily due to progressive fractional crystallization of different proportions of olivine, plagioclase and clinopyroxene as a magma cools. The slightly different trajectories are a consequence of the different proportions of fractionating phases which are controlled by initial (and subsequent changing) liquid composition, temperature and pressure. Langmuir (1992) and Grove et al. [1992] discuss these processes in greater detail. In some MORB suites, linear elemental trends may be due to mixing of more mafic and evolved end-member magmas. Mixing of magmas [*Rhodes, et al.*, 1979] and crustal assimilation [*Michael and Schilling*, 1989] are additional processes that are likely to occur in reservoirs beneath MORs but generally only play subtle roles in controlling MORB major element trends. Such processes may however, have much greater effects on trace element abundances and isotopic ratios [e.g. *Perfit, et al.*, in press]. The composite magma chamber model presented by Sinton and Detrick [1992] can explain many of chemical systematics observed in MORB suites. The model proposes there are melt-rich zones (magma lens) separate from crystal-rich zones (mush) which are variably effected by fractional crystallization, magma mixing and *in situ* crystallization. Slow spreading ridges which do not have steady-state magma bodies are dominated by mixing and processes such as *in situ* crystallization that occur in the mush zone, whereas magmatically robust fast spreading ridges are most heavily influenced by fractional crystallization. The model explains why MORB may vary with tectonomagmatic evolution at individual ridge segments; particularly at intermediate rates where magma lenses may be small and intermittent.

Suites of MORB glasses often define distinctive LLDs that match those determined by experimental crystallization of MORB at low to moderate pressures [*Juster, et al.*, 1989; *Grove, et al.*, 1992]. At the EPR and JdFR, the major element data are best explained by low-pressure (~1 kb) fractional crystallization whereas MAR data requires higher pressure crystallization (~5-10 kb) (Figure 22). This is consistent with other evidence suggesting that magmas at fast and intermediate spreading ridges evolve in a shallow magma lens or chamber and that magmas at slow spreading ridges evolve at significantly greater depths; possibly in the mantle lithosphere or at the crust-mantle boundary. Estimated depths of crystallization correlate with observations of increased depths of magma lens or fault rupture depth with decreasing spreading rate [*Phipps Morgan and Chen*, 1993a].

Although most of the geochemical variances observed within individual flow units can be explained by small to modest amounts of crystal fractionation during shallow-level storage and eruption, it is clear that this cannot account for all the observed variations. In some instances, such as along the northern part of the Cleft segment of the JdFR, some pillow mounds appear to be composite structures with the most recent eruptives having a slightly different composition (possibly due to magma mixing at depth). More detailed studies of submarine volcanic constructs like Serocki volcano might reveal similar composite histories. Overall, the chemical variability observed in eruptive units that are closely related in time and space, suggests subaxial magma bodies are not homogeneous on a small scale (< few km) and that fractional crystallization can occur during the diking/eruption process; particularly if there is significant along-axis propagation/flow. From a petrogenetic point of view, the different extents of chemical variability are probably a consequence of a number of interrelated processes : 1) style of eruptive processes, 2) stage of tectonomagmatic development of the ridge, 3) structure of the subaxial magma body. Thus, rapid eruption from a shallow, relatively well-mixed magma lens produces aphyric, homogeneous sheet flows (BBQ, Cleft) ; extended along-axis dike propagation associated with the eruption of pillow mounds produces more heterogeneous and crystalline

flows (CoAxial); and longer time scale, sequential eruptions from deeper, non-steady-state, magma bodies generate the most chemically and morphologically variable eruptive units (Serocki).

Explanations for differences in the major element compositions of parental or primary magmas are beyond the scope of this paper but it is important to note that lavas from slow spreading ridges commonly have greater Na_2O, Al_2O_3 and lower FeO and CaO/Al_2O_3 contents at given MgO values than lavas from fast spreading ridges [*Klein and Langmuir*, 1987; *Klein and Langmuir*, 1989; *Langmuir, et al.*, 1992; *Shen and Forsyth*, 1995; *Niu and Hekinian*, 1997]. Variations in these chemical characteristics are generally expressed as differences in Na_8, Fe_8, Al_8, etc. which are the values of these oxides calculated at an MgO content of 8.0 wt%. Generally higher Na_2O and Al_2O_3 concentrations at 8% MgO in MAR lavas in comparison to JdFR and EPR lavas is shown in Figure 21. Although the original hypothesis that global variations in MORB major element chemistry are a consequence of total extents of mantle melting and mean pressure of extraction due to variations in mantle temperature [*Klein and Langmuir*, 1987; *Langmuir, et al.*, 1992] and can explain many of the general chemical features observed in MORB, there are quite a few questions regarding MORB petrogenesis that have yet to be answered. What style of melting (e.g. fractional or equilibrium) occurs beneath ridges? What is the scale of compositional heterogeneity of MORB mantle? What does the melting regime look like? Is there a systematic relationship between spreading rate and extent of melting? How much melting occurs and over what depth intervals? Integration of detailed geochemical results with constraints placed on physical parameters by geophysical and experimental data show great promise for resolving some of these questions.

5.3. Volume-Rate Relationships

Although the data are somewhat limited, calculated volumes of individual flow units show an inverse exponential relationship to spreading rate (Figure 24a), contrary to what might be expected. The largest eruptive units are cones in the axis of the northern MAR whereas the smallest units are sheet/lobate flows on the EPR. Eruption intervals (time in years between successive eruptions) and number of eruptions over 10 ka per unit length of ridge crest have been calculated based on the calculated flow volumes and assuming these flows are responsible for constructing layer 2A of the oceanic crust. Results of these first-order calculations are presented in Figures 24b and 24c. The calculations assume layer 2A is 500 m thick (depth), spreading has occurred at a constant rate for 10 ka (width), and the length of ridge in question is 1 km (length). They suggest fast spreading ridges erupt on yearly to decadal time scales; intermediate on 10 to 100 year intervals; and slow spreading ridges on 1000 to 10 ka

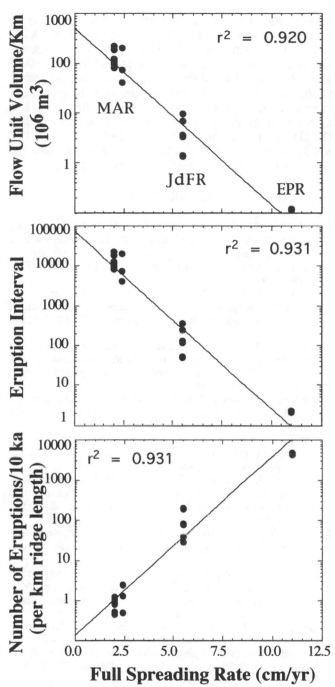

Figure 24. (a) Estimated eruption volumes of single eruptive units vs. spreading rate. The volumes plotted in this figure are from Table 3. The best fit lines are exponential curves. The data are obviously limited and, in some cases (e.g. MAR), poorly constrained, but show an inverse relationship between flow volume per unit length of ridge and spreading rate. Using some assumptions regarding spreading rate and thickness of crust formed over 10 ka, eruption intervals (b) and number of eruptions (c) per unit length of ridge can be predicted. The data suggest that fast ridges have frequent small eruptions whereas slow ridges are characterized by infrequent large eruptions.

intervals. This means that within a 10 ka time interval there may be thousands of small flows that build the crust out to ~500 m from the axis of fast spreading ridges but only a few (<10) large flows near the axis of slow spreading ridges. Decreasing the thickness of the extrusive crust by a factor of two results in halving the number of eruptions and doubling the eruption interval for each flow size, but does not change the gross relationships that are shown.

6. DISCUSSION AND CONCLUSIONS

Many features on the MOR that vary with spreading rate - the width of the neovolcanic zone, the overall morphology of the ridge crest, the size and frequency of eruptions, the morphology of lava flows - all appear to be fundamentally related to whether or not a steady-state magma reservoir can be sustained at a given spreading rate. This single factor has a tremendous control on the volcanic output over time. Where a steady-state magma reservoir exists, volcanism can keep-pace with or dominate over tectonism; where magma reservoirs are intermittent, tectonism tends to dominate. This variation in the degree of magma input also has a profound influence on the resulting structure of the upper oceanic crust, both in terms of the character of volcanic constructs and the degree of subsequent structural reorganization. On ridge segments where magma supply is high and diking events are frequent, magmatism can be seen as short-circuiting the tectonic cycle of faulting, since the build-up of tensional stresses perpendicular to the ridge by plate spreading will almost always be interrupted and reset by intrusions. On the other hand, plate spreading inevitably leads to faulting in the absence of robust magma input. The critical spreading rate, above which a steady-state magma reservoir can form and below which one cannot, appears to be about 50 mm/yr [*Phipps Morgan and Chen*, 1993b]. Near this critical spreading rate magma reservoirs may be quasi-steady-state.

What conclusions can be drawn from our very limited knowledge of individual volcanic events on the MOR? There is much that we still do not know, but one thing is clear: we are still missing most eruptions that occur. Crisp [1984] estimated that 75% of the new magma reaching Earth's surface each year is emplaced at the MOR crest, on average ~3 km³/yr extrusive and ~18 km³/yr intrusive. The few eruptions that have been documented represent only a tiny proportion of these volumes.

We can also conclude that the frequency of volcanic eruptions does not increase linearly with increasing spreading rate, and in fact, the relationship is distinctly non-linear (Table 1, Figure 24). If the relationship were linear, then the ratio of volcanism to tectonism would stay the same on all ridges, regardless of spreading rate, but this is clearly not the case. Eruptions occur often on fast ridges because the magma supply is continuous, whereas eruptions are infrequent on slow ridges because the magma supply is intermittent. Again, the key reason that this relationship is non-linear is that the continuity of the magma supply depends on whether or not a steady-state magma reservoir can be sustained at a given spreading rate and this, in turn, can be related to variations in the extents of melting of the mantle.

Another tentative conclusion is that the average volume of volcanic eruptions on the MOR also appears to be a function of spreading rate. Quantitative data about the volumes of individual eruptions are limited, but they suggest that slow ridges tend to have large eruptions and fast ridges tend to have small eruptions (Table 2). This appears to be related to the frequency of the eruptions at different spreading rates, and also to the differences in magmatic feeder systems. Fast ridges typically have short-lived fissure eruptions because their narrow feeder dikes freeze rapidly, whereas slow ridges often have point-source eruptions, with feeders that are thermally more stable and so can supply magma for much longer periods. The estimated volumes in Table 2 are based on the dimensions of single eruptive units described in the literature, and show that the relationship between spreading rate and eruption volume is also non-linear, and close to exponential (Figure 24a).

Another relationship that is clear is that the dominant morphology of erupted lava changes with spreading rate (mainly sheet flows on fast ridges and mainly pillows at slow ridges), and since this primarily reflects the rate of lava extrusion, we can conclude that dikes intrude at fast spreading ridges with a higher magma pressure than at slow-spreading ridges. This is probably due to the robust and continuous magma supply at fast-spreading ridges. Also the high frequency of intrusions tends to keep the compressive stress perpendicular to the ridge relatively high, which subsequent dikes can only overcome with high internal pressure. Dikes intruded where the ambient compressive stress is high would tend to produce narrow dike-induced grabens at the surface, consistent with observation and interpretation of structures associated with eruptive sites on the JdFR and possibly the EPR.

Acknowledgments. We thank all of our colleagues who have collaborated with us in discovering many of the features we have discussed and formulating the ideas we have presented, and who have participated in numerous cruises with us over the past decade. Specifically, we are grateful to R. Batiza, J.F. Bender, M.H. Edwards, R.W. Embley, J.H. Fink, D.J. Fornari, C.G. Fox, P.J. Fox, S.J. Goldstein, T.K.P. Gregg, R.M. Haymon, E.E.E. Hooft, K.C. Macdonald, K.H. Rubin, J.R. Reynolds, J.M. Sinton, K.W.W. Sims, D.K. Smith, and M.C Smith who have frequently discussed their ideas on seafloor volcanic processes, MORB geochemistry, radiometric dating and geophysics with us. We also thank K. Sapp and M.C Smith who helped prepare many of the figures and R. Milam who completed the final layout. D.K. Smith, D.J. Fornari, R.W. Embley, M.C. Smith provided us with preprints of their papers and gave us permission to use some of their figures. The manuscript was improved by suggestions from D.J. Fornari, and reviews by E.M. Klein and S.L Goldstein. We also wish to thank: W.R. Buck and J.A. Karson for inviting us to participate in the RIDGE Theoretical Institute and for putting up with our propensity to miss deadlines. Support for the field programs and shore-based research was provided by the following grants to MRP from the National Science Foundation (NSF): OCE9100503, OCE9402360 and OCE9530299 . WWC and MRP also gratefully acknowledge support from the NOAA Vents Program. PMEL contribution 1957.

REFERENCES

Abdallah, A., V. Courtillot, M. Kasser, A. Y. Le Dain, J. C. Lepine, B. Robineau, J. C. Ruegg, P. Tapponnier, and A. Tarantola, Relevance of Afar seismicity and volcanism to the mechanics of accreting plate boundaries, *Nature*, *282*, 17-23, 1979.

Allan, J. F., R. Batiza, M. R. Perfit, D. J. Fornari, and R. O. Sock, Petrology of lavas from the Lamont seamount chain and adjacent East Pacific Rise, *J. Petrol.*, *30*, 1245-1298, 1989.

Auzende, J. M., et al., Recent tectonic, magmatic, and hydrothermal activity on the East Pacific Rise between 17°S and 19°S: Submersible observations, *J. Geophys. Res.*, *101*, 17995-18010, 1996.

Auzende, J. M., and J. Sinton, NAUDUR explorers discover recent volcanic activity along the East Pacific Rise, *Eos Trans. Am. Geophys. Union*, *75*, 601 and 604-605, 1994.

Baker, E. T., and S. R. Hammond, Hydrothermal venting and the apparent magmatic budget of the Juan de Fuca Ridge, *J. Geophys. Res.*, *97*, 3443-3456, 1992.

Baker, E. T., J. W. Lavelle, R. A. Feely, G. J. Massoth, S. L. Walker, and J. E. Lupton, Episodic venting of hydrothermal fluids from the Juan de Fuca Ridge, *J. Geophys. Res.*, *94*, 9237-9250, 1989.

Baker, E. T., G. J. Massoth, and R. A. Feely, Cataclysmic hydrothermal venting on the Juan de Fuca Ridge, *Nature*, *329*, 149-151, 1987.

Baker, E. T., G. J. Massoth, R. A. Feely, R. W. Embley, R. E. Thomson, and B. J. Burd, Hydrothermal event plumes from the CoAxial seafloor eruption site, Juan de Fuca Ridge, *Geophys. Res. Lett.*, *22*, 147-150, 1995.

Baker, E. T., G. J. Massoth, J. E. Lupton, S. L. Walker, D. A. Tennant, C. Wilson, and N. Garfield, Time-series sampling of hydrothermal event plume(s) from the 1996 Gorda Ridge eruption, *EOS Trans. Am. Geophys. Union*, *77*, 46, Fall Meeting suppl., F1, 1996.

Baker, E. T., G. J. Massoth, J. E. Lupton, S. L. Walker, D. A. Tennant, C. Wilson, and N. Garfield, Patterns of event and chronic hydrothermal venting following a magmatic intrusion: New perspectives from the 1996 Gorda Ridge eruption, *Deep-Sea Res.*, in press.

Baker, E. T., and T. Urabe, Extensive distribution of hydrothermal plumes along the superfast spreading East Pacific Rise, 13°13'-18°40'S, *J. Geophys. Res.*, *101*, 8685-8695, 1996.

Ballard, R. D., W. B. Bryan, J. R. Heirtzler, G. R. Keller, J. G. Moore, and T. H. van Andel, Manned submersible observations in the Famous area, *Science*, *190*, 103-108, 1975.

Ballard, R. D., J. Francheteau, T. Juteau, C. Rangan, and W. Normark, East Pacific Rise at 21°N: the volcanic, tectonic, and hydrothermal processes of the central axis, *Earth Planet. Sci. Lett.*, *55*, 1-10, 1981.

Ballard, R. D., R. Hekinian, and J. Francheteau, Geological setting of hydrothermal activity at 12°50'N on the East Pacific Rise: a submersible study, *Earth Planet. Sci. Lett.*, *69*, 176-186, 1984.

Ballard, R. D., R. T. Holcomb, and T. H. van Andel, The Galapagos Rift at 86°W: 3. Sheet flows, collapse pits, and lava lakes of the rift valley, *J. Geophys. Res.*, *84*, 5407-5422, 1979.

Ballard, R. D., and J. G. Moore, *Photographic Atlas of the Mid-Atlantic Ridge Rift Valley*, 114 pp., Springer-Verlag, New York, 1977.

Ballard, R. D., and T. H. van Andel, Morphology and tectonics of the inner rift valley at 36°50'N on the Mid-Atlantic Ridge, *Geol. Soc. Am. Bull.*, *88*, 507-530, 1977.

Ballard, R. D., T. H. van Andel, and R. T. Holcomb, The Galapagos Rift at 86°W: 5. Variations in volcanism, structure, and hydrothermal activity along a 30-kilometer segment of the rift valley, *J. Geophys. Res.*, *87*, 1149-1161, 1982.

Barone, A. M., and W. B. F. Ryan, Single plume model for asynchronous formation of the Lamont seamounts and adjacent East Pacific Rise terrain, *J. Geophys. Res.*, *95*, 10801-10827, 1990.

Barr, S. M., Seamount chains formed near the crest of Juan de Fuca Ridge, northeast Pacific Ocean, *Mar. Geol.*, *17*, 1-19, 1974.

Batiza, R., P. J. Fox, P. R. Vogt, S. C. Cande, N. R. Grindlay, W. G. Melson, and T. O. O'Hearn, Morphology, abundance, and chemistry of near-ridge seamounts in the vicinity of the Mid-Atlantic Ridge ~26°S, *J. Geology*, *97*, 209-220, 1989.

Batiza, R., and Y. Niu, Petrology and magma chamber processes at the East Pacific Rise – 9°30'N, *J. Geophys. Res.*, *97*, 6779-6797, 1992.

Batiza, R., Y. Niu, J. L. Karsten, W. Boger, E. Potts, L. Norby, and R. Butler, Steady and non-steady state magma chambers below the East Pacific Rise, *Geophys. Res. Lett.*, *23*, 221-224, 1996.

Batiza, R., and D. Vanko, Volcanic development of small oceanic central volcanoes on the flanks of the East Pacific Rise inferred from narrow-beam echo-sounder surveys, *Mar. Geol.*, *54*, 53-90, 1983.

Bennett, J. T., S. Krishnaswami, K. K. Turekian, W. G. Melson, and C. A. Hopson, The uranium and thorium decay series nuclides in Mt St. Helens effusives, *Earth Planet. Sci. Lett.*, *60*, 1982.

Bergman, E. A., and S. C. Solomon, Earthquake swarms on the Mid-Atlantic Ridge: Products of magmatism or extensional tectonics?, *J. Geophys. Res.*, *95*, 4943-4965, 1990.

Björnsson, A., G. Johnsen, S. Sigurdsson, G. Thorbergsson, and E. Tryggvason, Rifting of the plate boundary in north Iceland, 1975-1978, *J. Geophys. Res.*, *84*, 3029-3038, 1979.

Bonatti, E., and C. G. A. Harrison, Eruption styles of basalt in oceanic spreading ridges and seamounts: Effect of magma temperature and viscosity, *J. Geophys. Res.*, *93*, 2967-2980, 1988.

Brandsdottir, B., and P. Einarsson, Seismic activity associated with the September 1977 deflation of the Krafla central volcano in northeastern Iceland, *J. Volcanol. Geotherm. Res.*, *6*, 197-212, 1979.

Brown, J. R., and J. A. Karson, Variations in axial processes on the Mid-Atlantic Ridge: The median valley of the MARK area, *Mar. Geophys. Res.*, *10*, 109-138, 1988.

Bryan, W. B., S. E. Humphris, G. Thompson, and J. F. Casey, Comparative volcanology of small axial eruptive centers in the MARK area, *J. Geophys. Res.*, *99*, 2973-2984, 1994.

Bryan, W. B., and J. G. Moore, Compositional variations of young basalts in the Mid-Atlantic Ridge rift valley near 36°49'N, *Bull. Geol. Soc. Amer.*, *88*, 556-570, 1976.

Bryan, W. B., and G. Thompson, Basalts from DSDP Leg 37 and the FAMOUS area: compositional and petrogenetic comparisons, *Can. J. Earth Sci.*, *14*, 875-885, 1977.

Bryan, W. B., G. Thompson, and J. N. Ludden, Compositional variation in normal MORB from 22°-25°N: Mid-Atlantic Ridge and Kane fracture zone, *J. Geophys. Res.*, *86*, 11, 815-11, 836, 1981.

Bryan, W. B., G. Thompson, and P. J. Michael, Compositional variation in a steady-state zoned magma chamber: Mid-Atlantic Ridge at 36° 50'N, *Tectonophysics*, *55*, 63-85, 1979.

Butterfield, D. A., K. K. Roe, J. Ishibashi, and J. E. Lupton, Time series chemistry of hydrothermal fluids from the southern East Pacific Rise, 14-18°S, *EOS Trans. Am. Geophys. Union*, *78*, 46, Fall Meeting suppl., F706, 1997.

Canales, J. P., J. J. Dañobeitia, R. S. Detrick, E. E. E. Hooft, R. Bartolomé, and D. F. Naar, Variations in axial morphology along the Galapagos spreading center and the influence of the Galapagos hotspot, *J. Geophys. Res.*, *102*, 27341-27354, 1997.

Cann, J. R., and M. R. Strens, Modeling periodic megaplume emission by black smoker systems, *J. Geophys. Res.*, *94*, 12227-12237, 1989.

Carbotte, S., and K. C. Macdonald, Comparison of seafloor tectonic fabric at intermediate, fast, and superfast spreading ridges: Influence of spreading rate, plate motions, and ridge segmentation on fault patterns, *J. Geophys. Res.*, *99*, 13609-13631, 1994.

Carbotte, S. M., J. C. Mutter, and L. Xu, Contribution of volcanism and tectonism to axial and flank morphology of the southern East Pacific Rise, 17°10'-17°40'S, from a study of layer 2A geometry, *J. Geophys. Res.*, *102*, 10165-10184, 1997.

Chadwick, W. W., Jr., and R. W. Embley, Lava flows from a mid-1980s submarine eruption on the Cleft Segment, Juan de Fuca Ridge, *J. Geophys. Res.*, *99*, 4761-4776, 1994.

Chadwick, W. W., Jr., and R. W. Embley, Graben formation associated with recent dike intrusions and volcanic eruptions on the mid-ocean ridge, *J. Geophys. Res.*, in press.

Chadwick, W. W., Jr., R. W. Embley, and C. G. Fox, Evidence for volcanic eruption on the southern Juan de Fuca Ridge between 1981 and 1987, *Nature*, *350*, 416-418, 1991.

Chadwick, W. W., Jr., R. W. Embley, and C. G. Fox, SeaBeam depth changes associated with recent lava flows, CoAxial segment, Juan de Fuca Ridge: Evidence for multiple eruptions between 1981-1993, *Geophys. Res. Lett.*, *22*, 167-170, 1995.

Chadwick, W. W., Jr., R. W. Embley, and T. M. Shank, The 1996 Gorda Ridge eruption: Geologic mapping, sidescan sonar, and SeaBeam comparison results, *Deep-Sea Res.*, in press.

Choukroune, P., J. Francheteau, and R. Hekinian, Tectonics of the East Pacific Rise near 12°50'N: A submersible study, *Earth Planet. Sci. Lett.*, 68, 115-127, 1984.

Christeson, G. L., G. M. Purdy, and G. J. Fryer, Seismic constraints on shallow crustal emplacement processes at the fast spreading East Pacific Rise, *J. Geophys. Res.*, 99, 17957-17973, 1994.

Christie, D., and J. Sinton, Evolution of abyssal lavas along propagating segments of the Galapagos spreading center, *Earth Planet. Sci. Let.*, 56, 321-335, 1981.

Clague, D. A., F. A. Frey, G. Thompson, and S. Rindge, Minor and trace-element geochemistry of volcanic rocks dredged from the Galapagos spreading center: Role of crystal fractionation and mantle heterogeneity, *J. Geophys. Res.*, 86, 9469-9482, 1981.

Cochran, J. R., et al., The Southeast Indian Ridge between 88°E and 118°E: Gravity anomalies and crustal accretion at intermediate spreading rates, *J. Geophys. Res.*, 102, 15463-15487, 1997.

Condomines, M., Comment on: "The volume and residence time of magma beneath active volcanoes determined by decay series disequilibria methods", *Earth Planet. Sci. Lett.*, 122, 251-255, 1994.

Condomines, M., C. Hemond, and C. J. Allègre, U-Th-Ra radioactive disequilibria and magmatic processes, *Earth Planet. Sci. Lett.*, 90, 243-262, 1988.

Crane, K., and R. D. Ballard, Volcanics and structure of the Famous Narrowgate Rift: Evidence for cyclic evolution: AMAR 1, *J. Geophys. Res.*, 86, 5112-5124, 1981.

Crisp, J. A., Rates of magma emplacement and volcanic output, *J. Volcanol. Geotherm. Res.*, 20, 177-211, 1984.

Davis, E., Evidence for extensive basalt flows on the sea floor, *Geol. Soc. Am. Bull.*, 93, 1023-1029, 1982.

Davis, E., and H. Villinger, Tectonic and thermal structure of the Middle Valley sedimented rift, northern Juan de Fuca Ridge, in *Proceedings of the Ocean Drilling Program, Initial Reports, v. 139*, edited by E. E. Davis, et al., pp. 9-41, Ocean Drilling Program, College Station, TX, 1992.

Davis, E. E., and J. L. Karsten, On the cause of the asymmetric distribution of seamounts about the Juan de Fuca Ridge: Ridge-crest migration over a heterogeneous asthenosphere, *Earth Planet. Sci. Lett.*, 79, 385-396, 1986.

Detrick, R. S., P. Buhl, E. Vera, J. Mutter, J. Orcutt, J. Madsen, and T. Brocher, Multi-channel seismic imaging of a crustal magma chamber along the East Pacific Rise, *Nature*, 326, 35-41, 1987.

Detrick, R. S., A. J. Harding, G. M. Kent, J. A. Orcutt, J. C. Mutter, and P. Buhl, Seismic structure of the southern East Pacific Rise, *Science*, 259, 499-503, 1993.

Detrick, R. S., J. C. Mutter, P. Buhl, and I. I. Kim, No evidence from multichannel reflection data for a crustal magma chamber in the MARK area on the Mid-Atlantic Ridge, *Nature*, 347, 61-64, 1990.

Duncan, R. A., and L. G. Logan, Radiometric dating of young MORB using the ^{40}Ar-^{39}Ar incremental heating method, *Geophys. Res. Lett.*, 21, 1927-1930, 1994.

Dziak, R. P., and C. G. Fox, Long-term seismicity and ground deformation at Axial volcano, Juan de Fuca Ridge, *EOS Trans. Am. Geophys. Union*, 78, 46, Fall Meeting suppl., F641, 1997.

Dziak, R. P., C. G. Fox, R. W. Embley, J. E. Lupton, G. Johnson, W. W. Chadwick Jr., and R. A. Koski, Detection of and response to a probable volcanogenic T-wave event swarm on the western Blanco Transform Fault Zone, *Geophys. Res. Lett.*, 23, 873-876, 1996.

Dziak, R. P., C. G. Fox, and A. E. Schreiner, The June-July 1993 seismo-acoustic event at CoAxial segment, Juan de Fuca Ridge: Evidence for a lateral dike injection, *Geophys. Res. Lett.*, 22, 135-138, 1995.

Einarsson, P., and B. Brandsdottir, Seismological evidence for lateral magma intrusion during the July 1978 deflation of the Krafla volcano in NE-Iceland, *J. Geophys.*, 47, 160-165, 1980.

Einsele, G., Basaltic sill-sediment complexes in young spreading centers: Genesis and significance, *Geology*, 13, 249-252, 1985.

Embley, R. W., and W. W. Chadwick Jr., Volcanic and hydrothermal processes associated with a recent phase of seafloor spreading at the northern Cleft segment: Juan de Fuca Ridge, *J. Geophys. Res.*, 99, 4741-4760, 1994.

Embley, R. W., W. W. Chadwick Jr., and J. Getsiv, The CoAxial segment: A site of recent seafloor eruptions on the Juan de Fuca Ridge, *EOS Trans. Am. Geophys. Union*, 76, 46, Fall Meeting suppl., F410, 1995a.

Embley, R. W., W. W. Chadwick Jr., I. R. Jonasson, D. A. Butterfield, and E. T. Baker, Initial results of the rapid response to the 1993 CoAxial event: Relationships between hydrothermal and volcanic processes, *Geophys. Res. Lett.*, 22, 143-146, 1995b.

Embley, R. W., W. W. Chadwick Jr., M. R. Perfit, and E. T. Baker, Geology of the northern Cleft segment, Juan de Fuca Ridge: Recent lava flows, sea-floor spreading, and the formation of megaplumes, *Geology*, 19, 771-775, 1991.

Embley, R. W., W. W. Chadwick Jr., T. Shank, and D. Christie, Geology of the 1996 Gorda Ridge eruption from analysis of multibeam, towed camera, sidescan, and ROV data, *EOS Trans. Am. Geophys. Union*, 77, 46, Fall Meeting suppl., F1, 1996.

Embley, R. W., J. E. Lupton, G. Massoth, T. Urabe, T. Shibata, V. Tunnicliffe, D. Butterfield, O. Okano, M. Kinoshita, and K. Fujioka, Geologic, chemical, and biologic evidence for recent volcanism at 17°22'-35'S, East Pacific Rise, *Earth Planet. Sci. Lett.*, in press.

Embley, R. W., J. E. Lupton, G. Massoth, T. Urabe, V. Tunnicliffe, D. A. Butterfield, T. Shibata, O. Okano, M. Kinoshita, and K. Fujioka, Geologic, chemical, and biologic evidence for recent volcanism at 17.5°S, *EOS Trans. Am. Geophys. Union*, 78, 46, Fall Meeting suppl., F705, 1997.

Embley, R. W., K. M. Murphy, and C. G. Fox, High resolution studies of the summit of Axial volcano, *J. Geophys. Res.*, 95, 12785-12812, 1990.

Feely, R. A., E. T. Baker, K. Marumo, T. Urabe, J. Ishibashi, J. Gendron, G. T. Lebon, and K. Okamura, Hydrothermal plume particles and dissolved phosphate over the superfast-spreading southern East Pacific Rise, *Geoch. Cosmoch. Acta*, 60, 2297-2323, 1996.

Fink, J. H., and R. W. Griffiths, Radial spreading of viscous-gravity currents with solidifying crust, *J. Fluid Mech.*, 221, 485-509, 1990.

Fink, J. H., and R. W. Griffiths, A laboratory analog study of the surface morphology of lava flows extruded from point and line sources, *J. Volcanol. Geotherm. Res.*, *54*, 19-32, 1992.

Fornari, D. J., R. Batiza, and M. A. Luckman, Seamount abundances and distribution near the East Pacific Rise 0°-24°N based on Seabeam data, in *Seamounts, Islands, and Atolls*, edited by: B. Keating, et al., pp. 13-21, American Geophysical Union, Washington, DC, 1987.

Fornari, D. J., and R. W. Embley, Tectonic and volcanic controls on hydrothermal processes at the mid-ocean ridge: An overview based on near-bottom and submersible studies, in *Seafloor Hydrothermal Systems: Physical, Chemical, Biological, and Geological Interactions, Geophys. Monograph 91*, edited by S. E. Humphris, et al., pp. 1-46, American Geophysical Union, Washington, DC, 1995.

Fornari, D. J., R. M. Haymon, M. R. Perfit, T. K. P. Gregg, and M. H. Edwards, Axial summit trough of the East Pacific Rise 9°N to 10°N: Geological characteristics and evolution of the axial zone on fast-spreading mid-ocean ridges, *J. Geophys. Res.*, in press-a.

Fornari, D. J., M. R. Perfit, J. F. Allan, R. Batiza, R. Haymon, A. Barone, W. B. F. Ryan, T. Smith, T. Simkin, and M. A. Luckman, Geochemical and structural studies of the Lamont seamounts: Seamounts as indicators of mantle processes, *Earth Planet. Sci. Lett.*, *89*, 63-83, 1988.

Fornari, D. J., W. B. F. Ryan, and P. J. Fox, The evolution of craters and calderas on young seamounts: Insights from Sea Marc I and SeaBeam sonar surveys of a small seamount group near the axis of the East Pacific Rise at 10°N, *J. Geophys. Res.*, *89*, 11069-11083, 1984.

Fornari, D. J., T. Shank, K. L. Von Damm, T. K. P. Gregg, M. Lilley, G. Levai, A. Bray, R. M. Haymon, M. R. Perfit, and R. Lutz, Time-series temperature measurements at high-temperature hydrothermal vents, East Pacific Rise 9°49-51'N: Monitoring a crustal cracking event, *Earth Planet. Sci. Lett.*, in press-b.

Fox, C. G., Evidence of active ground deformation on the mid-ocean ridge: Axial Seamount, Juan de Fuca Ridge, April-June, 1988, *J. Geophys. Res.*, *95*, 12813-12822, 1990.

Fox, C. G., Five years of ground deformation monitoring on Axial Seamount using a bottom pressure recorder, *Geophys. Res. Lett.*, *20*, 1859-1862, 1993.

Fox, C. G., W. W. Chadwick Jr., and R. W. Embley, Detection of changes in ridge-crest morphology using repeated multibeam sonar surveys, *J. Geophys. Res.*, *97*, 11149-11162, 1992.

Fox, C. G., and R. P. Dziak, Seismo-acoustic detection of volcanic activity on the Gorda Ridge, February-March 1996, *EOS Trans. Am. Geophys. Union*, *77*, 46, Fall Meeting suppl., F1, 1996.

Fox, C. G., and R. P. Dziak, Hydroacoustic detection of volcanic activity on the Gorda Ridge, February-March 1996, *Deep-Sea Res.*, in press.

Fox, C. G., R. P. Dziak, H. Matsumoto, and A. E. Schreiner, Potential for monitoring low-level seismicity on the Juan de Fuca Ridge using military hydrophone arrays, *Mar. Tech. Soc. J.*, *27*, 22-30, 1994.

Fox, C. G., and H. Matsumoto, Monitoring northeast Pacific seismicity using autonomous hydrophone moorings, *EOS Trans. Am. Geophys. Union*, *76*, 46, Fall Meeting suppl., F412, 1995.

Fox, C. G., H. Matsumoto, and T. K. Lau, Monitoring East Pacific Rise seismicity using autonomous hydrophone moorings, *EOS Trans. Am. Geophys. Union*, *78*, 46, Fall Meeting suppl., F705, 1997.

Fox, C. G., K. M. Murphy, and R. W. Embley, Automated display and statistical analysis of interpreted deep-sea bottom photographs, *Mar. Geol.*, *78*, 199-216, 1988.

Fox, C. G., W. E. Radford, R. P. Dziak, T. K. Lau, H. Matsumoto, and A. E. Schreiner, Acoustic detection of a seafloor spreading episode on the Juan de Fuca Ridge using military hydrophone arrays, *Geophys. Res. Lett.*, *22*, 131-134, 1995.

Fox, P. J., N. R. Grindlay, and K. C. Macdonald, The Mid-Atlantic Ridge (31°S-34°30'S): Temporal and spatial variations of accretionary processes, *Mar. Geophys. Res.*, *13*, 1-20, 1991.

Francheteau, J., and R. D. Ballard, The East Pacific Rise near 21°N, 13°N and 20°S: Inferences for along-strike variability of axial processes of the mid-ocean ridge, *Earth Planet. Sci. Lett.*, *64*, 93-116, 1983.

Francheteau, J., T. Juteau, and C. Rangan, Basaltic pillars in collapsed lava-pools on the deep ocean-floor, *Nature*, *281*, 209-211, 1979.

Frey, F. A., N. Walker, D. Stakes, S. R. Hart, and R. Nielsen, Geochemical characteristics of basaltic glasses from the AMAR and FAMOUS axial valleys, Mid-Atlantic Ridge (36°-37°N): Petrogenetic implications, *Earth Planet. Sci. Lett.*, *115*, 117-136, 1993.

Fujimoto, H., N. Seama, J. Lin, T. Matsumoto, T. Tanaka, and K. Fujioka, Gravity anomalies of the Mid-Atlantic Ridge north of the Kane Fracture Zone, *Geophys. Res. Lett.*, *23*, 3431-3434, 1996.

Garcia, M. O., K. H. Rubin, M. D. Norman, J. M. Rhodes, D. W. Graham, K. Spencer, and D. Muenow, Petrology and geochronology of basalt breccia from the 1996 earthquake swarm of Loihi Seamount, Hawaii: Magmatic history of its 1996 eruption, *Bull. Volcanol.*, in press.

Gente, P., J. M. Auzende, V. Renard, Y. Fouquet, and D. Bideau, Detailed geological mapping by submersible of the East Pacific Rise axial graben near 13°N, *Earth Planet. Sci. Lett.*, *78*, 224-236, 1986.

Gente, P., C. Mével, J. M. Auzende, J. A. Karson, and Y. Fouquet, An example of a recent accretion on the Mid-Atlantic Ridge: The Snakepit neovolcanic ridge (MARK area, 23°22'N), *Tectonophysics*, *190*, 1-29, 1991.

Gill, J., and M. Condomines, Short-lived radioactivity and magma genesis, *Science*, *257*, 1368-1376, 1992.

Gill, J., D. M. Pyle, and R. W. Williams, *Igneous Rocks*, 207-257 pp., Oxford Science Publications, Oxford, U.K., 1992.

Gill, J., R. Williams, and K. Bruland, Eruption of basalt and andesite lava degasses ^{222}Rn and ^{210}Po., *Geophys. Res. Lett.*, *12*, 17-20, 1985.

Goldstein, S. J., M. T. Murrell, and D. R. Janecky, Th and U isotopic systematics of basalts from the Juan de Fuca and Gorda Ridges by mass spectrometry, *Earth Planet. Sci. Lett.*, *96*, 134-146, 1989.

Goldstein, S. J., M. T. Murrell, D. R. Janecky, J. R. Delaney, and D. A. Clague, Geochronology and petrogenesis of MORB from the Juan de Fuca and Gorda ridges by ^{238}U-^{230}Th disequilibrium, *Earth Planet. Sci. Lett.*, *109*, 255-272, 1992.

Goldstein, S. J., M. T. Murrell, and R. W. Williams, [231]Pa and [230]Th chronology of mid-ocean ridge basalts, *Earth Planet. Sci. Lett.*, *115*, 151-159, 1993.

Goldstein, S. J., M. R. Perfit, R. Batiza, D. J. Fornari, and M. T. Murrell, [226]Ra and [231]Pa systematics of axial MORB, crustal residence ages, and magma chamber characteristics at 9°-10°N, East Pacific Rise, *Mineralogical Magazine*, *58a*, 335-336, 1994a.

Goldstein, S. J., M. R. Perfit, R. Batiza, D. J. Fornari, and M. T. Murrell, Off-axis volcanism at the East Pacific Rise detected by uranium-series dating of basalts, *Nature*, *367*, 157-159, 1994b.

Goodfellow, W. D., and J. M. Franklin, Geology, mineralogy, and chemistry of sediment-hosted clastic massive sulfides in shallow cores, Middle Valley, northern Juan de Fuca Ridge, *Econ. Geol.*, *88*, 2037-2068, 1993.

Green, J., and N. M. Short, *Volcanic landforms and surface features*, 519 pp., Springer, Berlin, Heidelberg, New York, 1971.

Gregg, T. K. P., and W. W. Chadwick Jr., Submarine lava-flow inflation: A model for the formation of lava pillars, *Geology*, *24*, 981-984, 1996.

Gregg, T. K. P., and J. H. Fink, Quantification of submarine lava-flow morphology through analog experiments, *Geology*, *23*, 73-76, 1995.

Gregg, T. K. P., D. J. Fornari, and M. R. Perfit, Lava pillars: "Rosetta stones" of deep sea eruption dynamics, *EOS Trans. Am. Geophys. Union*, *77*, 46, Fall Meeting suppl., F664, 1996a.

Gregg, T. K. P., D. J. Fornari, M. R. Perfit, R. M. Haymon, and J. H. Fink, Rapid emplacement of a mid-ocean ridge lava flow on the East Pacific Rise at 9° 46'-51'N, *Earth Planet. Sci. Lett.*, *144*, E1-E7, 1996b.

Griffiths, R. W., and J. H. Fink, Solidification and morphology of submarine lavas: A dependence on extrusion rate, *J. Geophys. Res.*, *97*, 19729-19737, 1992.

Grindlay, N. R., P. J. Fox, and P. R. Vogt, Morphology and tectonics of the Mid-Atlantic Ridge (25°-27°30'S) from SeaBeam and magnetic data, *J. Geophys. Res.*, *97*, 6983-7010, 1992.

Grove, T. L., R. J. Kinzler, and W. B. Bryan, Fractionation of mid-ocean ridge basalts, in *Mantle flow and melt generation at mid-ocean ridges, Geophys. Monograph 71*, edited by J. Phipps-Morgan, D. K. Blackman, and J. Sinton, pp. 281-310, Am. Geophys. Union, Washington, D.C., 1992.

Hall, L. S., and J. M. Sinton, Geochemical diversity of the large lava field on the flank of the East Pacific Rise at 8°17'S, *Earth Planet. Sci. Lett.*, *142*, 241-251, 1996.

Hammond, S. R., Offset caldera and crater collapse on Juan de Fuca ridge-flank volcanoes, *Bull. Volcanol.*, *58*, 617-627, 1997.

Harding, A. J., G. M. Kent, and J. A. Orcutt, A multichannel seismic investigation of upper crustal structure at 9°N on the East Pacific Rise: Implications for crustal accretion, *J. Geophys. Res.*, *98*, 13925-13944, 1993.

Harp, K. S., Geochemical variations in a single basaltic flow at 9°30'N on the East Pacific Rise, in *Proc. ODP, Sci. Results*, *142*, edited by R. Batiza Storms, M.A., and Allan, J.F., pp. 9-21, Ocean Drilling Program, College Station, Texas, 1995.

Hart, S. R., A large-scale isotope anomaly in the Southern Hemisphere mantle, *Nature*, *309*, 753-757, 1984.

Haymon, R., D. J. Fornari, M. H. Edwards, S. Carbotte, D. Wright, and K. C. Macdonald, Hydrothermal vent distribution along the East Pacific Rise crest (9°09'-54'N) and its relationship to magmatic and tectonic processes on fast-spreading mid-ocean ridges, *Earth Planet. Sci. Lett.*, *104*, 513-534, 1991.

Haymon, R. M., et al., Volcanic eruption of the mid-ocean ridge along the East Pacific Rise crest at 9° 45-52'N: Direct submersible observations of seafloor phenomena associated with an eruption event in April, 1991, *Earth Planet. Sci. Lett.*, *119*, 85-101, 1993.

Head, J. W., L. Wilson, and D. K. Smith, Mid-ocean ridge eruptive vents: Evidence for dike widths, eruption rates, and evolution of eruptions from morphology and structure, *J. Geophys. Res.*, *101*, 28265-28280, 1996.

Hekinian, R., J. M. Auzende, J. Francheteau, P. Gente, W. B. F. Ryan, and E. S. Kappel, Offset spreading centers near 12°53' N on the East Pacific Rise: submersible observations and composition of the volcanics, *Marine Geophys. Res.*, *7*, 359-377, 1985.

Hekinian, R., et al., East Pacific Rise near 13°N: geology of new hydrothermal fields, *Science*, *219*, 1321-1324, 1983.

Hekinian, R., G. Thompson, and D. Bideau, Axial and off-axial heterogeneity of basaltic rocks from the East Pacific Rise at 12°35'N-12°51'N and 11°26'N-11°30'N, *J. Geophys. Res.*, *94*, 17,437 - 17,463, 1989.

Hekinian, R., and D. Walker, Diversity and spatial zonation of volcanic rocks from the East Pacific Rise near 21°N, *Contrib. Mineral. Petrol.*, *96*, 265-280, 1987.

Hey, R., P. D. Johnson, F. Martinez, J. Korenaga, M. L. Somers, Q. J. Huggett, T. P. LeBas, R. I. Rusby, and D. F. Naar, Plate boundary reorganization at a large-offset, rapidly propagating rift, *Nature*, *378*, 167-170, 1995.

Holden, J. F., B. C. Crump, M. Summit, and J. A. Baross, Microbial blooms at the CoAxial segment, Juan de Fuca Ridge deep-sea hydrothermal vent site following a magma intrusion, *EOS Trans. Am. Geophys. Union*, *76*, 43, Fall Meet. Suppl., F411, 1995.

Hon, K., J. Kauahikaua, R. Denlinger, and K. Mackay, Emplacement and inflation of pahoehoe sheet flows: Observations and measurements of active lava flows on Kilauea volcano, Hawaii, *Geol. Soc. Am. Bull.*, *106*, 351-370, 1994.

Hooft, E. E. E., and R. S. Detrick, Relationship between axial morphology, crustal thickness, and mantle temperature along the Juan de Fuca and Gorda Ridges, *J. Geophys. Res.*, *100*, 22499-22508, 1995.

Hooft, E. E. E., R. S. Detrick, and G. M. Kent, Seismic structure and indicators of magma budget along the southern East Pacific Rise, *J. Geophys. Res.*, *102*, 27319-27340, 1997.

Hooft, E. E. E., H. Schouten, and R. S. Detrick, Constraining crustal emplacement processes from the variation of seismic layer 2A thickness at the East Pacific Rise, *Earth Planet. Sci. Lett.*, *142*, 289-310, 1996.

Humler, E., J. L. Thirot, and J. P. Montagner, Global correlations of mid-ocean ridge basalt chemistry with seismic tomographic images, *Nature*, *364*, 225-228, 1993.

Humphris, S. E., W. B. Bryan, G. Thompson, and L. K. Autio, Morphology, geochemistry, and evolution of Serocki volcano, in *Proc. Ocean Drilling Prog., Scientific Results*, *106/109*, edited by R. S. Detrick, et al., pp. 67-84, Ocean Drilling Program, College Station, Texas, 1990.

Johnson, P. D., Recent structural evolution of the East Pacific Rise 29°S large-scale dueling propagator system, M.A. Thesis, University of Hawaii, 1996.

Juniper, S. K., P. Martineu, J. Sarrazin, and Y. Gélinas, Microbial-mineral floc associated with nascent hydrothermal activity on CoAxial segment, Juan de Fuca Ridge, *Geophys. Res. Lett.*, 22, 179-182, 1995.

Juster, T. C., T. L. Grove, and M. R. Perfit, Experimental constraints on the generation of FeTi basalts, andesites and rhyodacites at the Galapagos Spreading Center, 85° and 95° W, *J. Geophys. Res.*, 94, 9251-9274, 1989.

Kappel, E. S., and W. R. Normark, Morphometric variability within the axial zone of the southern Juan de Fuca Ridge: Interpretation from SeaMARC II, SeaMARC I, and deep sea photography, *J. Geophys. Res.*, 92, 11,291-11,302, 1987.

Kappel, E. S., and W. B. F. Ryan, Volcanic episodicity and a non-steady state rift valley along northeast Pacific spreading centers: evidence from SeaMARC I, *J. Geophys. Res.*, 91, 13925-13940, 1986.

Karson, J. A., and P. A. Rona, Block-tilting, transfer faults, and structural control of magmatic and hydrothermal processes in the TAG area, Mid-Atlantic Ridge 26°N, *Geol. Soc. Am. Bull.*, 102, 1635-1645, 1990.

Karson, J. A., et al., Along-axis variations in seafloor spreading in the MARK area, *Nature*, 328, 681-685, 1987.

Karsten, J. L., J. R. Delaney, J. M. Rhodes, and R. A. Liias, Spatial and temporal evolution of magmatic systems beneath the Endeavour segment, Juan de Fuca Ridge: Tectonic and petrologic constraints, *J. Geophys. Res.*, 95, 19,235-19,256, 1990.

Karsten, J. L., S. R. Hammond, E. E. Davis, and R. G. Currie, Detailed geomorphology and neotectonics of the Endeavour segment, Juan de Fuca Ridge: New results from Seabeam swath mapping, *Geol. Soc. Am. Bull.*, 97, 213-221, 1986.

Kent, G. M., A. J. Harding, and J. A. Orcutt, Distribution of magma beneath the East Pacific Rise between the Clipperton Transform and the 9°17'N deval from forward modeling of common depth point data, *J. Geophys. Res.*, 98, 13945-13969, 1993.

Klein, E. M., and C. H. Langmuir, Global correlations of ocean ridge basalt chemistry with axial depth and crustal thickness, *J. Geophys. Res.*, 92, 8089-8115, 1987.

Klein, E. M., and C. H. Langmuir, Local versus global variations in ocean ridge basalt composition: A reply, *J. Geophys. Res.*, 94, 4241-4252, 1989.

Kong, L. S., R. S. Detrick, P. J. Fox, L. A. Mayer, and W. B. F. Ryan, The morphology and tectonics of the MARK area from SeaBeam and Sea MARC I observations (Mid-Atlantic Ridge 23°N), *Mar. Geophys. Res.*, 10, 59-90, 1988.

Kuo, B. Y., and D. W. Forsyth, Gravity anomalies of the ridge-transform system in the south Atlantic between 31 and 34.5°S: Upwelling centers and variation in crustal thickness, *Mar. Geophys. Res.*, 10, 205-232, 1988.

Lagabrielle, Y., M. H. Cormier, V. Ballu, and J. M. Auzende, From perfect dome to large collapse caldera: Tectonic/magmatic evolution of the EPR axial domain at 17°-19°S from submersible observations, *EOS Trans. Am. Geophys. Union*, 77, 46, Fall Meeting suppl., F660, 1996.

Langmuir, C. H., Petrology database, vol. 2 and vol. 3, in *East Pacific Rise Data Synthesis Final Report*, edited by S. Tighe, Joint Oceanographic Institutions, Inc., Washington, D.C., 1988.

Langmuir, C. H., J. F. Bender, and R. Batiza, Petrologic and tectonic segmentation of the East Pacific Rise, 5° 30'–14° 30' N, *Nature*, 322, 422-429, 1986.

Langmuir, C. H., E. M. Klein, and T. Plank, Petrological systematics of mid-ocean ridge basalts: constraints on melt generation beneath ocean ridges, in *Mantle flow and melt generation at mid-ocean ridges, Geophys. Monograph 71*, edited by J. Phipps-Morgan, D. K. Blackman, and J. Sinton, pp. 183-280, Amer. Geophys. Union, Washington, D.C., 1992.

le Roex, A. P., F. A. Frey, and S. H. Richardson, Petrogenesis of lavas from the AMAR valley and Narrowgate region of the FAMOUS valley, 36°-37°N on the Mid-Atlantic Ridge, *Contrib. Mineral. Petrol.*, 124, 167-184, 1996.

Leybourne, M. I., and N. A. Van Wagoner, Heck and Heckle seamounts, northeast Pacific Ocean: high extrusion rates of primitive and highly depleted mid-ocean ridge basalt on off-ridge seamounts, *J. Geophys. Res.*, 96, 16,275-16,293, 1991.

Lin, J., G. M. Purdy, H. Schouten, J. C. Sempéré, and C. Zervas, Evidence from gravity data for focused magmatic accretion along the Mid-Atlantic Ridge, *Nature*, 344, 627-632, 1990.

Lister, C. R. B., On the intermittency and crystallization mechanisms of sub-seafloor magma chambers, *Geophys. J. R. Astr. Soc.*, 73, 351-365, 1983.

Lonsdale, P., Abyssal pahoehoe with lava coils at the Galapagos Rift, *Geology*, 5, 147-152, 1977a.

Lonsdale, P., Structural geomorphology of a fast-spreading rise crest: The East Pacific Rise near 3°25'S, *Mar. Geophys. Res.*, 3, 251-293, 1977b.

Lonsdale, P., Segmentation of the Pacific-Nazca spreading center, 1°N-20°S, *J. Geophys. Res.*, 94, 12197-12225, 1989.

Lonsdale, P., and K. Becker, Hydrothermal plumes, hot springs, and conductive heat flow in the southern trough of Guaymas Basin, *Earth Planet. Sci. Lett.*, 73, 211-225, 1985.

Lonsdale, P. F., and L. A. Lawver, Immature plate boundary zones studies with a submersible in the Gulf of California, *Geol. Soc. Am. Bull.*, 91, 555-569, 1980.

Lowell, R. P., and L. N. Germanovich, Dike injection and the formation of megaplumes at ocean ridges, *Science*, 267, 1804-1807, 1995.

Lundstrom, C. C., J. Gill, Q. Williams, and M. R. Perfit, Mantle melting and basalt extraction by equilibrium porous flow, *Science*, 270, 1958-1961, 1995.

Lupton, J. E., E. T. Baker, N. Garfield, R. Greene, and T. Rago, Successful tracking of a hydrothermal event plume with a RAFOS neutrally-buoyant drifter, *EOS Trans. Am. Geophys. Union*, 77, 46, Fall Meeting suppl., F1, 1996.

Lupton, J. E., J. C. Ishibashi, and D. A. Butterfield, Gas chemistry of hydrothermal fluids along the southern East Pacific Rise, 13.5-18.5°S, *EOS Trans. Am. Geophys. Union*, 78, 46, Fall Meeting suppl., F706, 1997.

Luyendyk, B. P., and K. C. Macdonald, A geological transect across the crest of the East Pacific Rise at 21°N latitude made from the deep submersible Alvin, *Mar. Geophys. Res.*, 7, 467-488, 1985.

Macdonald, K. C., Mid-ocean ridges: Fine scale tectonic, volcanic and hydrothermal processes within the plate boundary zone, *Rev. Earth Planetary Sci.*, *10*, 155-190, 1982.

Macdonald, K. C., The crest of the Mid-Atlantic Ridge: Models for crustal generation processes and tectonics, in *The Geology of North America: The Western North Atlantic Region*, edited by P. R. Vogt, and B. E. Tucholke, pp. 51-68, Geological Society of America, Boulder, CO, 1986.

Macdonald, K. C., and P. J. Fox, The axial summit graben and cross-sectional shape of the East Pacific Rise as indicators of axial magma chambers and recent volcanic eruptions, *Earth Planet. Sci. Lett.*, *88*, 119-131, 1988.

Macdonald, K. C., P. J. Fox, R. T. Alexander, R. Pockalny, and P. Gente, Volcanic growth faults and the origin of Pacific abyssal hills, *Nature, 380*, 125-129, 1996.

Macdonald, K. C., P. J. Fox, L. J. Perram, E. M. F., R. M. Haymon, S. P. Miller, S. M. Carbotte, M.-H. Cormier, and A. N. Shor, A new view of the mid-ocean ridge from the behavior of ridge-axis discontinuities, *Nature, 335*, 217-225, 1988.

Macdonald, K. C., R. Haymon, and A. Shor, A 220 km^2 recently erupted lava field on the East Pacific Rise near lat 8°S, *Geology, 17*, 212-216, 1989.

Macdonald, K. C., R. M. Haymon, S. M. White, and P. J. Sharfstein, Evidence for extensive lava flows on the southern East Pacific Rise, *EOS Trans. Am. Geophys. Union, 78*, 46, Fall Meeting suppl., F705, 1997.

Macdonald, K. C., D. S. Scheirer, and S. Carbotte, It's only topography: Part I, *GSA Today, 3*, 1 and 24-25, 1993.

Macdougall, J. D., Using short-lived U and Th series isotopes to investigate volcanic processes, in *Annual Review of Earth and Planetary Sciences*, *23*, edited by G. W. Wetherill, A. L. Albee, and K. C. Burke, pp. 143-167, Annual Rev. Inc., Palo Alto, 1995.

Mastin, L. G., and D. D. Pollard, Surface deformation and shallow dike intrusion processes at Inyo Craters, Long Valley, California, *J. Geophys. Res.*, *93*, 13221-13235, 1988.

McConachy, T. F., R. D. Ballard, M. J. Mottl, and R. P. v. Herzen, Geologic form and setting of a hydrothermal vent field at 10°56'N East Pacific Rise: A detailed study using Angus and Alvin, *Geology, 14*, 295-298, 1986.

Melson, W. G., and T. O'Hearn, "Zero age" variations in the composition of abyssal volcanic rocks along the axial zone of the Mid-Atlantic Ridge, in *The Geology of North America, The Western North America Region*, edited by P. R. Vogt, and B. E. Tucholke, pp. 117-136, Geol. Soc. Amer., Boulder, CO, 1986.

Michael, P. J., and J.-G. Schilling, Chlorine in mid-ocean ridge magmas: evidence for assimilation of seawater-influenced components, *Geochim. et Cosmochim. Acta, 53*, 3131-3143, 1989.

Moore, J. G., Mechanism of formation of pillow lava, *Am. Sci., 63*, 269-277, 1975.

Morton, J. L., M. L. Holmes, and R. A. Koski, Volcanism and massive sulfide formation at a sedimented spreading center, Escanaba Trough, Gorda Ridge, northeast Pacific Ocean, *Geophys. Res. Lett., 14*, 769-772, 1987.

Morton, J. L., R. A. Zierenberg, and C. A. Reiss, Geologic, hydrothermal, and biologic studies at Escanaba Trough: An introduction, in *Geologic, Hydrothermal, and Biologic Studies at Escanaba Trough, Gorda Ridge, Offshore Northern California, U.S. Geol. Surv. Bull. 2022*, edited by J. L. Morton, R. A. Zierenberg, and C. A. Reiss, pp. 1-20, 1994.

Mutter, J. C., S. M. Carbotte, W. Su, L. Xu, P. Buhl, R. S. Detrick, G. M. Kent, J. A. Orcutt, and A. J. Harding, Seismic images of active magma systems beneath the East Pacific Rise between 17°05' and 17°35'S, *Science, 268*, 391-395, 1995.

Mutter, J. C., and J. A. Karson, Structural processes at slow-spreading ridges, *Science, 257*, 627-634, 1992.

Niu, Y., and R. Batiza, Magmatic processes at a slow spreading segment: 26° S Mid-Atlantic Ridge, *J. Geophys. Res., 99*, 19,719-19,740, 1994.

Niu, Y., and R. Batiza, Trace element evidence from seamounts for recycled oceanic crust in the Eastern Pacific mantle, *Earth Planet. Sci. Lett., 148*, 471-483, 1997.

Niu, Y., and R. Hekinian, Spreading-rate dependence of the extent of mantle melting beneath ocean ridges, *Nature, 385*, 326-329, 1997.

Normark, W. R., et al., Submarine fissure eruptions and hydrothermal vents on the southern Juan de Fuca Ridge: Preliminary observations from the submersible *Alvin*, *Geology, 14*, 823-827, 1986.

Normark, W. R., J. L. Morton, and S. L. Ross, Submersible observations along the southern Juan de Fuca Ridge: 1984 *Alvin* program, *J. Geophys. Res., 92*, 11283-11290, 1987.

Perfit, M. R., D. J. Fornari, A. Malahoff, and R. W. Embley, Geochemical studies of abyssal lavas recovered by DSRV ALVIN from the eastern Galapagos Rift - Inca Transform - Ecuador Rift: Trace elements and petrogenesis, *J. Geophys. Res., 88*, 10551-10572, 1983.

Perfit, M. R., D. J. Fornari, M. C. Smith, J. F. Bender, C. H. Langmuir, and R. M. Haymon, Small-scale spatial and temporal variations in mid-ocean ridge crest magmatic processes, *Geology, 22*, 375-379, 1994.

Perfit, M. R., D. J. Fornari, M. C. Smith, C. H. Langmuir, J. F. Bender, and R. M. Haymon, Fine-scale petrological variations along the East Pacific Rise crest 9°17'N to 9°54'N: results from ALVIN diving and rock coring during the Adventure program, *Eos Trans. Am. Geophys. Union, 72*, 491, 1991.

Perfit, M. R., W. I. Ridley, and I. R. Jonasson, Geologic, petrologic and geochemical relationships between magmatism and massive sulfide mineralization along the eastern Galapagos Spreading Center, *Rev. Economic Geol.*, in press.

Perfit, M. R., M. C. Smith, K. Sapp, D. J. Fornari, T. Gregg, M. H. Edwards, W. I. Ridley, and J. F. Bender, Geochemistry and Morphology of the Crestal Plateau of the East Pacific Rise ~9° 50', *Eos Trans. Am. Geophys. Union, 76*, F694, 1995.

Peterson, D. W., and R. I. Tilling, Transition of basaltic lava from pahoehoe to aa, Kilauea volcano, Hawaii: Field observations and key factors, *J. Volcanol. Geotherm. Res., 7*, 271-293, 1980.

Phipps Morgan, J., and Y. J. Chen, Dependence of ridge-axis morphology on magma supply and spreading rate, *Nature, 364*, 706-708, 1993a.

Phipps Morgan, J., and Y. J. Chen, The genesis of oceanic crust: Magma injection, hydrothermal circulation, and crustal flow, *J. Geophys. Res., 98*, 6283-6297, 1993b.

Pollard, D. D., P. T. Delaney, P. T. Duffield, E. T. Endo, and A. T. Okamura, Surface deformation in volcanic rift zones, *Tectonophysics*, *94*, 541-584, 1983.

Renard, V., R. Hekinian, J. Francheteau, R. D. Ballard, and H. Backer, Submersible observations at the axis of the ultra-fast spreading East Pacific Rise (17°30' to 21°30'S), *Earth Planet. Sci. Lett.*, *75*, 339-353, 1985.

Reynolds, J. A., C. H. Langmuir, J. F. Bender, K. A. Kastens, and W. B. F. Ryan, Spatial and temporal variability in the geochemistry of basalts from the East Pacific Rise, *Nature*, *359*, 493-499, 1992.

Rhodes, J. M., Homogeneity of lava flows: chemical data for historic Mauna Loan eruptions, *J. Geophys. Res.*, *88*, 1983.

Rhodes, J. M., M. A. Dungan, D. P. Blanchard, and P. E. Long, Magma mixing at mid-ocean ridges: Evidence from basalts drilled near 22° on the Mid-Atlantic Ridge, *Tectonophysics*, *55*, 35-61, 1979.

Rowland, S. K., and G. P. L. Walker, Pahoehoe and aa in Hawaii: Volumetric flow rate controls the lava structure, *Bull. Volcanol.*, *52*, 615-628, 1990.

Rubin, A. M., Dike-induced faulting and graben subsidence in volcanic rift zones, *J. Geophys. Res.*, *97*, 1839-1858, 1992.

Rubin, A. M., and D. D. Pollard, Dike-induced faulting in rift zones of Iceland and Afar, *Geology*, *16*, 413-417, 1988.

Rubin, K. H., and J. D. Macdougall, ^{226}Ra excesses in mid-ocean ridge basalts and mantle melting, *Nature*, *335*, 158-161, 1988.

Rubin, K. H., and J. D. Macdougall, Dating of neovolcanic MORB using (226Ra/^{230}Th) disequilibrium, *Earth Planet. Sci. Lett.*, *101*, 313-322, 1990.

Rubin, K. H., J. D. Macdougall, and M. R. Perfit, ^{210}Po-^{210}Pb dating of recent volcanic eruptions on the seafloor, *Nature*, *368*, 841-844, 1994.

Rubin, K. H., M. C. Smith, M. R. Perfit, D. M. Christie, and L. F. Sacks, Geochronology and petrology of lavas from the 1996 North Gorda Ridge eruption, *Deep-Sea Res.*, in press.

Ruegg, J. C., J. C. Lepine, A. Tarantola, and M. Kasser, Geodetic measurements of rifting associated with a seismo-volcanic crisis in Afar, *Geophys. Res. Lett.*, *6*, 817-820, 1979.

Scheirer, D. S., and K. C. Macdonald, Variation in cross-sectional area of the axial ridge along the East Pacific Rise: Evidence for the magmatic budget of a fast spreading center, *J. Geophys. Res.*, *98*, 7871-7885, 1993.

Scheirer, D. S., and K. C. Macdonald, Near-axis seamounts on the flanks of the East Pacific Rise, 8°N to 17°N, *J. Geophys. Res.*, *100*, 2239-2259, 1995.

Scheirer, D. S., K. C. Macdonald, D. W. Forsyth, and Y. Shen, Abundant seamounts of the Rano Rahi field near the southern East Pacific Rise, 15°S to 19°S, *Mar. Geophys. Res.*, *18*, 13-52, 1996.

Schilling, J.-G., R. H. Kingsley, and J. D. Devine, Galapagos hot spot-spreading center system, 1, Spatial, petrological and geochemical variations (83°W-101°W), *J. Geophys. Res.*, *87*, 5593-5610, 1982.

Schilling, J.-G., M. Zajac, R. Evans, T. Johnston, W. White, D. Devine, and R. Kingsley, Petrologic and geochemical variations along the Mid-Atlantic Ridge from 29°N to 73°N, *Am. J. Sci.*, *283*, 510-586, 1983.

Schreiner, A. E., C. G. Fox, and R. P. Dziak, Spectra and magnitudes of T-waves from the 1993 earthquake swarm on the Juan de Fuca Ridge, *Geophys. Res. Lett.*, *22*, 139-142, 1995.

Scott, D. R., and D. J. Stevenson, A self consistent model of melting, magma migration and bouyancy-driven circulation beneath mid-ocean ridges, *J. Geophys. Res.*, *94*, 2973-2988, 1989.

Sempéré, J. C., et al., The Southeast Indian Ridge between 88°E and 118°E: Variations in crustal accretion at constant spreading rate, *J. Geophys. Res.*, *102*, 15489-15505, 1997.

Sempéré, J. C., J. Lin, H. S. Brown, H. Schouten, and G. M. Purdy, Segmentation and morphotectonic variations along a slow-spreading center: The Mid-Atlantic Ridge (24°00'N-30°40'N), *Mar. Geophys. Res.*, *15*, 153-200, 1993.

Sempéré, J. C., G. M. Purdy, and H. Schouten, Segmentation of the Mid-Atlantic Ridge between 24°N and 30°40'N, *Nature*, *344*, 427-431, 1990.

Sempéré, J. C., B. P. West, and L. Géli, The Southeast Indian Ridge between 127° and 132°40'E: Contrasts in segmentation characteristics and implications for crustal accretion, in *Tectonic, Magmatic, Hydrothermal and Biological Segmentation of Mid-Ocean Ridges, Spec. Pub. 118*, edited by C. J. MacLead, P. Tyler, and C. L. Walker, pp. 1-15, Geological Society of London, 1996.

Shen, Y., and D. W. Forsyth, Geochemical constraints on initial and final depths of melting beneath mid-ocean ridges, *J. Geophys. Res.*, *100*, 2211-2237, 1995.

Shen, Y., D. W. Forsyth, D. S. Scheirer, and K. C. Macdonald, Two forms of volcanism: Implications for mantle flow and off-axis crustal production on the west flank of the southern East Pacific Rise, *J. Geophys. Res.*, *98*, 17875-17889, 1993.

Shen, Y., D. S. Scheirer, D. W. Forsyth, and K. C. Macdonald, Trade-off in production between adjacent seamount chains near the East Pacific Rise, *Nature*, *373*, 140-143, 1995.

Sigurdsson, O., Surface deformation of the Krafla fissure swarm in two rifting events, *J. Geophys.*, *47*, 154-159, 1980.

Sims, K. W. W., D. J. DePaolo, M. T. Murrell, W. S. Baldridge, S. J. Goldstein, and D. A. Clague, Mechanisms of magma generation beneath Hawaii and mid-ocean ridges: U/Th and Sm/Nd isotopic evidence, *Science*, *267*, 508-512, 1995.

Sims, K. W. W., S. J. Goldstein, D. J. Fornari, M. R. Perfit, M. T. Murrell, and S. R. Hart, U-series analyses of young lavas from 9°-10° N East Pacific Rise: Constraints on magma transport and storage times beneath the ridge axis, *Eos Trans. Am. Geophys. Union*, *78*, F792, 1997a.

Sims, K. W. W., S. J. Goldstein, M. R. Perfit, D. J. Fornari, S. R. Hart, M. T. Murrell, and M. C. Smith, Spatial and temporal variations in mid-ocean ridge basalt geochemistry along the East Pacific Rise: U-series disequilibria, radiogenic isotopes and elemental variability in a closely spaced, two-dimensional grid of samples from 9°48'-52'N., in *Seventh Annual Goldschmidt Conf., LPI 921*, pp. 194, Lunar and Planetary Institute, Houston, Texas, 1997b.

Sinton, J. M., and R. S. Detrick, Mid-ocean ridge magma chambers, *J. Geophys. Res.*, *97*, 197-216, 1992.

Sinton, J. M., D. S. Wilson, D. M. Christie, R. N. Hey, and J. R. Delaney, Petrologic consequences of rift propagation on oceanic spreading ridges, *Earth Planet. Sci. Let.*, *62*, 193-207, 1983.

Sinton, J. S., S. M. Smaglik, J. J. Mahoney, and K. C. Macdonald, Magmatic processes at superfast spreading mid-ocean ridges: Glass compositional variations along the East Pacific Rise, 13°-23°S, *J. Geophys. Res.*, *96*, 6133-6155, 1991.

Sleep, N. H., Tapping of melt by veins and dikes, *J. Geophys. Res.*, *93*, 10,255-10,272, 1988.

Smith, D. K., and J. R. Cann, Hundreds of small volcanoes on the median valley floor of the Mid-Atlantic Ridge at 24-30°N, *Nature*, *348*, 152-155, 1990.

Smith, D. K., and J. R. Cann, The role of seamount volcanism in crustal construction at the Mid-Atlantic Ridge (24°-30°N), *J. Geophys. Res.*, *97*, 1645-1658, 1992.

Smith, D. K., and J. R. Cann, Building the crust at the Mid-Atlantic Ridge, *Nature*, *365*, 707-715, 1993.

Smith, D. K., et al., Mid-Atlantic Ridge volcanism from deep-towed side-scan sonar images, 25°-29°N, *J. Volcanol. Geotherm. Res.*, *67*, 233-262, 1995a.

Smith, M. C., M. R. Perfit, and A. L. Heatherington, High-resolution petrochemical mapping of individual eastern Pacific lava flows: Variances and genetic fingerprints, *Eos Trans. Am. Geophys. Union*, *77*, F706, 1996.

Smith, M. C., M. R. Perfit, A. L. Heatherington, I. R. Jonasson, W. I. Ridley, R. W. Embley, and W. W. Chadwick, Elemental and isotopic constraints on the recent spatial and temporal variations in MORB: CoAxial Segment, Juan de Fuca Ridge, *Eos Trans. Am. Geophys. Union*, *76*, F410, 1995b.

Smith, M. C., M. R. Perfit, and I. R. Jonasson, Petrology and geochemistry of basalts from the southern Juan de Fuca Ridge: Controls on the spatial and temporal evolution of mid-ocean ridge basalt, *J. Geophys. Res.*, *99*, 4787-4812, 1994.

Smith, M. C., M. R. Perfit, I. R. Jonasson, and W. I. Ridley, Preliminary petrologic investigations of the new eruption on the CoAxial segment of the Juan de Fuca Ridge, *EOS Trans. Am. Geophys. Union*, *74*, 43, Fall Meeting suppl., 620, 1993.

Solomon, S. C., P. Y. Huang, and L. Meinke, The seismic moment budget of slow spreading ridges, *Nature*, *334*, 58-60, 1988.

Stakes, D. S., J. W. Shervais, and C. A. Hopson, The volcanic-tectonic cycle of the FAMOUS and AMAR valleys, Mid-Atlantic Ridge (36°47'N): Evidence from basalt glass and phenocryst compositional variations for a steady state magma chamber beneath the valley midsections, AMAR 3, *J. Geophys. Res.*, *89*, 6995-7028, 1984.

Swanson, D. A., Pahoehoe flows from the 1969-1971 Mauna Ulu eruption, Kilauea volcano, Hawaii, *Geol. Soc. Am. Bull.*, *84*, 615-626, 1973.

Sykes, L. R., Mechanisms of earthquakes and nature of faulting on the mid-ocean ridges, *J. Geophys. Res.*, *72*, 2131-2153, 1967.

Sykes, L. R., Earthquake swarms and seafloor spreading, *J. Geophys. Res.*, *75*, 6598-6611, 1970.

Takahashi, T. J., and J. D. Griggs, Hawaiian volcanic features: A photoglossary, in *Volcanism in Hawaii, U.S. Geol. Surv. Prof. Pap.*, *1350*, edited by R. W. Decker, T. L. Wright, and P. H. Stauffer, pp. 845-902, 1987.

Thompson, G., W. B. Bryan, R. Ballard, K. Hamuro, and W. G. Melson, Axial processes along a segment of the East Pacific Rise, 10-12°N, *Nature*, *318*, 429-433, 1985.

Thompson, G., W. B. Bryan, and S. E. Humphris, Axial volcanism on the East Pacific Rise, 10-12°N, in *Magmatism in the Ocean Basins, Geological Soc. Spec. Pub. 42*, edited by A. D. Saunders, and M. J. Norry, pp. 181-200, 1989.

Tolstoy, M., A. J. Harding, and J. A. Orcutt, Crustal thickness on the Mid-Atlantic Ridge: Bull's-eye gravity anomalies and focused accretion, *Science*, *262*, 726-729, 1993.

Toomey, D. R., S. C. Solomon, and G. M. Purdy, Tomographic imaging of the shallow crustal structure of the East Pacific Rise at 9°30'N, *J. Geophys. Res.*, *99*, 24,135-24,157, 1994.

Tryggvason, E., Widening of the Krafla fissure swarm during the 1975-1981 volcano-tectonic episode, *Bull. Volcanol.*, *47*, 47-69, 1984.

Turner, R. J. W., D. E. Ames, J. M. Franklin, W. D. Goodfellow, C. H. B. Leitch, and T. Höy, Character of active hydrothermal mounds and nearby altered hemipelagic sediments in the hydrothermal areas of Middle Valley, northern Juan de Fuca Ridge: Data on shallow cores, *Can. Mineral.*, *31*, 973-995, 1993.

Urabe, T., et al., The effect of magmatic activity on hydrothermal venting along the superfast-spreading East Pacific Rise, *Science*, *269*, 1092-1095, 1995.

van Andel, T. H., and R. D. Ballard, The Galapagos rift at 86°W: 2. Volcanism, structure, and evolution of the rift valley, *J. Geophys. Res.*, *84*, 5390-5406, 1979.

Vera, E. E., and J. B. Diebold, Seismic imaging of oceanic layer 2A between 9°30'N and 10°N on the East Pacific Rise from two-ship wide-aperture profiles, *J. Geophys. Res.*, *99*, 3031-3041, 1994.

Volpe, A. M., and S. J. Goldstein, ^{226}Ra-^{230}Th disequilibrium in axial and off-axis mid-ocean ridge basalts, *Geochim. Cosmochim. Acta*, *57*, 1233-1242, 1993.

Weaver, J. S., and C. H. Langmuir, Calculations of phase equilibrium in mineral-melt systems, *Computers and Geosciences*, *16*, 1-19, 1990.

Wentworth, C. K., and G. A. Macdonald, Structures and forms of basaltic rocks in Hawaii, *U.S. Geol. Surv. Bull. 994*, 95 pp., 1953.

White, S. M., K. C. Macdonald, D. S. Scheirer, M. H. Cormier, and R. M. Haymon, Near-axis volcanism at the SEPR: Is the neovolcanic zone 28 kilometers wide?, *EOS Trans. Am. Geophys. Union*, *78*, 46, Fall Meeting suppl., F689, 1997.

Wolfe, E. W., M. O. Garcia, D. B. Jackson, R. Y. Koyanagi, C. A. Neal, and A. T. Okamura, The Puu Oo eruption of Kilauea volcano, episodes 1-20, January 3, 1983, to June 8, 1984, in *Volcanism in Hawaii, U.S. Geol. Surv. Prof. Pap. 1350*, edited by R. W. Decker, T. L. Wright, and P. H. Stauffer, pp. 471-508, 1987.

Wright, D. J., R. M. Haymon, and D. J. Fornari, Crustal fissuring and its relationship to magmatic and hydrothermal processes on the East Pacific Rise crest (9° 12' to 54'N), *J. Geophys. Res.*, *100*, 6097-6120, 1995.

Zierenberg, R. A., R. A. Koski, J. L. Morton, R. M. Bouse, and W. C. Shanks, Genesis of massive sulfide deposits on a sediment-covered spreading center, Escanaba Trough, southern Gorda Ridge, *Econ. Geol.*, *88*, 2069-2098, 1993.

Zierenberg, R. A., J. L. Morton, R. A. Koski, and S. L. Ross, Geologic setting of massive sulfide mineralization in Escanaba Trough, in *Geologic, Hydrothermal, and Biologic Studies at Escanaba Trough, Gorda Ridge, Offshore Northern*

California, U.S. Geol. Surv. Bull. 2022, edited by J. L. Morton, R. A. Zierenberg, and C. A. Reiss, pp. 171-197, 1994.

Zonenshain, L. P., M. I. Kazmin, A. P. Lisitsin, Y. A. Bogdanov, and B. V. Baranov, Tectonics of the Mid-Atlantic rift valley between the TAG and MARK areas (26- 24°N): Evidence for vertical tectonism, *Tectonophysics*, *159*, 1-23, 1989.

Michael R. Perfit, Department of Geology, University of Florida, Gainesville, FL 32611

William W. Chadwick, Jr., Oregon State University, Hatfield Marine Science Center, Newport, OR 97365

Geological Consequences of Dike Intrusion
at Mid-Ocean Ridge Spreading Centers

Daniel Curewitz and Jeffrey A. Karson

Division of Earth and Ocean Sciences, Duke University, Durham NC

Dike intrusion is a fundamental process of crustal accretion at mid-ocean ridge spreading centers. Although many studies of ophiolites and spreading centers treat dikes as passive infillings of tensile fractures, observations and mechanical models from subaerial rift zones demonstrate that widespread, temporally and spatially complex mechanical effects accompany dike intrusion. These effects can potentially have a profound influence on the processes attending seafloor spreading. Specifically, dike intrusion should change the magnitude and orientation of the local stress field as a dike tip propagates along a spreading center. Consequently, dike emplacement could cause significant short-term (days to months), long-term (years to decades), and permanent changes in the structure of the axial crust. The mechanical effects of intrusion of even a single dike from a subaxial magma chamber should influence the entire narrow (<5 km) axial rise and axial depression at fast-spreading ridges. A dike of the same size at a typical slow-spreading ridge should affect only a portion of the wide (>10 km) median valley. At fast-spreading ridges the effects of intrusion are associated with recent microearthquake swarms, graben subsidence, fissures, eruptions, and hydrothermal megaplume events. In contrast, relatively infrequent dike intrusions with only local effects on ridge-axis morphology and geology may be reflected by scattered volcanic edifices and large-scale faulting at slow-spreading ridges. We suggest that repeated dike intrusion will create systematically contrasting crustal assemblages that should reflect the varying relationship between faulting and magmatism at different mid-ocean ridge spreading centers.

1. INTRODUCTION

Plate separation at mid-ocean ridges is accommodated by a combination of magmatic construction and mechanical extension. Globally the relation between these processes varies substantially and the end-members are commonly discussed in terms of fast- and slow-spreading ridges, the former exemplified by the East Pacific Rise (>100 mm/yr - full rate) and the latter by the central Mid-Atlantic Ridge (<30 mm/yr - full rate). One of the most important types of magmatic construction is the intrusion of basaltic or diabasic dikes. These represent conduits that tap basaltic

melt generated in the mantle and commonly stored in crustal magma chambers before they erupt on the seafloor. The continual intrusion of dikes during plate separation is generally considered to be one of the hallmarks of seafloor spreading.

Since the recognition of ophiolite complexes as subaerially exposed fragments of oceanic crust and upper mantle, dense dike swarms and sheeted dike complexes have been considered to be fundamental components of the oceanic crust [*Moores and Vine*, 1971; *Kidd*, 1977]. Individual dikes and sheeted dike complexes have been found by submersible studies of "tectonic windows" into upper crustal structures [*ARCYANA*, 1978; *Auzende et al.*, 1989b; *Francheteau et al.*, 1990; 1992; *Karson et al.*, 1992; *Hekinian et al.*, 1996; *Lawrence et al.*, 1998; *Karson*, this volume]. A sheeted dike complex was also penetrated at Ocean

Faulting and Magmatism at Mid-Ocean Ridges
Geophysical Monograph 106
Copyright 1998 by the American Geophysical Union

Drilling Program Hole 504B, the only deep crustal penetration into the oceanic crust [*Anderson et al.*, 1985b; *Becker*, 1989; *Alt*, 1993; 1995].

Recently, a dike intrusion event on the Juan de Fuca Ridge was inferred from a migrating swarm of earthquakes detected by a submarine seismic network [*Dziak et al.*, 1995; *Fox et al.*, 1995]. This event was spatially and temporally associated with fresh lava flows [*Chadwick et al.*, 1995], hydrothermal megaplumes [*Baker et al.*, 1995; *Lowell and Germanovich*, 1995], and graben subsidence [*Chadwick and Embley*, in press]. Studies of the Juan de Fuca dike intrusion event and similar fissure eruptions, megaplumes, and associated seafloor features along other fast-spreading ridge segments [*Haymon et al.*, 1991; 1993; *Baker*, 1989; 1994; *Fornari et al.*, in press; *Gregg et al.*, in press] have begun to provide new insights into the surface expression of dike intrusion along mid-ocean ridge spreading centers.

Although investigations of dikes and dike intrusion events on the seafloor are in their infancy, the study of these features in subaerial rift zones and exhumed dike complexes is a mature area of research [e.g. *Kidd*, 1977; *Halls and Fahrig*, 1985; *Delaney et al.*, 1986; *Decker et al.*, 1987; *Walker*, 1987; *Gudmundsson*, 1988; 1995; *Rubin and Pollard*, 1988; *Adamides*, 1990; *Malpas et al.*, 1990; *Rubin*, 1992; 1993c; *van Everdingen*, 1995; *Hofton and Foulger*, 1996a, b]. These observations and theoretical studies provide a basis for understanding dike intrusions in the oceanic crust and anticipating their role in seafloor spreading.

In this paper we explore the effects of dike intrusion in seafloor spreading environments by combining the substantial body of research on dike intrusion in subaerial rifts with available constraints on the geology of mid-ocean ridges and oceanic crust. We begin with a very brief review of key aspects of the mechanical and thermal parameters attending dike intrusion. These indicate that dike intrusion cannot be regarded simply as the passive filling of an existing crack with magma, as it has in nearly all studies of oceanic crust. Instead, dike intrusion is a forceful magmatic injection that can result in dike-tip propagation and dike lengthening, as well as substantial changes in surrounding crustal materials. Applying these results to mid-ocean ridges we consider the short-term (hours to months) and long-term (years to decades) effects of dike intrusion in different oceanic spreading environments. We summarize these effects by describing the time sequence of processes that collectively represent a dike intrusion event. Finally, we discuss systematic differences in crustal architecture and hydrothermal processes in end-member spreading environments that may be directly related to variations in the frequency and mechanical impact of dike intrusion in the context of these specific settings.

2. PHYSICAL ASPECTS OF DIKE INTRUSION

The following brief review of dike intrusion processes is not intended to be a complete summary of previous work. Such a summary would be well beyond the scope of this paper, but readers are referred to key citations for more details. We emphasize the mechanical and thermal effects known from theoretical and observational studies of dikes that we believe to be most relevant to dike intrusion along spreading centers.

2.1. Mechanics

Dike intrusion events have been studied in Iceland [*Bjornsson et al.*, 1979; *Opheim and Gudmundsson*, 1989; *Hofton and Foulger*, 1996a,b], Afar [*Abdallah et al.*, 1979], the Juan de Fuca Ridge [*Dziak et al.*, 1995], and Hawaii [*Klein et al.*, 1987]. These studies reveal several systematic relationships that arise during and immediately after dike injection.

During intrusion, seismicity migrates at velocities of several kilometers per day and marks the progress of the dike as it propagates along the rift zone [*Brandsdottir and Einarsson*, 1979; *Klein et al.*, 1987; *Rubin and Pollard*, 1988; *Dziak et al.*, 1995]. Geodetic measurements have shown that subsidence and fault-slip occur ahead of and above the dike tip-line (the three-dimensional termination of the dike) and that uplift and inhibition of fault slip occur on either side of the dike [*Rubin and Pollard*, 1988; *Rubin*, 1992; *Hofton and Foulger*, 1996a, b]. The areal extent of surface deformation related to dike intrusion is variable, but measurements show that uplifted flanks of rift zones extend tens of kilometers to either side of a dike injected into the shallow crust [*Rubin*, 1992; *Hofton and Foulger*, 1996a]. Normal faulting, fissuring, and subsidence related to dike-tip propagation are more limited in areal extent, and are generally confined to graben directly above or ahead of the dike [*Rubin and Pollard*, 1988; *Rubin*, 1992].

Modeling of geodetic data is used to infer dike dimensions and the orientation and magnitude of stresses imposed by dike intrusion [*Rubin and Pollard*, 1987; 1988; *Rubin*, 1992; *Hofton and Foulger*, 1996a, b]. These models suggest that dikes in subaerial rift zones are "blade-like", with horizontal width to length ratios of 10^{-2} to 10^{-4} (lengths of several tens of kilometers and widths less than 10 m). Dike heights are estimated at several kilometers and generally less than 10 km. These dikes are treated as internally pressurized Mode I cracks in an elastic medium, and mean normal stress changes that arise as a result of dike intrusion are proportional to dike length, dike height, and internal driving pressure (magma pressure) [*Pollard and Segall*, 1987; *Rubin*, 1992; 1995]. Tensile stress is concentrated in a region around the dike tip-line. The dimensions of the tip-line region and the magnitude of tensile stresses concentrated there are determined by the length of the dike (perpendicular to the tip-line) and internal driving pressure in the dike [*Delaney et al.*, 1986; *Rubin*, 1993c; *Rubin*, 1995]. Compressive stress is concentrated to either side of the dike in proportion to the internal driving pressure and the height of the dike [*Rubin and Pollard*, 1988; *Rubin*, 1992, A. Rubin, pers. commun.].

The internal magma pressure in a basaltic dike is related to four parameters: 1) excess pressure at the magma source, 2) density contrast between the magma and country rock (buoyancy), 3) lithostatic stress, specifically the increase of lithostatic stress with depth [*Parsons et al.*, 1992], and 4) tectonic stress [*Rubin and Pollard*, 1987; 1988; *Rubin*, 1995]. Additional changes in magma pressure may arise from magmatic processes such as the escape of volatiles, vesiculation within the dike, or flow of pore fluids resulting from changes in dike-tip geometry [*Rubin*, 1993c]. *Rubin and Pollard* [1987] use an internal magma pressure of between 2 and 10 MPa in their analysis of dike intrusion in Hawaii, and this value is used in several other analyses [*Rubin*, 1993c; 1995; *Hofton and Foulger*, 1996a, b]. We use 10 MPa throughout this paper, broadly in accord with theoretical constraints [*Parsons et al.*, 1992], calculated lava extrusion rates [*Gregg and Fink*, 1992], observed flow morphologies [*Head et al.*, 1996], and interpretations of geodetic uplift data in rift zones [*Rubin and Pollard*, 1987; 1988; *Rubin*, 1992; *Hofton and Foulger*, 1996a, b].

The dependence of compressive stress perturbation on dike height is a function of its shape. Dikes are typically much longer than they are high, therefore most areas near a dike are closer to its upper and lower tip-lines, and farther from the tip-lines at its horizontal extremes. Therefore, the magnitude of the mean normal stress perturbation at most points is more strongly influenced by the distance from the upper and lower edges of the dike, or a function of dike height. [A. Rubin, pers. commun.]. Thus, for a dike with dimensions of 1 m wide, 60 km long, 1.5 km high [after *Dziak et al.*, 1995], stress perturbations (expressed as surface deformation) that are a significant fraction of the internal magma pressure should extend out to horizontal distances (normal to the dike) of approximately 1.5 km (Figure 1). Here we use a value of 2 MPa as a lower limit of significant compressional stress.

The magnitude and areal extent of stress concentrations that can result from dike intrusion may also depend upon host rock properties, for instance, variations in elastic parameters caused by the presence of fluid-filled fractures and pore spaces [*O'Connell and Budiansky*, 1977; *Agnon and Lyakhovsky*, 1995]. Increased fracture density in the host rock decreases the effective elastic moduli [*Jaeger and Cook*, 1969], and may limit the areal extent and magnitude

of stress perturbations caused by intrusion. Fluid saturation in these fractures could counteract this effect depending on the compressibility of the fluid and the connectivity of the fracture network [*O'Connell and Budiansky*, 1977]. In general, the presence of fractures in country rock intruded by dikes should decrease the magnitude and therefore the lateral extent of significant stress changes and may promote distributed horizontal shortening rather than uplift of the surrounding rock.

2.2. Thermal Effects of Dike Intrusion

The intrusion of a 1-m-wide basaltic dike (temperature ~1200°C) into the shallow oceanic crust could have significant thermal effects on the rocks within 2 m of its margin [*Delaney*, 1987]. The dike should solidify within about 10 days of the cessation of magma flow [*Turcotte and Schubert*, 1982]. During the initial period of cooling the host rock within 2 m of the dike (and any fluids in fractures and pores) is conductively and advectively heated to temperatures of 350° to 400°C [*Lowell and Germanovich*, 1995; *Cherkaoui et al.*, 1997]. The circulation of hydrothermal fluids and concomitant advective cooling should limit the width of the hydrothermally altered zone of country rock near the dike margins to less than about two dike widths [*Lowell and Germanovich*, 1995].

Using conductive heat loss alone, a 1-m-wide dike should cool to ambient shallow crustal temperatures within 2 years [*Delaney*, 1987]. If advective cooling as a result of fluid circulation is taken into account, the time to cool to ambient temperatures decreases to less than a year, depending on the thermal properties of the fluid and the permeability of the host rock [*Cherkaoui et al.*, 1997]. Observations [*Embley et al.*, 1994] and models [*Cherkaoui et al.*, 1997] of hydrothermal flow at recent dike intrusion sites suggest that thermally driven circulation at seafloor black smoker vents ceases within three months of dike injection.

2.3. Sequence of Events During Dike Intrusion

Using the relationships outlined above, we now list the sequence of mechanical and thermal effects that would be expected to attend a dike intrusion event in oceanic crust. We do this by describing the effects that would occur on various time scales at a reference point along the path of dike propagation (see Figure 2).

During a dike intrusion event, the dike tip propagates at velocities of several kilometers per day [*Rubin and Pollard*, 1988; *Dziak et al.*, 1995]. Therefore, dike-normal tensile stress and extension at any given point should last only for a few minutes or hours, depending upon the propagation velocity of the dike and the size of the dike-tip region

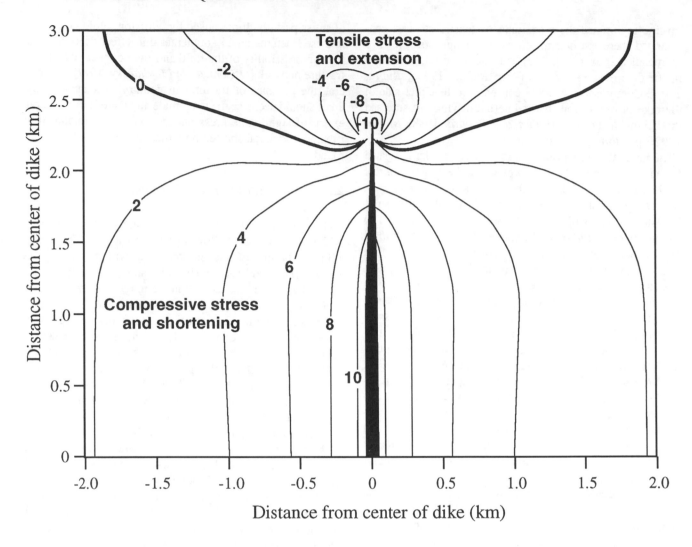

Figure 1. Schematic cross-section showing contours of mean normal stress that arise as a result of dike intrusion (after *Pollard and Segall*, 1987; *Rubin and Pollard*, 1988). Stress, in MPa, (compressive stress, positive values; tensile stress, negative values) is a function of magma pressure and dike geometry. In this example, internal magma pressure is 10 MPa above lithostatic, and compressive stress increases of at least 2MPa extend out to distances that scale with the height of the dike. Note that the diagram may be viewed as a vertical section through the crust for vertical dike propagation, or as a horizontal section through the crust for lateral dike propagation.

(Figure 2A, B). During this period of time, tensile fractures parallel to the plane of the dike would tend to open, while slip would be promoted on favorably oriented faults and fractures [*Rubin and Pollard*, 1988; *Mastin and Pollard*, 1988; *Rubin*, 1992].

Once the dike has propagated past the reference point, dike-tip tensile stresses are succeeded by dike-normal compressive stresses [*Delaney et al.*, 1986; *Pollard and Segall*, 1987; *Rubin and Pollard*, 1988]. These initially increase rapidly as the dike reaches its full width (Figure 2C). At

the same time, open cracks parallel to the dike margins would tend to close while fractures and faults should cease to slip or reverse their sense of slip depending upon their orientation [*Rubin and Pollard*, 1988; *Mastin and Pollard*, 1988; *Rubin*, 1992].

While magma flows in a dike, hydrothermal circulation and elevated pore pressures may arise as a result of heating of the surrounding rock and fluids in fracture networks and pore spaces [*Lowell and Germanovich*, 1995; *Cherkaoui et al.*, 1995] (Figure 2D). Hydrothermal activity should cease

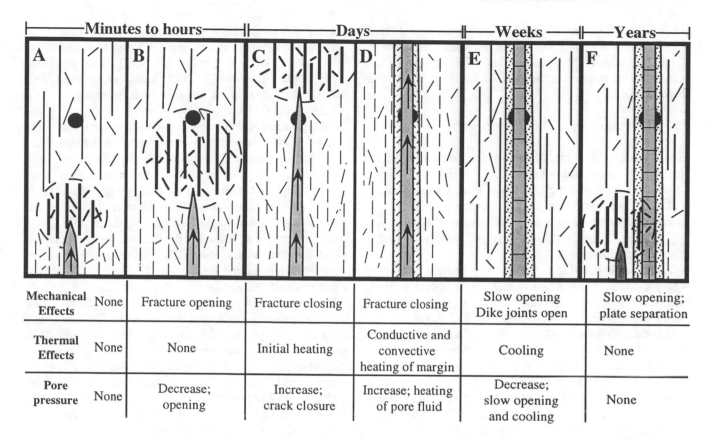

	Minutes to hours		Days		Weeks	Years
	A	B	C	D	E	F

	A	B	C	D	E	F
Mechanical Effects	None	Fracture opening	Fracture closing	Fracture closing	Slow opening Dike joints open	Slow opening; plate separation
Thermal Effects	None	None	Initial heating	Conductive and convective heating of margin	Cooling	None
Pore pressure	None	Decrease; opening	Increase; crack closure	Increase; heating of pore fluid	Decrease; slow opening and cooling	None

Figure 2. Schematic diagram (plan view) showing the effects of a lateral dike intrusion event at a stationary reference point in the crust (black dot). Heavy lines represent open fractures; dashed lines - actively closing fractures; thin lines - closed fractures; arrows in dikes - magma transport direction; dashed pattern - hydrothermal circulation and metamorphism near dike margin; stippled pattern - cemented hydrothermal breccia along dike margin. A. Dike tip still far from reference point. B. Zone of increased tensile stress around dike tip reaches reference point. The crust undergoes dilation, dike-parallel joints open and fractures with other orientations open or slip as the dike tip approaches the reference point. C. Dike tip propagates through the reference point and dike-normal compressive stress arising from dike inflation and magma pressure begins to "squeeze" the crust on either side of the dike. Dike-parallel joints tend to close and fractures in other would respond to this new stress field. D. Magma transport continues through the dike over a period of days or weeks. Fracture networks along the margin of the dike may host vigorous hydrothermal circulation, permeability may increase due to elevated pore pressure. Compressive stress may vary with fluctuations in magma pressure. E. Magma ceases flowing through the dike and over a period of weeks to months the dike cools to ambient temperatures. Hydrothermal minerals precipitate in the fractures adjacent to the dike, forming a selvedge of cemented hydrothermal breccia along the dike margin. Cooling and contraction of the dike and surrounding rocks results in the formation or re-opening of dike-parallel tensile cracks. F. Dike is fully solidified and cooled, plate separation relieves the dike-related compressive stress, and dike-parallel fractures open. The crust near spreading centers, including previously intruded dikes, should experience repeated dike intrusion events.

and elevated pore pressure should decay to background levels after several months [*Cherkaoui et al.*, 1997]. Hydrothermal precipitates in vein networks along dike margins, as well as altered dike rock and country rock may provide a record of elevated hydrothermal activity near dikes

[*Bird et al.*, 1985; *Rose and Bird*, 1994] (Figure 2E).

Over several years, continued plate separation, opening of thermal contraction joints, and slip along faults should effectively reduce the stresses imposed by dike intrusion to ambient values that reflect the regional state of stress

[*Hofton and Foulger*, 1996a, b]. In the axial region of mid-ocean ridge spreading centers, the sequence of events described above may be repeated over years or decades (see section 3.4) if more dikes intrude the same crustal volume (Figure 2F).

2.4. Geological Record of Dike Intrusion.

Intrusion of a dike can cause irreversible changes in the crust that are preserved in the geologic record of rift zones. Below we summarize the permanent effects accompanying addition of the dike to the crust. We use these relationships later in the paper as a basis for discussing the geological implications of repeated dike intrusion in different oceanic spreading environments.

Mafic dikes exposed in subaerial rift zones are commonly highly elongate and narrow, generally in accord with dimensions inferred from geodetic data [*Gudmundsson*, 1983; *Rubin and Pollard*, 1987; *Walker*, 1987; *Rubin*, 1995]. Relative dimensions of dikes in oceanic crust are likely to be similar based on limited seafloor observations [*Francheteau et al.*, 1990; 1992; *Karson et al.*, 1992; *Lawrence et al.*, 1998; *Karson*, this volume], one geophysical study [*Dziak et al.*, 1995], and studies of ophiolites [*Kidd*, 1977; *Malpas et al.*, 1990; *Nehlig*, 1994]. They commonly consist of massive, relatively fine-grained basaltic rock [*Wada*, 1994; *Gillis*, 1995 and references therein], with margin-perpendicular contraction joints [*van Everdingen*, 1995] and margin-parallel chemical and textural zonations [*Platten and Watterson*, 1987]. Dikes can partition the crust into hydrologically and mechanically isolated blocks [*Ryan et al.*, 1996]. They commonly act as barriers to horizontal groundwater flow, with spring discharge sites and aquifer characteristics reflecting the geometry of dike complexes [*Takasaki and Mink*, 1985; *Ingebritsen and Scholl*, 1993; *Scholl et al.*, 1996].

Fault slip, subsidence, and concentrated crustal fissuring are likely to accompany each successive intrusion and may result in the development of a well-defined rift-graben [*Rubin and Pollard*, 1988; *Opheim and Gudmundsson*, 1989; *Rubin*, 1992; *Wright et al.*, 1995a, b; *Hofton and Foulger*, 1996a, b; *Chadwick and Embley*, in press]. In competent rock units sets of dike-parallel joints may form near dike margins [*Delaney et al.*, 1986; *Reches and Fink*, 1988], or pre-existing fractures could be reactivated [*Gudmundsson*, 1986; 1995] (Figure 3A, B). The cumulative effect of superimposed dike intrusions in oceanic crust could be to develop highly fractured and faulted basaltic material in the axial region. Older parcels of basaltic lava and dikes have a greater potential for being affected by subsequent dike intrusions and therefore should have higher fracture densities. These general relationships have been observed in subaerial and oceanic rift zones [*Francheteau et al.*, 1990; 1992; *Karson et al.*, 1992; *Gudmundsson*, 1995], ophiolites [*Nehlig*, 1994; *van Everdingen*, 1995], and in deep drilling of oceanic crust [*Anderson et al.*, 1985b; *Becker*, 1989].

Hydrothermal alteration of dike rock and host rock, and hydrothermal precipitates in vein networks along dike margins reflect elevated hydrothermal activity near dikes (Figure 3C, D). Studies of dikes and sheeted dike complexes in ophiolites [*Adamides*, 1990], young oceanic crust [*Gillis*, 1995], and continental rifts [*Bird et al.*, 1985; *Rose and Bird*, 1994] show high-temperature (300° to 600°C) alteration near dikes. This takes the form of relatively narrow (1 to 10 dike widths) dike-parallel zones of intense hydrothermal metamorphism and veining that cut across pre-existing structures and metamorphic zones.

The cumulative effects of repeated intrusion of dikes with a strong preferred orientation may be to impart significant anisotropy of permeability to the affected volume. This permeability anisotropy has been inferred in many areas of current and ancient rifting and dike intrusion, including hot spots [*Ryan*, 1988; 1990], exhumed ophiolites [*Nehlig*, 1994; *van Everdingen*, 1995], and mid-ocean ridges [*Haymon et al.*, 1991; *Wright et al.*, 1995a; *Haymon*, 1996; *Wilcock and McNabb*, 1996].

3. FRAMEWORK OF DIKE INTRUSION AT SPREADING CENTERS

After considering dike intrusion into oceanic crust in a generic sense, we now consider the specific consequences of dike intrusion in two end-member spreading environments. We use available constraints from geological and geophysical studies of fast- and slow-spreading end-members of mid-ocean ridges to contrast the role of dike intrusion in these settings.

3.1. Fast-Spreading Ridges

Fast-spreading ridges (FSRs) are characterized by an axial rise (~5 km wide, ~2 km high) with a central axial depression (<200 m deep, <1 km wide) [*Macdonald*, 1982; *Scheirer and Macdonald*, 1993], and small (<9 km long, <40 m high) normal fault scarps [*Cowie et al.*, 1993; *Malinverno and Cowie*, 1993]. Earthquakes are rare and generally less than M = 4.0, accounting for less than 1% of the strain accommodation estimated in these environments [*Cowie et al.*, 1993]. Volcanic activity is concentrated in the axial depression [*Haymon et al.*, 1991; 1993; *Chadwick and Embley*, in press; *Fornari et al.*, in press] and

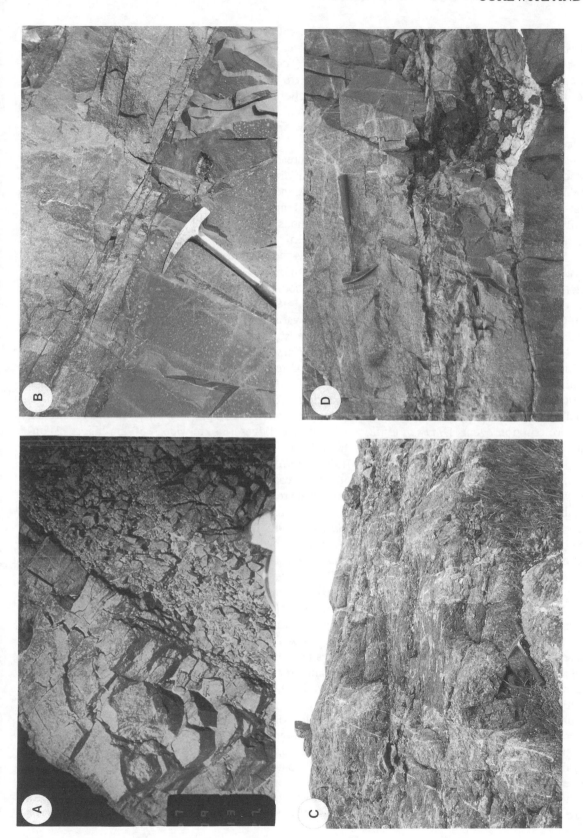

Figure 3. A. Diabase dike intruding altered and brecciated basaltic lava. Note intense fracturing in basalts and both margin-parallel and high-angle joints in the dike. Mid-Atlantic Ridge at 22°N. Dike is ~ 1.5 m wide. B. Chilled margin of diabase dike (beneath hammer head) intrudes an older, fractured diabase dike. Note closely spaced margin-parallel joints in the older dike. Tasilaq Fjord, East Greenland. C. Sheeted dike complex with about 6 parallel dikes. Note pale weathering veins and altered selvages of each dike margin. Blow me down Mountain massif, Bay of Islands Ophiolite Complex, Newfoundland. D. Margin-parallel fractures, net veins, and pale weathering hydrothermal breccia at margin between two diabase dikes. Fladø, East Greenland.

commonly takes the form of smooth-surfaced sheet and lobate flows that are areally extensive (~10 km long, 200 m wide) and generally <10 m thick [*Bonatti and Harrison,* 1988; *Gregg and Fink,* 1995; *Gregg et al.,* in press]. Fissure eruptions have been documented in several areas [*Haymon et al.,* 1991; 1993; *Gregg et al.,* in press] and resurfacing of the axial depression by lava flows has been documented using successive SeaBeam surveys and submersible investigations [*Chadwick et al.,* 1995].

Most hydrothermal activity is concentrated within the axial graben at FSRs [*Haymon et al.,* 1991; 1993; *Chadwick and Embley,* in press; *Fornari et al.,* in press]. Hydrothermal vents are generally less than 10 to 20 m in diameter and less than 10 m high [*Embley and Chadwick,* 1994; *Koski et al.,* 1994]. Maps of hydrothermal fields along FSRs show that active vents are concentrated near the tips of fresh, axis-parallel, elongate lava flows and pillow mounds, while extinct hydrothermal vents are commonly found to either side of freshly erupted lavas [*Haymon et al.,* 1991; *Embley and Chadwick,* 1994; *Koski et al.,* 1994; *Fornari and Embley,* 1995]. Individual hydrothermal vents appear to be active for only 10 to 50 years [*Koski et al.,* 1994], and sites of hydrothermal activity appear to migrate or relocate on similar time scales. This apparent reconfiguration of hydrothermal plumbing is thought to be related to dike intrusion and fissure eruption events [*Haymon et al.,* 1991; 1993; *Chadwick et al.,* 1995; *Fornari et al.,* in press].

Geological investigations at Hess Deep [*Auzende et al.,* 1989b; *Francheteau et al.,* 1990; 1992; *Karson et al.,* 1992], ODP Hole 504b [*Anderson et al.,* 1985b; *Alt et al.,* 1993], and ophiolites [*Kidd,* 1977] show that the vertical dimension of sheeted dike complexes ranges from 0.5 to 1.5 km in fast-spread crust. The height of individual dikes has not been directly determined; however, limitations on dike heights can be inferred from other relationships. Seismic studies of the fast-spreading EPR show the thickness of crust above an axial magma chamber is about 1.5 km [*Sinton and Detrick,* 1992]. Dikes derived from axial magma chambers beneath the axial rise of FSRs should be limited to this height. Furthermore, since there are few dikes in gabbroic rocks underlying sheeted dike complexes, most dikes must be intruded above the dike/gabbro contact.

Collectively, these observations suggest that typical dike heights at FSRs are commonly 1.5 km. Using this height and the relationship between the vertical dike dimension and the areal extent of dike-induced horizontal stress described above, we infer the impact of dike intrusion. The across-axis width of the area affected by a 1.5-km-high dike with internal pressure of 10 MPa should be ~3.0 km to ~4.0 km, with stresses of <2 MPa defining the limit of the affected area. The axial region of FSRs is typically 3 to 5 km wide [*Scheirer and Macdonald,* 1993], therefore dike intrusions are likely to have significant mechanical effects over most or all of this area.

3.2. Slow-Spreading Ridges

Slow-spreading ridges (SSRs) are characterized by a wide (~1500 m deep, >10 km wide) median valley [*Macdonald,* 1982; *Sempéré and Macdonald,* 1987], bounded by large (>50 km long, >200 m high) normal fault scarps that form mountainous rift-valley walls [*Searle and Laughton,* 1977; *Purdy et al.,* 1991; *Sempéré et al.,* 1993; *Shaw and Lin,* 1993]. Earthquakes reach M = 6.0 along SSRs, and have been estimated to account for up to 20% of the horizontal strain accommodation in these environments [*Huang and Solomon,* 1988]. On the scale of 10^6 yr, faulting accommodates variable amounts of plate separation along different ridge segments, probably as a function of magma budget [*Karson et al.,* 1987]. Estimates of plate separation accommodated by faulting vary from ~5% at Reykjanes Ridge to ~100% at the Mid-Atlantic Ridge at the Kane Transform [*Karson and Winters,* 1992].

Volcanic activity takes place on seamounts and along axial volcanic ridges (<5-km-wide, <50-km-long, <1-km-high) in the floor of the median valley [*Smith and Cann,* 1992; *Smith et al.,* 1995; *Head et al.,* 1996]. Lava flows in these areas commonly consist of highly fractured pillow lavas that are consolidated into mounds and elongate ridges, and only rarely take the form of sheet flows [*Bonatti and Harrison,* 1988; *Head et al.,* 1996 and references therein]. Volcanic mounds are commonly highly sedimented [*Ballard and Moore,* 1976; *Stakes et al.,* 1984; *Brown and Karson,* 1988; *Eberhart et al.,* 1988], and have a wide range of inferred ages based on interbedding with dated hydrothermal deposits [*Lalou, et al.,* 1993; 1995].

Hydrothermal activity at SSRs is found near neovolcanic ridges (possible dike intrusion sites), or along major fault zones bounding the median valley (away from any young volcanic rocks) [*Karson and Brown,* 1988; *Karson and Rona,* 1990; *Fouquet et al.,* 1993; 1994; *German et al.,* 1995; *Langmuir et al.,* 1997]. Hydrothermal mounds attain maximum dimensions of up to hundreds of meters in diameter and up to 50 m high [*Karson and Brown,* 1988; *Rona et al.,* 1993], and activity at individual vents or mounds spans up to 10^5 years [*Lalou et al.,* 1993; 1995]. Hydrothermal vents are generally much larger, longer-lived, and more isolated than in hydrothermal systems at FSRs [*Fornari and Embley,* 1995]. These characteristics suggest that structural rather than magmatic processes control hydrothermal activity in these settings [*German et al.*, 1995; *Wilcock and Delaney*, 1996; *Curewitz and Karson*, 1997].

Studies of slow-spread crust show that dikes are found intruding lower crustal gabbroic rocks and serpentinized upper mantle rocks [*Karson*, this volume], indicating that dike intrusion could take place in the lower reaches of the crust or even in the upper mantle, unless the plutonic rocks were uplifted prior to intrusion. In addition, magma chambers have not been identified at SSRs and are likely to be rare or absent [*Sinton and Detrick*, 1992]. Thus, shallow crustal intrusion as at FSRs is probably not common at SSRs. In some ridge segments, earthquakes occur at upper mantle depths suggesting a relatively cool, brittle, axial crust 5-7 km thick [*Toomey et al.*, 1985]. These observations suggest that dike heights in slow-spreading environments may reach 5.0 km or more (the full thickness of the crust); however, they provide no direct evidence of the crustal depth of dike intrusion. Models of magma source depths associated with axial seamount elevations at SSRs [*Smith and Cann*, 1992] suggest that dike heights range from 1.5 to 5.0 km, and most commonly 1.5 to 3.0 km. The area affected by a 1.5 to 3.0-km-high dike with internal pressure of 10 MPa should be 3.0 to 6.0 km in across-axis width, with stresses of <2 MPa defining the limit of the affected area. Thus, dike intrusion is likely to have significant mechanical effects over an area that is generally smaller than the >10-km-wide median valleys of SSRs.

3.3. The Effects of Fractured Oceanic Crust

The areal extent of significant dike-imposed stresses in different spreading environments are estimated above by assuming that the surrounding rock behaves elastically. However, many studies show that the basaltic material in the upper 2 km of oceanic crust is highly fractured [*Batiza*, 1990; *Fryer and Wilkins*, 1990; *Karson et al.*, 1992; *Christeson et al.*, 1994; *Wright et al.*, 1995a, b], and therefore cannot behave perfectly elastically. Instead, dike intrusion should tend to close fractures oriented parallel to the plane of the dike, open fractures oriented perpendicular to the dike, and cause slip on fractures at an angle to the dike. Considering the expected strong preferred orientation of cracks parallel to the ridge axis [*Wright et al.*, 1995a], dike-normal compressive stresses of 2 MPa should cause significant fracture closure. This value is based on laboratory experiments at low confining pressure that have shown that 60% of the fractures in a cracked laboratory specimen close under loads of this magnitude [*Brown and Scholz*, 1986]. Using measurements or estimates of the

elastic moduli of axial crust at mid-ocean ridge spreading centers [*Anderson et al.*, 1985b; *Becker et al.*, 1989; *Alt et al.*, 1993; *Wilcock and McNabb*, 1996] and dike dimensions described above, dike-normal compressive stress greater than 2 MPa would typically be expected up to 1.5 km away from a dike in fast-spread crust and up to 3.0 km away from a dike in slow-spread crust. Therefore, significant horizontal compressive and tensile stresses that result from dike intrusion should affect the entire axial graben and most of the axial rise at FSRs, but may have only local effects in the median valleys of SSRs (Figure 4).

3.4. Frequency of Dike Intrusion Events at Different Spreading Rates

If all plate separation at mid-ocean ridges is assumed to be accommodated by dike intrusion (an obvious oversimplification) then for a given dike width the average frequency of dike intrusion can be estimated. Typical dike widths observed in ophiolites and in seafloor exposures are about 1 m, so for a fast-spreading rate of 120 mm/yr a dike would be intruded about every 8 years. Similarly, for a slow-spreading rate of 20 mm/yr a dike would be intruded about every 50 years. An alternate calculation based upon plate separation rate, the width of the zone of active plate separation, and the elastic characteristics of the crust [*Gudmundsson*, 1988b; 1995] results in an estimate of 1 dike intruded every 90 to 150 years in Iceland. Applying this calculation to mid-ocean ridges results in estimates of 1 dike every 1.5 to 10 years at FSRs, and 1 dike every 45 to 100 years at SSRs, in accord with the simpler calculation.

If the importance of mechanical accommodation of plate separation expressed as faulting is taken into account, estimates of dike intrusion frequency change. At FSRs, where mechanical extension may comprise less than 10% of the total plate separation [*Bicknell et al.*, 1987; *Cowie et al.*, 1993], 8 years per dike intrusion is probably a reasonable value. This estimate is also in accord with the periodicity of hydrothermal megaplumes associated with dike intrusions at FSRs [*Baker*, 1994; *Chadwick et al.*, 1995].

In contrast, at SSRs, a substantial amount of plate separation is taken up by faulting [*Karson and Winters*, 1992; *Mutter and Karson*, 1992; *Shaw*, 1992; *Shaw and Lin*, 1993; *McAllister and Cann*, 1996]. Recent investigations have revealed the existence of oceanic core complexes and low-angle detachment faults with apparent horizontal displacements of >10 km, and slip histories spanning ~1 Ma [*Karson et al.*, 1987; *Cann et al.*, 1997; *Tucholke et al.*,

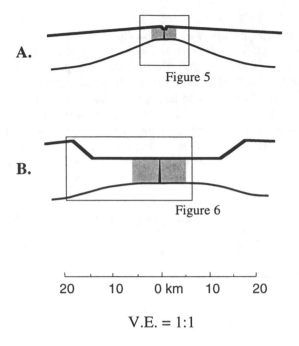

A.

Figure 5

B.

Figure 6

20   10   0 km   10   20

V.E. = 1:1

Figure 4. Schematic representation of end-member spreading environments with highly generalized morphology. The approximate area affected by the intrusion of a dike (black wedge) is shaded gray. Dike height controls the cross-axis width of the area effected. The height is determined by the thickness of crust above the magma supply: typically the depth to the axial magma chamber (FSRs) or the thickness of the crust (SSRs). A. Fast-spreading ridge with < 1-km-wide, 40-m-deep axial graben, and a 5 to 8 km wide axial rise. B. Slow-spreading ridge with 10- to 20-km-wide, 1.5-km-deep axial valley. Theoretically, a single dike imposes stress that affects the entire axial graben in fast-spreading environments. In contrast, in slow-spreading environments dikes are likely to have only a local effect within the broad rift valley. The inset boxes correspond to Figures 5 and 6.

1997]. Additionally, direct observations of rock units exposed in "tectonic windows" into oceanic crust have revealed that sheeted dike complexes, common in ophiolites and at FSRs, may be rare or absent in slow-spread crust [*Karson,* this volume]. These observations suggest that at SSRs dikes accommodate much less of the total plate separation than at FSRs. Therefore, the time interval between dike intrusions could be much longer than 50 years in magma starved areas of SSRs where mechanical extension apparently dominates over magmatic construction. In other words, magma budget and dike intrusion frequency are not proportional to spreading rate.

4. DISCUSSION

Dike intrusion events are likely to have greater mechanical and geological impact on the crust at FSRs than at SSRs for two reasons: 1) the axial region at FSRs is much narrower than at SSRs relative to expected dike heights and corresponding areas affected by dike intrusion, and 2) dike intrusions per unit of plate separation should be more frequent at FSRs than at SSRs [*Chadwick and Embley,* in press]. The intrusion of even a single 1-m-wide dike and its attendant changes in the local stress regime should affect an area larger than the width of the axial region at FSRs (Figure 5), but much narrower than typical rift valleys of SSRs (Figure 6). A comparison of geological traits of end-member spreading environments (Table 1) shows that there are significant variations in the shape, size, and lifespan of hydrothermal vents in these different settings. There are also significant differences in the morphology and distribution of lava flows and other volcanic edifices; and fault structures and fault distribution are markedly dissimilar in these different spreading environments. We suggest that these variations are at least in part a result of differences in the frequency and relative impact of dike intrusions between FSRs and SSRs.

4.1. Consequences of Dike Intrusion at Fast-Spreading Ridges

4.1.1. Hydrothermal activity. Frequent dike intrusions along individual FSR segments could result in the redistribution of permeability with each successive dike intrusion event, and may account for the general pattern of widely scattered, short-lived hydrothermal vents on FSRs. Propagation of a dike along the axial rift may open fractures in areas ahead of the dike and close fractures to either side of it. Redistribution of hydrothermal activity may occur if new vent sites form as dike-tip tensile stresses open new pathways for hydrothermal flow ahead of and above the dike, while old vents are shut off by dike-normal compression to either side of the dike. These effects could explain the observed distribution of active and extinct hydrothermal vents with respect to the position of recently intruded dikes and erupted lavas (Figure 5).

4.1.2. Volcanism. Frequent dike intrusions shallow in the crust at FSRs are suggested by widespread, relatively recent fissure eruptions along many ridge segments. Smooth-surfaced, lobate lava flow morphology may reflect high lava eruption rates (> 10^1 m^3/s) [*Gregg and Fink,* 1995; *Head et al.,* 1996], and may indicate relatively wide (>1.5 m) dikes and relatively high magma pressures

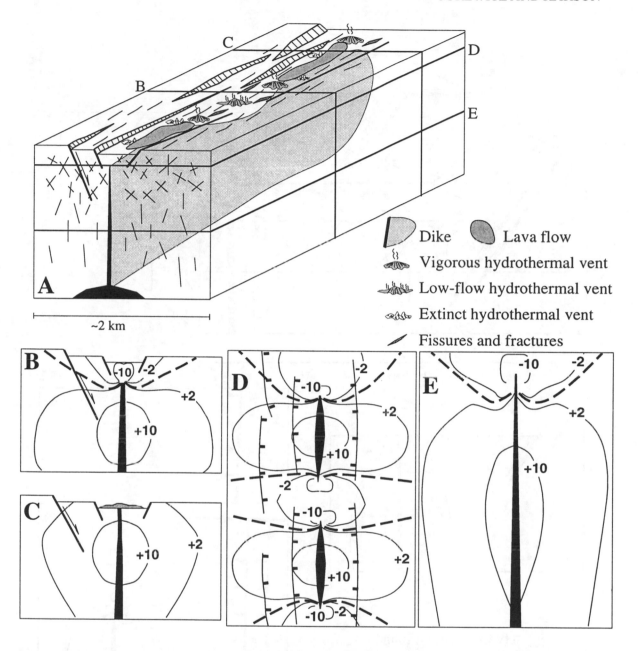

Figure 5. A. Schematic block-diagram showing a fast-spreading ridge intruded by a 1-m-wide, 1.5-km-high dike that is several times longer than it is high, corresponding to the inset in Figure 2. The spatial distribution of vigorous, low-flow, and extinct hydrothermal systems across the axial valley may reflect the influence of dike-induced stress changes affecting the entire area. B. Vertical section across an area where a dike does not reach the surface; tensile stress (in MPa) is concentrated in the overlying crust, compressive stress concentrated lower in the crust. C. Vertical section across an area where the dike reaches the surface and erupts as a lava flow; compressive stress field reaches the surface. D. Horizontal section through the axial rise at a depth of approximately 100 m. Dike geometry gives rise to heterogeneous stress distribution due to vertical and lateral propagation. As dike tips approach one another, tip-line stress fields interfere and give rise to heightened mechanical effects. E. Horizontal section through the axial graben at a depth of approximately 1 km; compressive stress normal to the dike margin dominates, except at the dike tip (stress contours after *Pollard and Segall*, 1987; *Rubin and Pollard*, 1988).

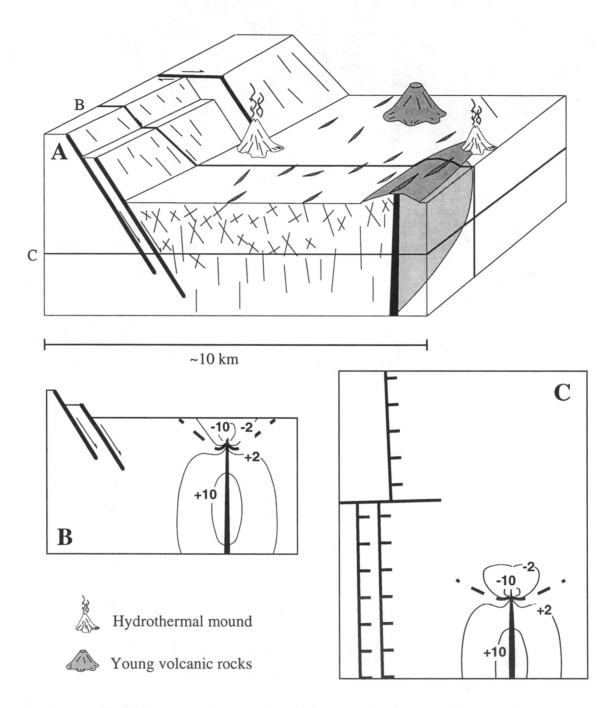

Figure 6. A. Schematic block-diagram showing a slow-spreading ridge intruded by a 1-m-wide, 5.0-km-high dike that is several times longer than it is high, corresponding to the inset in Figure 2. Hydrothermal mounds occur in the neovolcanic zone and along faults bounding the median valley. B. Vertical section showing the mean normal stress increases (in MPa) due to dike intrusion. C. Horizontal section at ~1 km depth. Only local areas within such broad median valleys are likely to be affected by dike intrusion even for dikes with such extreme heights (stress contours modified after *Pollard and Segall*, 1987; *Rubin and Pollard*, 1988).

Table 1. Characteristics of fast- and slow-spreading mid-ocean ridges showing contrasting rift dimensions, spatial distribution of volcanic activity, inferred average dike intrusion frequency, and hydrothermal vent size and life-span

	Fast-Spreading Ridges	Slow-Spreading Ridges
Spreading Rate	>100 mm/yr	< 30 mm/yr
Rift Width	<1 km wide	> 10 km
Neo-Volcanic Zone	~5 km wide	1 - 2 km wide
Intrusion Frequency[1.]	~ 8 yr	> 50 yr
Vent Size	<1,000 m^3	> 750, 000 m^3
Vent Life-span	< 100 yr	Up to 100,000 yr

1. Assumes all plate separation accommodated by intrusion.

[*Smith et al.*, 1995; *Head et al.*, 1996], associated with robust magmatic systems along FSRs.

4.1.3. Faulting. Dike intrusion plays an important role in the development and modification of rift morphology and topography in subaerial rifts [*Reches and Fink*, 1988, *Rubin and Pollard*, 1988; *Rubin*, 1992] and presumably also along FSRs. Faults that initially form during dike intrusion may slip farther during subsequent dike intrusions in the same area [*Rubin and Pollard*, 1988; *Opheim and Gudmundsson*, 1989]. Thus, repeated dike intrusion beneath the axial rise could result in substantial cumulative structural relief, much of which would likely be buried beneath lavas [*Chadwick and Embley*, in press]. Continual intrusion could result in highly fractured and fissured volumes of older crust and dikes. The rapid thickening of low-velocity seismic layer 2A in young oceanic crust is commonly thought to represent the growth of the extrusive layer [*Christeson et al.*, 1994; *Carbotte et al.*, 1997]. Instead, this may primarily represent increased fracture density in older crust resulting from repeated dike intrusion.

The cumulative effects of continual dike intrusion in the <5-km-wide plate boundary zone of FSRs may be to construct packages of crust with broadly consistent sheeted dike complexes, fault structures, and fracture density and orientation. This uniform architecture has been inferred along many widely separated FSR segments [*Scheirer and Macdonald*, 1993; *Karson*, this volume], and may correspond closely to the ophiolite template for oceanic crust.

4.2. Consequences of Dike Intrusion at Slow-Spreading Ridges

4.2.1. Hydrothermal Activity. The distribution and life-span of hydrothermal systems at SSRs may reflect infrequent dike intrusions that have geographically limited effects on the scale of the >10-km-wide median valley

(Figure 6). The life-span of some hydrothermal systems at SSRs is much longer than estimated intrusion recurrence intervals and cooling times for individual dikes. The location of hydrothermal vents along major fault zones suggests that they are likely hosted by persistent fracture networks created and maintained by fault propagation and interaction [*Curewitz and Karson*, 1997]. Communication with a deep crustal thermal cracking front [*Wilcock and Delaney*, 1996] through these fracture systems may provide the heat necessary for continued hydrothermal activity. Therefore, structural controls [*Curewitz and Karson*, 1997] may override the relatively short-term, spatially limited mechanical and thermal effects of dike intrusion at many hydrothermal sites on SSRs. In contrast, hydrothermal systems found on axial volcanic ridges of SSRs are more likely to be affected by recent dike intrusion events in a manner similar to that described for FSRs.

4.2.2. Volcanism. The distribution and morphology of eruptive vents and lava flows at SSRs may reflect the relative infrequency and spatially variable nature of dike intrusion in the median valley. Scattered axial volcanic ridges and seamounts are consistent with the expectation that magma chambers which could feed dike intrusions are neither spatially nor temporally persistent. This is also suggested by the wide range in apparent age of extruded lavas within individual ridge segments. The predominance of pillow lavas over sheet or lobate flows at SSRs may indicate eruption rates of less than 10^{-1} m^3/s [*Gregg and Fink*, 1995] and eruption through <1-m-wide dikes [*Head et al.*, 1996], consistent with relatively low-pressure magmatic systems at SSRs.

Different segments of SSRs have contrasting lava flow morphology and distribution, possibly depending upon the magma supply [*Bonatti and Harrison*, 1988; *Head et al.*, 1996]. Where there is a relatively constant magma supply, for example the Reykjanes Ridge, the crust may more closely resemble that found at FSRs with a thick extrusive layer, a well defined axial volcanic ridge, and ample evidence for fissure eruptions [*Murton and Parson*, 1993]. In contrast, where the magma supply is limited the extrusive layer is highly discontinuous and fewer dikes are expected. Tectonic windows show that sheeted dike complexes are relatively uncommon in slow-spread crust [*Karson*, this volume], suggesting that dike intrusion may not play a major role in constructing slow-spread crust.

4.2.3. Faulting. Studies of dike intrusion in subaerial settings show that incremental dike intrusion in rift zones can either promote or inhibit slip on faults even kilometers away [*Bursik and Sieh*, 1989; *Opheim and Gudmundsson*, 1989; *Hofton and Foulger*, 1996a, b]. Thus it seems very likely that faulting on the walls of the rift valleys of SSRs could be strongly influenced by subaxial dike intrusion. In particular, the relief on median valley bounding faults may in part reflect the mechanical effects of repeated dike intrusion. Magmatically active segments are likely to be the sites of localized graben subsidence associated with dike intrusions. However, these same segments may have muted axial relief as dike intrusion could accommodate extension otherwise expressed as normal faulting. In contrast, the extreme axial relief along large-offset (>10 km) normal faults found on some segments could reflect periods of limited magmatism and infrequent dike intrusion.

Distinct from FSRs, where the main features of the axial region may be related to dike intrusion, many features of SSRs appear to be controlled by tectonic processes while dike intrusion may only affect local areas within the rift. Thus, the interplay between faulting and sporadic magmatism may result in the highly variable crustal architecture observed among different SSR segments [Karson, this volume], and geologic structures that deviate substantially from the ophiolite template are expected.

4.3. Dike Intrusion and Megaplumes

Megaplumes at mid-ocean ridges consist of large volumes of hydrothermal fluid expelled into the ocean over short periods of time (<1 day). Megaplumes have been observed several times at FSRs; as yet no megaplume events have been observed at SSRs. Estimates of the volume of fluids expelled as megaplumes are on the order of $3x10^7$ m^3 [Baker et al., 1989] to $4x10^{10}$ m^3 [Baker et al., 1995]. Baker et al. [1989] suggest that the fluid flux that forms a megaplume may represent the equivalent of several years of normal hydrothermal fluid flow expelled over a single day, resulting in mass discharges 100 to 1000 times that of steady-state hydrothermal systems measured on FSRs. Water temperature, suspended particulate matter, and chemical composition all indicate that the high-temperature fluid in megaplumes is compositionally similar to fluids in known steady-state hydrothermal systems. This suggests that megaplumes represent the expulsion of mature hydrothermal fluids that have already reacted extensively with the crust [Baker et al, 1989; Baker, 1994; Baker et al, 1995; Lupton et al, 1995]. Thus, megaplumes represent very rapid mass transfer from the porous upper crust to the overlying water column. Both the volume and rate of fluid expulsion compare well with the expected mechanical effects of dike intrusion.

The association of megaplume events with earthquake swarms and fissure eruptions suggests that they are somehow caused by dike intrusion events [Dziak et al., 1995;

Embley et al., 1995; Fox et al., 1995; Lowell and Germanovich, 1995]. Published models of hydrothermal circulation and megaplume formation [Strens and Cann, 1982; Cann and Strens, 1989; Lowell and Germanovich, 1995] have attributed megaplumes to increased fracture permeability combined with increased heat flow driving hydrothermal circulation. They are thought to represent individual tectonic extension events associated with plate separation, and heat extraction from newly intruded dikes. Megaplume formation mechanisms that are proposed include hydraulic fracturing of the crust, and rapid, thermally driven outflow of resident hydrothermal fluids accompanied by the heating of ocean bottom water due to extrusion and cooling of molten lava. These models basically treat dike intrusion as a passive response to an extensional event arising from external forces.

We suggest that dike-perpendicular compression that arises as a dike propagates along a spreading axis is a potentially important and previously unexplored factor in megaplume formation. The temporal and spatial effects of dike intrusion outlined above are consistent with the time-period and volumetric requirements of megaplume formation. With dike intrusion into an area of established hydrothermal activity, extension ahead of the dike tip may open new pathways for fluid flow ahead of the propagating dike. At the same time, dike-normal compression behind the propagating dike tip could squeeze shut the fracture network which provides the pathways for fluid flow. Fracture networks along the dike path should be sequentially opened and shut as the dike propagates through the crust. Fluids residing in these fracture networks would be forced out of areas to either side of the dike, and swept ahead into the dike-tip region, laterally into the surrounding crust, or vertically out of the crust and into the axial graben.

The closure of fracture systems and the expulsion of fluids within these fracture systems could contribute a significant volume of fluid to a megaplume. Using our current understanding of fracture density in oceanic crust [Anderson et al., 1985b; Alt et al., 1993], it is likely that most of the dike-related compression in the upper kilometer of crust is accommodated by fracture closure [after Jaeger and Cook, 1969; O'Connell and Budiansky, 1977]. Delaney and McTigue [1994] calculated the theoretical effects of the intrusion of vertical planar sheet into the crust and found that the volume ratio of uplifted rock to that of the injected "dike" is 0.75. The difference between the uplift volume and the injection volume is therefore the amount of crack closure accommodated in the crust during dike intrusion. This calculation assumes a Poisson ratio of 0.25, which is reasonable for non-fractured rock, but is probably higher than that of fractured rock. With a lower Poisson ratio, the difference between uplift and injection volume increases, and the volume of cracks closed and fluid expelled by dike intrusion increases proportionally.

Using these relationships, the volume of fractures closed by the 1993 CoAxial dike intrusion event at the Juan de Fuca Ridge (assumed dike dimensions: 1-m-wide, 60-km-long, 1.5-km-high) ranges from 9×10^7 m^3 (100% of dike volume taken up by fracture closure) to 2.25×10^7 m^3 (25% of dike volume taken up by fracture closure). These volumes are comparable to estimates of the hydrothermal fluid expelled in the megaplume event. Thus, fracture closure and concomitant fluid expulsion related to dike injection are potentially important driving forces behind megaplume formation.

5. CONCLUSIONS

Applying the lessons learned from theory and observations of subaerial dikes to mid-ocean ridges suggests that dike intrusion associated with seafloor spreading is likely to be much more complex and interesting than most ophiolite and seafloor geological studies would suggest. Contrary to the prevailing view, dike intrusion along mid-ocean ridge spreading centers is not a passive fracture-filling event, but rather a dynamic process by which magma is forcefully injected into the crust. Dike intrusion events result in a range of short- and long-term effects to the surrounding crust. Although thermal effects are relatively short-lived and restricted to areas only a few dike widths from the margins, significant transient and permanent mechanical effects could occur at distances of kilometers away from typical, 1-m-wide mid-ocean ridge dikes. The specific effects of dike intrusion are governed by the spreading center framework, especially the depth to the magma supply and height of the dike, physical properties of shallow crustal rocks, width of the plate boundary zone, and frequency of dike intrusion relative to plate separation. These parameters vary substantially between the generalized fast- and slow-spreading end-members of ridges, therefore the intrusion of an individual dike or cumulative effects of repeated dike intrusions affecting a parcel of crust as it moves away from the axial region is likely to vary substantially with magma budget. Although a dike intrusion is likely to have only relatively minor, local effects along a slow-spreading ridge, it could potentially have a marked effect on the morphology, structure, and hydrothermal processes across the entire axial rise of a typical fast-spreading ridge. Megaplumes may be one of the most dramatic direct results of the mechanical and thermal perturbations caused by the injection of a dike into fractured oceanic crust.

Acknowledgments. We thank P. T. Delaney, D. J. Fornari, A. Gudmundsson, A. M. Rubin, and W. R. Buck for their invaluable input, advice, and editorial comments. Portions of this research arise from investigations supported by NSF Grants EAR-9508250 and OCE-9202661

REFERENCES

Abdallah, A., V. Courtillot, M. Kasser, A. Y. LeDain, J. C. Lepine, B. Robineau, J. C. Ruegg, P. Tapponier, and A. Tarantola, Relevance of Afar seismicity and volcanism to the mechanics of accreting plate boundaries, *Nature, 282,* 17-23, 1979.

Adamides, N. G., Hydrothermal circulation and ore deposition in the Troodos Ophiolite, Cyprus, in *Ophiolites -- Oceanic Crustal Analogues,* J. Malpas, E. Moores, A. Panayiotou, and C. Xenophontos, (eds), Geol. Surv. Dep., Minist. Agric. Natl. Resour., Nicosia, Cyprus, 685-704, 1990.

Agnon, A., and V. Lyakhovsky, Damage distribution and localization during dyke intrusion, in *Physics and Geochemistry of Dykes,* G. Baer and A. Heimann (eds), Balkema Press, 1-14, 1995.

Alt, J. C., Subseafloor processes in mid-ocean ridge hydrothermal systems, in *Seafloor Hydrothermal Systems,* Humphris, S. E., R. A. Zierenberg, L. S. Mullineaux, and R. E. Thompson, (eds), AGU, Washington, D.C., 85-114, 1995

Alt, J.C., H. Kinoshita, L. B. Stokking, et al., *Proceedings of the Ocean Drilling Program, Initial Reports,* vol. 148, Ocean Drill. Prog., College Station, TX (Ocean Drilling Program), 1993.

Anderson, R. N., M. D. Zoback, S. H. Hickman, and R. L Newmark, Permeability versus depth in the upper oceanic crust: in situ measurements in DSDP Hole 504B, eastern equatorial Pacific, *J. Geophys. Res., 95,* 3659-3669, 1985b.

ARCYANA, FAMOUS: Photographic Atlas of the Mid-Atlantic Ridge: Rift and Transform Fault at 3000 m Depth, Bordas, Paris, 128 pp., 1978.

Auzende, J. M., D. Bideau, E. Bonatti, M. Cannat, J. Honnorez, Y. Lagabrielle, J. Malavieille, V. Mamaloukas-Frangoulis, and C. Mével, The MAR-Vema Fracture Zone intersection surveyed by deep submersible *Nautile, Terra Nova, 2,* 68-73, 1989b.

Baker, E. T., J. W. Lavelle, R. A. Feely, G. J. Massoth, S. L. Walker, and J. E. Lupton, Episodic venting of hydrothermal fluids from the Juan de Fuca Ridge, *J. Geophys. Res., 94,* 9237-9250, 1989.

Baker, E. T., A 6-year time series of hydrothermal plumes over the Cleft segment of the Juan de Fuca Ridge, *J. Geophys. Res., 99,* 4889-4904, 1994.

Baker, E. T., G. J. Massoth, R. A. Feely, R. W. Embley, R. E. Thompson, and B. J. Burd, Hydrothermal event plumes from the CoAxial seafloor eruption site, Juan de Fuca Ridge, *Geophys. Res. Lett., 22,* 2, 147-150, 1995.

Ballard, R. D. and J. G. Moore, Photographic Atlas of the Mid-Atlantic Ridge Rift Valley, Springer-Verlag, NY, 114 pp., 1977.

Batiza, R., Volcanic processes, in *Proceedings of a Workshop on the Physical Properties of the Ocean Floor*, G. M. Purdy and G. J. Fryer, (eds), 43-54, 1990.

Becker, K. et al., Drilling deep into young oceanic crust, Hole 504B, Costa Rica Rift, *Rev. Geophys., 27*, 79-102, 1989.

Bicknell, J. D., J. C. Sempéré, K. C. Macdonald, and P. J. Fox, Tectonics of a fast spreading center: A Deep-Tow and Sea Beam survey of the East Pacific Rise at 19°30'S, *Mar. Geophys. Res., 9*, 25-45, 1987.

Bird, D. K., M. T. Rosing, C. E. Manning, and N. M. Rose, Geologic field studies of the Miki Fjord area, East Greenland, *Bull. Geol. Soc. Denmark, 34*, 219-236, 1985

Bjornsson, A., G. Johansen, S. Sigurdsson, G. Thorbergsson, and E. Tryggvason, Rifting of the plate boundary in north Iceland 1975-1978, *J. Geophys. Res., 84*, 3029-3038, 1979.

Bonatti, E., and C. G. A. Harrison, Eruption styles of oceanic basalt in oceanic spreading ridges and seamounts: Effect of magma temperature and viscosity, *J. Geophys. Res., 93*, 2967-2980, 1988.

Brandsdottir, B., and P. Einarsson, Seismic activity associated with the September 1977 deflation of Krafla volcano in North-Eastern Iceland, *J. Volc. Geoth. Res., 6*, 197-212, 1979.

Brown, J. R., and J. A. Karson, Variations in axial processes on the Mid-Atlantic Ridge: The neovolcanic zone in the MARK Area, *Mar. Geophys. Res., 10*, 109-138, 1988.

Brown, S. R., and C. H. Scholz, Closure of rock joints, *J. Geophys. Res., 91*, 4939-4948, 1986.

Bursik, M., and K. Sieh, Range front faulting and volcanism in the Mono Basin, Eastern California, *J. Geophys. Res., 94*, 15587-15609, 1989.

Cann, J.R., D. K. Blackman, D. K. Smith, E. McAllister, B. Janssen, S. Mello, E. Avgerinos, A. R. Pascoe, and J. Escartín, Corrugated slip surfaces formed at ridge-transform intersections on the Mid-Atlantic Ridge, *Nature, 385*, 329-332, 1997.

Cann, J. R., and M. R. Strens, Modeling periodic megaplume emission by black smoker systems, *J. Geophys. Res., 94*, 12227-12237, 1989.

Carbotte, S. M., J. C. Mutter, and L. Xu., Contribution of volcanism and tectonism to axial and flank morphology of the southern East Pacific Rise, 17°10'-17°40'S, from a study of layer 2A geometry, *J. Geophys. Res., 102*, 10165-10184, 1997.

Chadwick, W. W., and Embley, R. W., Graben formation associated with recent dike intrusions and volcanic eruptions on the mid-ocean ridge, *J. Geophys. Res.,* in press.

Chadwick, W. W., R. W. Embley, and C. G. Fox, SeaBeam depth changes associated with recent lava flows, CoAxial segment, Juan de Fuca Ridge: Evidence for multiple eruptions between 1981-1993, *Geophys. Res. Lett., 22*, 167-170, 1995.

Cherkaoui, A. S. M., W. S. D. Wilcock, and E. T. Baker, Modeling thermal fluxes following the 1993 diking event on the CoAxial Segment of the Juan de Fuca Ridge, *Trans. Amer. Geophys. Union, Eos, 76*, 46, F411, 1995.

Cherkaoui, A. S. M., W. S. D. Wilcock, and E. T. Baker, Thermal fluxes associated with the 1993 diking event on the CoAxial segment, Juan de Fuca Ridge: A model for the convective cooling of the dike, *J. Geophys. Res., 102*, 24887-24902, 1997.

Christeson, G. L., G. M. Purdy, and G. J. Fryer, Seismic constraints on shallow crustal emplacement processes at the fast spreading East Pacific Rise, *J. Geophys. Res., 99*, 17957-17973, 1994.

Cowie, P. A., C. H. Scholz, M. Edwards, and A. Malinverno, Fault strain and seismic coupling on mid-ocean ridges, *J. Geophys. Res., 98*, 17911-17920, 1993.

Curewitz, D., and J. A. Karson, Structural settings of hydrothermal flow: Permeability maintained by fault propagation and interaction, *J. Volcanol. Geotherm. Res., 79*, 149-168, 1997

Decker, R. W., T. L. Wright, and P. H. Stauffer (eds) *Volcanism in Hawaii*, U. S. Geol. Surv. Prof. Pap. 1350, 1667 pp., 1987.

Delaney, P. T., Heat transfer during emplacement and cooling of mafic dykes, in *Mafic Dyke Swarms*, Halls, H. C., and Fahrig, W. F.(eds), Geol. Assoc. Can., Spec. Pub., 34, 31-46, 1987.

Delaney, P. T., D. D. Pollard, J. I. Ziony, and E. H. McKee, Field relations between dikes and joints: Emplacement processes and paleostress analysis, *J. Geophys. Res., 91*, 4920-4938, 1986.

Delaney, P. T., and D. F. McTigue, Volume of magma accumulation or withdrawal estimated from surface uplift or subsidence, with application to the 1960 collapse of Kilauea Volcano, *Bull. Volcanol., 56*, 417-424, 1994.

Dziak, R. P., C. G. Fox, and A. E. Schreiner, The June-July 1993 seismo-acoustic event at CoAxial segment, Juan de Fuca Ridge: Evidence for a lateral dike injection, *Geophys. Res. Lett., 22*, 135-138, 1995.

Eberhart, G. L., P. A. Rona, and J. Honnorez, Geologic controls on hydrothermal activity in the Mid-Atlantic Ridge rift valley: Tectonics and volcanics. *Mar. Geophys. Res., 10*, 233-259, 1988.

Embley, R. W., and W. W. Chadwick, Volcanic and hydrothermal processes associated with a recent phase of seafloor spreading at the northern Cleft segment: Juan de Fuca Ridge, *J. Geophys. Res., 99*, 4741-4760, 1994.

Embley, R. W., W. W. Chadwick, Jr., E. T. Baker, D. Massoth, D. Butterfield, R. Koski, I. R. Jonasson, J. R. Delaney, R. E. McDuff, M. D. Lilley, and J. Holden, Comparison of two recent eruptive sites on the Juan de Fuca ridge: Geological setting and time-series observations, *Trans. Amer. Geophys. Union, Eos, 75*, 617, 1994.

Embley, R. W., W. W. Chadwick, I. R. Jonasson, D. A. Butterfield, and E. T. Baker, Initial results of the rapid response to the 1993 CoAxial event: Relationships between hydrothermal and volcanic processes, *Geophys. Res. Lett., 22*, 143-146, 1995.

Fornari, D. J., and R. W. Embley, Tectonic and volcanic controls on hydrothermal processes at the mid-ocean ridge: An overview based on near-bottom and submersible studies, in *Seafloor Hydrothermal Systems,* Humphris, S. E., R. A. Zierenberg, L. S. Mullineaux, and R. E. Thompson (*eds*), AGU, Washington, D.C., 466 p., 1995.

Fornari, D. J., R. M. Haymon, M. R. Perfit, T. K. P. Gregg, and M. H. Edwards, Axial summit trough of the East Pacific Rise 9°N to 10°N: Geological characteristics and evolution of the axial zone on fast-spreading mid-ocean ridges, *J. Geophys. Res.,* in press.

Fouquet, Y., J.-L. Charlou, I. Costa, J.-P. Donval, J. Radford-Konery, H. Pelle, H. Ondreas, N. Laurenco, M. Segonzac, and M. K. Tivey, A detailed study of the Lucky Strike hydrothermal site and the discovery of a new hydrothermal site: Menez Gwen; preliminary results of the DIVA1 cruise (5-29 May, 1994), *Inter-Ridge News, 3,* 14-17, 1994.

Fouquet, Y., A. Wafik, P. Cambon, C. Mevel, G. Meyer, and P. Gente, Tectonic setting and mineralogical and geochemical zonation in the Snake Pit sulfide deposit (Mid-Atlantic Ridge at 23°N), *Econ. Geol., 88,* 2018-2036, 1993.

Fox, C. G., W. E. Radford, R. P. Dziak, T.-K. Lau, H. Matsumoto, and A. E. Schreiner, Acoustic detection of a seafloor spreading episode on the Juan de Fuca Ridge using military hydrophone arrays, *Geophys. Res. Lett., 22,* 131-134, 1995.

Francheteau, J., R. Armijo, J. L. Cheminee, R. Hekinian, P. Lonsdale, and N. Blum, 1 Ma East Pacific Rise oceanic crust and uppermost mantle exposed by rifting in Hess Deep (equatorial Pacific Ocean), *Earth Planet. Sci. Lett., 101,* 281-295, 1990.

Francheteau, J., R. Armijo, J. L. Cheminee, R. Hekinian, P. Lonsdale, and N. Blum, Dike complex of the East Pacific Rise exposed in the walls of Hess Deep and the structure of the upper oceanic crust, *Earth Planet. Sci. Lett., 111,* 109-121, 1992.

Fryer, G. J., and Wilkins, R. H., Fractures, cracks, and large-scale porosity, in *Proceedings of a Workshop on the Physical Properties of the Ocean Floor,* G. M. Purdy and G. J. Fryer, (*eds*), 121-134, 1990.

German, C. R., L. M. Parson, HEAT Scientific team., Hydrothermal exploration near the Azores Triple Junction: Tectonic control of venting at slow-spreading ridges?, *Earth Planet. Sci. Lett., 138,* 93-104, 1995.

Gillis, K., Controls on hydrothermal alteration in a section of fast-spreading oceanic crust, *Earth Planet. Sci. Lett., 134,* 473-489, 1995.

Gregg, T. K. P., and J. H. Fink, Quantification of submarine lava-flow morphology through analog experiments, *Geology, 23,* 73-76, 1995.

Gregg, T. K. P., D. J. Fornari, M. R. Perfit, R. M. Haymon, and J. Fink, Rapid emplacement of a mid-ocean ridge lava flow on the East Pacific Rise at 9°46'-51'N, *Geology,* in press.

Gudmundsson, A., Form and dimensions of dykes in eastern Iceland, *Tectonophys., 95,* 295-307, 1983.

Gudmundsson, A., Mechanical aspects of postglacial volcanism and tectonics of the Reykjanes Peninsula, Southwest Iceland, *J. Geophys. Res., 91,* 12711-12721, 1986.

Gudmundsson, A., Effect of tensile stress concentration around magma chambers on intrusion and extrusion frequencies, *J. Volcanol. Geotherm. Res., 35,* 179-184, 1988.

Gudmundsson A., Infrastructure and mechanics of volcanic systems in Iceland, *J. Volcanol. Geotherm. Res., 64,* 1-22, 1995.

Halls, H. C., and W. F. Fahrig (*eds*) *Mafic Dyke Swarms,* Geol. Assoc. Can., Spec. Pub., 34, 503 pp., 1987.

Haymon, R. M., The response of ridge crest hydrothermal systems to segmented, episodic magma supply, in *Tectonic, Magmatic and Hydrothermal and Biological Segmentation of Mid-Ocean Ridges,* C. J. MacLeod, P. A. Tyler, and C. L. Walker (*eds*), *Geol. Soc. London Spec. Pub., 118,* 157-168, 1996.

Haymon, R. M., D. J. Fornari, M. H. Edwards, S. Carbotte, D. Wright, and K. C. Macdonald, Hydrothermal vent distribution along the East Pacific Rise crest (9°09' - 54'N) and its relationship to magmatic and tectonic processes on fast-spreading mid-ocean ridges, *Earth Planet. Sci. Lett., 104,* 513-534, 1991.

Haymon, R. M., D. J. Fornari, K. L. Von Damm, M. D. Lilley, M. R. Perfit, J. M. Edmond, W. C. Shanks III, R. A.Lutz, J. M. Grebmeier, S. Carbotte, D. Wright, E. McLaughlin, M. Smith, N. Beedle, and E. Olson, Volcanic eruption of the mid-ocean ridge along the East Pacific Rise crest at 9°45'-52'N: Direct submersible observations of seafloor phenomena associated with an eruption event in April, 1991, *Earth Planet. Sci. Lett., 119,* 85-101, 1993.

Head, J. W., L. Wilson, and D. K. Smith, Mid-ocean ridge eruptive vent morphology and structure: Evidence for dike widths, eruption rates, and evolution of eruptions and axial volcanic ridges, *J. Geophys. Res., 101,* 28265-28280, 1996.

Hekinian, R., J. Francheteau, R. Armijo, J. P. Cogne, M. Constantin, J. Girardeau, R. Hey, D. F. Naar, R. Searle, Petrology of the Easter microplate region in the South Pacific, *J. Volcanol. Geotherm. Res., 72,* 259-289, 1996.

Hofton, M. A., and G. R. Foulger, Postrifting anelastic deformation around the spreading plate boundary, north Iceland 1. Modeling of the 1987-1992 deformation field using a viscoelastic Earth structure, *J. Geophys. Res., 101,* 25403-25421, 1996.

Hofton, M. A., and G. R. Foulger, Postrifting anelastic deformation around the spreading plate boundary, north Iceland 2. Implications of the model derived from the 1987-1992 deformation field, *J. Geophys. Res., 101,* 25423-25436, 1996.

Huang, P. Y., and S. C. Solomon, Centroid depths of mid-ocean ridge earthquakes: Dependence on spreading rate, *J. Geophys. Res., 93,* 13445-13477, 1988.

Ingebritsen, S. E., and M. A. Scholl, The hydrogeology of Kilauea volcano, *Geothermics, 22,* 255-270, 1993.

Jaeger, J. C., and N. G. W. Cook, *Fundamentals of Rock Mechanics,* Chapman and Hall LTD, Science Paperbacks, London, 515 pp., 1969.

Karson, J. A., Accommodation Zones and Transfer Faults: Integral Components of Mid-Atlantic Ridge Extensional

Systems, in *Ophiolites Genesis and Evolution of Oceanic Lithosphere*, Tj. Peters, A. Nicolas and R.G. Coleman (eds), Kluwer, Dordrecht, 21-37, 1991.

Karson, J. A., Faulting and Magmatism at mid-ocean ridges: A perspective from tectonic windows in the oceanic lithosphere, in *Faulting and Magmatism at Mid-Ocean Ridges*, Buck, W.R., P. T. Delaney, and J. A. Karson, eds., American Geophysical Union, this volume.

Karson, J. A., and J. R. Brown, Geologic setting of the Snake Pit Hydrothermal site: An active vent field on the Mid-Atlantic Ridge, *Mar. Geophys. Res., 10*, 91-107, 1988.

Karson, J. A., S. D. Hurst, S. D., R. M. Lawrence, and SMARK Cruise Participants, Upper crustal construction and faulting at a segment-scale half graben on the Mid-Atlantic Ridge at 22°30'N (SMARK Area), *Trans. Amer. Geophys. Union, Eos, 77*, S271, 1996.

Karson, J. A., S. D. Hurst, and P. Lonsdale, Tectonic rotations of dikes in fast-spread oceanic crust exposed near Hess Deep, *Geology, 20*, 685-688, 1992.

Karson, J. A., and P. A. Rona, 1990, Block-tilting, transfer faults, and structural control of magmatic and hydrothermal processes in the TAG area, Mid-Atlantic Ridge 26°N, *Geol. Soc. Amer. Bull., 102*, 1635-1645, 1990.

Karson, J. A., G. Thompson, S. E. Humphris, J. M. Edmond, W. B. Bryan, J. R. Brown, A. T. Winters, R. A. Pockalny, J. F. Casey, A. C. Campbell, G. Klinkhammer, M. R. Palmer, R. J. Kinzler, and MM Sulanowska, Along-axis variations in seafloor spreading in the MARK area, *Nature, 328*, 681-685, 1987.

Karson, J. A., and A. T. Winters, 1992, Along-axis variations in tectonic extension and accommodation zones in the MARK area, Mid-Atlantic Ridge 23°N latitude, in *Ophiolites and Their Modern Oceanic Analogues*, L. M. Parson, B. J. Murton, and O. Browning, (eds), Geological Society Special Publication, 107-116, 1992.

Kidd, R. G. W., A model for the formation of the upper oceanic crust, *Geophys. J. R. Astr. Soc., 50*, 149-183, 1977.

Klein, F. W., R. Y. Koyanagi, J. S. Nakata, W. R. Tanigawa, The seismicity of Kilauea's magma system. in *Volcanism in Hawaii*, Decker, R. W., T. L. Wright, and P. H. Stauffer (eds) U. S. Geol. Surv. Prof. Pap. 1350, 1019-1185, 1987.

Koski, R. A., I. R. Jonasson, D. C. Kadko, V. K. Smith, and F. L. Wong, Compositions, growth mechanisms, and temporal relations of hydrothermal sulfide-sulfate-silica chimneys at the northern Cleft segment, Juan de Fuca Ridge, *J. Geophys. Res., 99*, 4813-4832, 1994.

Lalou, C., J.-L. Reyss, E. Brichet, M. Arnold, G. Thompson, Y. Fouquet, and P. A. Rona, New age data for Mid-Atlantic Ridge hydrothermal sites: TAG and Snakepit chronology revisited, *J. Geophys. Res., 98*, 9705-9713, 1993.

Lalou, C., J.-L. Reyss, E. Brichet, P. A. Rona, and G. Thompson, Hydrothermal activity on a 10^5-year time scale at a slow-spreading ridge, TAG hydrothermal field, Mid-Atlantic Ridge 26° N, *J. Geophys. Res., 100*, 17855-17862, 1995.

Langmuir, C., S. Humphris, D. Fornari, C. Van Dover, K. Von Damm, M. K. Tivey, D. Colodner, J.-L. Charlou, D.

Desonie, C. Wilson, Y. Fouquet, G. Klinkhammer, and H. Bougault, Hydrothermal vents near a mantle hot spot: The Lucky Strike vent field at 37°N on the Mid-Atlantic Ridge, *Earth Planet. Sci. Lett., 148*, 69-91, 1997.

Lawrence, R.M., J. A. Karson, and S. D. Hurst, Dike orientations and fault-block rotations in slow-spread oceanic crust at the SMARK Area, Mid-Atlantic Ridge at 22°40'N, *J. Geophys. Res., 103*, 663-676, 1998.

Lin, J., and E. A. Bergman, Rift grabens, seismicity and volcanic segmentation of the Mid-Atlantic Ridge: Kane to Atlantis fracture zones (abstract), *Trans. Amer. Geophys. Union, Eos, 71*, 1572, 1990.

Lowell, R. P., and L. N. Germanovich, Dike injection and the formation of megaplumes at ocean ridges, *Science, 267*, 1804-1807, 1995.

Lowell, R. P., P. A. Rona, and R. P. von Herzen, Seafloor hydrothermal systems, *J. Geophys. Res., 100*, 327-352, 1995.

Lupton, J. E., E. T. Baker, G. J. Massoth, R. E. Thomson, B. J. Burd, D. A. Butterfield, R. W. Embley, and G. A. Cannon, Variations in water-column ^{3}He/heat ratios associated with the 1993 CoAxial event, Juan de Fuca Ridge, *Geophys. Res. Lett., 22*, 155-158, 1995.

Macdonald, K. C., Mid-ocean ridges: Fine-scale tectonic, volcanic and hydrothermal processes within the plate boundary zone, *Ann. Rev. Earth Planet. Sci., 10*, 155-190, 1982.

Mastin, L. G., and D. D. Pollard, Surface deformation and shallow dike intrusion processes at Inyo Craters, Long Valley, California, *J. Geophys. Res., 93*, 13221-13235, 1988.

Malinverno, A, and P. A. Cowie, Normal faulting and topographic roughness of mid-ocean ridge flanks, *J. Geophys. Res., 98*, 17921-17939, 1993

Malpas, J., E. Moores, A. Panayiotou, and C. Xenophontos, (eds), *Ophiolites -- Oceanic Crustal Analogues*, Geol. Surv. Dep., Minist. Agric. Natl. Resour., Nicosia, Cyprus, 733 pp., 1990.

McAllister, E. and J.R. Cann, Initiation and evolution of boundary -wall faults along the Mid-Atlantic Ridge at 25-29°N, in *Tectonic, Magmatic, Hydrothermal, and Biological Segmentation of Mid-Ocean Ridges*, edited by C. J. MacLeod, C. J. Tyler, and C. L. Walker, *Geol. Soc. Lond., Spec. Publ.* 118, 1996.

Moores, E. M. and F. J. Vine, The Troodos massif, Cyprus, and other ophiolites as oceanic crust: evaluation and implications, *Phil. Trans. R. Soc. Lond. A, 268*, 443-466, 1971.

Murton, B.J. and L.M. Parson, Segmentation, volcanism and deformation of oblique spreading centres: A quantitative study of the Reykjanes Ridge, *Tectonophys., 222*, 237-257, 1993.

Mutter, J. C., and J. A. Karson, Structural processes at slow-spreading ridges, *Science, 257*, 627-634, 1992.

Nehlig, P., Fracture and permeability analysis in magma-hydrothermal transition zones in the Samail ophiolite (Oman), *J. Geophys. Res., 99*, 589-601, 1994.

O'Connell, R. J., and B. Budiansky, Viscoelatic properties of fluid-saturated cracked solids, *J. Geophys. Res., 82, 36,* 5719-5735, 1977.

Opheim, J. A., and A. Gudmundsson, Formation and geometry of fractures, and related volcanism, of the Krafla fissure swarm, northeast Iceland, *Geol. Soc. Amer. Bull., 101,* 1608-1622, 1989.

Parsons, T., N. H. Sleep, and G. A. Thompson, Host rock rheology controls on the emplacement of tabular intrusions: Implications for underplating of extending crust, *Tectonics, 11,* 1348-1356, 1992.

Platten, I. A., and J. Watterson, Magma flow and crystallization in dyke fissures, in *Mafic Dyke Swarms,* Halls, H. C., and Fahrig, W. F. (*eds*), Geol. Assoc. Can., Spec. Pub., 34, 65-73, 1987.

Pollard, D. D., P. T. Delaney, W. A. Duffield, E. T. Endo, and A. T. Okamura, Surface deformation in rift zones, *Tectonophys., 94,* 541-584, 1983.

Pollard, D. D., and P. Segall, Theoretical displacements and stresses near fractures in rock: With applications to faults, joints, veins, dikes, and solution surfaces, in *Fracture Mechanics of Rock,* B. K. Atkinson, ed., Academic Press, London, England, 1987.

Purdy, G. M., J.C. Sempere, H. Schouten, D. DuBois, and R. Goldsmith, Bathymetry of the Mid-Atlantic Ridge 24°-31°N: A map series: *Mar. Geophys. Res., 12,* 247-252, 1991.

Reches, Z., and J. Fink, The mechanism of intrusion of the Inyo Dike, Long Valley Caldera, California, *J. Geophys. Res., 93,* 4321-4334, 1988.

Rona, P. A., Y. A. Bogdanov, E. G. Gurvich, N. A. Rimski-Korsakov, A. M. Sagalevitch, M. A. Hannington, and G. Thompsen, , Relict Hydrothermal zones in the TAG hydrothermal field, mid-Atlantic Ridge 26°N, 45°W, *J. Geophys. Res., 98,* 9715-9730, 1993

Rose, N. M., and D. K. Bird, Hydrothermally altered dolerite dykes in East Greenland: Implications for Ca-metasomatism of basaltic protoliths, *Contrib. Mineral. Petrol., 116,* 420-432, 1994.

Rubin, A. M., Dike-induced faulting and graben subsidence in volcanic rift zones, *J. Geophys. Res., 97,* 1839-1858, 1992.

Rubin, A. M., Tensile fracture of rock at high confining pressure: Implications for dike propagation, *J. Geophys. Res. 98,* 15919-15935, 1993c.

Rubin, A. M., Propagation of Magma-filled cracks, *Ann. Rev. Earth Planet. Sci., 23,* 287-336, 1995.

Rubin, A. M., and D. D. Pollard, Origin of blade-like dikes in volcanic rift zones, in *Volcanism in Hawaii,* Decker, R. W., T. L. Wright, and P. H. Stauffer (*eds*), U. S. Geol. Surv. Prof. Pap. 1350 v. 2, 1449-1470, 1987.

Rubin, A. M., and D. D. Pollard, Dike-induced faulting in rift zones of Iceland and Afar, *Geology, 16,* 413-417, 1988.

Ryan, M. P., The mechanics and three-dimensional internal structure of active magmatic systems: Kilauea Volcano, Hawaii, *J. Geophys. Res, 93,* 4213-4248, 1988.

Ryan, M. P., The physical nature of the Iceland magma transport system, in *Magma Transport and Storage,* M. P. Ryan ed., John Wiley & Sons Ltd., West Sussex, England, 1990.

Ryan, M. P., J. Yang, and R. N. Edwards, The three-dimensional structure of basaltic rift zone hydrothermal convection systems, *Trans. Amer. Geophys. Union, Eos, 77,* 17, S257-S258, 1996.

Scheirer, D. S., and K. C. Macdonald, Variation in cross-sectional area of the axial ridge along the East Pacific Rise: Evidence for the magma budget of a fast spreading center, *J. Geophys. Res., 98,* 7871-7885, 1993.

Scholl, M. A., S. E. Ingebritsen, C. J. Janik, and J. P. Kauahikaua, Use of precipitation and groundwater isotopes to interpret regional hydrology on a tropical volcanic island: Kilauea volcano area, Hawaii, *Water Resour. Res., 32,* 3525-3537, 1996.

Searle, R. C., and A. S. Laughton, Sonar studies of the Mid-Atlantic Ridge and Kurchatov fracture zone, *J. Geophys. Res., 82,* 5313-5328, 1977.

Sempéré, J.-C., J. Lin, H. S. Brown, H. Schouten, and G. M. Purdy, Segmentation and morphotectonic variations along a slow-spreading center: The Mid-Atlantic Ridge (24°00'N-30°40'N), *Mar. Geophys. Res., 15,* 153-200, 1993.

Sempéré, J.-C., and K. C. Macdonald, Marine tectonics: Processes at mid-ocean ridges, *Rev. Geophys., 25,* 1313-1347, 1987.

Shaw, P. R., Ridge segmentation, faulting and crustal thickness in the Atlantic Ocean, *Nature, 358,* 490-493, 1992.

Shaw, P. R., and J. Lin, Causes and consequences of variations in faulting style at the Mid-Atlantic Ridge, *J. Geophys. Res., 98,* 21839-21851, 1993.

Sinton, J. M., and R. S. Detrick, Mid-ocean ridge magma chambers, *J. Geophys. Res., 97,* 197-216, 1992.

Smith, D. K., and J. R. Cann, The role of seamount volcanism in crustal construction at the Mid-Atlantic Ridge (24°-30°N), *J. Geophys. Res., 97,* 1645-1658, 1992.

Smith, D. K., S. E. Humphris, and W. B. Bryan, A comparison of volcanic edifices at the Reykjanes Ridge and the Mid-Atlantic Ridge at 24°-30°N, *J. Geophys. Res., 100,* 22485-22498, 1995.

Solomon, S. C., P. Y. Huang, and L. Meinke, The seismic moment of slowly spreading ridges, *Nature, 334,* 58-60, 1988.

Stakes, D. S., J. W. Shervais, and C. A. Hopson, The volcano-tectonic cycle of the FAMOUS and AMAR Valleys, Mid-Atlantic Ridge (36° 47' N): evidence from basalt glass and phenocryst compositional variations for a steady-state magma chamber beneath the valley mid-sections, AMAR 3, *J. Geophys. Res., 89,* 6995-7028, 1984.

Strens, M. R., and J. R. Cann, A model of hydrothermal circulation in fault zones at mid-ocean ridge crests, *Geophys. J. R. Astr. Soc., 71,* 225-240, 1982.

Takasaki, K. J., and J. K. Mink, Dike-impounded ground-water reservoirs, Island of Oahu, *U. S. Geol. Surv. Water Supply Pap. 2217,* 77 pp., 1985.

Toomey, D. R., S. C. Solomon, G. M. Purdy, M. H. Murray, Microearthquakes beneath the median valley of the Mid-Atlantic Ridge near 23°N: Hypocenters and focal mechanisms, *J. Geophys. Res., 90,* 5443-5458, 1985.

Tucholke, B. E., J. Lin, M. C. Kleinrock, M. A. Tivey, T. B. Reed, J. Goff, and G. E. Jaroslow, Segmentation and crustal structure of the western Mid-Atlantic Ridge flank, 25°25'-27°10'N and 0-29 my., *J. Geophys. Res., 102*, 10203-10224, 1997.

Turcotte, D. L., and G. Schubert, *Geodynamics Applications of Continuum Physics to Geological Problems*, John Wiley and Sons, Inc., New York, 450 p. 1982

van Everdingen, D. A., Fracture characteristics of the sheeted dike complex, Troodos ophiolite, Cyprus: Implications for permeability of oceanic crust, *J. Geophys. Res., 100*, 19957-19972, 1995.

Wada, Y., On the relationship between dike width and magma viscosity, *J. Geophys. Res., 99*, 17743-17755, 1994.

Walker, G. P. L., The dike complex of Koolau volcano, Oahu: Internal structure of a Hawaiian rift zone, in *Volcanism in Hawaii*, Decker, R. W., T. L. Wright, and P. H. Stauffer (*eds*) U. S. Geol. Surv. Prof. Pap. 1350, 961-993, 1987.

Wilcock, W. S. D., and A. McNabb, Estimates of crustal permeability on the Endeavour segment of the Juan de Fuca mid-ocean ridge, *Earth Planet. Sci. Lett., 138*, 83-91, 1996.

Wright, D. J., R. M. Haymon, and D. J. Fornari, Crustal fissuring and its relationship to magmatic and hydrothermal processes on the East Pacific Rise crest (9°12' to 54'N), *J. Geophys. Res., 100*, B4, 6097-6120, 1995.

Wright, D. J., R. M. Haymon, and K. C. Macdonald, Breaking new ground: Estimates of crack depth along the axial zone of the East Pacific Rise (9°12'-54'N), *Earth Planet. Sci. Lett., 134*, 441-457, 1995.

Daniel Curewitz and Jeffrey A. Karson, Division of Earth & Ocean Sciences, Duke University Box 90229, Durham, NC 27708
(e-mail: *curewitz@geo.duke.edu, jkarson@geo.duke.edu*)

Formation and Development of Fissures at the East Pacific Rise: Implications for Faulting and Magmatism at Mid-Ocean Ridges

Dawn J. Wright

Department of Geosciences, Oregon State University, Corvallis, OR

Fissures control the gross permeability of ocean crust, providing critical pathways for hydrothermal fluids and magma and lending important clues to the cycling of volcanic and hydrothermal processes and the spatial and temporal stability of ridge segments. And yet due to limitations in mapping technology that existed until the late 1980s, the critically important parameters of fissure abundance, spacing, length, width, and depth have rarely been reported anywhere on the seafloor, with the exception of the fast-spreading EPR. However, with the increasing availability of high resolution mapping tools, research agendas at various international workshops are now citing the determination of these parameters as a high priority for morphotectonic studies of the mid-ocean ridge that seek to more fully understand the nature of extensional failure of the ocean crust. This chapter reviews the fundamentals of fissure formation and development, as well as the only known studies to have focussed specifically on fissuring at mid-ocean ridges. Also discussed are the distinctions between tectonic and eruptive fissuring, the implications of fissures changing into faults within the neovolcanic zone, and the use of fissures for mapping out fourth-order ridge segment boundaries and assessing the cycling of magmatic and hydrothermal processes therein. The chapter concludes by posing research questions to guide future studies of fissuring in the neovolcanic zone, as well as faulting and eruption on the mid-ocean ridge crest and flanks.

INTRODUCTION

Fissures (on a "micro-scale" of 10s of meters along-strike) and faults (on a "macro-scale" of kilometers to 10s of kilometers) provide essential pathways into the upper oceanic crust through which magma and hydrothermal fluids may migrate (faults may even be primary conduits for fluid flow *deep* into the crust). The nature and distribution of crustal cracking provide insights into the spatial and temporal stability of mid-ocean ridge (MOR) segment and have important implications for magmatic and hydrothermal venting along the MOR crest, as well as hydrothermal circulation on- and off-axis. Crustal cracking on the order of 10's to 100's of meters in length controls the gross permeability of the crust and, ultimately, fluid flux

pathways, alteration of associated igneous rock, and the physical properties of ocean crust. A fundamental question to be considered is the role of fissuring (tensile failure) versus faulting (shear failure). Fault populations on the MOR have been the subject of considerable study within the past few years (e.g., *Edwards et al.* [1991], *Carbotte and Macdonald* [1992 and 1994], *Cowie et al.* [1994], *Macdonald et al.* [1996], *van Wyk de Vries and Merle* [1996]). Similarly, in connection to faulting and magmatism on the MOR, much attention has been given to the study of subsurface dikes (i.e., magma-filled cracks; see *Gudmundsson* [1990], *Karson et al.* [1992], *Rubin* [1992], *Johnson and Salem* [1994], *Lowell and Germanovich* [1995], and *Dziak et al.* [1995]). However, in order to fully understand the nature of mechanical failure in an extensional environment, one must also consider the location, distribution, and geometry of fissures. In the past this has been difficult on the MOR because of the lack of good quality data at the "micro-scale." *Crane* [1987] was the first to tabulate fissure abundance and length along the axial zone

Faulting and Magmatism at Mid-Ocean Ridges
Geophysical Monograph 106
Copyright 1998 by the American Geophysical Union

of the MOR (using the side-looking sonars of the deep-towed Sea MARC I with selectable swath widths of 1.5, 3, and 6 km over the EPR at 10°35' to 13°10'N). No further studies focusing primarily on fissures were performed on any part of the MOR until the 1989 *Argo I* survey of the EPR at 9°12' - 54'N provided unprecedented coverage of the axial zone (up to 80% *visual* coverage with a 10-16 m swath and 90% sonar coverage with a 300 m swath) [*Haymon et al.*, 1991]. This finally provided a data set that was of sufficient density, continuity, and geographical precision to meaningfully map and analyze fissure distributions at a fine-scale [*Wright et al.*, 1995a], as well as hydrothermal vents, lava flows, and biological communities [*Haymon et al.*, 1991]. Indeed, given the near-bottom mapping technology, fissures offer several advantages in the field over faults: (1) large numbers may be present within a small area; (2) they are likely to be confined to a single flow morphology or lithology; and (3) they are commonly well-exposed over their entire length. This chapter reviews current ideas on fissure formation and distribution, primarily on the fast-spreading East Pacific Rise (EPR) where most of the existing data have been gathered, and discusses the important implications of ridge-crest fissuring for axial faulting, magmatism, segmentation, and the cycling of tectonic, magmatic, and hydrothermal processes.

FISSURE FORMATION AND DISTRIBUTION

Background From Fracture Mechanics

For the MOR environment a fissure is here defined as an open, Mode I tension crack initiating at the surface (Figure 1). Unlike a dike, which is modeled by *Rubin* [this volume] as a pressurized, static, magma-filled crack in an elastic solid, a fissure is not filled with magma, but may serve as an important conduit for upwardly-propagating magma and hydrothermal fluids. This may be possible if the fissure is tectonic in origin and intersecting a horizontal lava tube or interconnected voids in the uppermost (~0-50 m) of the seafloor. *Haymon et al.* [1993], *Perfit et al.* [1993], and *Goldstein et al.* [1994] have documented the existence of such lava tubes and voids at the EPR, 9°N. A distinction that I will return to later in more detail is that of tectonic fissures (formed near the axis because the lithosphere is stretched as it accelerates from zero to full spreading velocity) from eruptive fissures (assocated with dikes propagating to the vicinity of the surface). Detailed discussion of the mechanics of fissure formation in volcanic environments are available only from studies on Iceland, Hawaii, and Afar (e.g., *Pollard et al.* [1983], *Rubin and Pollard* [1987], *Gudmundsson and Bäckström* [1991], and *Hayward and Ebinger* [1996]), where quantitative field measurements and resulting theoretical models have allowed for a more thorough treatment of the problem than attempts on the seafloor (which are not unlike mapping in a dark football stadium with a flashlight [*Gregg*, 1997]). The

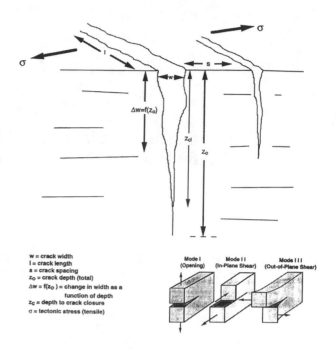

w = crack width
l = crack length
s = crack spacing
z_0 = crack depth (total)
$\Delta w = f(z_0)$ = change in width as a function of depth
z_c = depth to crack closure
σ = tectonic stress (tensile)

Mode I (Opening)   Mode II (In-Plane Shear)   Mode III (Out-of-Plane Shear)

Figure 1. Illustration of the relevant parameters involved in considering the effects of cracking on the properties of an igneous crustal plate. Symbols are defined in the illustration. The fracture mechanics models reviewed in the text are primarily concerned with the parameters w (crack width), z_0 (crack depth), and σ (remote tensile stress). In terms of the loading applied to a crack, fissures are characterized as Mode I , where the principal load is applied normal to the crack plane.

Iceland, Hawaii, and Afar results are directly applicable to the MOR environment because the same mechanical processes are at work. The following is a brief review of the mechanics of fissure formation in volcanic rift zones. For a thorough introduction to fracture mechanics see *Jaeger and Cook* [1979] or *Anderson* [1995]; for applications to dike propagation see *Rubin* [this volume].

According to the First Law of Thermodynamics, when a system goes from a nonequilibrium state to equilibrium, there will be a net decrease in energy [*Anderson*, 1995]. In 1920 A. A. Griffith applied this idea to the formation of a crack which, he surmised, takes place when the energy stored in the structure is sufficient to overcome the surface energy of the material [*Griffith*, 1920]. Since cracking involves the breaking of bonds, the stress on the atomic level must reach or exceed the cohesive stress [*Anderson*, 1995]. From this Griffith developed a theory of strength of solids based on the assumption that the local stress intensification necessary to break atomic bonds is provided by minute internal and surface flaws already existing in the material [*Griffith*, 1920; *Griffith*, 1924]. The Griffith theory assumes that crack initiation occurs from the points of highest tensile stress on the surfaces of these microscopic flaws or "Griffith's cracks" in brittle material (in a biaxial stress field), and this has since been elucidated by the theoretical and laboratory studies of *Bieniawski* [1967] and *Huang et al.* [1993]. Joints, lava flow contacts, and tension cracks may be regarded as the macroscopic analogy to "Griffith's cracks" [*Gudmundsson and Bäckström*, 1991]. If σ_1 is the greatest compressive stress, σ_3 the least compressive stress, and T_0 the tensile strength of the rock, then the two-dimensional Griffith crack initiation is [*Griffith*, 1924; *Jaeger and Cook*, 1979]:

$$\text{If } \sigma_1 < -3\sigma_3, \text{ then } \sigma_3 = -T_0 \tag{1}$$

$$\text{If } \sigma_1 > -3\sigma_3, \text{ then } (\sigma_1 - \sigma_3)^2 = 8T_0(\sigma_1 + \sigma_3) \tag{2}$$

Formula (1) applies to the tensile regime whereas formula (2) applies to compression. When considering the formation and development of fissures on the MOR several mechanical factors need to be taken into account, the most important being elastic moduli, tensile stress, and tensile strength of the host rock.

Elastic Moduli. For the complete specification of a linear elastic material, any two of the elastic moduli (λ, Lamé's constant; G, the shear modulus (rigidity); ν, Poisson's ratio; E, Young's modulus; or K, the bulk modulus) must be known [*Jaeger and Cook*, 1979]. For MOR fracture mechanics models Poisson's ratio is used either with Young's modulus or with the shear modulus. The dynamic Poisson's ratio may be calculated from seismic compressional- (V_p) and shear-wave (V_s) velocities. From these dynamic moduli, the static moduli, which should be used in the crack models presented below, can often be inferred. However, because direct shear-wave velocity measurement in young ocean crust is difficult, especially in the uppermost part of the crust, the dynamic Poisson's ratio is largely unknown [*Shaw*, 1994]. One is therefore compelled to rely on a static Poisson's ratio determined from laboratory measurements of basalt as an approximation of the *in-situ* value. The commonly assumed Poisson's ratio for oceanic crust, based on the laboratory measurements of *Christensen* [1978], is 0.3 (yielding $V_p/V_s = 1.9$). However, the amplitude modeling of the on-bottom seismic refraction data of *Christeson et al.* [1994; 1997] near 9°30'N indicates a Poisson's ratio at the seafloor ranging from 0.43 to 0.49 (i.e., $V_p/V_s > 3$). This value is similar to other determinations of Poisson's ratio for young oceanic crust, which fall in the range of 0.39 to 0.46 [*Diachok et al.*, 1984; *Harding et al.*, 1989; *Vera et al.*, 1990]. These higher values of Poisson's ratio are more common for material dominated by thin cracks (i.e., aspect ratios much less than one) than by material dominated by more equidimensional voids, such as vesicles or interpillow voids [*Jaeger and Cook*, 1979; *Shearer*, 1988].

The dynamic Young's modulus, E_d, is given by *Jaeger and Cook* [1979] as:

$$E_d = V_p^2 \, \rho \, \frac{(1+\nu)(1-2\nu)}{(1-\nu)} \tag{3}$$

where V_p is P-wave velocity, ν is Poisson's ratio and ρ is rock density. For the purposes of this chapter, V_p is the average P-wave velocity and ρ is the average rock density of the uppermost 1 km of crust at the EPR, 9°12'-54'N. The relationship between E_d and the static Young's modulus (E_s) is somewhat complex and depends on the rock in question [*Cheng and Johnston*, 1981; *Eissa and Kazi*, 1988]. In general, because cracks propagate much more slowly than the velocities of seismic waves, the static Young's modulus (as well as the static Poisson's ratio) should be used [*Gudmundsson*, 1983; *Gudmundsson*, 1990]. In laboratory measurements the E_d/E_s ratio is commonly around 2.0 [*Cheng and Johnston*, 1981], but for *in-situ* measurements this ratio ranges from 1.5 to 9.1 for common extrusive rocks [*Gudmundsson*, 1990; *Link*, 1968]. Because *in-situ* measurements in the Tertiary and Quaternary lava piles of Iceland suggest a ratio of 2.0 [*Forslund and Gudmundsson*, 1991; *Gudmundsson*, 1988], *Wright et al.* [1995b] adopted the same E_d/E_s ratio for their estimates of fissure depth at the EPR 9"N. A static shear modulus, G_s, is given by *Jaeger and Cook* [1979]:

$$G_s = \frac{E}{2(1+\nu)} \tag{4}$$

Tensile Stress. Most MOR fissures are known to be vertical at the surface and must be generated by absolute tensile stresses (i.e., they are Mode I cracks; Figure 1). Under most loading conditions, absolute tensile stresses attain their peak values at the surface, so it is likely that all fissures form as tensile fractures at or near the surface, while with downward propagation some may subsequently change into normal faults [*Opheim and Gudmundsson*, 1989; *Wright et al.*, 1995b]. *Gudmundsson* [1983] has shown that it is possible to estimate tensile stress during crack formation from Young's modulus, Poisson's ratio, and the length/width ratios of the cracks. On the EPR at 9°N *Wright et al.* [1995b] found that the length/width ratio for a small population of fissures resulted in a tensile stress estimate of 30 ± 10 MPa, which is similar to estimates used by *Lachenbruch* [1973] and *Macdonald et al.* [1991]. In Iceland's Krafla fissure swarm, estimates range from 12 to 134 MPa, with an average of 20-30 MPa [*Opheim and Gudmundsson*, 1989]. Therefore, any crack formation model for the MOR must be able to generate tensile stresses on at least the order of ~20-40 MPa, or, alternatively, explain how existing crack width:length ratios might be overestimated.

Tensile Strength. No *in situ* values of tensile strength have been published for the MOR, but estimates suggest that it is very low. The best determinations currently available for volcanic rift zone environments range from 1 MPa to 6 MPa, with an inaccuracy of a factor 2, based on the work of *Haimson and Rummel* [1982] in Iceland. Fortunately this range covers the possible inaccuracy in ignoring the potential effects of pore-fluid pressure on crack formation which have not been rigorously quantified for the MOR. The general effect of pore-fluid pressure is to increase the *probability* of failure in a rock [*Helm*, 1994]. Pore-fluid pressure affects only the normal stress, not the shear stress (A. Gudmundsson, personal communication, 1995).

Fissure Length, Width and Depth

On the MOR fissure length is determined fairly easily from side-scan sonar surveys (e.g., *Edwards et al.* [1991], *Carbotte and Macdonald* [1992], and *Macdonald et al.* [1992]) and fissure width can, in principle at least, be determined from near-bottom camera observations (e.g., *Gente et al.* [1986] and *Wright et al.* [1995a]), although this has rarely been reported. *Nur* [1982] found that in tension crack systems the length of cracks should normally be equal to, or greater than, their depths. This is assuming, however, that the applied tensile stress increases with depth, which may not be entirely appropriate for MORs. *Wright et al.* [1995b] employed a slightly different model assuming that the width of a crack may be controlled by either its depth or its length (Figure 1). In order to know which is the minimum or width-controlling dimension, their model drew on the reasoning of *Gudmundsson and Bäckström* [1991], and *Gudmundsson* [1992]: (1) use the Griffith crack criterion [*Griffith*, 1924] to estimate the maximum possible depth of absolute tension in the crust; (2) compare this depth with an estimate of average length for the cracks under consideration; (3) if cracks are generally longer than this depth, then crack depth is the minimum and thus the width-controlling dimension, and the model applies. Returning to formula (2), substituting $-T_0$ for σ_3 gives:

$$\sigma_1 \leq 3T_0 \qquad (5)$$

If $\sigma_1 = \rho g z$ then $\qquad (6)$

$$T_0 = \frac{\rho g z}{3} \qquad (7)$$

and solving for depth z gives:

$$z_{max} = \frac{3T_0}{\rho \ g} \qquad (8)$$

where z_{max} is the maximum depth of absolute tension in the crust. *Gudmundsson and Bäckström* [1991] report an average z_{max} of 500 m for Holocene fissures in the rift zone of Iceland and find that most large-scale tension cracks should develop into normal faults at crustal levels deeper than 500-800 m. *Wright et al.* [1995b] calculated a z_{max} of 400 ± 100 m for the EPR 9°12'-54'N.

The length of fissures on the EPR at 9°N ranges from ~30 to ~650 m with an average length of ~170 m, based on the *Argo I* sonar data of *Wright et al.* [1995a]. Many of the cracks in general are rubble-filled, and many, if not most, of the shorter cracks are part of arrays, with the distances between the ends of the cracks being much shorter than the lengths of the cracks themselves. These arrays behave in a mechanical fashion essentially as a single crack. It is therefore surmised that those cracks are generally longer than z_{max}, the inferred maximum depth of absolute tension. So *Wright et al.* [1995b] employed the following fracture mechanics model to estimate crack depth from crack width:

$$z_0 = \frac{E \ w}{\sigma \ V_1(d/b)} \qquad (9)$$

where $V_1(d/b)$ is the stress function of *Tada et al.* [1973], which has 1% accuracy for any d/b:

$$V_1(d/b) = \frac{1.46 + 3.42 \ [1 - \cos\left(\frac{\pi}{2}\right)\frac{d}{b}]}{[\cos\left(\frac{\pi}{2}\right)\frac{d}{b}]^2} \qquad (10)$$

In the stress function above, d is 400 m, in accordance with the earlier estimate of z_{max}, and b is the brittle thickness of the crust, which would be the depth to the 600° isotherm on a fast-spreading axis. This translates to ~1200 m for the EPR at 9-10°N [*Lin and Parmentier*, 1989]. The average depth to the axial magma chamber (AMC) reflector for the EPR at 9-10°N is just below this at ~1600 m [*Kent et al.*, 1993]. This model predicts depths between ~60-275 ± 40 m, including cracks that may penetrate the Layer 2A/2B boundary north of 9°42'N, where crack widths are the greatest [*Wright et al.*, 1995b]. These cracks are also located where the 1991 eruption of the ridge crest occurred at ~9°45'-52'N [*Haymon et al.*, 1993]. In general, these crack depths are comparable to the depths directly observed in the eroded Tertiary and Pleistocene lava piles of the rift zone in southwest Iceland [*Gudmundsson*, 1987; *Forslund and Gudmundsson*, 1991], and to those inferred for the Holocene pahoehoe lava flows of the rift zone in northeast Iceland [*Gudmundsson and Bäckström*, 1991], which range in depth from 200-400 m.

Models of Fissure Development

The development of fissures in volcanic rift zones on land has been attributed mainly to tensile stresses generated either by magmatic intrusions or by divergent plate motion. Current models of fissure formation are similar for the MOR where most workers agree that the presence of fissures is due to either: (a) lithospheric stretching (amagmatic, tensile cracking near-axis as the ridge accelerates from zero velocity to its full spreading rate, resulting in tectonic fissures); (b) magmatic intrusion (extensional cracking on-axis in the crust overlying dikes, resulting in eruptive fissures); or (c) thermal contraction of aging crust where tension cracks result from the shrinkage of cooling rock. Models have been proposed in the literature for both lithospheric stretching and magmatic intrusion, and are evaluated below.

A simple plate motion model (Figure 2) assumes that tensile stress due to far-field plate motion builds up gradually within the rift zone until new fissures form, old ones propagate, and tensile stress is temporarily relaxed [*Björnsson*, 1985]. This model is also very attractive, as there is undoubtedly relative tensile stress that builds up within the MOR rift zone. However, because we are now aware of the importance of diking within the neovolcanic zone, particularly in the crust directly above a magma reservoir, a model invoking diking seems more appropriate, particularly within an axial summit trough. A plate motion model would be especially appropriate outside of an axial summit trough, within ~1-2 km of the axis.

In the dike model developed by *Pollard et al.* [1983], dike emplacement produces significant changes in the local stress field of crustal rock directly over the top of the dike. Contours of maximum principal stress outline a region of tensile stress that spreads outward and upward from the top of the dike (Figure 3). The point immediately over the dike at the surface is stress free and a pair of tensile maxima occurs on either side of this spot. The *Pollard et al.* [1983] model of deformation for the 1976 Krafla event in Iceland, places a dike (of thickness somewhat greater than 2 m) at a depth of 250 m. The ambient stress at this depth must be near the level required for tensile failure of the crust (~30 MPa) in order for fissures to form above the dike at the surface. The Krafla event was unusually voluminous, resulting in a zone of fissuring and faulting ~6 km wide [*Sigurdsson*, 1980], much larger than a typical neovolcanic zone on the MOR. The numerical modelling of *Head et al.* [1996] reveals that dike widths for the MOR should range from as little as ~0.2 m up to ~3 m. This is based on observations in analogous settings, including an average thickness of less than 2 m for Pleistocene dike swarms in Iceland [*Gudmundsson*, 1995], 1.5 m for the Troodos ophiolite [*Kidd*, 1977], 2-3 m for the seafloor [*Francheteau et al.*, 1990; *Hurst et al.*, 1994; *Gregg et al.*, 1996]. Indeed, the experiments of *Mastin and Pollard* [1988] demonstrate that the top of a dike must be shallower than about ten times its thickness at its top in order to generate surface

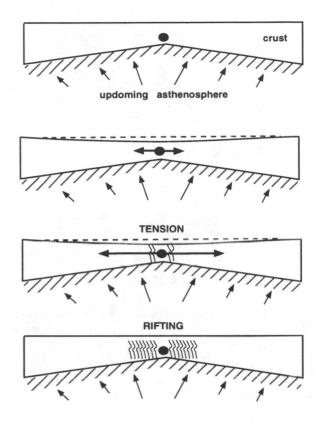

Figure 2. Simple plate motion model of fissure development after *Björnsson* [1985], based on the 1975 rifting episode in the Krafla region of NE Iceland. Tension is gradually built up in the axial rift zone by ridge push forces and then is released every few years to few centuries in a rifting episode, accompanied by the intrusion of magma into fissures in the crust. Arrows are proportional to deviatoric stress in the crust.

fractures, which start to develop into normal faults only when the dike top is at a depth of about five times the thickness of the dike at its top. So a typical 2- to 4-m-thick dike on the slow-spreading MAR or in Iceland would have to be within 20-40 m of the surface to generate tension cracks [*Gudmundsson*, 1983; *Gudmundsson*, 1990]. *Head et al.* [1996] find that dikes in the range of 0.2-3 m thickness will cause fissuring at the surface only if they penetrate to depths of less than a few tens of meters. Dikes greater in thickness may be able to produce fissuring from much greater depths, given that the ambient stress at depth is near the level required for tensile failure.

Gudmundsson [1988] proposed a plate motion model in combination with dike intrusion, in which most dikes do not reach the surface but build up temporarily high horizontal compressive stresses at some depth in the crust. During particular rifting events, the uppermost 1-2 km of the crust might be free of tectonic compressive stress, whereas such stress would still be maintained in deeper layers, and relatively large tensile stress would be needed to relax the compressive stress [*Gudmundsson*, 1988]. The

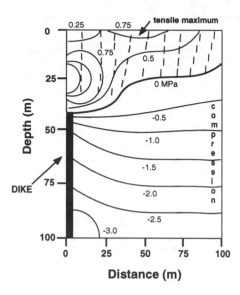

Figure 3. Dike model of fissure development after *Pollard et al.* [1983], showing idealized contours of the maximum principal stress near a vertical crack cutting a vertical plane. Dike is ~100 m high with depth to center of ~75 m and subject to a driving pressure of 1 MPa. Short dashed lines are trajectories of the minimum principal stress and indicate potential planes of secondary cracking.

result is that high absolute tensile stress may be generated in the uppermost part of the crust during relaxation of horizontal compressive stress at deeper crustal levels. So in the *Gudmundsson's* [1988] model, the lithospheric stretching of plate motion is primarily responsible for surficial crack formation, but earlier or current dikes may build up horizontal compressive stress at deeper levels in the crust which would still need to be relaxed. This model still requires quantitative derivation and will first be applied to crack formation in the Krafla fissure swarm (A. Gudmundsson, personal communication, 1995).

Fissure Distribution

Several studies have included mapping of the distribution and orientation of fissuring on MOR crests spreading at different rates (e.g., *Normark* [1976], *Luyendyk and Macdonald* [1977], *Gente et al.* [1986], *Kappel and Ryan* [1986], *USGS* [1986], *Crane* [1987], *Edwards et al.* [1991], *Embley et al.* [1991], *Hey et al.* [1992], and *Wright et al.* [1995a] which derives fissure distribution from the first segment-scale, continuous visual survey ever provided for a MOR crest). All of these studies agree on the following spatial and temporal aspects of fissure distribution: (1) most fissuring occurs primarily on the crest of the MOR in a 1- to 2-km wide region within the neovolcanic zone; (2) lengths and widths of fissures are

generally 10-500 m and 0.2-3 m respectively; (3) orientations are largely sub-parallel to the adjacent ridge axis; and, (4) spacings are usually <100 m. Where the ridge has recently erupted, axial fissures formed by these mechanisms may be filled or covered by new, uncracked lava flows. Such contact relationships have been commonly observed in areas where very recent eruptions have been documented such as at the CoAxial site at 46°-47°N on the JdFR [*Embley et al.*, 1993] and at the EPR at 9°-10°N [*Haymon et al.*, 1993].

The high rates of processes on the super fast-spreading Southern East Pacific Rise (SEPR), at sites such as 17-19°S [*Detrick et al.*, 1993; *Auzende et al.*, 1996; *Haymon et al.*, 1997], beg the question of whether volcanotectonic events become larger and more intense, or simply much more frequent [*Mottl et al.*, 1996]. And perhaps the along-strike demarcation between portions of ridge segments that are controlled more by tectonic than magmatic processes will not be as clear. How closely will the along-strike variability in crustal fissuring and other fine-scale features of the seafloor correspond to any 4th-order morphotectonic segmentation of the ridge crest? What will be the spatial and temporal constraints on volcanic-hydrothermal-tectonic cycling? Analyses of fissure data collected in 1993 and 1996 [*Auzende et al.*, 1996; *Haymon et al.*, 1997; *Wright et al.*, in prep.], in relation to the hydrothermal and magmatic data are still in progress, but we hope to provide more insight by late 1998.

FISSURING AT DEPTH: IMPLICATIONS FOR FAULTING WITHIN THE NEOVOLCANIC ZONE

Oceanic crust is produced at the ridge in a neovolcanic zone 1-2 km wide and is subsequently fissured and tectonized by normal faults within 2-3 km of the ridge axis [*Macdonald*, 1982]. The process by which the faults grow along-strike is poorly known, but is thought to involve the formation of a set of open tension cracks which then link and coalesce to form the slip surface [*Scholz*, 1989; *Cartwright et al.*, 1995]. *Cowie and Scholz* [1992a] model fault growth in terms of an inelastic deformation zone at the fault tip where cracking and frictional wear of the rock produce the mature fault surface. Their model predicts a linear relation between fault length and maximum displacement, and a tapered displacement profile instead of the elliptical one produced by a simple linear elastic fracture mechanics model. The proportionality constant between fault length and displacement is a function of the rock's shear strength divided by its shear modulus. *Walsh and Watterson* [1988] and *Marrett and Allmendinger* [1991] have found a non-linear relationship between fault length and maximum displacement. *Cowie and Scholz* [1992b] note that this non-linearity may be due to the small scale over which the observations were made or inappropriately combining of data from different environments (i.e., with different rock properties). Studies are in progress on the Mid-Atlantic Ridge (MAR) (e.g., *Shaw and Kleinrock* [submitted]) and on the Juan de Fuca Ridge (JdFR) (e.g.,*Wright , Embley, and Chadwick* [in prep.]) with data that span a much greater scale range for faults in a single tectonic environment and rock type to resolve these divergent results.

Compared with fissures, most MOR normal faults are large and grade into tension fractures at their ends [*Shaw and Lin*, 1993; *Kleinrock and Humphris*, 1996]. This would indicate that MOR fissures have to attain a certain minimum length and/or depth in order to develop into normal faults. In Iceland, *Opheim and Gudmundsson* [1989] define normal faults as fissures having vertical displacement in excess of 1 m. Their results indicate that a fissure has to attain a length of several hundred meters before any significant vertical displacement occurs. Similarly on the EPR at 9°N *Wright et al.* [1995b] found that a fissure has to attain a width of ~5-10 m and a subsequent depth of ~400 m before potentially changing into a normal fault, at which point the tension crack would tend to close. However, such a fissure may have to be carried out of the neovolcanic zone by divergent plate motion before changing to a normal fault. *Edwards et al.* [1991] and *Carbotte and Macdonald* [1994] have shown that normal faulting is not common on fast-spreading ridges within ± 3 km of the ridge axis because the lithosphere is too thin and weak to support it. Furthermore, *Chen and Morgan* [1990] point out that near-axis faulting may be impeded at fast-spreading ridges because the crust is decoupled from extensional stresses by an axial magma chamber (AMC). On slow-spreading ridges the lithosphere is sufficiently thick and strong enough to support normal faulting right along the axis, and such faults may extend all the way to the base of the crust (e.g., *Macdonald and Luyendyk* [1977], *Huang and Solomon* [1988], and *Kong et al.* [1992]).

TECTONIC VERSUS ERUPTIVE FISSURING: LINKAGES TO RIDGE SEGMENTATION AND MAGMATISM

The finest scale of ridge segmentation (fourth order) probably corresponds to individual fissure eruption events. For example, at this very fine scale on the EPR, excellent correlations can be seen within individual fourth order segments between lava age, density and width of fissuring, and hydrothermal vent abundance [*Haymon et al.*, 1991; *Wright et al.*, 1995a]. See *Langmuir et al.* [1986] or *Macdonald et al.* [1991] for definitions of ridge segmentation, orders 1-4.

For example, at the EPR 9°N, *Wright et al.* [1995a] found the widest cracks to be in the youngest lavas suggesting that these cracks are eruptive in origin and control the locations of high-temperature hydrothermal venting during the early stages of the volcanic-hydrothermal-tectonic cycle. Once

dikes are emplaced in crust that is already subject to external tensile stress (related to divergent plate motion), tensile stress concentration in portions of the crust directly above the tops of these dikes should already be high enough to form the wide cracks. Also, the thickness of these dikes must be at least 2-3 m. Dike thickness is proportional to dike dimension and driving pressure (the difference between the magma pressure in the dike and the regional least compressive stress): the thicker the dike the greater the driving pressure, and the wider the fissure forming above it [*Rubin and Pollard*, 1987]. So at the very outset these fissures are wide and deep, remaining so until they are carried away from the neovolcanic zone by divergent plate motion. If their widths increased with time within the neovolcanic zone so would their depths, and eventually, at crustal depths of 500-800 m [*Gudmundsson and Bäckström*, 1991], the cracks would change into normal faults and thereby tend to close at the surface. And although all fissures provide pathways for seawater to enter the crust, which might potentially cool the hydrothermal and magmatic systems, most hydrothermal heat loss in the axial zone of fast-spreading centers probably occurs early along the wide, deep, eruptive fissures.

In contrast to these sparse eruptive fissures in younger lavas, narrower, shallower cracks occurring in older, colder crust on the EPR probably accumulate with time in response to tensile stresses. These shallower cracks do not tap magma and typically are not hydrothermally active. They are primarily tectonic in origin, forming in response to crustal extension.

The direct proportionality of crack depth to crack width presented in *Wright et al.* [1995b] adds to other observational evidence indicating that wide fissures in relatively young lava flows are primarily eruptive and are thus deep enough to facilitate the flux of magma and fluids from the tops of the feeder dikes during an eruption event (Figure 4). *Alvin* observations of the 1991 eruption site on the EPR at 9°N confirm that many wide fissures there may be eruptive in origin, and numerous high-temperature, vapor-rich vents are localized along the margins of these wide fissures [*Haymon et al.*, 1993]. *Haymon et al.* [1993] proposed the intrusion of dikes to ~200 m beneath the floor of the axial summit caldera during the eruption, which drove the phase separation of hydrothermal fluids near the tops of the dikes and a large flux of vapor-rich fluids through the overlying rubbly lavas. A possible origin for these dikes is the inflation of the AMC due to the injection of melt from the upper mantle. In the magmatic pressure change model of *Gudmundsson* [1987], magma accumulation and slight increases in magmatic pressure within the AMC prior to an eruption cause uplift and bending of the overlying crust, thereby generating tensile stress in the upper part of the crust sufficiently large enough to form cracks. During uplift and bending, the potential tensile stress is much higher than during ordinary divergent plate motion, resulting in cracks at the leading edge of these dikes that may be exceptionally wide and deep.

In addition to observations of fissure width and depth, *Wright et al.* [1995a] found that fissure density appears to increase near the centers of fourth-order segments but to decrease near the tips of segments (Figure 5). This is contrary to cracking patterns observed at a second-order scale, where crack density increases toward segment tips [*Sempéré and Macdonald*, 1986; *Macdonald et al.*, 1991]. The cracking pattern of second-order segments is associated with recent crustal extension or far-field tensile stress as the plates separate [*Sempéré and Macdonald*, 1986; *Macdonald et al.*, 1991]. Stresses result in a crack propagation force [*Macdonald et al.*, 1991] that promotes increased cracking at segment tips. This crack propagation force increases as the segment lengthens [*Macdonald et al.*, 1991]. The general paucity of fissures at the ends of fourth-order segments suggests that a different mechanism of fissuring is at work

at this scale. The pattern of fissuring at this scale is driven by the propagation of dikes along-strike away from small, shallow, localized sites of melt injection from the magma lens, and that fourth-order segment boundaries are thus null points between cracking fronts at the tips of advancing dikes (Figure 6).

IMPLICATIONS OF FISSURING FOR VOLCANIC-TECTONIC CYCLING AT FAST-SPREADING RIDGES

Related to the segmentation of the MOR is the *cycling* of processes at the ridge axis. The structure of the ridge crest is related to both magmatic and tectonic processes that may be cyclic *within* a 2nd-, 3rd- or fourth-order ridge segment. Volcanic-tectonic cycles have been proposed in previous studies of the JdFR [*Lichtman and Eissen*, 1983; *Embley et al.*, 1991], the EPR (e.g., *Gente et al.* [1986] and *Haymon et al.* [1991]) the MAR (e.g., *Eberhart et al.* [1988] and *Fouquet et al.* [1993]), and the Galapagos Rift (e.g., *Embley et al.* [1988]).

On the fast-spreading EPR, fine-scale volcanic, tectonic, and hydrothermal characteristics of the axial zone strongly reflect the processes of ridge segmentation on both a second- and a fourth-order scale. The distribution of fine-scale features have spatially and temporally constrained the model of *Haymon et al.* [1991] in which the individual fourth-order segments are in different phases of a volcanic-hydrothermal-tectonic cycle that begins with cracking/diking and eruptive fissuring, followed by magmatic drainback, gravitational collapse, possible development of an ASC, and then cooling of the heat source, initiation of tectonic fissuring, and

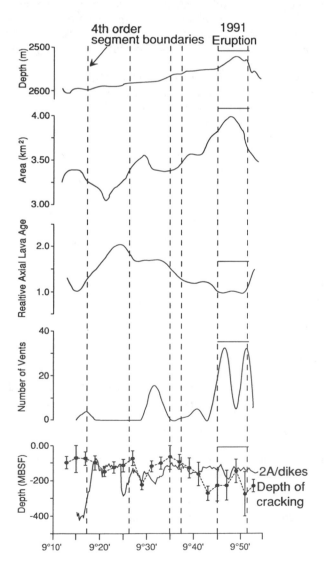

Figure 4. Correlation of along-strike crack depth with along-strike axial topography, axial cross-sectional area, crack density, relative axial lava age, and hydrothermal vent abundance for the EPR crest, 9°12'-54'N. Vertical dashed lines mark the latitudes of 4th-order ridge axis discontinuities as determined by *Haymon et al.* [1991]. Profiles from top to bottom: (a) Seafloor topography from the Sea Beam bathymetry of *Macdonald et al.* [1992]. Vertical exaggeration is 93x. (b) Smoothed along-strike variation in ridge axis cross-sectional area digitized from *Scheirer and Macdonald* [1993]. (c) Along-strike variation in crack density (i.e., the number of fissures per square km of seafloor imaged with the ARGO I video camera) after *Wright et al.* [1995a]. (d) Along-strike variation in relative axial lava age from *Wright et al.* [1995a]. Lava age criterion based on *Haymon et al.* [1991]. (e) Along-strike variation in the number of hydrothermal vents actively discharging hydrothermal fluids in 1989 after *Haymon et al.* [1991] and *Wright et al.* [1995a]. Fluid count includes high-temperature vents (black, white, gray smokers and smoke plumes), and low-temperature vents (milky or cloudy water). (f) Along-strike variation in depth of Layer 2A/2B boundary (solid line) and estimated depth of cracking from *Wright et al.* [1995b]. MBSF = meters below seafloor. Vertical exaggeration is 25x.

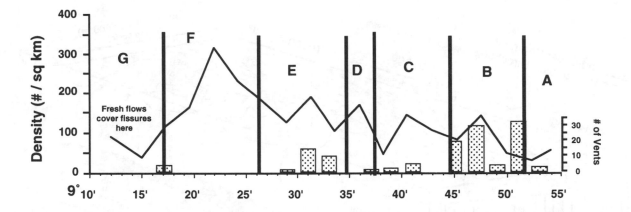

Figure 5. Along-strike variation in fissure density as a function of latitude for the EPR at 9°12'-54'N from *Wright et al.* [1995a]. Dashed lines mark the latitude of fourth-order boundaries at discontinuities with a significant cross-strike offset and/or along-strike overlap of the ridge axis. Note the peaks in fissure density near the middle of most of the fourth-order segments. The number of vents actively discharging hydrothermal fluids per minute of latitude in 1989 is overlain for comparison. Active vent count includes high-temperature vents (black, white, gray smokers and smoke plumes) and low-temperature vents (milky or cloudy water; see Table 3 of *Haymon et al.* [1991]).

waning of hydrothermal and magmatic activity (Figure 7). At the end of the cycle, hydrothermal activity ceases, and cold tectonic cracking as well as mass wasting that widens the ASC, are the dominant processes modifying the axial zone.

Positive correlation of fissure density with relative ages of axial lavas reveals the tendency of the crust in the axial zone to accumulate more cracks with time rather than to widen existing cracks [*Wright et al.*, 1995a]. This key observation implies that the occasional formation of wide, presumably deep cracks is intimately related to episodes of dike intrusion, eruption of new, uncracked flows, and hydrothermal venting. When intrusions and eruption cease, crustal cooling and extension continue to produce additional smaller, shallower cracks in the frozen volcanic carapace overlying the sheeted dikes. These cold, amagmatic cracks increase in number with time and normally are not sites of active hydrothermal discharge.

A second important observation is that along-strike variability in crustal fissuring and other fine-scale features of the seafloor correspond to the fourth-order morphotectonic segmentation of the ridge crest [*Haymon et al.*, 1991]. The tendency of fourth-order segment boundaries to coincide with along-strike changes in the mean density of fissuring, in relative ages of axial lavas, and in hydrothermal vent abundances supports the premise of *Haymon et al.* [1991] that each fourth-order segment is in its own phase of a volcanic-hydrothermal-tectonic cycle. Such a cycle would begin with cracking/diking and eruptive fissuring, followed by magmatic drainback, gravitational collapse, possible development of an ASC, and then cooling of the heat source, initiation of tectonic fissuring, and waning of hydrothermal and magmatic activity. At the end of the cycle,

hydrothermal activity ceases, and cold tectonic cracking as well as mass wasting that widens the ASC, are the dominant processes modifying the axial zone [*Haymon et al.*, 1991; *Wright et al.*, 1995a].

CONCLUSION: DIRECTIONS FOR FUTURE RESEARCH

Detailed studies of both axial and near-axial regions of the EPR and SEPR, as well as the MAR, JdFR, the Indian Ocean Ridges, and the Galapagos Rift will continue to provide many insights into MOR geological processes. However, a number of first-order geological problems remain to be solved. Among these is the determination of the influence of axial tectonic activity (which includes crustal cracking) on the evolution of crustal morphology and physical properties. Specific questions include the following:

• On the "micro-scale" of fissures, what can their distributions within the axial zone allow us to infer about the timing and spatial extent of volcanic-tectonic cycles along the ridge axis not only at fast-spreading ridges but at super-fast, intermediate, and slow-spreaders? And what is the relative importance of volcanic, tectonic, and hydrothermal activity within these cycles in controlling seafloor morphology?

• Does the abundance of fissuring vary inversely along-strike with lava age and hydrothermal vent abundance at intermediate and slow spreading rates? This relationship was predicted for fast-spreading rates by the volcanic-hydrothermal-tectonic cycle for fourth-order segments proposed by *Haymon et al.* [1991] and quantified by *Wright et al.* [1995a] (Figure 7).

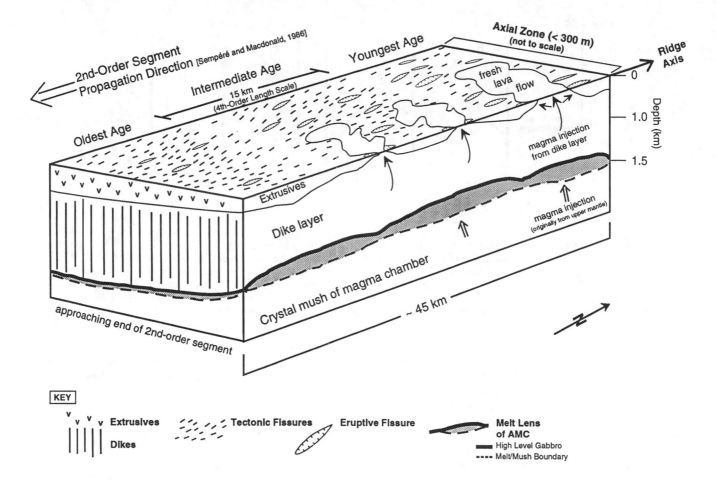

Figure 6. Schematic block diagram showing variations in the density and width of cracks observed along the EPR crest at 9°12'-54'N from *Wright et al.* [1995a]. This diagram distinguishes eruptive fissures from tectonic fissures and distinguishes fissure density distribution at a second-order scale from fissure density distribution at a fourth-order scale. Note that there are two levels at which magma injection occurs, as shown in the diagram: shallow injection from the dike layer and deep injection from the upper mantle.

• *Wright et al.* [1995a] proposed for the fast-spreading EPR that, counter-intuitively, the widest cracks do *not* occur in the oldest lavas but rather in the youngest. Again, how well does this correlation hold at intermediate-spreading ridges such as the JdFR? *Wright et al.* [1995a and 1995b] found that the tendency of young lavas to host sparse, wide fissures has important implications for fissures being deep enough to tap melt and channel eruptions, given a reasonably shallow magma chamber. This is especially pertinent to the southern JdFR where multichannel seismic reflection data have been used to resolve a weak reflection, centered beneath the ridge axis, that is interpreted be the top of an axial magma chamber at 2.3-2.5 km depth beneath the seafloor [*Morton et al.*, 1987].

• On the "macro-scale" of faults, how do they grow in length, and are there systematic changes in fault geometry with distance along- or across-strike?

• Will observations of fault lengths and displacements spanning the diverse tectonic environments of the JdFR and MAR (e.g., *Kleinrock and Humphris* [1996] and *Wright, Embley, and Chadwick* [in prep]) verify or disprove the linear relationship between fault length and maximum displacement predicted by the model of *Cowie and Scholz* [1992b]?

• On the crest of the JdFR is a dike model of crack formation as in *Pollard et al.* [1983] more appropriate, or a plate motion model [*Björnsson*, 1985], and what are the implications of this for the segmentation of an intermediate-rate spreading center? More observations of fissure length, width, and spatial distribution on the JdFR are needed.

• If the elevated centers of ridge segments are indeed the primary loci of upwelling magma from the mantle, is there a progressive change in fissure and fault distribution with

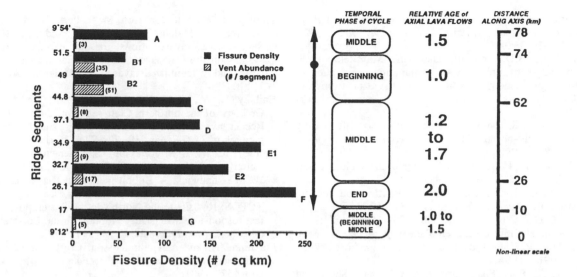

Figure 7. Illustration of the spatial and temporal constraints on a volcanic-hydrothermal-tectonic cycle for the EPR at 9°12'-54'N from *Wright et al.* [1995a]. The weighted density of fissures (per square kilometer of seafloor imaged) in each fourth-order ridge segment is represented by solid bars. The number of active high- and low-temperature vents per ridge segment is shown by the striped bars with the actual number in parentheses. The circle represents the 9°50'N axial high of the survey area, and the arrows extending above and below the circle represent both an increase in along-strike depth and in axial lava age. Moving to the right, the next column identifies the temporal phase of a segment cycle and its approximate along-strike extent. The phase of the cycle marked "BEGINNING" in parentheses at the bottom of the column refers to the fresh, unfissured flow spanning the boundary between segments F and G. Whether the source of the fresh flow is in segments G or F is uncertain; it is speculated that the source may be a local axial high, at ~9°15'N in segment G. The next column to the right identifies relative lava ages inferred according to the criteria of *Haymon et al.* [1991] and averaged within each ridge segment. The column on the far right scales the entire length of the study area, showing the along-strike distance in kilometers.

distance from the ridge segment high? In other words, is the along-strike variability in the scale of fissuring and faulting related to the magmatic segmentation of the ridge?

The critically important parameters of fissure abundance, spacing, length, width, and depth have rarely been reported anywhere on the seafloor, with the exception of the fast-spreading EPR, and have only recently been perceived as important in marine geology [*Johnson*, 1990]. The research agendas being set at various international workshops (e.g., *Purdy and Fryer* [1990], *Dziewonski and Lancelot* [1995], and *Mottl et al.* [1996]) are now including the determination of these parameters as a high priority for morphotectonic studies of MOR crust, certainly for any study that hopes to more fully understand the nature of extensional failure of the ocean crust. Clearly we need to continue "breaking new ground" with more fine-scale surveys and quantitative studies of axial and off-axial fissures as they relate to faulting and magmatism along the global MOR.

Acknowledgements. Many thanks to Rachel Haymon, Ken Macdonald, Agust Gudmundsson, Allan Rubin, Dan Scheirer, and Amy Clifton for helpful and encouraging discussions regarding the ideas expressed in this paper. The critical reviews of Agust Gudmundsson and an anonymous referee significantly improved the manuscript. This work was supported by National Science Foundation grant OCE-9521039.

REFERENCES

Anderson, T.L., *Fracture Mechanics: Fundamentals and Applications*, 688 pp., CRC Press, Inc., Boca Raton, Florida, 1995.

Auzende, J.-M., V. Ballu, R. Batiza, D. Bideau, J.-L. Charlou, M.-H. Cormier, Y. Fouquet, P. Geistodoerfer, Y. Lagabrielle, J. Sinton, and P. Spadea, Recent tectonic, magmatic, and hydrothermal activity on the East Pacific Rise between 17°S and 19°S: Submersible observations, *J. Geophys. Res., 101*, 17,995-18,010, 1996.

Bieniawski, Z.T., Mechanism of brittle fracture of rock. Part 1, Theory of the fracture process., *Int. J. Rock Mech. Min. Sci., 4*, 395-406, 1967.

Björnsson, A., Dynamics of crustal rifting in Iceland, *J. Geophys. Res., 90*, 10,151-10,162, 1985.

Carbotte, S., and K. C. Macdonald, East Pacific Rise 8°-10°30'N: Evolution of ridge segments and discontinuities from SeaMARC II and three-dimensional magnetic studies, *J. Geophys. Res., 97*, 6959-6982, 1992.

Carbotte, S.M., and K.C. Macdonald, Comparison of seafloor tectonic fabric at intermediate, fast, and super fast spreading ridges: Influence of spreading rate, plate motions, and ridge segmentation on fault patterns, *J. Geophys. Res.*, *99* , 13,609-13,631, 1994.

Cartwright, J.A., B.D. Trudgill, and C.S. Mansfield, Fault growth by segment linkage: An explanation for scatter in maximum displacement and trace length data from the Canyonlands Grabens of SE Utah, *J. Struct. Geol.*, *17*, 1319-1326, 1995.

Chen, Y., and W.J. Morgan, Rift valley/no rift valley transition at mid-ocean ridges, *J. Geophys. Res.*, *95* , 17,571-17,581, 1990.

Cheng, C. H., and D. H. Johnston, Dynamic and static moduli, *Geophys. Res. Lett.*, *8*, 39-42, 1981.

Christensen, N. I., Ophiolites, seismic velocities and oceanic crustal structure, *Tectonophysics, 47*, 131-157, 1978.

Christeson, G.L., G.M. Purdy, and G.J. Fryer, Seismic constraints on shallow crustal emplacement processes at the fast-spreading East Pacific Rise, *J. Geophys. Res.*, *99* , 17,957-17,996, 1994.

Christeson, G.L., P.R. Shaw, and J.D. Garmany, Shear and compressional wave structure of the East Pacific Rise, 9°-10°N, *J. Geophys. Res.*, *102* , 7821-7836, 1997.

Cowie, P. A., and C. H. Scholz, Displacement length scaling relationship for faults: Data synthesis and discussion, *J. Struct. Geol.*, *14*, 1149-1156, 1992a.

Cowie, P. A., and C. H. Scholz, Physical explanation for the displacement length relationship of faults using a post-yield fracture mechanics model, *J. Struct. Geol.*, *14*, 1133-1148, 1992b.

Cowie, P.A., A. Malinverno, W.B.F. Ryan, and M.H. Edwards, Quantitative fault studies on the East Pacific Rise: A comparison of sonar imaging techniques, *J. Geophys. Res.*, *99*, 15,205-15,218, 1994.

Crane, K., Structural evolution of the East Pacific Rise axis from 13°10'N to 10°35'N: Interpretations from SeaMARC I data, *Tectonophysics, 136*, 65-124, 1987.

Detrick, R.S., A.J. Harding, G.M. Kent, J.A. Orcutt, and J.C. Mutter, Seismic structure of the southern East Pacific Rise, *Science, 259* , 499, 1993.

Diachok, O. I., R. L. Dicus, and S. C. Wales, Elements of a geoacoustic model of the upper crust, *J. Acoust. Soc. Am.*, *75*, 324-334, 1984.

Dziak, R.P., C.G. Fox, and A.E. Schreiner, The June-July 1993 seismo-acoustic event at CoAxial segment, Juan de Fuca Ridge: Evidence for a lateral dike injection, *Geophys. Res. Lett.*, *22* , 135-138, 1995.

Dziewonski, A., and Y. Lancelot, *Multidisplinary Observatories on the Deep Seafloor*, 229 pp., INSU/CNRS/IFREMER/ USSAC/ODP, Marseille, France, 1995.

Eberhart, G. L., P. A. Rona, and J. Honnorez, Geologic controls of hydrothermal activity in the Mid-Atlantic Ridge rift valley: Tectonics and volcanics, *Mar. Geophys. Res.*, *10*, 233-259, 1988.

Edwards, M. H., D. J. Fornari, A. Malinverno, W. B. F. Ryan, and J. Madsen, The regional tectonic fabric of the East Pacific Rise from 12°50'N to 15°10'N, *J. Geophys. Res.*, *96*, 7995-8018, 1991.

Eissa, E. A., and A. Kazi, Relation between static and dynamic Young's modulus, *Int. J. Rock Mech. Min. Sci. Geomech. Abstr.*, *25*, 479-482, 1988.

Embley, R. W., I. R. Jonasson, M. R. Perfit, J. M. Franklin, M. A. Tivey, A. Malahoff, M. F. Smith, and T. J. G. Francis, Submersible investigations of an extinct hydrothermal system on the Galapagos Ridge: Sulfide mounds, stockwork zone, and differentiated lavas, *Can. Mineral.*, *26*, 517-539, 1988.

Embley, R. W., W. Chadwick, M. R. Perfit, and E. T. Baker, Geology of the northern Cleft segment, Juan de Fuca Ridge: Recent lava flows, sea-floor spreading, and the formation of megaplumes, *Geology, 19*, 771-775, 1991.

Embley, R. W., I. R. Jonasson, S. Petersen, G. Massoth, D. Butterfield, V. Tunnicliffe, T. Parker, K. Juniper, J. Sarrazin, W. W. Chadwick, A. Bobbitt, and J. Getsiz, Investigations of recent eruptive sites on the Juan de Fuca Ridge using the ROPOS: Revisits to the North Cleft 1980s eruption area and a fast response to a June/July eruption on the CoAxial Segment (abstract), *Eos, Trans. AGU, 74*, 561, 1993.

Forslund, T., and A. Gudmundsson, Crustal spreading due to dikes and faults in southwest Iceland, *J. Struct. Geol.*, *13*, 443-457, 1991.

Fouquet, Y., A. Wafik, P. Cambon, C. Mevel, G. Meyer, and P. Gente, Tectonic setting and mineralogical and geochemical zonation in the Snake Pit sulfide deposit (Mid-Atlantic Ridge at 23°N), *Econ. Geol.*, *88*, 2018-2036, 1993.

Francheteau, J., R. Armijo, J.L. Cheminee, R. Hekinian, P. Lonsdale, and N. Blum, 1 Ma East Pacific Rise oceanic crust and uppermost mantle exposed by rifting in Hess Deep (equatorial Pacific Ocean), *Earth Planet. Sci. Lett.*, *101*, 281-295, 1990.

Gente, P., J. M Auzende, V. Renard, Y. Fouqet, and D. Bideau, Detailed geological mapping by submersible of the East Pacific Rise axial graben near 13°N, *Earth Planet. Sci. Lett.*, *78*, 224-236, 1986.

Goldstein, S.J., M.R. Perfit, R. Batiza, D.J. Fornari, and M.T. Murrell, Off-axis volcanism at the East Pacific Rise detected by Uranium-series dating of basalts, *Nature, 367*, 157-159, 1994.

Gregg, T.K.P., Investigating submarine volcanic eruption dynamics: Comparative field studies and developing methodologies, *RIDGE Events, 8* , 1-4, 1997.

Gregg, T.K.P., D.J. Fornari, M.R. Perfit, R.M. Haymon, and J.H. Fink, Rapid emplacement of a mid-ocean ridge lava flow on the East Pacific Rise at 9°46'-51'N, *Earth Planet. Sci. Lett.*, *144*, E1-E8, 1996.

Griffith, A. A., The phenomena of rupture and flow in solids, *Philos. Trans. R. Soc. London, 221*, Ser. A, 163-197, 1920.

Griffith, A. A., Theory of rupture, in *Proceedings of the First International Congress on Applied Mechanics*, edited by C. B. Biezeno and J. M. Burgers, 55-63, Waltman, Delft, 1924.

Grindlay, N. R., P. J. Fox, and P. R. Vogt, Morphology and tectonics of the Mid-Atlantic Ridge (25°-27°30'S) from Sea Beam and magnetic data, *J. Geophys. Res.*, *97*, 6983-7010, 1992.

Gudmundsson, A., Form and dimension of dykes in eastern Iceland, *Tectonophysics*, *95*, 295-307, 1983.

Gudmundsson, A., Geometry, formation and development of tectonic fractures on the Reykjanes Peninsula, southwest Iceland, *Tectonophysics, 139*, 295-308, 1987.

Gudmundsson, A., Effect of tensile stress concentration around magma chambers on intrusion and extrusion frequencies, *J. Volcanol. Geotherm. Res.*, *35*, 179-194, 1988.

Gudmundsson, A., Emplacement of dikes, sills, and crustal magma chambers at divergent plate boundaries, *Tectonophysics, 176,* 257-275, 1990.

Gudmundsson, A., Formation and growth of normal faults at the divergent plate boundary in Iceland, *Terra Nova, 4,* 464-471, 1992.

Gudmundsson, A., Infrastructure and mechanics of volcanic systems in Iceland, *J. Volcanol. Geotherm. Res., 64,* 1-22, 1995.

Gudmundsson, A., and K. Bäckström, Structure and development of the Sveinagja Graben, Northeast Iceland, *Tectonophysics, 200,* 111-125, 1991.

Haimson, B. C., and F. Rummel, Hydrofracturing stress measurements in the Iceland research drilling project drill hole at Reydarfjördur, Iceland, *J. Geophys. Res., 87,* 6631-6649, 1982.

Harding, A. J., J. A. Orcutt, M. E. Kappus, E. E. Vera, J. C. Mutter, P. Buhl, D. R. S., and T. M. Brocher, Structure of young oceanic crust at 13°N on the East Pacific Rise from expanding spread profiles, *J. Geophys. Res., 94,* 1989.

Haymon, R. M., D. J. Fornari, M. H. Edwards, S. Carbotte, D. Wright, and K. C. Macdonald, Hydrothermal vent distribution along the East Pacific Rise crest (9°09'-54'N) and its relationship to magmatic and tectonic processes on fast-spreading mid-ocean ridges, *Earth Planet. Sci. Lett., 104,* 513-534, 1991.

Haymon, R., D. Fornari, R. Lutz, M. Perfit, K. Macdonald, K. Von Damm, M. Lilley, W. C. I. Shanks, D. Nelson, M. Edwards, M. Kappus, D. Colodner, D. Wright, D. Scheirer, M. Black, H. Edmonds, E. Olson, and T. Geiselman, Dramatic short-term changes observed during March '92 dives to April '91 eruption site on the East Pacific Rise (EPR) crest, 9°45-52'N (abstract), *Eos Trans. AGU, 43,* 524, 1992.

Haymon, R. M., D. J. Fornari, K. L. Von Damm, M. D. Lilley, M. R. Perfit, J. M. Edmond, W. C. Shanks III, R. A. Lutz, J. B. Grebmeier, S. Carbotte, D. Wright, E. McLaughlin, E. Smith, N. Beedle, and E. Olson, Volcanic eruption of the mid-ocean ridge along the East Pacific Rise crest at 9°45-52'N: 1. Direct submersible observations of seafloor phenomena associated with an eruption event in April, 1991, *Earth Planet. Sci. Lett., 119,* 85-101, 1993.

Haymon, R., K.C. Macdonald, S. Baron, L. Crowder, J. Hobson, P. Sharfstein, S. White, B. Bezy, E. Birk, F. Terra, D. Scheirer, D. Wright, L. Magde, C. Van Dover, S. Sudarikov, and G. Levai, Distribution of fine-scale hydrothermal, volcanic and tectonic features along the EPR crest, 17°15'-18°30'S: Results of near-bottom acoustic and optical surveys (abstract), *Eos Trans. AGU, 78,* F705, 1997.

Hayward, N.J., and C.J. Ebinger, Variations in the along-axis segmentation of the Afar Rift system, *Tectonics, 15,* 244-257, 1996.

Head, J.W., III, L. Wilson, and D.K. Smith, Mid-ocean ridge eruptive vent morphology and substructure: Evidence for dike widths, eruption rates, and evolution of eruptions and axial volcanic ridges, *J. Geophys. Res., 101,* 28,265-28,280, 1996.

Helm, D.C., Hydraulic forces that play a role in generating fissures at depth, *Bull. Assoc. Engr. Geol., 31,* 293-304, 1994.

Hey, R.N., J.M. Sinton, M.C. Kleinrock, R.N. Yonover, K.C. Macdonald, S.P. Miller, R.C. Searle, D.M. Christie, T.M. Atwater, N.H. Sleep, H.P. Johnson, and C.A. Neal, ALVIN investigation of an active propagating rift system, Galapagos 95.5°W, *Mar. Geophys. Res., 14,* 207-226, 1992.

Huang, F., Z. Wang, and Y. Zhao, The development of rock fracture from microfracturing to main fracture formation, *Int. J. Rock Mech. Min. Sci. & Geomech. Abstr., 30,* 925-928, 1993.

Huang, P.Y., and S.C. Solomon, Centroid depths of mid-ocean ridge earthquakes: Dependence on spreading rate, *J. Geophys. Res., 93,* 13,445-13,477, 1988.

Hurst, S.D., J.A. Karson, and K.L. Verosub, Paleomagnetism of tilted dikes in fast spread oceanic crust exposed in the Hess Deep Rift: Implications for spreading and rift propagation, *Tectonics, 13,* 789-802, 1994.

Jaeger, J. C., and N. G. W. Cook, *Fundamentals of Rock Mechanics,* 3rd edition, Chapman and Hall, London, 1979.

Johnson, H. P., Tectonic regulation of the physical properties of the sea floor: The view from the Juan de Fuca Ridge, in *Proceedings of a Workshop on the Physical Properties of Volcanic Seafloor,* edited by G. M. Purdy and G. J. Fryer, 55-72, Woods Hole Oceanographic Institution, Woods Hole, MA, 1990.

Johnson, H.P., and B.L. Salem, Magnetic properties of dikes from the oceanic upper crustal section, *J. Geophys. Res., 99,* 21,733-21,740, 1994.

Kappel, E. S., and W. B. F. Ryan, Volcanic episodicity and a non-steady state rift valley along Northeast Pacific spreading centers: Evidence from Sea MARC I, *J. Geophys. Res., 91,* 13,925-13,940, 1986.

Karson, J.A., S.D. Hurst, and P. Lonsdale, Tectonic rotations of dikes in fast-spread oceanic crust exposed near Hess Deep, *Geology, 20,* 685-692, 1992.

Kent, G. M., A. J. Harding, and J. A. Orcutt, Distribution of magma beneath the East Pacific Rise between the Clipperton Transform and the 9°17'N deval from forward modeling of common depth point data, *J. Geophys. Res., 98,* 13,945-13,970, 1993.

Kidd, R.G.W., A model for the process of formation of the upper oceanic crust, *Geophys. J. R. Astron. Soc., 50,* 149-183, 1977.

Kleinrock, M.C., and S.E. Humphris, Structural asymmetry of the TAG Rift Valley: Evidence from a near-bottom survey, *Geophys. Res. Lett., 23,* 3439-3442, 1996.

Kong, L.S., S.C. Solomon, and G.M. Purdy, Microearthquake characteristics of a mid-ocean ridge along-axis high, *J. Geophys. Res., 97,* 1659-1686, 1992.

Lachenbruch, A. H., A simple mechanical model for oceanic spreading centers, *J. Geophys. Res., 78,* 3395-3417, 1973.

Langmuir, C. H., J. F. Bender, and R. Batiza, Petrological and tectonic segmentation of the East Pacific Rise, 5°30'N-14°30'N, *Nature, 322,* 422-429, 1986.

Lichtman, G. S., and J.-P. Eissen, Time and space constraints on the evolution of medium-rate spreading centers, *Geology, 11,* 592-595, 1983.

Lin, J., and E. M. Parmentier, Mechanisms of lithospheric extension at mid-ocean ridges, *Geophys. J., 96,* 1-22, 1989.

Link, H., On the correlation of seismically and statically determined moduli of elasticity of rock masses, *Felsmech. Ingenieurgeol., 4,* 90-110, 1968.

Lowell, R.P., and L.N. Germanovich, Dike injection and the formation of megaplumes at ocean ridges, *Science, 267,* 1804-1807, 1995.

Luyendyk, B. P., and K. C. Macdonald, Physiography and

structure of the inner floor of the FAMOUS rift valley: Observations with a deep-towed instrument package, *Geol. Soc. Am. Bull., 88,* 648-663, 1977.

Macdonald, K. C., Mid-ocean ridges: Fine scale tectonic, volcanic and hydrothermal processes within the plate boundary zone, *Ann. Rev. Earth Planet. Sci., 10,* 155-190, 1982.

Macdonald, K.C., and B.P. Luyendyk, Déep-tow studies of the structure of the Mid-Atlantic Ridge crest near lat 37° N, *Geol. Soc. America Bull., 88,* 621-636, 1977.

Macdonald, K., J.-C. Sempéré, and P. J. Fox, East Pacific Rise from Siqueiros to Orozco Fracture Zones: Along-strike continuity of axial neovolcanic zone and structure and evolution of overlapping spreading centers, *J. Geophys. Res., 89,* 6049-6069, 1984.

Macdonald, K. C., P. J. Fox, L. J. Perram, M. F. Eisen, R. M. Haymon, S. P. Miller, S. M. Carbotte, M.-H. Cormier, and A. N. Shor, A new view of the mid-ocean ridge from the behaviour of ridge-axis discontinuities, *Nature, 335,* 217-225, 1988.

Macdonald, K. C., D. S. Scheirer, and S. M. Carbotte, Mid-ocean ridges: discontinuities, segments and giant cracks, *Science, 253,* 986-994, 1991.

Macdonald, K. C., P. J. Fox, S. Carbotte, M. Eisen, S. Miller, L. Perram, D. Scheirer, S. Tighe, and C. Weiland, The East Pacific Rise and its flanks, 8°-17°N: History of segmentation, propagation and spreading direction based on SeaMARC II and Sea Beam studies, *Mar. Geophys. Res., 14,* 299-334, 1992.

Macdonald, K.C., P.J. Fox, R.T. Alexander, R. Pockalny, and P. Gente, Volcanic growth faults and the origin of Pacific abyssal hills, *Nature, 380,* 125-129, 1996.

Mammerickx, J., R.N. Anderson, H.W. Menard, and S.M. Smith, Morphology and tectonic evolution of the East Central Pacific, *Geol. Soc. Am. Bull., 86,* 111-118, 1975.

Mastin, L.G., and D.D. Pollard, Surface deformation and shallow dike intrusion processes at Inyo Craters, Long Valley, California, *J. Geophys. Res., 93* , 13,221-13,235, 1988.

Marrett, R., and R. W. Allmendinger, Estimates of strain due to brittle faulting: Sampling of fault populations, *J. Struct. Geol., 13,* 735-738, 1991.

Morton, J. L., N. H. Sleep, W. R. Normark, and D. H. Tompkins, Structure of the southern Juan de Fuca Ridge from seismic reflection records, *J. Geophys. Res., 92,* 11,315-11,326, 1987.

Mottl, M., E. Baker, K. Macdonald, J. Sinton, and G. Wheat, *Processes and Fluxes on a Superfast Spreading Ridge: The Southern East Pacific Rise,* RIDGE, Durham, NH, 36 pp., 1996.

Normark, W. R., Delineation of the main extrusion zone of the East Pacific Rise at Lat. 21°N, *Geology, 4,* 681-685, 1976.

Nur, A., The origin of tensile fracture lineaments, *J. Struct. Geol., 4,* 31-40, 1982.

Opheim, J. A., and A. Gudmundsson, Formation and geometry of fractures, and related volcanism, of the Krafla Fissure Swarm, Northeast Iceland, *Geol. Soc. Am. Bull., 101,* 1608-1622, 1989.

Perfit, M.R., D.J. Fornari, M.C. Smith, J.F. Bender, C.H. Langmuir, and R.M. Haymon, Small-scale spatial and temporal variations in mid-ocean ridge crest magmatic processes, *Geology, 22* , 375-379, 1994.

Pockalny, R. A., and P. J. Fox, Estimation of crustal extension within the plate boundary zone of slowly spreading mid-ocean ridges (abstract), *Eos Trans. AGU, 70,* 1300, 1989.

Pollard D. D., P. T. Delaney, W. A. Duffield, E. T. Endo, and A. T. Okamura, Surface deformation in volcanic rift zones, *Tectonophysics, 94,* 541-584, 1983.

Purdy, G. M., and G. J. Fryer, *Proceedings of a Workshop on the Physical Properties of Volcanic Seafloor,* Woods Hole Oceanographic Institution, Woods Hole, MA, 279 pp., 1990.

Rubin, A.M., Dike-induced faulting and graben subsidence in volcanic rift zones, *J. Geophys. Res., 97,* 1839-1858, 1992.

Rubin, A. M., and D. D. Pollard, Origins of blade-like dikes in volcanic rift zones, in Volcanism in Hawaii, vol. 2, edited by R. W. Decker, T. L. Wright, and P. H. Stauffer, pp. 1449-1470, *U.S. Geol. Surv. Prof. Paper 1350,* Denver, Colo., 1987.

Scheirer, D. S., M. Cormier, K. Macdonald, R. Haymon, and S. White, Sojourn Leg 1: Completing the picture of the East Pacific Rise and its flanks in the MELT area (abstract), *Eos, Trans. AGU, 77,* F663, 1996.

Scheirer, D. S., and K. C. Macdonald, Variation in cross-sectional area of the axial ridge along the East Pacific Rise: Evidence for the magmatic budget of a fast-spreading center, *J. Geophys. Res., 98,* 7871-7886, 1993.

Scholz, C. H., Mechanics of faulting, *Ann. Rev. Earth Planet. Sci., 17,* 309-334, 1989.

Schouten, H., K. D. Klitgord, and J. A. Whitehead, segmentation of mid-ocean ridges, *Nature, 317,* 225-229, 1985.

Sempéré, J.-C., and K. C. Macdonald, Deep-tow studies of the overlapping spreading centers at 9°03'N on the East Pacific Rise, *Tectonics, 5,* 881-900, 1986.

Shaw, P. R., Age variations of oceanic crust Poisson's ratio: Inversion and a porosity evolution model, *J. Geophys. Res., 99,* 3057-3066, 1994.

Shaw, P.R., and J. Lin, Causes and consequences of variations in faulting style at the Mid-Atlantic Ridge, *J. Geophys. Res., 98,* 21,839-21,851, 1993.

Shaw, P.R., and M.C. Kleinrock, Growth and evolution of normal faults at the Mid-Atlantic Ridge at the TAG area, *Geophys. Res. Lett.,* submitted.

Shearer, P. M., Cracked media, Poisson's ratio and the structure of the upper oceanic crust, *Geophys. J. R. Astron. Soc., 92,* 357-362, 1988.

Sigurdsson, O., Surface deformation of the Krafla fissure swarm in two rifting events, *J. Geophys., 47,* 154-159, 1980.

Tada, H., P. C. Paris, and G. R. Irwin, *The Stress Analysis of Cracks Handbook,* 383 pp., Del Research Corporation, Hellertown, Pennsylvania, 1973.

U.S. Geological Survey Juan de Fuca Study Group, Submarine fissure eruptions and hydrothermal vents on the southern Juan de Fuca Ridge: Preliminary observations from the submersible *Alvin, Geology, 14,* 823-827, 1986.

van Wyk de Vries, B., and O. Merle, The effect of volcanic constructs on rift fault patterns, *Geology, 24,* 643-646, 1996.

Vera, E. E., J. C. Mutter, P. Buhl, J. A. Orcutt, A. J. Harding, M. E. Kappus, R. S. Detrick, and T. M. Brocher, The structure of 0- to 0.2-m.y.-old oceanic crust at 9°N on the East Pacific Rise from expanded spread profiles, *J. Geophys. Res., 95,* 15,529-15,556, 1990.

Walsh, J. J., and J. Watterson, Analysis of the relationship between displacements and dimensions of faults, *J. Struct. Geol., 10,* 239-247, 1988.

Wright, D.J., R.M. Haymon, and D.J. Fornari, Crustal fissuring and its relationship to magmatic and hydrothermal processes on the East Pacific Rise crest (9°12'-54'N), *J. Geophys. Res., 100 ,* 6097-6120, 1995a.

Wright, D.J., R.M. Haymon, and K.C. Macdonald, Breaking new ground: Estimates of crack depth along the axial zone of the East Pacific Rise (9°12'-54'N), *Earth Planet. Sci. Lett., 134,* 441-457, 1995b.

Dawn J. Wright, Department of Geosciences, 104 Wilkinson Hall, Oregon State University, Corvallis, OR, 97331-5506, dawn@dusk.geo.orst.edu.

Ultramafic-Mafic Plutonic Rock Suites Exposed Along the Mid-Atlantic Ridge (10°N-30°N). Symmetrical-Asymmetrical Distribution and Implications for Seafloor Spreading Processes.

Yves Lagabrielle[1], Daniel Bideau[2], Mathilde Cannat[3], Jeffrey A. Karson[4] and Catherine Mével[3]

Along specific portions of the axis of the Mid-Atlantic Ridge (MAR), the spreading of the lithosphere appears to occur without abundant magmatism. Horizontal extension of the ocean floor is then accommodated mostly by tectonic stretching. Such tectonically governed processes lead to the creation of an ocean basement composed of mantle-derived ultramafics and associated mafic plutonic rocks. In this paper, we present a review of the geological and tectonic settings of the most intensively surveyed areas where such mantle and deep crustal rocks have been recovered along the central Atlantic Ridge between 15°N and 35°30'N. The regions where detailed geological observations have been made are: 1. the intersection high north of Oceanographer fracture zone; 2. two intersection highs near small-offset axial discontinuities north of Hayes fracture zone; 3. the region south of the Kane fracture zone near 23°N (MARK area); 4. the northern and southern intersections of the 15°20'N fracture zone; and 5. off-axis oblique depressions representing the trace of migrating segment discontinuities in the MARK area. Mantle-derived serpentinites and associated gabbros are commonly exposed along only one side of the axial valley wall, that is, asymmetrically with respect to the spreading axis, while basaltic formations are found on the opposite wall. However, we emphasize that such an asymmetrical distribution in the topography, structure, and geology of the axial valley is not observed all along the studied areas of the MAR, for example in the 15°N region where serpentinized peridotites are documented across wide areas on both sides of the axis. Two major points are discussed. (1) The processes leading to the unroofing of mantle

[1] Centre National de la Recherche Scientifique, Institut Universitaire Européen de la Mer, Plouzané, France

[2] Institute Français de Recherche pour l'Exploitation de la Mer, Plouzané, France

[3] Centre National de la Recherche Scientifique, Université Pierre & Marie Curie, Paris, France

[4] Duke University, Department of Geology, Durham, North Carolina, USA

Faulting and Magmatism at Mid-Ocean Ridges
Geophysical Monograph 106
Copyright 1998 by the American Geophysical Union

ultramafics and associated rocks are not necessarily linked to a stage of amagmatic extension of a previously constructed thick crust; some appear to occur in areas of permanent low magma budget, at segment ends or within wider regions of starved magma production. Such regions may retain this character over millions of years as shown by the geology of the off-axis traces of axial discontinuities. (2) Since geological settings with a symmetrical distribution of deep crustal and mantle rocks on the valley walls are found, a model of asymmetrical extension and denudation along a single major detachment fault cannot be applied all along the ridge axis. This implies that more attention must be paid to the degree of across-axis asymmetry or symmetry of tectonic processes responsible for the uplift and exposure of mantle-derived and lower crustal rocks.

1. INTRODUCTION

Numerous geological, geophysical, and petrological data collected during the past 20 years along the axis of the Mid-Atlantic Ridge, have revealed a typical fabric composed of successive individual spreading segments. Each segment generally comprises a bathymetric high in the center and bathymetric deeps at both ends. The wavelength of the segmentation is of the order of 20 to 100 km. Off-axis investigations at 32°-29°N, 29°-28°N, 24°-20°N, 32°S-34°S [Fox et al., 1991; Rommevaux et al., 1994; Gente et al., 1994; Sempéré et al., 1995] have shown that the length of segments may vary over time, leading to a typical regional fabric consisting of a succession of highs separated by a network of depressions. The alignment of the depressions does not remain orthogonal to the ridge axis thus defining a more or less oblique pattern.

A negative residual gravity anomaly at the central part of the segments suggests relatively thick mafic crust and hot mantle at depth, while the segment tips show a positive residual gravity anomaly suggesting that they have a thin mafic or serpentinite crust and are characterized by a relatively starved magmatic activity [Kuo and Forsyth, 1988; Lin et al., 1990; Blackman and Forsyth, 1991; Grindlay et al., 1991; Lin and Phipps Morgan, 1992; Escartin and Lin, 1995; Shaw and Lin, 1996]. Petrological investigations also indicate that the center of the segments have a higher magma supply than the ends and that magma chambers at the segment tips are small and not continuous [Niu and Batiza, 1994]. Seismic experiments confirm that the crust is thicker (8 km) in the center and thinner (2-4 km) at the end of the spreading segments [Tolstoy et al., 1993; Wolfe et al., 1995].

Many authors have pointed out that the mechanisms responsible for the tectonic denudation of plutonic rock suites at segment tips are basically asymmetrical processes [Karson and Dick, 1983; Karson, et al., 1987; Karson, 1990; Mutter and Karson, 1992; Tucholke and Lin, 1994; Cannat, 1996]. Asymmetrical extension accounts indeed for the presence of common topographic features such as the inside and outside corners of first-order discontinuities with a typical asymmetrical elevation [Fox and Gallo, 1984; Severinghaus and Macdonald, 1988]. It accounts also for the asymmetrical distribution of ultramafic-plutonic rock suites at many segment tips. In most models of segment geology, the inside corner exposes mafic plutonic rocks and serpentinized mantle ultramafics, whereas the outside-corner wall of the rift valley and the valley itself expose basaltic formations [Karson and Dick, 1983; Karson et al., 1987; Karson and Rona, 1990; Karson, 1990; Mével et al., 1991; Cannat, 1993; Tucholke and Lin, 1994; Karson and Lawrence, 1997a]. Calculations suggest that normal faulting along low-angle faults that penetrate the lower crust an upper mantle at inside corners result in an extreme thinning of the crust and may lead to deep crustal and mantle-derived rock exposures [Escartin and Lin, 1995]. Thermo-mechanical analog models yield insights into the tectonic processes of thinning and related uplift of the lithosphere at slow-spreading centers [Shemenda and Grocholsky, 1994].

In this paper, we first review the detailed geology and structure of the regions where serpentinized peridotites and associated gabbros have been recovered. We show that the asymmetrical distribution in the topography, the structure and the geology of ridge segments is not observed all along the MAR where ultramafic rocks have been recovered. This implies that a simple model of asymmetrical extension and denudation cannot be applied all along the MAR and more generally along all slow-spreading ridge segments. We then address some basic questions related to the tectonic processes leading to the unroofing of ultramafic and plutonic rocks at active spreading centers, and to their links with the local- and regional-scale magmatic activity.

2. A REVIEW OF THE OCCURRENCES OF DEEP CRUSTAL AND MANTLE ROCKS ALONG THE MAR AXIS

Seafloor sampling along the MAR has revealed numerous occurrences of ultramafic-plutonic exposures.

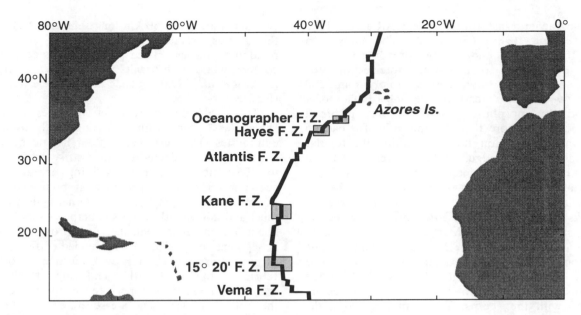

Figure 1. Simplified map of the axial zone of the Mid-Atlantic Ridge and location of the regions discussed in text.

Compilations of published occurrences obtained by dredging, drilling or direct sampling from diving submersibles have been presented over the years by *Fox and Heezen* [1965]; *Bonatti*, [1976]; *Dick et al.* [1984]; *Michael and Bonatti* [1985]; *Dick* [1989]; *Juteau et al.* [1990b]; *Lagabrielle and Cannat* [1990]; *Cannat et al.* [1992]; *Bonatti et al.* [1992]; *Cannat* [1993]; *Tucholke and Lin* [1994], *Cannat et al.* [1995]; *Karson and Lawrence* [1997a, 1997b]. All compilations show that mafic and ultramafic plutonic rocks are exposed within a variety of tectonic settings : along the walls of the axial valley, along the inactive and active branches of transform valley walls, and along oblique depressed regions of the ridge flanks which represent the off-axis track of axial discontinuities. In this paper, we focus on serpentinized peridotites and gabbros exposed along the axial valley walls and along oblique discontinuities. However, since crucial information related to the emplacement of mantle serpentinites and mafic plutonic rocks at the ridge axis has also been obtained along specific sections of fracture zones, additional data obtained along transform fault walls will be presented together with data from the axial zone.

The regions where detailed geological observations have been made during submersible dives or deep towed camera surveys are relatively few, from north to south, they are (Figure 1):

-the intersection high north of Oceanographer fracture zone [*OTTER Team*, 1984];

-two intersection highs near small offset axial discontinuities north of Hayes fracture zone [*Bideau et al.*, 1996; *Gracia et al.*, 1997];

-the region south of the Kane fracture zone near 23°N (Mid-Atlantic Ridge south of Kane or MARK area) [*Karson et al.*, 1987; *Brown and Karson*, 1988; *Mével et al.*, 1991; *Karson and Lawrence*, 1997a, 1997b; *Cannat et al.*, 1997a];

-the northern and southern intersections of the 15°20'N fracture zone [*Rona et al.*, 1987; *Bougault et al.*, 1993; *Cannat and Casey*, 1995; *Cannat et al.*, 1997b];

-parts of off-axis oblique depressions representing the trace of migrating segment discontinuities in the MARK area [*Cannat et al.*, 1995; *Durand et al.*, 1996; *Gente et al.*, 1996].

The geological data sets from the MARK area and from 15°N are the most substantial and include information from a wide array of research tools. In these two regions, ultramafic and gabbroic outcrops are not restricted to ridge-transform intersection massifs, but extend 45 km (MARK area) to 55 km (15°N region) along axis from the Kane or Fifteen Twenty fracture zones, respectively.

Dredges and drill holes have also documented numerous axial and off-axis occurrences of serpentinized peridotites and gabbros in various settings, mostly in the areas discussed in this paper. Additional important dredge sites are located near 26°N [*Tiezzi and Scott*, 1980] and 16°52'N [*Cannat et al.*, 1992].

2.1. The Eastern Ridge-Transform Intersection at the Oceanographer Fracture Zone

Deep-towed camera surveys and *Alvin* dives have been made at the eastern ridge-transform intersection of the Oceanographer transform with the MAR axis in 1980 [*OTTER Team*, 1984] (Figure 2). The study area includes the southern end of the spreading segment, north of the fracture zone, and the intersection area itself. The axial valley floor deepens rapidly from 3200 meters below sealevel (mbsl) to 4600 mbsl in the nodal basin and the morphology evolves into a broad asymmetric valley more than 15 km wide. The eastern wall of the intersection domain retains the regular tectonic fabric of the ridge, while the western wall, at the inside corner shows numerous oblique scarps. Dives in the axial valley recovered only recent pillow basalts from various constructions such as short ridges, "haystacks" and hummocky flows often disected by recent fissures and faults. Marks of active tectonics have been observed along the inside flank of the intersection. Vertical scarps exposing semi-consolidated carbonate sediments are regularly spaced along the slope. Fresh talus and down-slope trending ravines are associated with the scarps. The sampling along the western wall of the intersection domain is not so dense that one may accurately assess the distribution and spatial organisation of the rock types. Only extrusive or minor intrusive basalts and serpentinized peridotites were recovered during this field expedition and no gabbroic rocks were sampled. The serpentinized peridotites have been collected in place along few active scarps in the upper slope of the intersection flank around 3000-2800 mbsl while strongly altered basalts coated with manganese oxides and associated doleritic dikes cutting through them were sampled downslope around 3200 mbsl. As at many intersection massifs, mantle rocks are found at a shallower level than basaltic formation. The eastern flank of the intersection domain only exposes pillow basalts and shows the most active scarps at its base, close to the axial valley.

Despite limited submersible coverage, this region exhibits a number of characters which are found at other similar settings as shown by the examples presented below: asymmetrical cross-sectional topographic profile; asymmetrical distribution of the rock types with shallow exposures of mantle-derived rocks at the inside corner and relatively old basaltic formations cropping out along the opposite wall; recent volcanic rocks occurring on the inner floor of the axial valley which could be coeval with tectonic activity along both walls of the intersection (Figure 2).

2.2. Segment OH3, North of the Hayes Fracture Zone

The MAR axis between the Oceanographer and the Hayes fracture zones consists of 3 major second-order segments, 50 to 80 km long (segments OH1, OH2, OH3), separated by non-transform discontinuities [*Detrick et al.*, 1995] (Figure 3). Recent dives during the Oceanaut cruise of the submersible *Nautile* in 1995,

conducted along the 50 km long segment OH3, north of the Hayes fracture zone recovered serpentinized peridotites and associated plutonic rocks at both inside corners, while fresh basalts were sampled at the center [*Bideau et al.*, 1996]. Outside corners were not sampled (Figure 3).

The axial valley of segment OH3 presents a typical "hour glass" shape, with increasing width towards both extremities. The inner valley floor has no neovolcanic ridge. It shows a dome-like along-strike profile shoaling to 3050 mbsl. The most striking characteristic of segment OH3 is the presence at both ends of two asymmetrical topography. Inside corner high rise to 2200 mbsl and 2400 mbsl at the southern and northern ends, about 500 m deeper than the summit of the ridge crest at segment center. They form rounded edifices, 5-10 km across, contrasting strongly with the linear, north-south fabric of the axial valley and ridge flanks. These intersection massives appear to be disconnected from the rest of the walls, and are separated from the region of ridge-parallel fabric by an E-W trending depressed area. Two dives (OT14 and OT8) were conducted along the flanks of the northern and southern massifs respectively, during the Oceanaut cruise [*Bideau et al.*, 1996; *Gracia et al.*, 1997] (Figure 4).

2.2.1. Northern intersection massif.

The northern inside-corner massif extends from depths of 3500 mbsl to 2200 m, with an average slope of 25°. The floor of the axial valley at the foot of the massif consists of a ridge formed by pillow lavas partly hidden under a relatively thick sedimentary cover. Serpentinized peridotites and subordinate dolerites were sampled continuously from 3000 mbsl to 2500 mbsl along the basal part of the intersection massif. Visual observations during dive OT14 (Figure 4) suggest that the dolerites are shallow-level intrusions of basaltic melt within the peridotites. Outcrops of massive dolerites are surrounded by rubble of mixed serpentinites and dolerites (*Bideau et al.*, unpublished data). Gabbroic veinlets are also present in some serpentinite samples. At 3000 mbsl a major vertical active fault is observed along a sharp scarp, 10 m high, that cuts through pelagic sediments and ultramafic breccias.

2.2.2. Southern intersection massif.

Very fresh basaltic pillows and draped lavas are exposed on the floor of the axial valley at the base of the inside corner slope (Figure 4). Serpentinized peridotites are exposed from the foot of the valley wall at 3800 mbsl up to the summit of the massif at 2400 mbsl; these occur along the slopes as more or less sedimented talus and as outcrops of massive serpentinized peridotites near the top of the massif. Rare

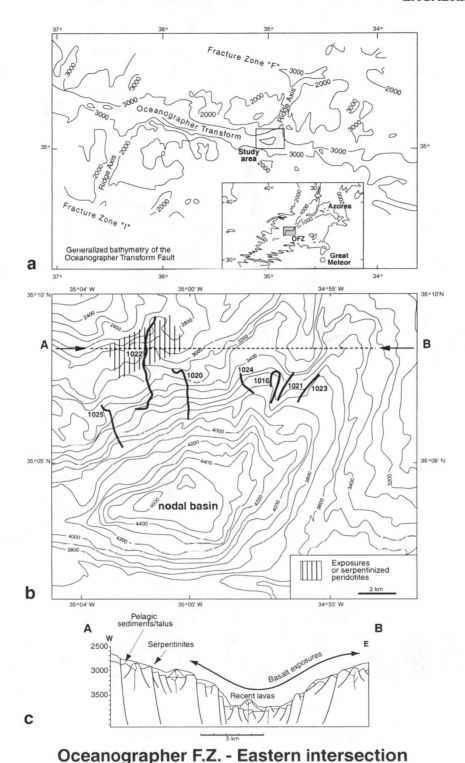

Oceanographer F.Z. - Eastern intersection

Figure 2. The Oceanographer transform fault after *OTTER* [1984]. (a) : Generalized bathymetry; (b) : detailed bathymetry of the eastern intersection with the ridge axis; (c) : corresponding cross section AB. The tracks of the Alvin dives are shown as well as the area of serpentinite exposures. Cross section AB only shows the surficial extension of exposures, but not the inferred internal structure as shown in the original figure by *OTTER* [1984].

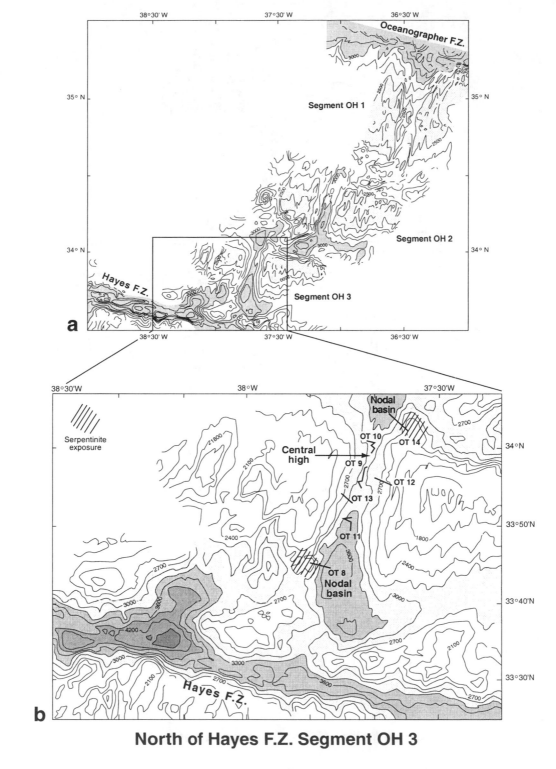

North of Hayes F.Z. Segment OH 3

Figure 3. The OH3 segment, north of the Hayes fracture zone, after *Bideau et al.*, [1996]. (a) : Location map of segment OH3 between the Oceanographer and the Hayes frature zones; (b) : detailed bathymetry of the OH3 segment, the location of Nautile dive tracks and the possible extension of serpentinite exposures at both intersection massifs are shown.

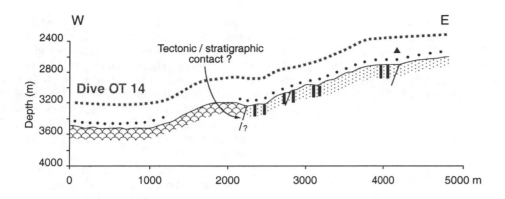

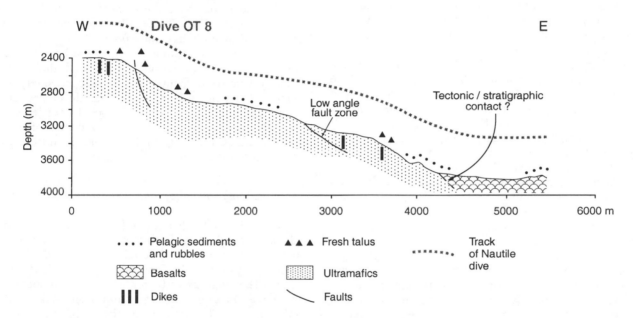

Figure 4. Geological cross-sections established along the tracks of Nautile dives OT 14 and OT 8 of the OCEANAUT cruise [*Bideau et al.*, 1996]. See location of dive tracks on Figure 3.

diabase samples were also recovered in two areas near 3400 mbsl. The geological relationships between the mafic and the ultramafic rocks are unclear, but the presence of gabbro and diabase dikelets within serpentinized peridotite samples suggest that mafic melts intruded the peridotites during their progressive uplift.

Vertical scarps are numerous, but the marks of recent tectonism have not been observed at the base of the slope. One main vertical fault scarp cross-cutting the ultramafics is present in the upper part of the massif. The most striking observation is a major, low-angle fault surface affecting the serpentinites at 3200 mbsl. Smooth serpentinites occur in N-S trending exposures that dip 40° to the east and show large dip-parallel striae, 0.5 to 1 m wide. Samples from the fault surface are fine-

grained mylonites confirming that the deformed zone exposes a major tectonic discontinuity. Discrete vertical faults and open fissures delineate small steps within the mylonitic outcrops.

2.3. The MARK Area

The Mid-Atlantic Ridge south of the Kane fracture zone (MARK area) has been the site of numerous cruises which collected diverse types of data from both surface-ship and direct seafloor investigations. In this region, the rift valley has an asymmetrical cross-sectional profile with a shallower and steeper western wall. Serpentinized upper mantle peridotites and associated mafic plutonic rocks are commonly exposed on the steepest and shallowest parts of the axial valley wall [*Karson et al.*, 1987; *Brown and Karson*, 1988; *Mével et al.*, 1989; *Cannat et al.*, 1990; *Juteau et al.*, 1990a; *Gente et al.*, 1991]. The 80 kilometers of ridge axis south of the Kane fracture zone has been divided into three sections on the basis of morphology, geology, and geophysics (Figure 5). Here we focus on the 35 km-long northern section and adjoining areas.

2.3.1. The western axial valley wall.

Sheared gabbros and minor foliated serpentinites are exposed along the northeast curved flank of the ridge transform intersection (RTI) massif (also refered to as the inside-corner massif), at the Kane fracture zone. The gabbros are commonly highly deformed and they are cut by relatively early low-angle normal faults and shear zones and by more recent vertical normal faults. This region has been the site of numerous submersible and drilling cruises providing a detailed description of the geology of the RTI massif (Figure 5). The surface geology is constrained by 16 Alvin and Nautile submersible dives [*Karson and Dick*, 1983; *Mével et al.*, 1991; *Karson and Winters*, 1992; *Auzende et al.*, 1993; *Karson and the SMARK cruise participants*, unpubl. data].

The gabbroic rocks extend 15 km along the ridge axis and 8-10 km normal to the ridge axis. During ODP Leg 153, ten holes were drilled at 4 Sites (921-924) on the western axial valley wall providing subsurface samples [*Shipboard Scientific Party*, 1997]. Here, the axial valley wall is essentially a dip slope (i.e., foliation-parallel exposure) on a major crustal detachment fault, itself cut by numerous late high-angle normal faults. Gabbros recovered from the drill sites are not part of a massive, continuous, mafic plutonic layer, but instead represent relatively small, variably deformed and metamorphosed gabbroic plutons, each probably less than one kilometer across. Detailed investigations indicate a complex history of successive intrusion and deformation of mafic plutons and isolated dikes, in a region characterized by very low or episodic magma budget.

A 20 km-long belt of serpentinites is present south of the gabbro exposures. The first investigation of the serpentinites was carried out by the submersible *Alvin*, in 1986, at 23°22'N [*Karson et al.*, 1987], followed by the Hydrosnake and the Gravinaute cruises of the *Nautile* [*Mével et al.*, 1989; *Durand et al.*, 1996], the SMARK cruise of the *Alvin* [*Karson and Lawrence*, 1997a,b]. The serpentinized peridotites were drilled during ODP legs 109 and 153 at Sites 670 and 920 [*Shipboard Scientific Party*, 1988; *Juteau et al.*, 1990a, b; *Shipboard Scientific Party*, 1997]. Dive observations show that the serpentinized peridotites crop out over a region at least 20 km long and 2 km wide, i.e. about half the length of the ridge segment. Two types of faults have been observed. Low-angle faults dipping about 40°

east are parallel to pervasive schistosity in the serpentinites. Such planes reflect deformation in the conditions of the greenschist metamorphic facies [*Mével et al.*, 1991]. These planes are intersected by more recent high-angle normal faults which form east-facing scarps 10 m to 100 m high. The high-angle faults are still active, as shown by the presence of fresh talus on top of unconsolidated sediments at the base of fault scarps [*Karson et al.*, 1987; *Mével et al.*, 1991; *Karson and Lawrence*, 1997b].

Recent pillow lavas are exposed on the adjacent axial valley floor . The contact between the basalts and the ultramafics is buried beneath talus of serpentinites and basalt fragments [*Mével et al.*, 1991]. At the upper part of the wall, above the serpentine belt, basalts are also present. They are older and more heavily sedimented than on the valley floor.

Mineralogical studies of the gabbroic dikelets trapped within the mantle suggest that the gabbroic rocks crystallized from melts of progressively more fractionated composition, as the mantle cooled during its upward displacement below the axial floor [*Cannat et al.*, 1997b]. This is similar to the model proposed by Cannat and Casey [1995] for gabbroic intrusives in serpentinized peridotites of the 15°20' region.

2.3.2. The axial valley and the eastern wall.

The axial valley has a continuous neovolcanic ridge (Snake Pit ridge) where recent volcanic construction and intense hydrothermal activity are focused (Figure 5) [*Karson et al.*, 1987; *Kong et al.*, 1988; *Gente et al.*, 1991]. The ridge extends more than 40 km south from the Kane fracture zone and culminates at a depth of 3490 mbsl. The flanks of the ridge are weakly tectonized and exhibit lava flow fronts [*Karson et al.*, 1987; *Brown and Karson*, 1988; *Karson and Brown*, 1988; *Mével et al.*, 1989; *Gente et al.*, 1991].

The east flank of the axial valley is known from bathymetry, from a few widely spaced camera runs and from submersible dives [*Karson and Dick*, 1983; *Karson et al.*, 1987; *Dubois et al.*, 1994a,b; *Durand et al.*, 1996]. Results of the Gravinaute cruise of the submersible Nautile have shown that only pillow lavas are exposed from the inner valley floor to the top of the eastern wall [*Durand et al.*, 1996]. The east flank is cut by fault scarps of less than 30 m height at the bottom, to more than 50 m high near the summit. Most of the scarps correspond to N-S trending high-angle normal faults that cut through pillow lavas or more massive basalts. The faults dip 60-80° both east and west, indicating that the architecture of the wall is more complex than a succession of simple steps. Basalts of the eastern wall extend north into the Kane fracture zone where they form the eastern side of the nodal basin.

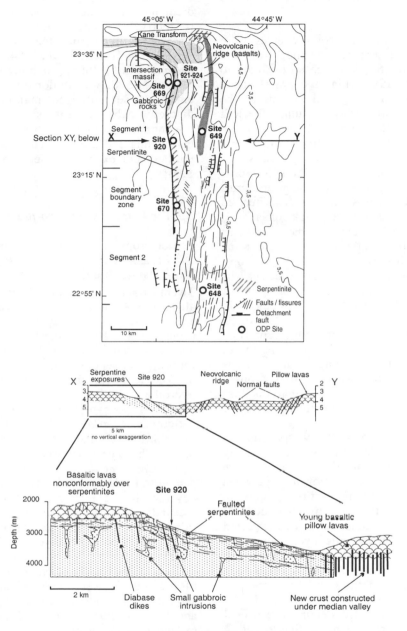

Kane F.Z. Mark Area

Figure 5. Geology of the axial region of the MAR south of the Kane fracture zone after *Detrick et al.* [1988], *Karson et al.* [1987], *Mével et al.* [1991], *Shipboard Scientific Party* [1997], and *Karson and Lawrence* [1997a]. The map shows the location of serpentinite exposures (hatched pattern : south of the intersection massif and on the north wall of the nodal basin) and gabbros exposures (grey pattern : along the flank of the intersection massif). Cross section XY shows the structural setting of serpentinite exposures in the vicinity of ODP site 920 after *Karson and Lawrence* [1977a]. The EW section below is after *Karson and Lawrence* [1977a]. It illustrates the possible geologic relations around Site 920 based on submersible and drilling results in the hypothesis of a seafloor spreading with a very low and intermittent magma supply.

2.3.3. Additional data from the southern wall of the Kane transform valley.

During the Kanaut cruise, 20 Nautile dives were performed in the Kane Fracture Zone, from its intersection with the Mid-Atlantic Ridge up to 80 km to the west [*Auzende et al.*, 1993] (Figure 6). Most of the dives have been conducted along four massifs of the southern wall of the fracture valley, including the RTI massif, separated by NS-oriented depressions at intervals of about 20 km. The dives show that serpentinized peridotites, gabbros and basalts are exposed along major normal and strike-slip faults. Moreover, the outcrops suggest that the magmatic crust in this area is very heterogeneous and is characterized by reduced thickness and by the lack of a well developed dike complex. The geology of the central part of the RTI massif on the southern wall of the Kane farcture zone is complex (Figure 6, X'Y' section). Outcrops on the lower part of the slope are metabasalts and dolerites with pervasive shear planes dipping to the NE. The middle slope has major EW-trending fault planes dipping 35° to 50°N (towards the transform valley) and exposing spectacular mylonitic zones in metagabbros and associated metaperidotites. The top of the massif lacks a typical dike complex and consists only of basaltic pillows, breccias and flows exposed above 2500 m. These basaltic formations are undeformed and were most probably emplaced after the ductile deformation of the deep rocks.

The three massifs explored to the west along the fracture wall show very similar although more heavily sedimented crustal sections. Crucial observations have been made there regarding the relationships between mafic rocks and serpentinized peridotites (Figure 6, section XY). The high-standing massifs do not correspond to thick volcanic constructions as expected by a model of magmatic-amagmatic cyclic evolution [*Pockalny et al.*, 1988]. Instead, serpentinites with isolated gabbroic dikes, typically 1 m thick, have been found demonstrating that melts have been trapped within the mantle at a relatively high level within the lithosphere. Massive gabbros also occur in restricted lenses, a few kilometers across at most. Sheeted dikes such as those found in some ophiolite complexes, were not observed along the transform valley wall, and where present, the basaltic cap is surprisingly thin (100-200 m).

2.4. Off-Axis Oblique Depressions in the MARK Area

Submersible sampling as well as systematic dredging have been conducted off axis along the oblique depressions which characterize the ridge flanks of the MARK area up to a few million years old [*Gente et al.*, 1991; 1995] (Figure 7). Submersible studies and dredging during the *Gravinaute* and *Seadmall* cruises, respectively, were conducted within the oblique depressions intersecting the MAR axis around 23°10'N and 22°10'N [*Dubois et al.*, 1994a,b; *Cannat et al.*, 1995; *Durand et al.*, 1996]. Sampling confirms that these regions are thin-crust domains made of tectonically emplaced serpentinized mantle rocks, intruded by discrete gabbroic intrusions and covered by a discontinuous, thin cap of basalts. These results suggest that mantle and deep crustal rocks can be continuously uplifted and exposed at segment tips [*Cannat et al.*, 1995].

2.5. The Fifteen Degrees North Region

The outcrops of mantle-derived ultramafic rocks in the 15°N region are the most extensive yet reported for the Mid-Atlantic Ridge (Figure 8). North of the Fifteen Twenty fracture zone, they form a belt at least 20 km along the west wall of the axial valley (Figures 8 and 9) and they also crop out at least locally on the east wall (Figures 8 and 10). South of the Fifteen Twenty fracture zone, ultramafic rocks with minor gabbros and basalts form a prominent, 30 km long, transverse ridge (Figure 8). The 15°N region is characterized by methane-rich hydrothermal plume signatures in the water column related to the hydrothermal alteration of serpentinized peridotites [*Charlou et al.*, 1991]. Geological maps and descriptions of the 15°N ultramafic and gabbroic outcrops have been published previously [*Bougault et al.*, 1993; *Casey et al.*, 1992; *Cannat and Casey*, 1995; *Cannat et al.*, 1997b]. The complete bathymetric coverage in the 15°N region was acquired during the recent Faranaut cruise covering 25 to 35 km on each ridge flank. A detailed account of the geology of this region, based on this bathymetric data and on the results of 23 dives of the submersible *Nautile* (Faranaut cruise), can be found in *Cannat et al.* [1997b]. A number of previous cruises recovered abundant gabbros and serpentinized peridotites in the region between 14°42'N and 15°53'N [*Rona et al.*, 1987; *Skolotnev et al.*, 1989] (*Silantyev et al.*, submitted).

Striking features of the bathymetry both north and south of the Fifteen Twenty fracture zone are wide domains of rugged morphology extending down to 14°30'N south of the Fifteen Twenty fracture zone, and up to 15°50'N north of the fracture zone [*Cannat et al.*, 1997b]. In these domains, ridge-parallel abyssal hills are rare, while short inward, outward, or oblique facing scarps are common. In other regions of the Atlantic, this type of rugged morphology occurs in the off-axis traces of axial discontinuities [*Sloan and Patriat*, 1992; *Shaw and Lin*, 1993; *Cannat et al.*, 1995; *Neuman and Forsyth*, 1995].

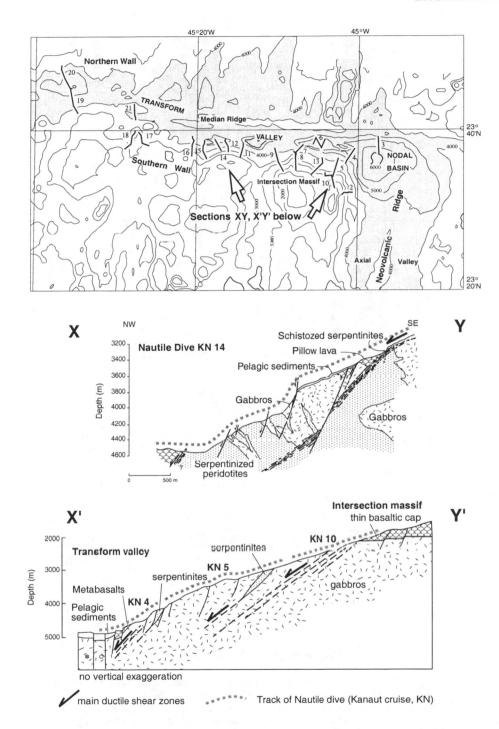

X
NW   SE   Y

Nautile Dive KN 14

Schistozed serpentinites
Pillow lava
Pelagic sediments
Gabbros
Gabbros
Serpentinized peridotites

0   500 m

X'   Intersection massif   Y'
                  thin basaltic cap

Transform valley   serpentinites   KN 10
                   serpentinites   KN 5
Metabasalts   KN 4   gabbros
Pelagic sediments

no vertical exaggeration

↙ main ductile shear zones   ······· Track of Nautile dive (Kanaut cruise, KN)

Kane F.Z. , southern wall and intersection massif

Figure 6. Geology of the southern wall of the Kane fracture zone and of the intersection massif after *Detrick et al.* [1988], *Karson et al.* [1987], *Mével et al.* [1991], and *Auzende et al.* [1993; 1994]. Geological sections XY and X'Y' are based on submersible observations made during the KANAUT cruise [*Auzende et al.*, 1993; 1994]. In both sections, the outcrop features suggest that the crust is highly heterogeneous and characterized discontinuous gabbroic intrusions and the lack of a well developed dike complex.

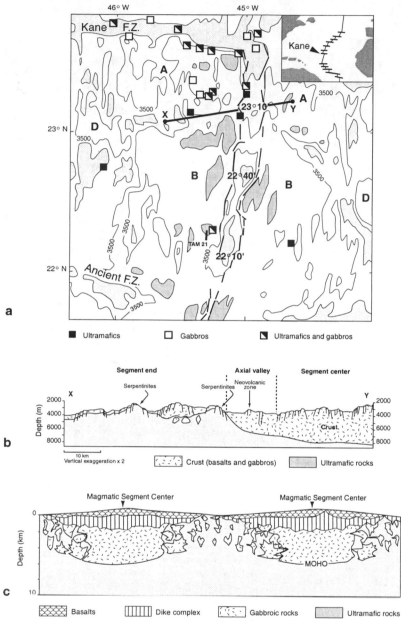

Mark area - South Kane

Figure 7. Simplified geology of the MAR axis and flanks south of the Kane fracture zone based on bathymetry, dredging, and drilling data compiled by *Cannat et al.* [1995] and *Durand et al.* [1996].

a : Simplified bathymetric map after *Gente et al.* [1991] showing dredges, drilling and diving sampling localities of ultramafic rocks and gabbros after a compilation by *Cannat et al.* [1995]. Heavy lines represent the major scarps of the axial valley. A, B, C, D refer to the polygonal domains bounded by oblique off-axis traces of segment discontinuities. Present-day non-transform discontinuities are labelled bt latitude(23°10'N to 22°10'N). Recent Nautile dive TAM 21 on a serpentine massif close to the 22°10'N discontinuity is indicated [*Gente et al., 1996*].

b : Simplified geological section XY modified from *Durand et al.* [1996]. This section crosses : 1) oblique discontinuity between domain A and B; 2) axial valley and the western serpentine belt; 3) segment center of east flank.

c : Sketch showing along-axis section of two idealized spreading segments, simplified after *Cannat et al.* [1995]. Magmatic crust is shown to be continuous at segment centers and becomes progressively thinner and more discontinuous at segment ends where mantle-derived rocks can be exposed.

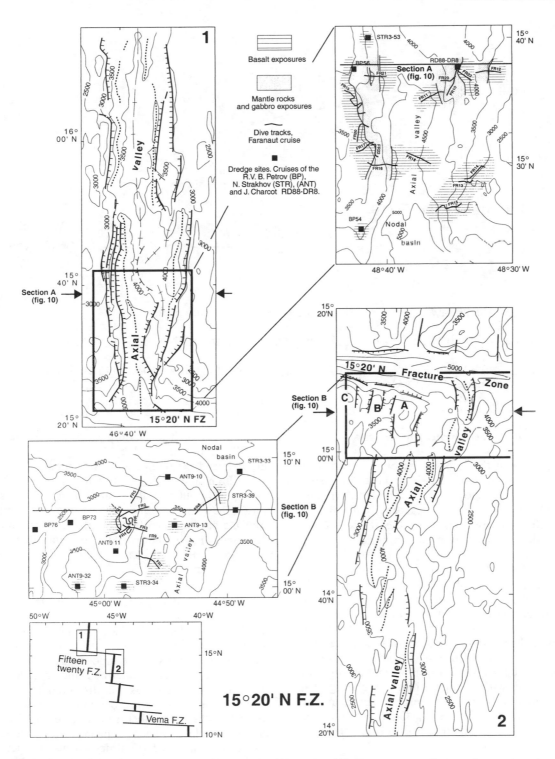

Figure 8. Simplified bathymetry and geology of the Fifteen Twenty Fracture zone intersection regions based on figures from *Cannat and Casey*, [1995] and *Cannat et al*. [1997b]. Detailed maps show the location of Nautile dive tracks conducted during the FARANAUT cruise [*Bougault et al.*, 1993] and the extent of ultramafic-gabbroic and basaltic rocks along the axial valley and walls. Location of dredge sites are shown according to *Bougault et al.* [1993], *Dick et al*. (unpublished data) and *Silantyev et al.* (submitted).

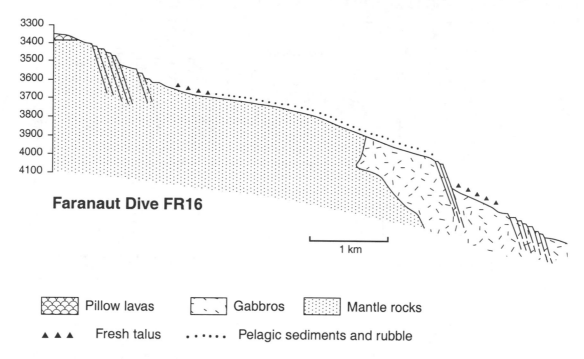

Faranaut Dive FR16

1 km

▨ Pillow lavas   ⬡ Gabbros   ▦ Mantle rocks

▲ ▲ ▲ Fresh talus   ⋯⋯ Pelagic sediments and rubble

Figure 9. Simplified geological section along the western flank of the axial valley at 15°30'N, according to observations made during FARANAUT dive 16.

2.5.1. South of the fifteen twenty fracture zone.

Between 15°N and the intersection with the fracture zone, the west ridge flank is the shallowest. It consists of a succession of rounded massifs (labelled A, B and C; Figure 8) and forms the eastern end of the anomalously elevated south wall of the fracture zone. Dredging has shown that serpentinized peridotites crop out on both the east and west walls of the axial valley, and on massifs A, B, C. (Figure 8) [*Skolotnev et al.*, 1989] (*Dick et al*, unpublished data; *Silantyev et al.*, submitted). Basalts crop out in the floor of the axial deeps, and form a small cap on top of massif A.

Dive observations show that the serpentinized peridotites form discontinuous outcrops alternating with serpentinite rubble. The ultramafic rocks are mostly serpentinized harzburgites with very small amounts of clinopyroxene and with compositions suggesting that they are residues from a high percentage of mantle melting [*Cannat and Casey*, 1995]. Gabbros have been sampled as dikes and dikelets in the serpentinites [*Cannat and Casey*, 1995].

Serpentinite exposures include massive or brecciated rocks and striated fault surfaces. Brecciation resulted from pervasive fracturing accompanied by the growth of carbonate minerals in the fractures network. The striated fault planes are cataclastic serpentinites, up to one meter thick, with millimeter-scale clasts in serpentine gouge. Most fault planes dip 30 to 50° to the NE or to the SW. Scarps formed by later, steeper faults in the serpentinized peridotites are also common, they are 1 to 30 m high and trend dominantly N-S to NNE-SSW. Basalts crop out as pillows and tubes with fresh glassy rims. They are commonly covered with sediments. Contacts between the basalts and the nearby ultramafics are concealed by talus and sediment. Basalts in the axial deep are at greater depths than neighboring serpentinite outcrops.

2.5.2. North of the fifteen twenty fracture zone.

The two ridge flanks are markedly asymmetrical. North of 15°50'N, the west rift valley wall is steeper than the east rift wall. The topography to the east of the axial valley is not well defined and includes a 4 to 8 km-wide hill to the east of the axial deep. South of 15°38'N, the axial valley becomes narrower and well defined by the 4000 mbsl isobath. South of 15°30'N, the east ridge flank forms a prominent ridge-transform intersection massif 2100 m deep near the fracture zone.

Serpentinized peridotites, with minor gabbro, form a continuous belt on the western wall of the axial deep between 15°29'N and 15°39'N (Figures 8, 9 and 10). Serpentinized peridotites and gabbros also crop out extensively along the complex eastern wall of the median valley near 15°36'N. Basalts crop out in the

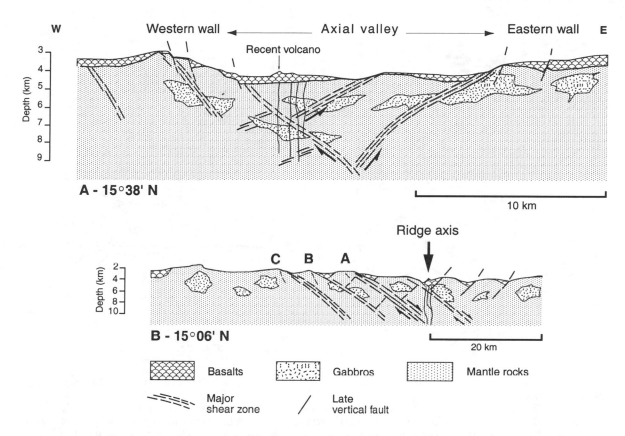

Figure 10. Idealized cross sections of the MAR north and south of the 15°20'N fracture zone [modified from *Cannat et al.*, 1997b) and (*Sylantiev et al.*, submitted). The major faults that accommodate the tectonic extension of the axial lithosphere are acting on both sides of the valley. At 15°38'N serpentinized peridotites and gabbros are exposed at three localities along the section on both sides of the valley. Some of the basalts of the axial valley may have been erupted over the serpentinites. Sections A, B and section C are located on Figure 8. Note that sections C is not at the scale of sections A and B.

axial valley and along the eastern ridge flank near the transform. Serpentinized peridotites have, however, been dredged off-axis on the ridge-transform intersection massif (at 15°25'N and 46°25'W; Silantyev et al., submitted). Outcrops of basalts also cap the ultramafic and gabbroic rocks on top of the western ridge flank and basalt fragments have been sampled above the eastern ridge flank ultramafic and gabbroic outcrops near 15°36'N.

The ultramafic rocks are serpentinized harzburgites with small amounts of clinopyroxene and rare serpentinized dunites. Their mineral compositions suggest that they are residues from a lower, although high, percentage of mantle melting than ultramafics sampled south of the Fifteen Twenty fracture zone [*Cannat and Casey*, 1995]. They have fine-grained and strongly foliated porphyroclastic textures suggestive of high strain in the ductile part of the axial lithosphere [*Cannat et al.*, 1992]. Thin gabbroic dikelets are present in about 50% of the ultramafic samples. They have crystallized from variably fractionated melts trapped in the axial mantle lithosphere [*Cannat and Casey*, 1995] (Figure 11).

Striated fault planes in the serpentinites have been observed, with shallow eastward dips, only along the west wall of the valley between 15°30' and 15°40'N. Ultramafic rocks are frequently brecciated into centimeter- to meter-sized fragments of serpentinized peridotite in a sheared matrix of serpentine, talc, and smaller ultramafic clasts. These breccias mark the planes of major N-S- striking faults in the serpentinites [*Cannat et al.*, 1997b]. Brecciated intervals have been observed in dives to be at least a few tens of meters thick, and they probably are thicker. They are overlain by unsedimented ultramafic debris with little to no manganese coating, suggesting that the outcrops have been subjected to recent tectonic movements and mass wasting.

Basalts appear as pillows, tubes, massive flows and breccia. Only one decimeter-wide vertical dike was observed in the pillows of the west flank. Most basalts are covered by a few decimeters of sediment, but the

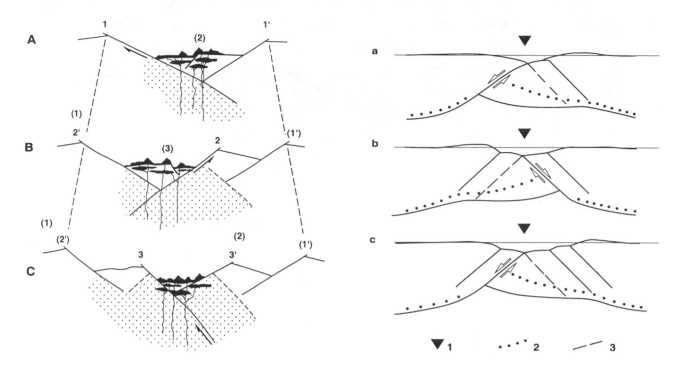

Figure 11. Comparison of two types of models of seafloor spreading with low magma supply to the crust and alternating faulting.
A, B, C are 3 sketches from *Cannat et al.* [1997b] showing the possible evolution of a system of alternating conjugate faults based on field data from the 15°20'N region. This model accounts for the emplacement of mantle rocks (dotted areas) on the seafloor. The active main fault is labelled with a number and the conjugate fault with an '. Inactive or future active faults are between brackets. Sketches a, b, c [after *Shemenda and Grocholsky*, 1994], refer to thermo-mechanical modelling with a specially fabricated hydrocarbon material, cooled from above in a tank. 1 : position of the active spreading center; 2 : position of the surface which was the "lithosphere" base in the previous stage of the process; 3: forming new detachment fault. Note the similarities between models based on field observations and physical modelling.

pillows in the axial deep at 15°30'N are only lightly dusted with sediment and have fresh glassy rims and intact glass horns that indicate recent emplacement. Field arguments suggest that at least some of the basalts of the axial valley may have been erupted over the serpentinites exposed at the base of the ridge flank. The basalt-ultramafic contact on top of the west ridge flank generally follows the 3500 m isobath and is therefore probably sub-horizontal. This suggests that this contact is not a fault, but more probably a stratigraphic contact.

3. DISCUSSION

3.1. Various Tectonic Settings

Our review shows that mantle-derived ultramafics and associated plutonic rocks are exposed within a range of different tectonic settings, but they are mainly related to environments characterized by low magma production or tectonic interactions between spreading and strike-slip processes (Figure 12). Mantle-derived ultramafics and associated plutonic rocks are not restricted to segment ends. They are also found along the walls of the axial valley relatively far from any discontinuity (northern segment of the MARK area) or within large regions of the ridge axis which have dramatically reduced magmatic sections (15°N region). Moreover, the distribution of mantle-derived rocks and related mafic plutonics along the axial valley is not necessarily asymmetrical, as emphasized by models of inside corner emplacement, but it can be symmetrical as in the case of the 15°N region where serpentinites are exposed along both walls of the valley (Figure 12). Finally, the variety in structural settings in which serpentinized peridotites are found suggests that variations in tectonic or magmatic processes might play an important role in the distribution of ultramafic exposures along a slow-spreading ridge.

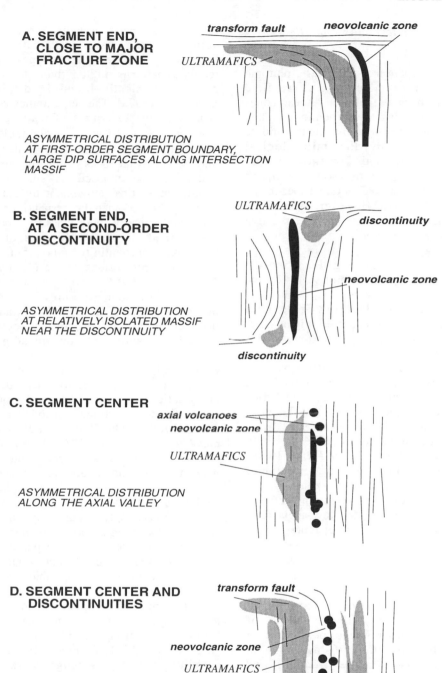

A. SEGMENT END, CLOSE TO MAJOR FRACTURE ZONE

transform fault

neovolcanic zone

ULTRAMAFICS

ASYMMETRICAL DISTRIBUTION AT FIRST-ORDER SEGMENT BOUNDARY, LARGE DIP SURFACES ALONG INTERSECTION MASSIF

B. SEGMENT END, AT A SECOND-ORDER DISCONTINUITY

ULTRAMAFICS

discontinuity

neovolcanic zone

ASYMMETRICAL DISTRIBUTION AT RELATIVELY ISOLATED MASSIF NEAR THE DISCONTINUITY

discontinuity

C. SEGMENT CENTER

axial volcanoes

neovolcanic zone

ULTRAMAFICS

ASYMMETRICAL DISTRIBUTION ALONG THE AXIAL VALLEY

D. SEGMENT CENTER AND DISCONTINUITIES

transform fault

neovolcanic zone

ULTRAMAFICS

axial volcanoes

SYMMETRICAL DISTRIBUTION ALONG THE AXIAL VALLEY

Figure 12. Series of sketches showing the different possible locations and structural settings of ultramafic and gabbroic outcrops along the MAR. Structural maps show idealized ridge segments with simplified tectonic pattern shown by thin lines. Areas where serpentinites and associated gabbros are exposed are shown with grey patterns. The neovolcanic zone is indicated by a continuous black pattern when it is supposed to be robust and by isolated spots representing seamounts in segments (or part of segments) characterized by permanent low magma production. A, B, C, D could refer respectively to the Kane F.Z.-ridge intersection, to the OH3 segment, to the South MARK area and to the eastern 15°20'N F.Z.-ridge intersection.

3.2. Significance of Low-Angle Faults and High-Angle Faults

In nearly all of the serpentinite and mafic plutonic terranes that have been examined by direct field observations, low-angle normal faults and shear zones are the main structures affecting the serpentinites. They have developed at elevated temperatures in the middle to lower crust as shown by the mineralogical assemblages found in deformed serpentinites and gabbros at various localities: at the southern tip of segment OH3, north of the Hayes Fracture zone; in the 15°N region; in the MARK area. However, most of the observed low-angle fault surfaces are no longer active along the valley wall, as indicated by their relatively high grade mineralogy (greenschist facies), the presence of pelagic sediments covering them, and the lack of associated fresh talus. Active fault scarps on the wall of the axial valleys are produced by high-angle faults which cut through the older low-angle fault planes. Low-angle surfaces in serpentinites originated at depth in the mantle and have been progressively unroofed at the ridge axis. When reaching the surface, the low- angle faults are subsequently offset by sub-vertical faults which accommodate the uplift of the axial valley walls. It is likely that the fault zone undergoes significant deformation forcing it to flatten as it moves progressively toward the surface and away from the axis in response to extension [Karson, 1990; Mutter and Karson, 1992; Tucholke and Lin, 1994]. However, recent paleomagnetic studies of serpentinites and gabbroic rocks in low-angle shear zones in the MARK area indicate that little if any tectonic rotation accompanied the footwall uplift of the low-angle fault zones [Karson and Lawrence, 1997 a, b; Lawrence et al., in press].

3.3. Crustal Stretching and Low Magma Budgets

The first discoveries of serpentinite exposures led many authors to propose a diapiric mode of emplacement of abyssal serpentinized peridotites in transform environments [Bonatti, 1976] as well as in environments away from discontinuities [Aumento and Loubat, 1971; Bonatti et al., 1971; Bonatti and Hamlyn, 1981; Francis, 1981]. In these models, it was supposed that weak, relatively low-density serpentinite moved vertically upward through crustal fractures where major faults penetrated into mantle peridotites. More recently, mapping and detailed submersible observations of numerous ultramafic exposures have not supported such a mechanism [Karson and Lawrence, 1997a, b; Cannat et al., 1997a], especially in areas where the serpentinites and gabbros form rectilinear exposures clearly linked to the axial valley wall. The serpentinites are not localized along individual fault zones, but instead they form wide belts. Serpentinite foliations and vein networks have systematic geometries and kinematic relations that suggest fairly consistent tectonic patterns rather than the chaotic or radial patterns classically found in diapirs [Karson and Lawrence, 1997a]. The serpentinites commonly crop out near extensive exposures of mafic plutonic rocks that have been exposed along major faults. If mafic lower crustal rocks are exposed by faulting, then it is not unreasonable that rocks originating in the upper mantle could also be exposed by similar faulting without significant vertical movement by diapirism. Finally, it should be noted that the serpentinite exposures of the MAR are typically solid, massive exposures of fractured and veined serpentinized peridotite and not the incoherent serpentinite mud found in the forearc serpentinite protrusions of the Mariana forearc [Fryer, 1992].

Thus serpentinite exposures of the MAR appear to be the result of major faulting. Two kinds of models have been proposed that explain the occurrence of these faulted exposures at slow-spreading centers. Some models view the emplacement of mantle-derived rocks at the seafloor as a purely tectonic phenomenon that would occur during a period of reduced magmatism or amagmatic extension that could be punctuated by magmatic constructional episodes [Karson and Dick, 1983; Karson, 1990; Mutter and Karson, 1992; Tucholke and Lin, 1994] (Figure 13A). These models assume that a normal, layered crust is created during stages of large magma supply and is subsequently sheared and thinned [Pockalny et al., 1988]. This implies a cyclic axial evolution, with alternating stages of relatively high and low magma supply. Such a mechanism of tectonic stretching has been proposed for instance to explain the occurrence of gabbroic rocks exposed along the southern wall of the Vema fracture zone (Figure 1) [Aubry et al., 1992] in a region of relatively thick crust [Auzende et al., 1989].

Other models propose steady-state, across-axis partitioning of faulting and magmatic construction leading to a strong outside-inside corner asymmetry [Dick et al., 1981; Karson, 1991; Cannat, 1993], permitting the magma budget to remain more or less constant (Figure 13B, 13C). Such models are required to account for the presence of exposures of deep mafic rocks, mostly gabbros, at high inside corners of segment tips, such as the MARK area. The abundance of basalts and the existence of a relatively thick crust (as deduced from gravimetric models) [Morris and Detrick, 1991; Rommevaux et al., 1994; Blackman and Forsyth, 1991] in the opposite, outside corner plate, suggest a relatively robust axial magmatism. The occurrence of numerous shear zones within the gabbros, and the direct observation of low-angle ductile fault planes exposed on the seafloor also provide demonstrative evidence that

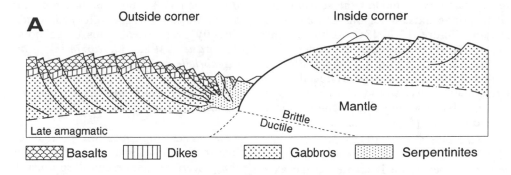

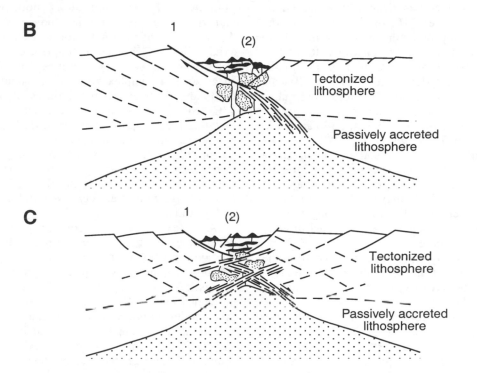

Figure 13. Three models of lithosphere evolution at the axis of slow-spreading ridge based on figures by *Tucholke and Lin*, [1994] and *Cannat et al.* [1997b].

Sketch A represents the late amagmatic stage of the model by *Tucholke and Lin*, [1994] across an inside- and outside-corner pair, during which the tectonic exposure of mantle rocks and is viewed as a rapid process, along a single detachment fault dipping steeply through the crust. In this model, the mantle is thought to be exposed during cyclic, relatively amagmatic stages. The two separating plates display a strong geological and structural asymmetry during magmatic phases, but less asymmetry during late amagmatic phases. Such a model could apply to regions where a typical dike complex and a continuous layer of gabbro have been tectonically denuded.

Sketch B [*Cannat et al.*, 1997b] concerns region of moderate to low rate of magma production. It could apply to the MARK area with an axial region markedly asymmetric and gabbros and serpentinites emplaced only on the western side of the axial valley.

Sketch C [*Cannat et al.* 1997b] could apply to the 15°30-15°38' region where ultramafics are exposed on both axial valley walls and where the walls have similar elevation. The plates will have similar geology characterized by a strongly reduced proportion of crustal components.

tectonic processes were responsible for upper lithosphere thinning.

In a more recent category of models the serpentinites are emplaced at the ridge axis in regions of long-lived low magma budget (Figure 13B, 13C). Little if any mafic material is added to the mantle undergoing serpentinization, uplift, and lateral spreading. Such a steady-state spreading of highly reduced and heterogeneous magmatic crust would characterize not only the segment tips, but also entire segments or wider domains of the ridge axis [*Hess*, 1962; *Brown and Karson*, 1988; *Cannat and Casey*, 1995; *Cannat et al.*, 1995; *Cannat*, 1993; 1996]. The idea of *Hess* [1962] that a crust of serpentinized peridotite is created at the axis of the MAR by upward motion along the ascending branch of a convecting cell is generally supported by these observations, even if the details of the mechanisms that develop beneath the valley floor have since received more elaborate explanations. Results of drilling faulted exposures during ODP Leg 153 in the MARK area suggest that deformation and small-scale magmatic construction are more or less continuous processes on the scale of tens to hundreds of thousands of years [*Karson and Lawrence*, 1997a, b; *Dilek et al.*, 1997; *Karson et al.*, 1997; *Cannat et al.*, 1997a; *Karson*, this volume]. In other words, crustal evolution along slow-spreading ridges does not necessarily require an absolute decoupling between magmatic and amagmatic phases in time and space. This may explain that mantle-derived rocks are preferentially found along the rift valley walls and yet have never been observed directly on the floor of the axial valley.

In summary, the two main observations from the MAR axis which support the hypothesis of the non-cyclic and magma-starved origin for mantle unroofing can be outlined as follows:

-In areas where mantle-derived ultramafics are exposed, the gabbros are observed as intrusive into the peridotites and do not form a continuous layer on a regional scale. Sheeted dike complexes are not present, suggesting that in these regions, a substantial magmatic crust did not develop. Intrusions of relatively small gabbros bodies and isolated diabase dike probably persist during the tectonic uplift of the ultramafic material to the surface;

-Oblique depressions where abundant ultramafic material is exposed, and which trace the axial discontinuities off-axis, are continuous geological features. They correspond to persistent conditions of low magma budget seafloor spreading where large magmatic fluctuations do not occur. These regions may move along the axis with segment growth or retreat, reflecting the changing distribution of mantle melting through time. These areas of magma-poor spreading show strong similarities with fossil oceanic lithosphere exposed in the Alpine and Appennine ophiolites. These ophiolites are characterized by sections showing features typical of very low magma budget; they systematically lack a dike complex, gabbro bodies are intrusive within the serpentinites, and pelagic sediments commonly rest directly over serpentinites or gabbros [*Lombardo and Pognante*, 1982; *Lagabrielle and Cannat*, 1990; *Lagabrielle and Lemoine*, 1997].

3.4. Symmetrical and Asymmetrical Distribution of Lithologies Along Axis

A remaining question is that of the geometry, distribution and kinematics of the faults which control the process of mantle and deep crust denudation. Under what conditions are serpentinites and associated mafic plutonic rocks exposed along both walls of axial valleys? As suggested by the study of the 15°N region, the occurence of ultramafic rocks on facing walls of the axial valley may be due to frequent changes of faulting polarity in the axial region: instead of slip on a single shear zone, a complex array of cross-cutting, conjugate faults and shear zones may be present (Figures 11 and 13C). Slip on major tectonic discontinuities could shift to faults located progressively inward toward the axial domain as spreading proceeds [*Cannat et al.*, 1997b]. Example of the 15°N region suggest that such mechanism concerns areas of very low magma supply to the crust. It leads to a poorly defined rift valley morphology, with relatively low relief and smooth axial valley walls.

A relevant model for progressively inward development of faults with alternation from one side of the axial valley to the other has been produced by thermo-mechanical experiments on a model lithosphere consisting of a special hydrocarbon undergoing horizontal tension [*Shemenda and Grocholsky*, 1994]. In this model (Figure 11), the faults that mark the spreading axis change polarity and jump suddenly and frequently from one side of the axis to the other. As shown in Figure 11, the tectonic fabric of the axial zone obtained from this model strongly resembles the fabric of the slow-spreading ridge axis with low magma budget such as the 15°N area. Recent multichannel seismic reflection data obtained in the Canary Basin on Jurassic to Cretaceous oceanic crust show a rather complex basement structure with opposite dipping reflectors that reach deep into the crust [*Ranero et al.*, 1997]. Such a geometry confirm that extensional activity at a slow-spreading center may provide a symmetrical tectonic pattern accros axis.

Finally, our review of currently available data indicates that a single and universal model for tectonic emplacement of mantle-derived serpentinites and associated mafic plutonic rocks cannot be proposed for the entire Mid-Atlantic Ridge and that the processes of mantle uplift and unroofing may change significantly along axis due to variations in the volume of magma which are extracted from the mantle and supplied to the crust. The occurrence of serpentinites exposed on most places along the MAR is not solely the expression of an extreme tectonic stretching, but also the consequence of drastically reduced magma supply due either to a local cause as expressed at segment discontinuities or to a more regional cause involving large domains of the MAR axis and the underlying mantle as in the 15°N region.

Acknowledgements. This study was partly funded by grants from the "Institut National des Sciences de l'Univers, Programme Géosciences Marines". We would like to thank the captains, officers, crew and the *Nautile* team aboard the R/V Nadir and l'Atalante who allowed precious observations during several cruises on the MAR. We wish to thank Jean-Marie Auzende and Henri Bougault, chief-scientists of cruises Vemanaute, Kanaut and Faranaut as well as Pascal Gente, Jerôme Dyment, Marcel Lemoine, Thierry Juteau, Yves Fouquet, Nathalie de Coutures and Astrid Rio for helpful discussions during the preparation of this manuscript. Brian Tucholke and John Mutter provided very careful reviews of the manuscript with constructive comments and suggestions. We wish to thank Bernadette Coléno and Kiram Lezzar for help in art graphics and editing.

REFERENCES

Aubry, M.P., W.A. Berggren, A. Schaaf, J.M. Auzende, Y. Lagabrielle, and V. Mamaloukas-Frangoulis, Paleontological evidence for early exposure of deep oceanic crust along the Vema Fracture Zone southern wall (Atlantic Ocean, 10°45'N), *Marine Geol.*, 107, 1-7, 1992.

Aumento, F., and H. Loubat, The Mid-Atlantic Ridge near 45°N, XVI, Serpentinized ultramafic intrusions, *Can. J. Earth Sci*, 8, 631-663, 1971.

Auzende, J.M., D. Bideau, E., Bonatti, M. Cannat, J. Honnorez,, Y. Lagabrielle, J. Malavieille, V. Mamaloukas-Frangoulis, and C. Mével, Direct observation of a section through slow-spreading oceanic crust, *Nature*, 337, 726-729, 1989.

Auzende, J.M., M. Cannat, P. Gente, J.P. Henriet,, T. Juteau, J.A. Karson,, Y. Lagabrielle, C. Mével, and M. Tivey, Affleurements de roches profondes de la croûte océanique et du manteau sur le mur sud de la zone de fracture Kane (Atlantique central): observations par submersible, *C. R. Acad. Sc. Paris*, 317, Serie II, 1641-1648, 1993.

Auzende, J.M., M. Cannat,, P. Gente, J.P. Henriet, T. Juteau, J.A. Karson, Y. Lagabrielle, C. Mével, and M. Tivey, Observation of sections of oceanic crust and mantle cropping out on the southern wall of Kane F.Z. (North Atlantic), *Terra Nova*, 6, 143-148, 1994.

Bideau, D., R. Hékinian, C. Bollinger, M. Constantin, E. Gracia, G. Guivel,, B. Sichler, R. Apprioual, and R. Le Gall, Submersible investigation of highly contrasted magmatic activities recorded on two segments of the Mid-Atlantic Ridge near 34°52'N and 33°55'N, *InterRidge News*, 5, 9-14, 1996.

Blackman, D.K., and D.W. Forsyth, Isostatic compensation of tectonic features of the Mid-Atlantic Ridge: 25-27°30'S, *J. Geophys. Res.*, 96, N.B7, 11741-11758, 1991.

Bougault, H., J.L. Charlou, Y. Fouquet, H.D. Needham, N. Vaslet, P. Appriou,, P. Jean Baptiste, P.A. Rona, L. Dmitriev, and Silantiev, S., Fast and slow spreading ridges: structure and hydrothermal activity, ultramafic topographic highs and CH4 output, *J. Geophys. Res.*, 98, 9643-9651, 1993.

Bonatti, E., J. Honnorez, and G. Ferrara, Peridotite-gabbro-basalt complex from the equatorial Mid-Atlantic Ridge, *Phil. Trans. R. Soc. London*, 268, 385-402, 1971.

Bonatti, E., Serpentinite intrusions in the oceanic crust, *Earth Planet. Sci. Lett.*, 32, 107-113, 1976.

Bonatti, E., and J. Honnorez, Sections of the Earth's crust in the Equatorial Atlantic, *J. Geophys. Res.*, 81, B23, 4104-4116, 1976.

Bonatti, E., and P. Hamlyn, Oceanic ultramafic rocks, in *The Sea (Vol. 7), The Oceanic Lithosphere*, edited by C. Emiliani, pp. 241-283, New York (Wiley)., 1981.

Bonatti, E., A. Peyve, P. Kepezhinskas, N. Kurentsova, M. Seyler, S. Skolotnev, and G. Udintsev, Upper mantle heterogeneity below the Mid-Atlantic Ridge, 0°-15°N, *J. Geophys. Res.*, Vol.97, N.B4, 4461-4476, 1992.

Brown, J.R., and J.A. Karson, Variations in axial processes on the MAR : the median valley of the MARK area, *Mar. Geophys. Res.*, 10, 109-138, 1988.

Cannat, M., Emplacement of mantle rocks in the seafloor at Mid-Ocean ridge, *J. Geophys. Res.*, 98, B3, 4163-4172, 1993.

Cannat, M., How thick is the magmatic crust at slow-spreading ocean ridges?, *J. Geophys. Res.*, 101, 2847-2857, 1996.

Cannat, M., T. Juteau, and E. Berger, Petrostructural analysis of the Leg 109 serpentinized peridotites, in *Proceedings of the O.D.P., Sci. Results*, 106/109, pp. 47-56., 1990.

Cannat, M., D. Bideau, and H. Bougault, Serpentinized peridotites and gabbros in the Mid Atlantic Ridge axial valley at 15°37'N and 16°52'N, *Earth Planet. Sci. Lett.*, 109, 87-106, 1992.

Cannat, M. and J. Casey, An ultramafic lift at the Mid-Atlantic Ridge: successive stages of magmatism in serpentinized peridotites from the 15°N region, in *Mantle and Lower crust exposed in oceanic ridges and in ophiolites*, edited by R.L.M. Vissers and A. Nicolas, pp. 5-34, Kluwer Academic Publishers, The Netherlands., 1995.

Cannat, M., C. Mével, M. Maia, C. Deplus, C. Durand, P. Gente, P. Agrinier, A. Belarouchi, G. Dubuisson, E. Humler, and J. Reynolds, Thin crust, ultramafic exposures, and rugged faulting patterns at the Mid-Atlantic Ridge (22°- 24°N), *Geology*, 23, 1, 49-52, 1995.

Cannat., M., F. Chatin,, H. Whitechurch, and G. Ceuleneer, Gabbroic rocks trapped in the upper mantle at the Mid-Atlantic Ridge, in *Proceedings of the Ocean Drill. Program, Initial Rep*, edited by J.A. Karson, M. Cannat, D.J. Miller, and D. Elthon, Vol.153, pp. 243-264, College Station, TX., 1997a.

Cannat, M., Y. Lagabrielle, N. de Coutures, H. Bougault, J. Casey, L. Dmitriev, and Y. Fouquet, Ultramafic and gabbroic exposures at the Mid-Atlantic Ridge: geological mapping in the 15°N region, *Tectonophysics*, 279, 193-213, 1997b.

Casey, J. F., M. Cannat, and the FARANAUT Scientific Party, Crustal architecture and mechanisms of ultramafic and gabbroic exposure along rift valley walls of the Mid-Atlantic Ridge north and south of the Fifteen Twenty transform, in *Am. Geophys. Union*, Fall Meeting (abstract), 73, p. 537, *EOS Transactions*, 1992.

Charlou, J.L., H. Bougault, P. Appriou, T. Nelsen, and P. Rona, Different TDM/CH4 hydrothermal plume signature: TAG site at 26°N, and serpentinized ultrabasic diapir at 15°05'N on the Mid-Atlantic Ridge, *Geoch. Cosm. Acta*, 55, 3209-3222, 1991.

Detrick, R.S., P.J. Fox, N. Schulz, R. Pockalny, L. Kong, L. Mayer, and W.B.F. Ryan, Geologic and tectonic setting of the MARK area, in *Proceedings. Ocean Drilling Program, Init. Repts. (PartA)*, 106/109, edited by R. Detrick., J. Honnorez., W.B. Bryan., T. Juteau. et al., pp. 15-22, Ocean Drilling Program, College Station, TX., 1988.

Detrick, R.S., H.D. Needham, and V. Renard, Gravity anomalies and crustal thickness variations along the Mid-Atlantic ridge between 33°N and 40°N, *J. Geophys. Res.*, 100, B3, 3767-3787, 1995.

Dick, H.J.B., Abyssal peridotites, very slow spreading ridges and ocean ridge magmatism, in *Magmatism in the Ocean Basins*, edited by A.D. Saunders and M.J. Norris, 42, pp. 71-105, *Geol. Soc. London Spec. Publ*, 1989.

Dick, H.J.B., G. Thompson, and W.B. Bryan, Low-angle faulting and steady-state emplacement of plutonic rocks at ridge-transform intersections, *Amer. Geophys. Union Trans., EOS*, 62, p. 406, 1981.

Dick, H.J.B., R.L. Fisher, and W.B. Bryan, Mineralogic variability of the uppermost mantle along mid-ocean ridges, *Earth Plan. Sci. Lett.*, 69, 88-106, 1984.

Dilek, Y., A. Coulton, and S.D. Hurst, Serpentinization and hydrothermal veining in peridotites at Site 920 in the MARK area, in *Proceedings of the Ocean Drilling Program, Scientific Results*, v. 153, edited by Karson, J.A., Cannat, M., Miller, D.J., and Elthon, D., Ocean Drilling Program, Texas A&M University, College Station, TX., 1997.

Dubois, J., and Scientific Party, The Gravinaut cruise : Seafloor gravity and electromagnetism at the MARK-Snake Pit area, *InterRidge News*, 3, 1, 3-6, 1994a.

Dubois, J., V. Ballu, P. Beuzart, S. Bonvalot, C. Deplus, M. Diament, A. Dubreule, C. Durand, M. F. Esnoult, A. Schultz, and P. Tarits, Sea-bottom gravity and electromagnetic measurements on MARK area, in *EGS 19, Grenoble 25-29 april 1994*, Vol.PO25, p. 49, 1994b.

Durand, C., V. Ballu, P. Gente, and J. Dubois, Horst and graben structures on the flanks of the Mid-Atlantic Ridge in the MARK area (23°22'N): submersible observation, *Tectonophysics*, 265, 275-297, 1996.

Escartin, J., and J. Lin, Ridge offsets, normal faulting, and gravity anomalies of slow spreading ridges, *J. Geophys. Res.*, 100, B4, 6163-6177, 1995.

Fox, P.J., and B.C. Heezen, Sands of the Mid-Atlantic Ridge, *Science*, 149, 1367-1370, 1965.

Fox, P.J., and D.G. Gallo, A tectonic model for ridge-transform-ridge plate boundaries: implications for the structure of the oceanic lithosphere, *Tectonophysics*, 104, 205-242, 1984.

Fox, P.J., N.R. Grindlay, and K.C. Macdonald, The Mid-Atlantic ridge (31°S-34°30'S): temporal and spatial variations of accretionary processes, *Mar. Geophys. Res.*, 13, 1-20, 1991.

Francis, T.J.G., Serpentinization faults and their role in the tectonics of slow-spreading ridges, *J. Geophys. Res.*, 86, 11616-11622, 1981.

Fryer, P., A synthesis of Leg 125 drilling of serpentinite seamounts on the Mariana and Izu-Bonin forearcs, in *Proceedings. Ocean Drill. Program, Initial Rep*, edited by P. Fryer, J.A. Pearce, L.B. Stokking, et al., pp. 593-614, College Station, TX., 1992.

Gente, P., C. Mével, J.M. Auzende, J.A. Karson, and Y. Fouquet, An example of a recent accretion on the Mid-Atlantic Ridge: The Snake Pit neovolcanic ridge (MARK Area, 23°22'N), *Tectonophysics*, 190, 1-29, 1991.

Gente, P., C. Durand, R. Pockalny, M. Maia,, C. Deplus, C. Mével, G. Ceuleneer,, M. Cannat,, C. Laverne, D. Aslanian, A. Campan, H. Leau, E. Marion, and M. Seyler, Structures obliques sur les flancs de la Dorsale Médio-Atlantique: traces fossiles de la propagation le long de l'axe de segments d'accrétion, *C. R. Acad. Sc. Paris*, 318, II, 1239-1246, 1994.

Gente, P., R.A. Pockalny, C. Durand, M. Maïa, C. Deplus, G. Ceuleneer, C. Mével, M. Cannat, and C. Laverne, Characteristics and evolution of the segmentation of the Mid-Atlantic ridge between 20°N and 24°N during the last 10 million years, *Earth Planet. Sci. Lett.*, 129, 55-71, 1995.

Gente, P., G. Ceulenner, O. Dauteuil, J. Dyment, C. Honsho,, C. Laverne, C. Le Turdu, M.C. Mitchell, M. Ravilly, and R. Thibaud, On- and off-axis submersible investigations on an highly magmatic segment of the Mid-Atlantic Ridge (21°40'N): the TAMMAR Cruise, *InterRidge News*, 5, 2, 27-31, 1996.

Gracia, E., D. Bideau, R. Hekinian,, Y. Lagabrielle, and L.M. Parson, Along-axis magmatic oscillations and exposure of ultramafic rocks in a second-order segment of the Mid-Atlantic Ridge (33°43'N to 34°07'N). *Geology*, 25, 12, 1059-1062.

Grindlay, N.R., P.J. Fox, and K.C. Macdonald, Second order ridge axis discontinuities in the South Atlantic : morphology, structure and evolution, *Mar. Geophys. Res.*, 13, 21-49, 1991.

Hess, H.H., History of ocean basins, in *Petrologic Studies*, Burlington Volume, A.E.J., 1962.

Juteau, T., E. Berger, and M. Cannat, Serpentinized, residual mantle peridotites from the MAR median valley, ODP hole 670a (21°10'N, 45°02'W, leg 109): primarily mineralogy and geothermometry, in *Proceedings of the Ocean Drilling Program, Scientific Results*, Vol.106/109, pp. 27-45, 1990a.

Juteau, T., M. Cannat, and Y. Lagabrielle, Serpentinized peridotites in the upper oceanic crust away from transform zones: a comparison of the results of previous DSDP and ODP legs, in *Proceedings of the Ocean Drilling Program, Scientific Results*, Vol.106/109, pp. 303-308, 1990b.

Karson, J.A., Seafloor spreading on the Mid-Atlantic Ridge: implications for the structure of ophiolites and oceanic lithosphere produced in slow-spreading environments, in *Proceedings of the Symposium "Troodos 1987"*, edited by J. Malpas, E.M. Moores, A. Panayiotou. et C. Xenophontos, pp. 547-555, Geological Survey Department, Nicosia, Cyprus., 1990.

Karson, J.A., and H.J.B. Dick, Tectonics of ridge-transform intersections at the Kane fracture zone, *Mar. Geophys. Res.*, 6, 51-98, 1983.

Karson, J.A., Faulting and Magmatism at Mid-Ocean Ridges: A Perspective from Tectonic Windows in the Oceanic Lithosphere, in *Proceedings of the 4th RIDGE Theoretical Institute*, edited by W.R. Buck, P.A. Delaney and J.A. Karson, American Geophysical Union, this volume.

Karson, J.A., G. Thompson, S.E. Humpris, J.M. Edmond, W.B. Bryan, J.R. Brown, A.T. Winters, R.A. Pockalny, J.F. Casey, C. Campbell, G. Klinkhammer, M.R. Palmer, R.J. Kinzler, and M. Sulanowska, Along-axis variations in seafloor spreading in the MARK area, *Nature*, 328, 681-685, 1987.

Karson, J.A., and J.B. Brown, Geologic setting of the Snake pit hydrothermal site : an active vent field on the Mid-Atlantic ridge, *Mar. Geophys. Res.*, 10, 91-107, 1988.

Karson, J.A., and P.A. Rona, Block-tilting, transfer faults, and structural control of magmatic and hydrothermal processes in the TAG area, Mid-Atlantic ridge 26°N, *Geol. Soc. Am. Bull.*, 102, 1635-1645, 1990.

Karson, J.A., and A.T. Winters, Along-axis variations in tectonic extension and accomodation zones in the MARK area, Mid-Atlantic ridge 23°N latitude, in *Ophiolites and their modern oceanic analogues*, edited by L.M. Parson, B.J. Murton, and P. Browning, pp. 107-116, *Geol. Soc. Special Publ*, 60, 1992.

Karson, J.A., and R.M. Lawrence, Tectonic setting of serpentinite exposures on the western median valley wall of the MARK area in the vicinity of ODP Site 920, in *Proceedings of the Ocean Drill. Program, Initial Rep*, edited by J.A. Karson, M. Cannat, D.J. Miller, and D. Elthon, Vol.153, pp. 5-21, College Station, TX., 1997a.

Karson, J.A., and R.M. Lawrence, Tectonic window into gabbroic rocks of the middle oceanic crust in the MARK area near ODP Sites 921-924, in *Proceedings of the Ocean Drill. Program, Initial Rep*, edited by J.A. Karson, M. Cannat, D.J. Miller, and D. Elthon, Vol.153, pp. 61-76, College Station, TX., 1997b.

Karson, J.A., M. Cannat, D.J. Miller, and D. Elthon, *Proceedings of the Ocean Drill. Program, Initial Rep*, Vol.153, College Station, TX., 1997.

Kong, L.S., R.S. Detrick, P.J. Fox, L.A. Mayer, and W.B.F. Ryan, The morphology and tectonics of the MARK area from Sea Beam and Sea MARC I observations (Mid-Atlantic Ridge 23°N), *Mar. Geophys. Res.*, 10, 59-90, 1988.

Kuo, B.Y., and D.W. Forsyth, Gravity anomalies of the ridge-transform system in the South Atlantic between 31 and 34.5°S: upwelling centers and variations in the crustal thickness, *Mar. Geophys. Res*, 10, 205-232, 1988.

Lagabrielle, Y., and M. Cannat, Alpine jurassic ophiolites resemble the modern central Atlantic basement, *Geology*, 18, 319-322, 1990.

Lagabrielle, Y., and M. Lemoine, Alpine, Corsican and Apennine ophiolites: the slow-spreading ridge model, *C. R. Acad. Sci*, 325, 909-920, 1997.

Lawrence, R.M., J.A. Karson, and S.D. Hurst, Dike orientations and fault-block rotations in slow-spread oceanic crust at the SMARK Area, Mid-Atlantic Ridge at 22°40'N, *J. Geophys. Res*, in press.

Lin, J., G.M. Purdy, H. Shouten, J.-C. Semperé, and C. Zervas, Evidence from gravity data for focused magmatic accretion along the Mid-Atlantic Ridge, *Nature*, 344, 627-632, 1990.

Lin, J., and J. Phipps Morgan, The spreading rate dependence of three-dimensional Mid-Ocean ridge gravity structure, *Geophys. Res. Lett.*, 19, N.1, 13-16, 1992.

Lombardo, B., and U. Pognante, Tectonic implications in the evolution of the western Alps ophiolite metagabbro. *Ofioliti*, 7, 2/3, 371-394, 1982.

Mével, C., J.M. Auzende, M. Cannat, J.P. Donval, J. Dubois, Y. Fouquet, P. Gente,, D. Grimaud, J.A. Karson, M. Segonzac, and M. Stievenard, La ride du Snake Pit (dorsale médio-Atlantique, 23°22'N): résultats préliminaires de la campagne Hydrosnake, *C. R. Acad. Sc. Paris*, 308, II, 545-552, 1989.

Mével, C., M. Cannat, P. Gente, E. Marion, J.M. Auzende, and J.A. Karson, Emplacement of deep crustal and mantle rocks on the west median valley wall of the MARK area (MAR, 23°N), *Tectonophysics*, 190, 31-53, 1991.

Michael, P.J., and E. Bonatti, Peridotite composition from the North Atlantic: regional and tectonic variations and implications for partial melting, *Earth Plan. Sci. Lett.*, 73, 91-104, 1985.

Morris, E., and R.S. Detrick, Three-dimensional analysis of gravity anomalies in the MARK area, Mid Atlantic Ridge, 23°N, *J. Geophys. Res*., 96, 4355-4366, 1991.

Mutter, J.C., and J.A. Karson, Structural processes at slow-spreading ridges, *Science*, 257, 627- 634, 1992.

Neuman, G.A., and D.W. Forsyth, High resolution statistical estimation of seafloor morphology: Oblique and orthogonal fabric on the flanks of the Mid-Atlantic Ridge, 34°-35.5°S, *Marine Geophys. Res.*, 17, 221-250, 1995.

Niu, Y., and R. Batiza, Magmatic processes at a slow spreading ridge segment: 26°S Mid-Atlantic Ridge, *J. Geophys. Res.*, Vol.99, B10, 19719-19740, 1994.

OTTER Group: Karson, J.A., P.J. Fox, H. Sloan, K.T. Crane, W.S.F. Kidd, E. Bonatti, J.B. Stroup, D.J. Fornari, D. Elthon, P. Hamlyn, J.F. Casey, D.G. Gallo, D. Needham, and R. Sartori, The geology of the Oceanographer transform: the ridge-transform intersection, *Mar. Geophys. Res.*, 6, 109-141, 1984.

Pockalny, R.A., R.S. Detrick, and P.J. Fox, Morphology and tectonics of the Kane transform from Sea Beam bathymetry data, *J. Geophys. Res.*, 93, B4, 3179-3193, 1988.

Ranero, C. R., E. Banda, and P. Buhl, The crustal structure of the Canary Basin: accretion processes at slow spreading centers, *J. Geophys. Res.*, 97, B5, 10185-10201, 1997.

Rommevaux, C., C. Deplus, P. Patriat, and J.C. Sempéré, Three-dimensional gravity study of the Mid-Atlantic ridge : evolution of the segmentation between 28° and 29°N during the last 10 My, *J. Geophys. Res.*, 99, B2, 3015-3029, 1994.

Rona, P.A., L. Widenfalk, and K. Boström, Serpentinized ultramafics and hydrothermal activity at the Mid-Atlantic ridge crest near 15°N, *J. Geophys. Res.*, 92, B2, 1417-1427, 1987.

Sempéré, J.C., P. Blondel, A. Briais, T. Fujiwara,, L. Géli, N. Isezaki, J.E. Pariso, L. Parson, P. Patriat, and C. Rommevaux, The Mid-Atlantic ridge between 29°N and 31°30'N in the last 10 Ma, *Earth Planet. Sci. Lett.*, 130, 45-55, 1995.

Severinghaus, J. P., and K. C. Macdonald, High inside corners at ridge-transform intersections, *Mar. Geophys. Res.*, 9, 353-367, 1988.

Shaw, P.R., and J. Lin, Causes and consequences of variations in faulting style at the Mid-Atlantic ridge, *J. Geophys. Res.*, 98, B12, 21839-21851, 1993.

Shaw, P. R., and J. Lin, Models of ocean ridge lithospheric deformation: Dependence on crustal thickness, spreading rate, and segmentation, *J. Geophys. Res.*, 101, B18, 17977-17993, 1996.

Shemenda, A.I., and A. L. Grocholsky, Physical modelling of slow seafloor spreading, *J. Geophys. Res.*, 99, B5, 9137-9153, 1994.

Shipboard Scientific Party Site 670, in *Proceedings of the Ocean Drilling Program, Initial Rep.*, edited by W.B. Bryan, T. Juteau, *et al.*, Vol.109, pp. 203-240, College Station, Texas., 1988.

Shipboard scientific Party Sites 920-924, in *Proceedings of the Ocean Drilling Program, Initial Rep*, edited by J.A. Karson, M. Cannat, D.J. Miller, and D. Elthon, Vol.153, College Station, TX., 1995.

Skolotnev, S. G., N. V. Tsukanov, A.A. Peyve, Raznitisin. and A. Yu, 15°20'N fracture zone bedrocks: General characteristics, in *Structure of the Cape Verde fracture zone, Central Atlantic*, edited by Y. M. Pushcharovsky, 439, pp. 40-60, Academy of Sciences of the USSR, Order of the Red Banner of Labour, Geological Institute Transactions., 1989.

Sloan, H., and P. Patriat, Kinematics of the north American-African plate boundary between 28° and 29°N during the last 10 Ma : evolution of the axial geometry and spreading rate and direction, *Earth Planet. Sci. Lett.*, 113, 323-341, 1992.

Tiezzi, L.J., and R.B. Scott, Crystal fractionation in a cumulate gabbro, Mid-Atlantic Ridge, 26°N, *J. Geophys. Res.*, 85, 5438-5454, 1980.

Tolstoy M., A.J. Harding, and J.A. Orcutt, Crustal thickness on the Mid-Atlantic Ridge: bull's eye gravity anomalies and focused accretion, *Science*, Vol.262, 726-729, 1993.

Tucholke, B.E., and J. Lin, A geological model for the structure of ridge segments in slow-spreading ocean crust, *J. Geophys. Res.*, 99, 11937-11958, 1994.

Wolfe, C.J., G.M. Purdy, D.R. Toomey. and S.C. Solomon, Microearthquake characteristics and crustal velocity structure at 29°N on the Mid-Atlantic Ridge: the architecture of a slow-spreading segment, *J. Geophys. Res.*, 100, 24449-24472, 1995.

Daniel Bideau, IFREMER, Département Géosciences Marines, Centre de Brest, BP 70, 29280 Plouzané Cedex, France.

Mathilde Cannat, CNRS, UPMC, Laboratoire de Pétrologie, 5, place Jussieu, Tour 26-00,E3, 75252 Paris Cedex 05, France.

Jeffrey A. Karson, 206 Bluestone Building, Department of Geology, Duke University, Durham, NC 27708-0227, USA.

Yves Lagabrielle, CNRS, UMR 6538 Domaines Océaniques, Institut Universitaire Européen de la Mer, Place Nicolas Copernic, 29285 Plouzané Cedex, France.
E-mail : yves.lagabrielle@univ-brest.fr

Catherine Mével, CNRS, UPMC, Laboratoire de Pétrologie, 5, place Jussieu, Tour 26-00,E3, 75252 Paris Cedex 05, France.

Internal Structure of Oceanic Lithosphere:
A Perspective from Tectonic Windows

Jeffrey A. Karson

Division of Earth and Ocean Sciences, Duke University, Durham, NC

Major faulted escarpments on the seafloor provide "tectonic windows" into oceanic crust and upper mantle. Direct observations in these settings reveal that the spatial arrangement, internal structure, and contacts between major rock units are significantly more complex than commonly anticipated on the basis of seismic studies and ophiolite analogs. From this perspective, a stratiform, ophiolite-like sequence of rock units, including basaltic volcanic rocks, sheeted diabase dike complex, isotropic and layered gabbroic and ultramafic rocks over upper mantle peridotites–all separated by generally horizontal contacts, may be much less common in the oceanic lithosphere than generally thought. Conversely, documented examples of large outcrop areas (tens of kilometers across) that lack the ophiolite-like sequence or that contain structures that do not conform to the ophiolite model call into question the basic assumptions made in the reconstruction and interpretation of ophiolite complexes. Historically, the stratiform ophiolite architecture has been the basis for inferences of the interaction between tectonism and magmatism at mid-ocean ridge spreading centers. A growing number of constraints on geological relations along seafloor escarpments hint at much broader range of interactions between tectonic deformation and magmatic construction.

Along slow-spreading ridges, the magma budget (volume of magma per unit plate separation) is highly variable, giving rise to a wide range of morphologic and geologic features along wide rift valleys. The diversity of crustal architectures and internal structures seen in tectonic windows is correspondingly large. Although it is possible that a relatively simple, layered ophiolite-like crust develops in regions of relatively high magma budget (Reykjanes Ridge, Azores region, etc.), more complex structures that differ from those of stratiform ophiolites are present where lower magma budgets prevail. Significant deviations from a simple layered structure occur near transform faults, but also occur along typical rift valley walls tens of kilometers away from obvious ridge-axis discontinuities. Under some circumstances, large-scale stretching and thinning of the lithosphere results in the formation of "oceanic core complexes" analogous to those of highly extended continental terranes. Tectonic windows into regions of variable magma budget commonly show

1. INTRODUCTION

Faulting and Magmatism at Mid-Ocean Ridges
Geophysical Monograph 106
Copyright 1998 by the American Geophysical Union

Fault zones that create major escarpments on the seafloor provide direct access to outcrops of the oceanic crust and uppermost mantle referred to as "tectonic windows" into the oceanic lithosphere. Although there are potential problems that: 1) some of the ophiolite rock units are missing, 2) contacts between major rock units are neither horizontal nor laterally continuous, and 3) internal structures of major rock units are highly variable. In several intensively studied locations, volcanic units appear to lie directly (unconformably) over variably deformed volumes of metadiabase, metagabbro, or serpentinite. These relations reflect a sputtering magma supply and/or heterogeneous magmatic accretion across axial valleys over periods of tens to hundreds of thousands of years. Structurally complex oceanic crust that lacks a simple layered structure is likely to be a typical product of many slow-spreading ridges.

At intermediate- to fast-spreading ridges the magma budget appears to be consistently higher and accordingly magmatic construction dominates. Spreading center morphology suggests only limited faulting; however, available geologic data suggest that the oceanic crust created in these environments may be more variable and geologically complex than generally thought. Although tectonic windows in fast-spread crust are rare, the crustal structures seen in them show variations on the ophiolite model. Surficial lavas are likely to blanket complicated structures that result from subsurface crustal collapse related to growth faulting, dike intrusion, and inflation/deflation of axial summit regions and underlying magma chambers. These mechanisms accommodate significant vertical mass transport at various crustal levels that are required to generate the observed thicknesses of major rock units.

Collectively, available geological observations from tectonic windows cast significant doubts on the prevailing global application of an ophiolite model for oceanic crust and the interpretation of marine seismic data based on this model. In addition, these observations indicate that tectonic windows into the oceanic lithosphere are a potentially rich source of information on the diverse geological processes attending seafloor spreading.

in reconstructing the geology of the oceanic lithosphere from such faulted outcrops, observations from these areas provide an extremely important perspective on the architecture of the oceanic crust and clues to the nature of seafloor spreading processes in time and space. These observations provide direct evidence of the nature of the rock units that make up the oceanic crust and the contacts between them. Models of seafloor spreading processes derived from indirect geophysical studies of the oceanic lithosphere, morphologic studies of active spreading centers, and ophiolites, must be reconciled with the results of these studies.

This paper provides a review and critical evaluation of the internal structure of the oceanic lithosphere based on direct observations of rock units exposed at tectonic windows. The paper begins with a brief review of the development of current views of oceanic lithosphere. This is followed by a description of the tectonic settings, morphologies, and fault structures of tectonic windows that have been investigated to date. After a brief review of the geology of major rock units, the specific relations at the most intensively investigated tectonic windows is presented. These relations have important implications for the relationship between magmatic construction and mechanical deformation across the spectrum of global spreading environments.

2. EVOLUTION OF THE CURRENT VIEW OF OCEANIC LITHOSPHERE

The current perception of the internal composition and structure of the oceanic lithosphere and crust evolved over the past 30 years through a combination of observations and inferences from several different perspectives. The single overriding notion that has influenced the perception of the oceanic lithosphere is that the oceanic crust and upper mantle have a persistent horizontally layered structure that in turn reflects processes at spreading centers. This view was initially based on indirect evidence that must be reconsidered in light of more recent seafloor investigations.

Early investigations of oceanic crust were based on dredged samples of a limited range of mafic to ultramafic lithologies [*Shand*, 1949; *Quon and Ehlers*, 1963]. It was very soon recognized that these lithologies were similar to those of ophiolite complexes found in mountain belts which had previously been interpreted (at least by some early workers) as oceanic basement on the basis of their overlying deep-water marine sedimentary units [*Lotti*, 1886; *Suess*, 1909; *Steinmann*, 1927].

Early photos of the seafloor revealed young-looking pillow basalts near the newly recognized Mid-Atlantic Ridge

(MAR), adding to the correlation with ophiolites and providing evidence of active volcanic and tectonic processes [*Heezen et al.*, 1959]. Most importantly, *Heezen* [1960] correlated bathymetric profiles of the Mid-Atlantic Ridge with the topography of the East African Rift and suggested similarities in extensional tectonics in oceanic and continental settings.

Based on additional sampling and thermal considerations, *Hess* [1962] suggested that the oceanic crust was capped by a thin rind of basalt but dominated by altered serpentinite grading downward into anhydrous peridotites of the upper mantle. These assertions presaged much more recent perceptions of similarities between continental and mid-ocean ridge tectonics [*Buck*, 1991; *Mutter and Karson*, 1992] and very low melt production along parts of slow-spreading ridges with oceanic crust dominated by serpentinites [i.e., *Brown and Karson*, 1988; *Cannat*, 1993; *Sleep and Barth*, 1997].

The views of Heezen and Hess were prophetic, but in ways that could not be anticipated in their day. In retrospect, it is remarkable how over only the past few years much more detailed information from the continental rifts and the seafloor have lead back to similar conclusions. These pioneers in the field of marine geology were among the first to grapple with the question of how tectonic extension and magmatic construction are related at divergent plate boundaries. Over the past two decades, detailed studies of the composition and structure of oceanic crust have been made by direct observations from submersibles and deep crustal drilling. The nature of major rock units and the contacts between them document relationships between magmatic and tectonic processes. In addition, they provide a basis for correlations with geophysical studies of crustal structure and guide the interpretation of active processes along spreading centers.

The current view of the geology of the oceanic crust and upper mantle has developed in parallel with studies of ophiolite complexes. In the early 1970's *Moores and Vine* [1971], *Dewey and Bird* [1971] and others correlated the layered geologic structure of ophiolite complexes with the seismic layers of the oceanic crust and upper mantle. From the internal structure and composition of ophiolite rock units and their spatial arrangements, these workers interpreted how tectonic and magmatic processes interacted at mid-ocean ridge spreading centers to create oceanic lithosphere. Reviewing these early papers, it is clear that these workers anticipated nearly all of the major geological questions that continue to drive investigations of spreading center processes today. These include the construction of volcanic units in the upper crust, the formation of sheeted dike complexes, the manifestations of tectonic extension at various crustal levels, the size, shape, and duration of magma chambers and formation of the lower crust, the definition of the Moho, and the generation, segregation, and transport of basaltic melt in the mantle beneath spreading

centers. Numerous studies over the following years provided a wealth of detailed geological and geochemical data that have refined and expanded the understanding of ophiolites (see *Coleman* [1977] and *Nicolas* [1989] for reviews and references). Despite the potential value of ophiolites in investigating the details of tectonic and magmatic processes in spreading environments, a number of significant problems have come to light concerning the direct application of ophiolite geology to the oceanic lithosphere (Table I). Nevertheless, ophiolites provide powerful, accessible analogs for oceanic lithosphere.

Following the Penrose Conference on ophiolites [*Penrose Conference Participants*, 1972] it was generally considered that "complete ophiolite complexes" (and by inference oceanic lithosphere) had a large-scale stratiform sequence of rock units: basaltic pillow lavas, sheeted diabase dike complex, layered and non-layered gabbroic to ultramafic rocks, interpreted as oceanic crust. These rock units lie above melt-depleted peridotites, interpreted as residual upper mantle (Figure 1). Other ophiolite complexes with similar rock units, but more complex and disrupted structures were generally considered to have been "dismembered" during obduction and tectonic transport. Some authors have pointed out that some ophiolites that lack a "complete" ophiolite assemblage and stratiform structure, but that appear to have formed in oceanic spreading environments, may have good analogs in slow-spread crust [*Lemoine et al.*, 1987; *Karson*, 1990; *Lagabrielle and Cannat*, 1990; *Alexander and Harper*, 1992; *Dilek et al.*, 1991].

The interpretation and reconstruction of ophiolite complexes have been strongly influenced by marine seismic refraction studies that documented a consistent, horizontally layered velocity structure for oceanic crust [*Raitt*, 1963; *Woollard*, 1975; *Houtz and Ewing*, 1976]. The seismic structure of the oceanic crust was interpreted in terms of two velocity layers: Layer 2 with compressional wave velocities of ~5.0-6.0 km/sec and Layer 3 with velocities of ~6.5-7.5 km/sec. Beneath the Moho discontinuity, upper mantle was characterized by velocities > 8.0 km/sec (Figure 1). Since these early studies the "seismic structure" of the oceanic crust has been investigated in great detail [for example, *Spudich and Orcutt*, 1980; *Bratt and Purdy*, 1984; *Purdy*, 1987; *Vera et al.*, 1990; *Harding et al.*, 1989;1993; *Christeson et al.*, 1992]. In addition, investigations in different tectonic settings have documented variations in seismic structure associated with specific tectonic settings [*White et al.*, 1984; *Calvert and Potts*, 1985; *Toomey et al.*, 1985; 1990; *Purdy and Detrick*, 1986; *Kong et al.*, 1992; *Tolstoy et al.*, 1993; *Detrick et al.*, 1987; 1990]. Almost without exception, the results of these studies have been interpreted in terms of an ophiolite-like geologic structure.

Laboratory seismic velocity measurements on oceanic and ophiolite rock samples helped support the correlation between oceanic seismic structure and ophiolite geology

Table 1. The Trouble With Ophiolites

1. Uncertain Plate Boundary Provenance
MOR Spreading Center?
- Spreading rate
- Magma budget
- Distance from transforms & other segment boundaries
- Distance from hot spots, cold spots, etc.
- Spreading history (symmetry, propagation, pole changes)

Other Spreading Regimes?
- Forearc – Arc – Backarc
- Hot spot or Ocean Island
- Magmatic rift or rifted margin

2. Obscure Post-Spreading Modifications
What Formed at a Spreading Center and What Came Later?
- Sedimentary rocks
- Magmatic intrusions / lavas
- Faults
- Alteration

3. Non-Unique Reconstructions
Suspect Paleo-Horizontal Markers?
- Marine sediments
- Lava flows
- Igneous layering
- Rock unit contacts
- "Petrologic Moho"

Suspect Paleo-Vertical Markers?
- Fissures, Gjar
- Dikes

Lack of Spreading Direction Reference Markers?
- Tilting of sedimentary bedding, lava flows, dikes, etc.
- Dip of faults
- Dike chilling statistics
- Igneous layering geometry
- Peridotite solid-state flow asymmetry

Assumed "Stratigraphic" Sequences?
- 1972 Penrose Conference view
- "Dismembered" ophiolites

4. Tenuous Correlations with Marine Geophysics
Scale of exposures relative to seismic imaging "footprint"?
Correlation between seismic velocity and rock types?

[*Christensen*, 1972; 1978; *Fox et al.*, 1973; *Petersen et al.*, 1974; *Salisbury and Christensen*, 1978; *Spudich and Orcutt*, 1986]. However, several studies have pointed out the strong influence of porosity and alteration on seismic velocities and that these parameters may overlay and dominate over the velocity variations from mineralogical variations of rock units [*Schreiber and Fox*, 1977; *Fox and Stroup*, 1981; *Karson and Fox*, 1986; *Carlson and Herrick*, 1990;

Wilkens et al., 1991]. Laboratory studies suggest that oceanic rocks might be differentiated on the basis of seismic properties measured on the scale of a few centimeters and at elevated confining pressures. However, marine seismic studies typically integrate velocities over length scales of at least a few kilometers and direct observations (see below) document significant heterogeneities in igneous mineralogy, metamorphism, deformation fabrics, fracture porosity, etc., well below the scale of the laboratory samples. Thus the fine-scale laboratory investigations of seismic velocities may only define end-members in a complex family of parameters that influence seismic velocities. Given these considerations, detailed correlations between seismic velocity intervals and major rock units are likely to be nonunique and should be used with extreme caution.

In many recent studies it is common to see marine geologists and geophysicists equate seismic velocity layers with major rock units, despite any documented correlations. Seismic studies and deep drilling in the eastern Pacific in the vicinity of Ocean Drilling Program (ODP) Site 504B demonstrate that the boundary between seismic layers 2 and 3 lies within the sheeted dike complex and appears to correspond to a change in alteration and porosity [*Detrick et al.*, 1994]. In addition, partially serpentinized peridotites with seismic velocities comparable to mafic rocks are known to crop-out on the seafloor near areas with "normal " seismic structures [*Brown and Karson*, 1988]. Seismic velocities and the major rock units thought to make up the oceanic crust cannot be matched in a unique way and therefore seismic crustal structure cannot predict geological structure at a high level of confidence. The only real exception to this would be that velocities of >8.0 km/sec are expected only in anhydrous peridotites. This ambiguity frustrates attempts to use seismic structure, including total "seismic" crustal thickness, as a means of determining the thickness of individual rock units or the total thickness of magmatic material in the oceanic crust.

In summary, two major, related assumptions discussed above have guided considerations of processes at spreading centers: 1) Following marine seismic studies, a generally continuous horizontally layered assemblage of rock units is inferred. 2) Based on analogies with ophiolite complexes, a specific arrangement of rock units and internal features is expected. These two perspective have strongly influenced one another: The reconstruction of layer thicknesses and internal structures in ophiolite complexes is based on the assumption of a horizontal, laterally continuous, layered structure which stems from the seismic structure of oceanic lithosphere. At the same time, the interpretation of the seismic structure of oceanic lithosphere is based on the internal geology of ophiolite complexes. Is this just internal consistency or circular reasoning?

Over the past two decades, submersible studies and crustal drilling along major tectonic escarpments of the seafloor have provided direct observational evidence for the composition, internal structure, and large-scale arrangement of rock

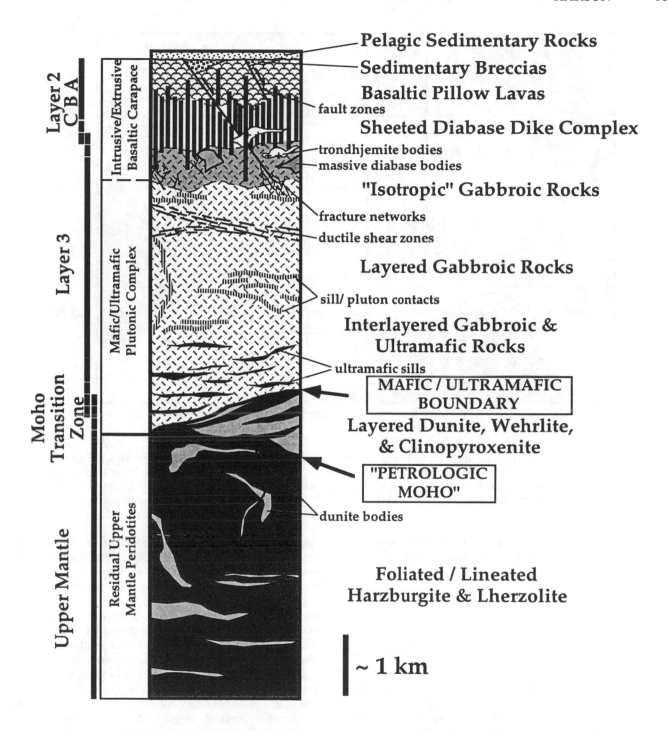

Figure 1. Prevailing view of the internal structure and composition of oceanic crust and correlations with seismic data. The generalized columnar section of rock units and structures is based on the reconstruction of ophiolite complexes with the assumption that contacts between major rock units are horizontal and that dikes are vertical [*e.g. Penrose Conference Participants,* 1972]. The internal structure and composition of rock units and the nature of the contacts that separate them vary significantly laterally. Relief of hundreds of meters along individual contacts is common. Acoustic velocities of rock samples from these units measured under laboratory conditions are commonly correlated with the generalized seismic velocity structure of the oceanic lithosphere.

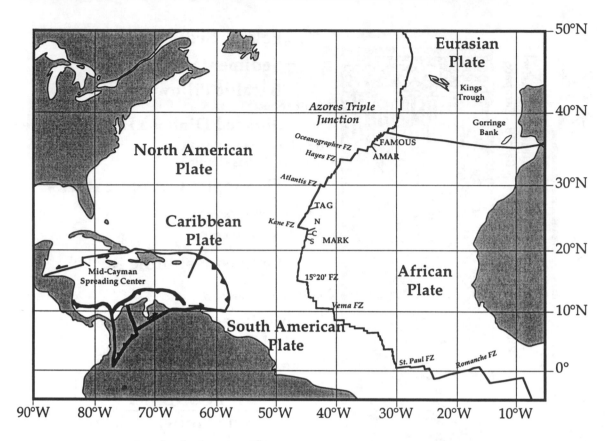

Figure 2. Locations of major tectonic windows into the oceanic lithosphere around the Central Atlantic Ocean. Locations on the MAR indicated by latitude or nearby fracture zones and also include King's Trough, Gorringe Bank, TAG= Trans-Atlantic Geotraverse Area, MARK= Mid-Atlantic Ridge at the Kane FZ (N, C, S= Northern, Central, and Southern rift segments, respectively), and Mid-Cayman Spreading Center in the Caribbean.

units of the oceanic crust. These observations challenge the two major assumptions discussed above and provide new approaches to investigating the nature of seafloor spreading.

3. TECTONIC WINDOWS INTO THE OCEANIC LITHOSPHERE

3.1 Settings of Tectonic Windows

Faulted escarpments that expose significant vertical sections of the oceanic lithosphere occur throughout the ocean basins [*Vine and Offset Drilling Working Group*, 1995]. They include deep gashes in mature oceanic lithosphere well away from active spreading centers and also escarpments near spreading centers and fracture zones. Possible tectonic windows into oceanic crust also exist in the inner walls of deep-sea trenches at subduction zones [for example *Hess*, 1962; *Fisher and Engel*, 1969; *Bloomer and Hawkins*, 1983; *Stern et al.*, 1996].

Major tectonic windows are most common in slow-spreading environments. Several different tectonic settings have been identified. These include major escarpments that define the median valley walls of the spreading centers and transform valley walls, as well as escarpments in relatively old oceanic lithosphere produced by intra-oceanic rifting, or thrusting (Figure 2).

Tectonic windows along median valley wall exposures are essentially parallel to structures expected to form along spreading centers, for example, normal faults, dikes, etc. These are well suited to reveal along-strike variations in spreading processes and to provide information on crustal accretion variables along spreading segments. Being situated in the active plate boundary zone, they may be modified before moving laterally away from the spreading center. Transform valley exposures should truncate seafloor spreading structures at a high angle and thus show spreading variations with time. Fracture zone walls cutting across lithosphere from zero to tens of millions of years old can potentially reveal long-term spreading histories. Active transform valley walls will typically pass spreading center terminations at ridge-transform intersections (RTIs) and are therefore susceptible to later modifications [*Karson and Dewey*, 1978; *Karson and Dick*, 1983].

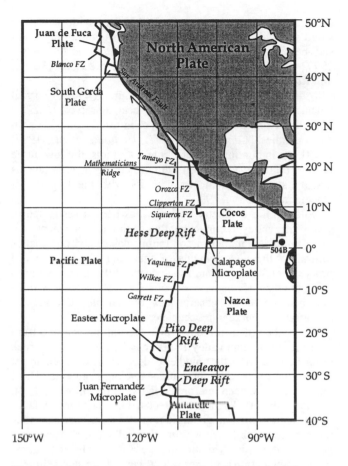

Figure 3. Locations of major tectonic windows into the oceanic lithosphere in the Eastern Pacific Region. Locations include the Blanco Transform, Mathematicians Ridge, Hess Deep Rift, Endeavor Deep Rift, Pito Deep Rift. ODP Site 504B is in ~6 Ma lithosphere south of the Costa Rica Rift.

Tectonic windows into lithosphere created at fast- and intermediate-spreading ridges are relatively rare (Figure 3). Unlike the major rift valleys of slow-spreading ridges, the axial grabens or depressions of faster spreading ridges are typically less than 1 km wide and only tens to 200 m high. Individual faults, typically with less than tens of meters of relief, expose only basaltic volcanic rocks. The faults and relief they create are still very much within the region of active volcanism [e.g., *Macdonald*, 1982] and therefore are likely to be intruded by dikes or buried by lavas before they move laterally away from the axis. Transform escarpments are typically only a few hundred meters high and in some cases expose plutonic rock units. The most impressive tectonic windows into fast-spread crust occur at propagating rift tips, commonly near microplate boundaries. Escarpments up to a few kilometers in relief provide excellent exposures, especially where rifting affects relatively old oceanic lithosphere.

In viewing the internal structure of the oceanic lithosphere it is important to recognize that the crustal structure in these areas may be influenced by the tectonic setting that created the window. Furthermore, faulting associated with the formation of major escarpments can also cause changes in the structure, porosity and alteration of rocks exposed along them. Thus, it is possible that rocks exposed in tectonic windows are not entirely representative of the more regional character of the oceanic lithosphere. Although these factors make it likely that sampling biases exist to some degree, the growing number of observations at tectonic windows does not show any consistent patterns with respect to major tectonic boundaries, supporting the view that these windows may provide views of the internal structure of the oceanic lithosphere that are significant on a regional scale.

3.2 Faults, Fault-Line Scarps, and Seafloor Escarpments

It is important to make the distinction between seafloor escarpments and the individual faults that create them. Contrary to many descriptions, the escarpments mapped with multibeam bathymetry and side-scan sonar systems generally do not correspond to individual faults. Direct observations show that the escarpments typically have a stair-step morphology created by steep "fault-line scarps" (tens to a few hundred meters high) separating more gently sloping terraces (tens to hundreds of meters wide). The fault-line scarps commonly have slopes of $30°-90°$ and expose massive to highly fractured basement rocks or indurated talus and debris slide deposits. Fault surface features typical of subaerial fault zones [*Hancock and Barka*, 1987] are exposed in many places. Locally, polished or slickensided fault surfaces dipping 30°-60° and commonly down-dip plunging slickenlines, striac, and grooves are well developed. Mass wasting and talus accumulation have modified these steep slopes according to the composition and fracture densities in the local rock units.

Fault geometry and mass wasting may modify escarpments in several ways. They may tend to increase or decrease slopes relative to the dips of the major fault zones. For example, slumping of the headwalls of fault scarps could either steepen or degrade local slopes. Talus accumulations at the base of scarps would tend to decrease slopes. Small-scale normal or reverse faulting could either increase or decrease the dip of major detachment fault surfaces [*Westway and Kusznir*, 1993] that accommodate much more displacement.

Despite these complexities, many interpretations and analyses of seafloor tectonics essentially equate escarpments with major faults [*Kong et al.*, 1988; *Goff*, 1991; *Shaw*, 1992; *Shaw and Lin*, 1993; *Cowie et al.*, 1993; *Goff et al.*, 1993; *Escartín and Lin*, 1995]. Although continuous, linear escarpments on the seafloor may serve as rough approximations of major fault zones, the slopes of escarpments are not the same as the dips of faults that create them and their relief is not the same as the vertical fault displacement. These differences can have important implications for estimating uplift rates, fault heaves, and seafloor tectonics in general. In the following section the geometry and kinematics of faulting associated with tectonic windows into the oceanic lithosphere are briefly outlined.

3.2.1 Median Valley Floor Escarpments. High-resolution surveys of faulted escarpments along the MAR axis has been reported in numerous studies [*Luyendyk and Macdonald*, 1977; *Macdonald and Luyendyk*, 1977; *Karson and Dick*, 1983; *OTTER Team*, 1984; *Brown and Karson*, 1988; *Kong et al.*, 1988; *Murton and Parson*, 1993; *Smith et al.*, 1995; *Allerton et al.*, 1996; *Kleinrock et al.*, 1996; *Lawson et al.*, 1996]. These studies show that closely spaced, small-relief, normal fault scarps and fissures occur across the median valley floor. Locally they cut volcanic features and elsewhere are partially buried by basaltic flows reflecting the continuous magmatic construction and stretching of the axial lithosphere. These small-offset faults appear to expose only volcanic rocks and thus yield only limited information about the internal structure of the uppermost oceanic crust. Despite exposures that in some places reach a few hundred meters vertically, very few studies have documented the internal structure and stratigraphy of lavas in these settings. Except near major transforms or obvious axial discontinuities where oblique- or even strike-slip faulting has been predicted, median valley faults are generally considered to be normal faults. Lack of mapped stratigraphy and poorly preserved kinematic indicators hamper quantifying fault displacements.

Fault scarps along the axes of intermediate- to fast-spreading ridges have relief of typically tens of meters [*CYAMEX Scientific Team*, 1981; *Luyendyk and Macdonald*, 1985; *Macdonald and Luyendyk*, 1985; *Kappel and Ryan*, 1986; *Macdonald*, 1998; *Perfit and Chadwick*, 1998]. Both inward and outward dipping faults are documented, with outward-dipping faults becoming less common at slower-spreading rates [*Macdonald and Luyendyk*, 1985; *Carbotte and Macdonald*, 1990]. In the axial region of spreading centers these features cut into young basaltic material and may be buried by later flows, either on- or off-axis [*Perfit and Chadwick*, 1998 and references therein]. Locally, growth fault morphologies are documented as a means of constructing abyssal hills [*Macdonald et al.*, 1996]. Overall, the limited relief created by faulting and continuing volcanic activity along intermediate- to fast-spreading ridge axes does not create substantial windows into the upper oceanic crust.

3.2.2 Median Valley Walls. The median valley walls that bound the rift valleys of slow-spreading ridges average about 1.5 km in relief [*Shih*, 1980], but vary from barely detectable to as much as 5 km in relief. Submersible observations show that these steep escarpments are not single faults, but rather complexly faulted areas in which spaced, steeply dipping normal faults dominate. The fault styles vary from segment to segment, and in some places change significantly over just a few kilometers along the ridge axis [*Karson et al.*, 1987; *Karson and Rona*, 1990; *Shaw and Lin*, 1993; *Cann et al.*, 1997; *Tucholke et al.*, 1997]. The

most common features are normal faults with relatively small vertical separations (<100 m). Most commonly, fault surfaces are more or less planar [*Macdonald and Luyendyk*, 1977], but listric forms have also been described [*Karson and Rona*, 1990].

In some places, extensive, low-angle (<30°), normal ("detachment") fault surfaces have been observed [*Dick et al.*, 1981; *Karson and Dick*, 1983] or inferred from side-scan sonar studies [*Cann et al.*, 1997; *Tucholke et al.*, 1997; 1998]. As in continental detachment systems, the low-angle faults are cut by later steep faults resulting in escarpments that are steeper than the local faults with the largest displacements [*Karson*, 1990].

Although the geometry of faults has been investigated in some rift valley tectonic windows, the displacements on these faults is a matter of continuing debate and one that is central to the understanding of the internal structure and process(es) of formation of the oceanic lithosphere. Quantifying the displacements of oceanic fault zones is very difficult because of the general lack of reliable structural or stratigraphic markers in oceanic crust.

Despite the rather limited vertical separations inferred from the relief of median valley faults (hundreds of meters), rock units derived from the middle to deep crust or even upper mantle are exposed along some spreading segments. These exposures have been interpreted in two different ways. First, the deep-level rocks might be exposed by very limited faulting if the basaltic and gabbroic upper to middle crustal rock units were initially very thin. Several workers have suggested that diminished melting near relatively cool transform intersections of spreading centers resulted in the formation of relatively thin igneous crust [*Francheteau, et al.*, 1976; *Stroup and Fox*, 1981]. This interpretation is supported in a general way by the thinning of the seismic crust near fracture zones [*White et al.*, 1984; *Detrick et al.*, 1993b]. Some studies suggest that even small, non-transform offsets along spreading centers are associated with thin crust, especially along inside plate corners [*Tucholke and Lin*, 1994; *Cannat et al.*, 1995c; *Escartín and Lin*, 1995]. Given the distribution of rock units and normal fault displacements inferred from scarp heights, this implies a drastic reduction in thickness of igneous crustal rock units individually and collectively. In this model, gabbroic rocks would have crystallized at depths of only 100-200 m and the total igneous crustal thickness could be as little as 500 m. A variation of this interpretation is that crust in these settings could be composed of serpentinite with variable proportions of gabbroic intrusions and a thin cap of basaltic lavas [*Brown and Karson*, 1988; *Cannat*, 1993; *Cannat et al.*, 1997b].

Alternatively, deep-level rocks could be exposed by various fault geometries that caused large-scale (kilometers) mechanical thinning of upper to middle crustal units. *White and Stroup* [1979] suggested that conjugate normal faults were responsible for extensive exposures of plutonic rock units on the walls of the Mid-Cayman Spreading Center.

Roughly symmetrical conjugate faulting is found in serpentinites in the 15°20'N area of the MAR [*Lagabrielle et al.*, 1998]. Large-scale conjugate faulting such as would be required to exhume deep-level rocks is not well documented, but is a potentially viable mechanism. Structures in gabbroic rocks on the MAR near the Kane Transform Ridge-Transform Intersection were interpreted as very extensive shear zones and detachment faults with large (kilometers) displacements [*Dick et al.*, 1981; *Karson and Dick*, 1983] in an oceanic core complex [*Karson*, 1990]. Recent bathymetric and sonar studies have shown that extensive lineated surfaces interpreted as detachment faults are widely developed over dome-like massifs in slow-spread crust [*Cann et al.*, 1997; *Tucholke et al.*, 1998; also see *RIDGE Multibeam Synthesis at: http://imager.ldeo.columbia.edu*] suggesting that core complex tectonics is widespread in this setting.

Following studies of continental core complexes, displacements on detachment fault surfaces could result in the exposure of deep-level rocks in two ways. First, if the faults were initially more steeply dipping and cut down into deep crustal materials, displacement and rotation could exhume rocks formed in the deep crust and upper mantle [*Spencer*, 1984; *Wernicke and Axen*, 1988; *Buck*, 1988]. Second, detachment faults could expose deep-level rocks if they cut into crustal materials that were previously thinned by ductile stretching [*Crittenden et al.*, 1980; *Davis and Lister*, 1988; *Lister and Davis*, 1989]. If large-scale rotations or ductile deformation occur in these settings significant complications could result in crustal structures viewed through tectonic windows into them.

Several observations suggest that very large-scale faulting has indeed occurred along slow-spreading ridges. The rock units exposed and the fault geometries observed in tectonic windows imply that individual major faults once cut from the surface to the upper mantle. Coarse-grained cumulate gabbros suggest crystallization at crustal depths of a few kilometers. Although geothermal gradients beneath median valley floors are poorly constrained, low-pressure granulite to amphibolite facies metagabbros formed there suggest exhumation from a few kilometers depth. Serpentinized upper mantle harzburgites imply depths of 4-6 km, corresponding to seismic crustal thicknesses. Fluid inclusion microanalysis of metagabbros from 3500 m water depth on the median valley wall detachment fault of the northern MARK Area indicate depths of formation corresponding to >1 kb [*Kelley and Delaney*, 1987]. Assuming a typical MAR bathymetry for this region, about 2 km of unroofing is implied. Faulting that cuts the full thickness of the crust (5-7 km) is corroborated by seismic studies along the MAR axis [*Toomey et al.*, 1985; 1988; *Kong et al.*, 1992; *Wolfe et al.*, 1995] as well as in older crust [*White et al.*, 1990; *Mutter and Karson*, 1992; *Louden et al.*, 1996]. Paleomagnetic studies of low-angle fault zones (30°-45° dips) along the MAR show that the fault-zones have not been rotated [*Hurst et al.*, 1997; *Lawrence*, 1998; *Lawrence et al.*, 1998] and that these faults are likely to have slipped in their present orientations. Finally, the flow-line parallel lengths of tectonic windows (up to ten kilometers) and of lineated surfaces interpreted as detachment faults (tens of kilometers) imply very large magnitudes and durations of displacement on major faults [*Karson*, 1990; *Cann et al.*, 1997b; *Tucholke et al.*, 1998].

Thus, although observations in tectonic windows suggest that relatively small-displacement (tens of meters) faults may expose plutonic rocks in thin or even discontinuous igneous crust along some slow-spreading ridge segments (see below), there is also strong evidence that deep-level rocks can be exhumed from substantial crustal depths (kilometers). It is likely that some combination of thin igneous crust and large-displacement faulting create the extensive plutonic rock exposures found along slow-spreading ridge median valley walls.

3.2.3 Fracture Zone Walls. *Bonatti and Honnorez* [1976] regarded the escarpments of the Vema Transform as essentially a continuous section through the oceanic crust and upper mantle based on the distribution of rock types dredged at different depths. *Francheteau et al.* [1976] challenged this interpretation by citing the step-faulted structure of the wall of Fracture Zone A in the FAMOUS Area that would result in repetition of relatively shallow crustal units from top to bottom of the escarpment. They further proposed other mechanisms by which coarse-grained mafic crustal rocks and mantle rocks might be exposed at the surface.

The stair-step morphologic patterns of fracture zone valley walls suggest vertical separations on faults similar to those on high-relief median valley walls near RTIs. In a few places kinematic indicators demonstrate lateral displacement on some faults [*Auzende et al.*, 1993; *Karson*, 1998], but the absolute or even relative magnitudes of these displacements are not known at present.

The preferential location of major detachment faults and core complexes along transform faults, especially at inside plate corners [*Severinghaus and Macdonald*, 1988], could permit relatively small-offset faults to expose deep-level rocks in previously stretched and thinned crust [*Tucholke and Lin*, 1994; *Tucholke et al.*, 1998]. Major scarps along transforms may provide windows into the internal structure of oceanic core complexes initially formed at RTI's [*Auzende et al.*, 1993; *Cann et al.*, 1997; *Karson*, 1998].

3.2.4 Other Types of Tectonic Windows. Oceanic crust and upper mantle rocks are exposed in a few places away from spreading center and fracture zones. Tectonic windows are also found off-axis in abandoned rifts and at propagating rift tips. Kings Trough [*Kidd et al.*, 1982] is an abandoned rift in the eastern Atlantic (Figure 2; [*Srivastava and Tapscott*, 1986]). Other examples occur at propagating rift tips along the edges of microplates in the eastern Pacific [*Francheteau et al.*, 1988; 1990; 1994; *Hooft et al.*, 1995] and provide some of the only know tectonic windows into lithosphere formed at the fast-spreading East Pacific Rise (EPR; Figure 3).

A potentially important type of tectonic window into fast- to intermediate-rate crust occurs at the abandoned EPR spreading center at the Mathematicians Ridge in the eastern Pacific (Figure 3-*Batiza and Vanko* [1986]). Tectonic extension apparently continued after magmatism ceased resulting in the exposure of deep crustal rocks. Spreading centers abandoned by ridge jumps and propagation events are fairly common on the seafloor and may provide more tectonic windows into crust with different spreading rates.

Finally, convergent plate boundaries appear to provide exposures of a diverse suite of mafic and ultramafic rock units. *Hess* [1962] described serpentinites from the wall of the Puerto Rico Trench. An island-arc ophiolite suite was inferred to crop-out on the inner wall of the Mariana Trench based on rocks recovered by dredging [*Bloomer and Hawkins*, 1983], drilling and submersible studies [*Fryer et al.*, 1985]. Recent submersible studies also document substantial exposures of back-arc basin plutonic mafic and ultramafic rocks in the rifted walls of the Mariana Trough [*Stern et al.*, 1996].

3.3 Major Rock Units of the Oceanic Lithosphere

The general characteristics of the major rock units found in the oceanic lithosphere are outlined in this section. The composition, internal structure, and outcrop character of each unit are discussed to give a sense of the nature of materials exposed in tectonic windows (Figures 4-7).

3.3.1 Surficial Rock Units. Basement rock units of the seafloor are commonly covered by some thickness of sedimentary material. This is composed of coarse clastic material that accumulates rapidly at the base of steep slopes in tectonically active areas and pelagic ooze that generally thickens with age to progressively bury the clastic material and basement outcrops [*Fox and Heezen*, 1965; *Dick et al.*, 1978; *Marks*, 1981; *Lagabrielle and Auzende*, 1982; *Barany and Karson*, 1989].

In young lithosphere, the most voluminous surface material is basaltic breccia produced by lithification of fragmental basaltic talus, scree, or rubble ranging from sand to boulder size material (Figure 4a). It is typically framework supported and cemented by carbonate or ferromanganese hydroxide. Widely exposed carbonate-matrix-supported breccias with basaltic clasts are probably debris slide deposits (Figure 4b).

Massive to crudely bedded volumes of breccia drape rugged seafloor topography. Although texturally very immature, clast diversity in individual deposits is very limited. The most common compositions are variably weathered glassy, aphanitic, and holocrystalline basalts. Breccias with clasts of serpentinite, gabbro, metagabbro, and a variety of fault rocks have also been found. Polymict breccias are rare. Breccia clasts within these deposits attest to down-slope recycling of clastic materials. Although the breccias commonly contain clasts of basement rock units that they

overlie, basaltic breccias commonly overlie plutonic rock units [*Karson and Dick*, 1983].

The breccia exposures are commonly deeply incised and sculpted by subsequent mass wasting. Even along median valley walls, they are sufficiently well indurated to support subvertical slump scars and fault-line scarps (Figure 4b). In some of these places breccias crop-out over vertical intervals of several meters to several tens of meters. Poorly developed bedding makes it difficult to determine if there are exposures of even thicker accumulations or just a veneer of debris adhering to underlying outcrops. This material is especially widespread along transform valley walls. Steep escarpments up to several tens of meters high in this material attest to its coherence. Considering how widespread this it is at spreading centers and along transform walls, it is likely that this type of colluvial material is one of the most common surficial deposits on Earth.

Other surficial material includes plastic to well indurated carbonate ooze [*Marks*, 1981]. Chalky deposits commonly directly overlie basaltic basement filling in fine-scale topography over lava flow surfaces. Fine lamination to thin bedding and extensive burrowing and bioturbation are also common. This bedding provides an important potential paleohorizontal marker with which to evaluate tectonic rotations [*Karson and Rona*, 1990].

Sedimentary rocks associated with ophiolite complexes provide an important means of interpreting the setting in which they have formed [*Moores*, 1982, and references therein]. Although ophiolite clast breccias are anticipated along oceanic fracture zones [*DeLong et al.*, 1979], the sedimentary breccias found in many ophiolites [e.g., *Abbate et al.*, 1980] may have formed along escarpments in any of the tectonic windows described here.

3.3.2 Basaltic Volcanic Units. Since the earliest submersible studies of the MAR during the FAMOUS program, numerous other investigations have provided detailed descriptions of the basaltic volcanic units [*Ballard et al.*, 1975; 1979; *Ballard and Moore*, 1977; *ARCYANA*, 1975; 1978; see *Perfit and Chadwick*, 1998 for a recent review] that typically make up the uppermost basement rock units of oceanic crust and ophiolites (Figure 4c, d). At slow-spreading ridges, surface mapping shows that basaltic units are constructed by the eruption many small (kilometer scale) lava flows [*Smith and Cann*, 1993]. At faster-spreading ridges fissure eruptions and sheet flows are also important [*Haymon et al.*, 1991; *Perfit et al.*, 1994].

Fault scarps expose truncated sections of lava stratigraphy reaching estimated thicknesses of as much as about 1 km. In addition, several DSDP and ODP drill holes have penetrated basaltic lava units up to about 0.5 km. In some drill holes, lavas are interlayered with thin beds of pelagic sedimentary material. At DSDP Site 395, basaltic pillow lavas are interlayered with sedimentary breccias bearing clasts of gabbroic rocks and serpentinites [*Purdy and Rabinowitz*, 1978].

Basaltic volcanic rocks are the most commonly exposed basement materials exposed on oceanic scarps. In all cases,

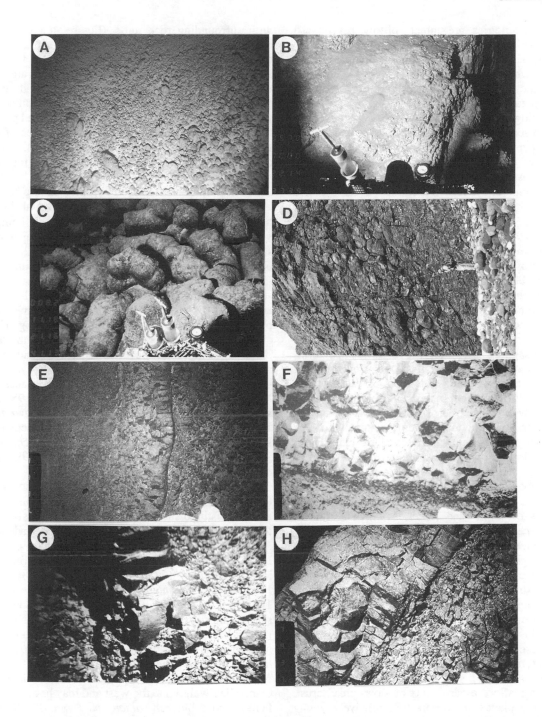

Figure 4. Outcrop photos of upper crustal rock units. (fov=field of view) a. Mixed rubble and pelagic sediment at the base of steep scarp of basaltic material (MARK Area, vertical incident ANGUS photo [*Karson et al.*, 1987]; fov ~5 m across). b. Steep fault scarp (1-2 m high) cutting consolidated, framework-supported basaltic breccia (MARK Area, *Alvin* Dive 1022). c. Basaltic pillow lavas with minor, light-colored, pelagic sediment (neovolcanic ridge of N. MARK Area, *Alvin* Dive 1021). d. Truncated section of basaltic pillow lavas (Oman Ophiolite, Wadi Shafran). e. Subvertical diabase dike (~ 0.5 m wide) in fractured basaltic material (SMARK Area, *Alvin* Dive 2571). f. Subhorizontal diabase sill (?), ~2 m thick, with vertical columnar jointing in fractured basaltic host rock (SMARK Area, *Alvin* Dive 2885). g. Vertical diabase dike (~0.5 m wide) in intensely fractured basaltic pillow lavas (SMARK Area, *Alvin* Dive 2887). h. Moderately dipping, fractured, diabase dike (~0.6 m wide) in fractured basaltic material (SMARK Area, *Alvin* Dive 2887).

where flow morphologies are intact, pillow basalts and tubes dominate (Figure 4c). Pillow basalts and tubes typically have dimensions of less than 2 m in diameter. They are inherently fractured and jointed on the scale of tens of centimeters. Although they appear to be common along the axes of some spreading segments, especially at fast- to intermediate-rate ridges, sheet flows or blocky flows have only rarely been seen in fault escarpments.

Especially deeper in the lava pile, the basaltic material is highly fragmented (Figure 4 e-h). Typically, fine-scale fracturing is so intense that original lava flow morphologies are not readily apparent. Flow structures appear to have been obscured by locally pervasive fractures spaced at a few centimeters. This may be an important feature indicating that large-scale cataclastic flow is an important, but underappreciated deformation mechanism in the upper oceanic crust. It is also possible that some of this fragmental material originated as blocky lava flows.

On submarine cliffs, lava flows are commonly truncated along young fault scarps. The headwalls of the faults remain steep but become scalloped in plan view due to the calving of masses of rock material from scoop-shaped scars. These masses disaggregate as they move downslope forming cobble- to gravel-size fragments. Downslope-trending corrugations produced by debris-slide tongues and intervening ravines are well-developed. In many places, the slide and debris chutes show evidence of erosion and scouring and are hundreds of meters long and ten to twenty meters wide. The distinctive wedge-shaped fragments created by radial joint patterns intersecting pillow rims are diagnostic.

Massive basaltic material occurs as rare lava flows or sills a few meters thick (Figure 4f) and as diabase dikes (Figures 4e,g,h). These and other massive basaltic units exhibit distinctive patterns of joints and fractures which are much more widely spaced than those of the surrounding pillow basalts. Consequently escarpments in these more massive materials are rougher and more angular. Downslope, talus derived from these units is coarser and less continuous than in basaltic exposures of similar age. Massive tabular fragments of individual dikes, sills or flows are common.

In ophiolite complexes, the base of the basaltic volcanic units typically is obscured by a downward increasing proportion of diabase dikes [*Moores and Vine*, 1971; *Abbots*, 1979; *Casey et al.*, 1981; *Pallister*, 1981; *Rosencrantz*, 1983; *Rothery*, 1983]. This type of contact has not yet been observed in seafloor escarpments of slow-spread crust. Instead, in several places, it appears that relatively young basaltic volcanic material directly (unconformably) overlies older, much more faulted and metamorphosed rock units. In tectonic windows of crust formed at faster spreading rates, basaltic unit are underlain by sheeted dike complexes or massive diabase units.

3.3.3 Diabase Dikes. Diabase dikes are commonly observed in basaltic lava units, sheeted dike complexes, and plutonic rock units (described below). Typically individual dikes are 0.5-1 m wide and have sharply defined contacts with surrounding rock units (Figure 4e,g,h). Faults and rock debris commonly limit the ability to trace dikes vertically or laterally across seafloor escarpments for more than a few tens of meters at most. Individual dikes are easily distinguished from surrounding rock units and even other dikes, because of their sharply defined chilled margins which are commonly marked by closely spaced parallel fractures and evenly spaced cross joints (the traces of columnar cooling joints) that terminate at the margins. Simple mechanical models of spreading centers predict that the dikes will be vertical and parallel to the ridge axis; however, considerable variations have been documented (see below).

In the basaltic units, dikes are planar overall but tend to have rather undulating margins. Rather than being strictly vertical, they tend to have sinuous traces upslope across steep escarpments. Where dikes occur in pillow lavas or other closely fractured materials, they commonly stand-out in relief on scarps creating promontories (where they intersect the scarp at a high angle) or smooth walls (where they are parallel). These features commonly become accentuated by the degradation and downslope channeling of material along their margins or derived from higher up the escarpments. Dikes typically occur as individuals and more rarely in swarms of several dikes in close proximity. Dikes have been observed in the basaltic units in many places along the MAR [*ARCYANA*, 1978; *Auzende et al.*, 1989a; *Zonenshain et al.*, 1989; *Karson and Rona*, 1990; *Lawrence et al.*, 1998] as well as in crust formed at intermediate to fast spreading rates [*Anderson et al.*, 1982; *Francheteau et al.*, 1992; 1994; *Karson et al.*, 1992; *Alt et al.*, 1993; *Juteau et al.*, 1995; *Naidoo*, 1998].

Dikes also occur in massive diabase, gabbroic, and serpentinite units. In these units the dikes have planar margins and are distinguished from surrounding massive rocks by their joint patterns and their tendency to degrade somewhat more rapidly, producing narrow, slot-like embayments. Dikes are described in gabbroic rocks in tectonic windows into slow-spread crust [*Auzende et al.*, 1978; *Stroup and Fox*, 1981; *Zonenshain et al.*, 1989; *Karson and Lawrence*, 1997a,b; *Lawrence et al.*, 1998]. They cut amphibolites and serpentinites in the southern wall of the Kane Transform [*Auzende et al.*, 1993]. Dikes also cut serpentinites in an number of other locations [*Auzende et al.*, 1978; *Cannat et al.*, 1997; *Karson and Lawrence*, 1997b].

At the TAG Area, ridge-parallel dikes on the eastern median valley wall dip to the west and may have been tilted in fault-bounded blocks [*Karson and Rona*, 1990]. Structural and paleomagnetic studies of dikes in upper crustal rock units of the southern MARK (SMARK) area show that they have highly variable orientations and that some have been tectonically rotated prior to the intrusion of later dikes, faulting, and uplift into the median valley walls [*Lawrence et al.*, 1998]. Near the Oceanographer Fracture Zone, dikes in basaltic pillow lavas have oblique trends that mimic the bathymetric lineaments which are intermediate between the ridge axis and the transform valley trends [*OTTER Team*,

1984]. Dikes at the Vema Transform [*Auzende et al.*, 1989a] and in exposures along the Kane Transform [*Auzende et al.*, 1993] strike at a high angle to the nearby active transform faults.

3.3.4 Sheeted Diabase Dike Complexes. Sheeted dike complexes are one of the defining characteristics of ophiolite complexes (Figures 5a-d) and are therefore expected in oceanic crust. Despite extensive exposures of basaltic and gabbroic rock units thought to bracket this distinctive unit vertically, exposures in slow-spread crust are very rare. In fact, a sheeted dike complex has been described in only two locations to date [Vema Transform- *Auzende et al.*, 1989a; Kings Trough- *Klitgord et al.*, 1990]. Sheeted dike complexes are thought to be an integral part of crust formed at fast- to intermediate spreading ridges. Major outcrops of sheeted dikes occur at Hess Deep [*Francheteau et al.*, 1992; *Karson et al.*, 1992] and Pito Deep [*Francheteau et al.*, 1994]. No sheeted dike complex has been identified at the Blanco Transform, even though a vertical crustal section of ~2 km appears to be exposed. Studies of cores, borehole video, and geophysical logs from Hole 504B also document a vertical section through a sheeted dike complex [*Anderson et al.*, 1985; *Alt et al.*, 1993; *Dilek*, 1998].

Where present, sheeted dike complexes on submarine escarpments have a distinctive "ribbed" outcrop fabric defined by the differential degradation of steeply dipping, subparallel dikes that are typically 0.5-1.5 m wide (Figures 5b-d). More closely jointed or fractured dikes tend to degrade more rapidly than massive dikes and to create steep ravines and embayments in steep cliffs. Steep walls along individual dike margins are commonly created by the removal of more fractured material. Grain-size variations and chilled margins are not visible in outcrops, but the margins of individual dikes are obvious from joint patterns. Exposures limited by debris slides, talus, and faults do not permit individual dikes to be traced vertically for more than a few tens of meters.

Despite, many current generalized models of seafloor spreading, the dikes in sheeted complexes are not always vertical and ridge-parallel. Some exposures of sheeted dike complexes seen in tectonic windows have essentially vertical, subparallel dikes (Figure 5b; *Auzende et al.*, 1989a; *Klitgord et al.*, 1990; *Francheteau et al.*, 1992). However, much more complex networks of variably dipping dikes (Figures 5c,d) in multiple, cross-cutting assemblages have also been described from Hess Deep [*Karson et al.*, 1992; *Hurst et al.*, 1994a; *Tougas*, 1998]. Paleomagnetic studies of oriented dike samples show that assemblages of these dikes have been tectonically rotated [*Hurst et al.*, 1994a]. *Pariso and Johnson* [1989] also document tectonic rotations of dikes drilled at ODP Site 504B. Locally, faults and hydrothermal veins cut the sheeted dike complex [*Agar*, 1991; *Karson et al.*, 1992; *Dilek*, 1998]. Samples from this unit show a wide range of alteration effects, generally with intensity and temperature of metamorphism increasing downward [*Alt et al.*, 1993; *Gillis*, 1995].

The upper and lower contacts of sheeted dike complexes are not well exposed, but appear to be gradational in the sense that the proportion of dikes decreases both upward into extrusive basaltic material and downward into gabbroic rocks. At Hess Deep, the upper contact of the sheeted dike complex appears to have significant relief. Sheeted dike swarms on the order of several tens of meters wide penetrate much as 200 m upward into the lavas. Elsewhere, the proportion of dikes in lavas decreases gradually upward over an interval of as much as 500 m referred to as a "transition zone" [*Francheteau et al.*, 1992]. A similar interval of mixed extrusive and intrusive material is identified at ODP Site 504B [*Anderson et al.*, 1982; *Dilek*, 1998].

The contact at the base of the sheeted dike complex is of special interest because it must lie near the contact between rapidly chilled upper crustal rocks and material that crystallized in an underlying magma chamber. Observations from ophiolite complexes suggest a number of possible relationships including "rooting" of dikes in coarse-grained gabbros, intrusion of the dikes by high-level gabbro plutons [*Rosencrantz*, 1983; *Rothery*, 1983; *Nicolas and Boudier*, 1992] or major low-angle fault zones [*Allerton and Vine*, 1987; *Varga and Moores*, 1990; *Hurst et al.*, 1994a]. The transition from dominantly sheeted dikes to gabbroic rocks appears to take place over a vertical distance of less than about 200 m at Hess Deep [*Karson et al.*, 1992]. Dikes and gabbros at this level have similar compositions [*Natland and Dick*, 1996] suggesting that the dikes are derived from nearby gabbro bodies.

Outcrop patterns suggest that as much as a few hundred meters of relief occurs on both the upper and lower contacts of oceanic sheeted dike complexes, but these depend on how bounding transitional units are defined. Thus, the estimates of the thickness of sheeted dike complexes are highly variable even over a few kilometers laterally. Estimates range from 300 m to 1200 m [*Auzende et al.*, 1989a; *Francheteau et al.*, 1992; *Karson et al.*, 1992]. Similar variations are noted in ophiolites [*Pallister*, 1981; *Rosencrantz*, 1983; *Rothery*, 1983].

Collectively, these results show that the internal structure and bounding contacts of sheeted dike complexes in tectonic windows into the oceanic crust are extremely variable. Where they are best developed they commonly have may similarities to those of ophiolite complexes. Some sheeted dike complexes have internal structures that are significantly more complex and suggest localized faulting and block rotations. It is not yet known if these reflect local spreading anomalies or an important aspect of seafloor spreading that has not been previously recognized. Sheeted dike complexes are missing from the crustal sections in some tectonic windows into slow-spread crust. In these areas, spreading must be accommodated by processes other than incremental dike intrusion.

3.3.5 Massive Diabase Units. This type of unit has been described in only two locations to date at the southern MARK (SMARK) area [*Karson et al.*, 1996; *Lawrence et*

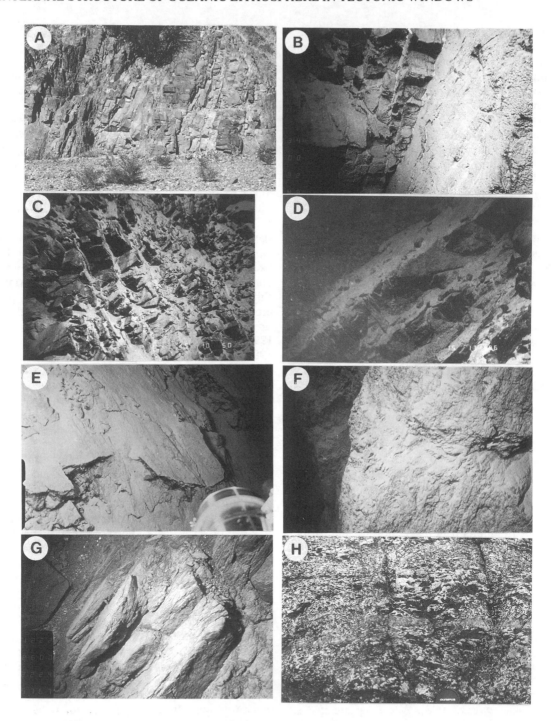

Figure 5. Outcrop photos of middle crustal rock units. a. Ophiolite sheeted diabase dike complex (Oman Ophiolite, Wadi Bani Suq-Daris; fov=~15m across). b. Sheeted dike complex with dikes ~0.5-1 m wide (Vema Transform, *Nautile* Dive VN #03). c. Sheeted dike complex with moderately dipping dikes (Hess Deep Rift, *Alvin* Dive 2219). d. Seafloor sheeted diabase dike complex with several, gently dipping dikes, each 0.5-1 m wide (Hess Deep Rift, *Alvin* Dive 2215). e. Massive diabase unit with spaced (~10 cm), moderately dipping, cataclastic shear zones (SMARK Area, *Alvin* Dive 2889). f. Outcrop of massive, densely fractured, gabbro to metagabbro (MARK Area, *Nautile* Dive HS14). g. Outcrop of strongly foliated metagabbro (MARK Area, *Nautile* Dive HS14). h. Heterogeneous grain size and texture in high-level ophiolite gabbro (Bay of Islands Ophiolite, Lewis Hills). Fov for all seafloor photos is ~5m across.

Figure 6. Outcrop photos of lower crustal and upper mantle rock units. a. Layered gabbros (Bay of Islands Ophiolite, Lewis Hills). b. Massive to laminated gabbro with plagioclase-rich lenses (Oman Ophiolite, Wadi Hamiliyaya). Note low-angle truncations at contacts suggesting that massive layers are sills. c. Ultramafic, wehrlite sills (dark) in light-colored gabbroic host (Oman Ophiolite, Samrah Oasis). d. Anastomosing ductile shear zones in wehrlite to feldspathic wehrlite (Bay of Islands Ophiolite, Lewis Hills). e. Ophiolite peridotite with foliation (trace ~horizontal in view), subordinate lineation, and multiple generations of cross-cutting syntectonic veins of dark pyroxenite and light-colored dunite (Bay of Island Ophiolite, Lewis Hills). f. Gabbroic pegmatite dike and later gabbroic dikes or sills in peridotite (Oman Ophiolite). g. Lumpy outcrop of massive, fractured serpentinite, fov ~1 m across (Vema Transform, *Nautile* Dive VN #01). h. Foliated serpentinite with typical phaccoidal fabric (central MARK Area, *Nautile* Dive HS13; fov=~3m).

al., 1998] and at the Blanco Transform [*Juteau et al.*, 1995; *Naidoo*, 1998]. Since some outcrops of massive, jointed rock units have been mapped on the seafloor using only deep-towed camera systems without ground-truth, it is possible that massive diabase is much more widespread. Areally extensive bodies of massive diabase to microgabbro have been described in some ophiolite complexes [*Rosencrantz*, 1983; *Robinson and Malpas*, 1990] where they appear to represent discrete, shallow-level intrusions or cupolas of larger plutons.

In the SMARK Area, closely spaced samples in extensive outcrop areas outline a massive diabase unit as much as a 600 m thick. Detailed sampling and examination of three-dimensional exposures revealed no chilled margins or the distinctive joint patterns found in sheeted dike complexes. Instead, the unit appears to consist of fairly homogeneous subophitic diabase (Figure 7a) that has been overprinted by closely spaced, moderately west-dipping, cataclastic shear zones (Figure 7b) and greenschist facies metamorphism. The shear zones impart a slabby or shingled form to the outcrops (Figure 5e). The deformation and metamorphism are not so pervasive that the protolith is not evident in nearly all samples. Locally, hydrothermal breccias and quartz veins are also present. Rarely, less altered and deformed diabase dikes cut the metadiabase unit.

In the Blanco Transform wall, a massive diabase unit is identified by side-scan sonar backscatter patterns and submersible sampling [*Juteau et al.*, 1995; *Naidoo*, 1998]. This unit is about 500 m thick and crops-out for at least several kilometers across the scarp face parallel to the spreading direction. Material in this unit appears to consist of a number of thick (tens of meters) massive, tabular, generally horizontal masses, but it is not clear if these are large flows, sills, or perhaps relatively fine-grained upper portions of magma chamber cupolas that intruded the upper crust.

3.3.6 Gabbroic Units. Gabbroic rocks are major components of ophiolite complexes [*Moores and Vine*, 1971; *Casey et al.*, 1981; *Pallister and Hopson*, 1981] and make up significant portions of the tectonic windows into the oceanic lithosphere. They represent coarse-grained igneous material that crystallized in or at the edges of magma chambers or as plutons in the middle to lower crust. Mapping with deep-towed cameras and submersibles shows that gabbroic rocks commonly occur as very extensive outcrop areas ranging from massive, jointed rock in rugged angular cliffs (Figure 5f) to strongly foliated, platy rocks commonly expressed as smooth dip-slopes (foliation essentially parallel to outcrop surface- Figure 5g).

Gabbroic rock exposures are typically massive to crudely banded in oceanic escarpments. Rugged, alpine-like exposures typical of gabbroic rocks have rarely been found in other lithologies. Joints, fractures, and veins are spaced at intervals of meters except near fault zones. Foliated metagabbros tend to have distinct outcrop patterns. Extensive faulted surfaces dipping 30°-50° have been followed upslope

as much as one hundred meters [*Karson and Dick*, 1983]. These are apparently the result of extensive sheet slides that exploit the fault surfaces and other low-angle foliation planes.

The depressions in escarpments and the base of fault-line scarps are piled with blocks of gabbro and finer grained debris from upslope. The talus at the base of outcrops of foliated material has a distinctive tabular nature reflecting the penetrative deformation fabric of the rock. Locally, huge boulders of massive gabbro up to a few tens of meters across have been observed and these can be difficult to distinguish from outcrops, especially if partially buried by rubble or pelagic ooze.

Drilling and closely spaced sampling in exposures of gabbroic rocks in a number of settings have provided samples that help constrain the scale, contacts, and composition of individual rock units [*Fox and Stroup*, 1981; *Malcolm*, 1981; *Elthon*, 1987; *Dick et al.*, 1991; 1998; *Natland and Dick*, 1996; *Pedersen et al.*, 1996; *Cannat et al.*, 1995a; *Casey*, 1997]. In general, these studies indicate that the gabbroic plutonic rock units of slow-spread crust are a highly heterogeneous collage of igneous units that have been assembled during deformation and metamorphism at spreading centers [*Dick et al.*, 1991; 1998; *Cannat et al.*, 1995b; *Karson and Lawrence*, 1997a]. Generally, less deformed gabbroic rock masses cut somewhat older, more deformed and metamorphosed units. Although the absolute ages of the rock bodies are not known, the geological relations suggest a more or less continuous process of crustal deformation, hydration, and igneous activity beneath ridge axes. Deformation and metamorphism tend to coincide spatially and extensional shear zones dipping toward the local spreading center appear to be very common (Figure 5g).

Collectively, samples recovered have an exceedingly wide range of textures and compositions. Individual samples commonly have very heterogeneous textures similar to the upper ("isotropic") gabbroic units of ophiolite complexes (Figure 5h; [*Casey et al.*, 1981; *Pallister and Hopson*, 1981; *Rosencrantz*, 1983]). Gabbroic rocks with igneous textures (Figure 7c) or identifiable protoliths include gabbro, gabbronorite, troctolite, leucogabbro to trondhjemite, and ferrogabbro to oxide gabbro. Although compositional layering is very common in middle to lower crustal sections of ophiolites (Figure 6a-c), this type of layering is generally not evident from surface exposures. Drilling during ODP Legs 118, 147, and 153 show that igneous and metamorphic compositional layering is common in oceanic gabbros and even in tectonic windows where it is not visible in outcrops [*Dick et al.*, 1991; *Gillis et al.*, 1993a; *Cannat et al.*, 1995a].

Heterogeneous greenschist facies alteration restricted to fractures and veins is nearly ubiquitous. Pervasive metamorphism to low-pressure granulite, amphibolite, and greenschist facies assemblages is common [*Gillis et al.*, 1993a; *Cannat et al.*, 1995a]. The granulite facies metagabbros tend to be weakly foliated with a pervasively to locally

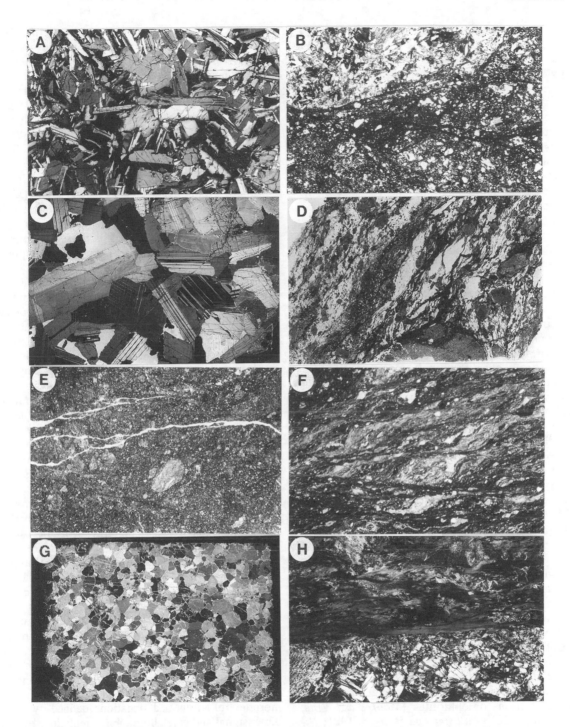

Figure 7. Photomicrographs of oceanic rocks. (fov = 4 mm for all) a. Typical subophitic texture in diabase with plagioclase laths partially enclosed by clinopyroxene (SMARK Area, *Alvin* Dive 2871). b. Dark, cataclastic band (upper part of view) cutting subophitic diabase (SMARK Area, *Alvin* Dive 2889). c. Coarse-grained gabbro (N. MARK Area, Dive 1008). d. Gneissic metagabbro with evidence of crystal-plastic deformation and recrystalliza- tion of plagioclase (Mid-Cayman Spreading Center). e. Cataclastic metagabbro (?) with mylonite clasts (N. MARK Area, *Alvin* Dive 1011). f. Schistose to cataclastic metagabbro with S/C fabric (N. MARK Area, *Alvin* Dive 1015). g. Peridotite with mosaic texture in olivine (Bay of Islands Ophiolite, Lewis Hills). h. Cataclastic to schistose shear zone (upper part of view) cutting typical "mesh-textured" serpentine (central MARK Area, ODP Site 920).

recrystallized, microstructures and essentially anhydrous assemblages (Figure 7d). Amphibolites vary from coarse-grained flazer gabbros and metagabbro gneisses to finely laminated mylonites and are very similar to those of ductile shear zones in some ophiolites (Figure 6d). Non-foliated amphibolites that may have a contact metamorphic origin are uncommon. Greenschist facies metabasites include mylonites, chlorite schists, and cataclasites (Figure 7e-f). Retrograde effects are evident in virtually all samples.

Hydrothermal quartz veins are fairly common and both discordant and concordant varieties have been found. These samples provide important constraints on the nature of hydrothermal circulation in the middle to deep crust and also on the depths of origin of the surrounding units. *Delaney et al.* [1985] described quartz-cemented breccias with quartz-mantled, angular, sulfide and altered mafic clasts. This particular sample came from a vein that was concordant with shear zones and foliation surfaces on the western median valley wall of the MARK Area near the Kane RTI. The textures suggest high flow rates sufficient to suspend and transport millimeter to centimeter size particles. Fluid inclusion microanalyses documented a decreasing fluid-temperature history during the formation of the breccias. Nearby metagabbros yielded fluid inclusions that give pressure constraints suggesting as much as 2 km of unroofing and >3 km of vertical uplift of these units [*Kelley and Delaney*, 1987], one of the few constraints on the magnitude of crustal uplift and denudation in the oceanic crust.

In slow-spreading environments, gabbroic assemblages appear to have been constructed by multiple, syntectonic intrusive events resulting in extremely complex patterns of igneous, deformational, and metamorphic features. This result is in accord with theoretical [*Sleep*, 1975; *Sleep and Barth*, 1997] and seismic studies [e.g. *Detrick et al.*, 1990] that indicate that large magma chambers do not develop along slow-spreading ridges.

Although exposures of gabbroic rocks are widespread in the tectonic windows of slow-spread crust, they are rare in intermediate- to fast-spread crust. Exposures of gabbroic rocks that might reveal the mode of formation of middle to lower crustal material in sustained magma chambers of fast-spreading ridges [e.g. *Sinton and Detrick*, 1992] are very rare. The best known exposures are the faulted and dismembered gabbroic rocks at Hess Deep [*Francheteau et al.*, 1990; *Gillis et al.*, 1993a].

3.3.7 Ultramafic Units. Variably serpentinized peridotites form the basal sections of ophiolites [*Moores and Vine*, 1971; *Casey et al.*, 1981; *Pallister and Hopson*, 1981; *Nicolas*, 1989 and references therein; Figure 6e, 7g]. Serpentinized peridotites have bulk densities and compressional wave velocities that overlap with those of other oceanic crustal rocks and therefore would not be distinguished in gravity or seismic velocity data [e.g., *Christensen*, 1972]. Shear wave velocity, Poisson's ratio, and magnetic properties of serpentinites are distinctive and may provide a means of mapping the extent of them in the oceanic crust. In hand specimen and outcrop character, oceanic serpentinized peridotites resemble those of Alpine Peridotites and the basal members of ophiolite complexes [*Coleman*, 1977; *Nicolas*, 1989]. They are interpreted as residual upper mantle material that has been depleted by partial melting and extraction of MORB liquids [*Michael and Bonatti*, 1985; *Dick*, 1989].

The common recovery of serpentinites at shallow structural levels by drilling [*Juteau et al.*, 1988], dredging [*Aumento and Loubat*, 1971; *Miyashiro et al.*, 1969; *Dick*, 1989], and submersible studies [e.g., *Lagabrielle et al.*, 1998] indicates that these altered upper mantle peridotites are major components of slow-spread crust [*Hess*, 1962; *Juteau et al.*, 1988; *Cannat*, 1993]. This upper mantle material may have been exposed on the seafloor by faulting [*Karson*, 1990; *Cannat et al.*, 1997b], serpentinite diapirism [*Bonatti*, 1976; *Francis*, 1981] or some combination of these two mechanisms.

Serpentinite outcrops superficially resemble lumpy, highly fragmented outcrops of pillow basalt but have distinctly different aspects when considered in detail. Steep headwall scarps are less common in serpentinites; low-relief, degraded rock masses are typical. Pervasive development of fractures, veins and other surfaces of low cohesion result in collapse of nearly all steep topography to produce an undulating rubbly surface. Like the basalts, smooth, downslope lineated talus ramps form aprons on major exposures. Serpentinites tend to have numerous large jagged blocks incorporated in the talus. Outcrops that do protrude from the debris are knobby on all scales (up to tens of meters) and in some places have jagged, serrated corners.

In areas where detailed sampling has been done, the peridotites are moderately to extensively serpentinized and commonly deeply weathered in surface exposures. The protoliths of the serpentinites appear to be dominantly harzburgites and lesser lherzolites with rare pyroxenites, wehrlites, and dunites. Outcrop morphologies are diverse, but most commonly include heterogeneous, lumpy-looking masses with numerous fractures (Figure 6g). Discrete serpentinite shear zones with phaccoidal mesoscopic fabrics (Figures 6h, 7h) occur in moderately dipping orientations with normal-slip kinematic indicators. Dense arrays of serpentine veins that mimic the foliations in shear zone or high-temperature fabrics are pervasive in many samples.

In many seafloor serpentinite exposures high-temperature deformation fabrics that were produced under mantle conditions are preserved. Typically these include strong metamorphic foliations and lineations defined by the preferred dimensional and crystallographic orientations of olivine and lesser orthopyroxene with traces of clinopyroxene and spinel. Where oriented samples have been collected, these foliations have moderate to gently dipping orientations [*Boudier*, 1978; *Cannat et al.*, 1990; *Ceuleneer and Cannat*, 1997], but may have been rotated significantly during faulting at crustal levels. Locally, millimeter to centimeter-scale veins with concentrations of spinel, clinopyroxene, or plagioclase have dunitic selvages and are interpreted as melt pathways [*Casey*, 1997]. Some samples have mylonitic fabrics indicative of extreme grain-size reduction and large strains developed in crustal shear zones [*Melson et al.*, 1972; *Jaroslow et al.*, 1996; *Karson*, 1998].

The serpentinized peridotites in some places are rather homogeneous over tens of meters, but more commonly appear to be invaded by irregular gabbroic dikes and plutons with dimensions of tens to hundreds of meters. Locally, foliated to mylonitic amphibolites and in the serpentinites record high-temperature deformation of the mafic inclusions.

In some exposures, fresh to moderately altered diabase dikes with chilled margins cut the serpentinites [*Cannat et al.*, 1997b; *Karson and Lawrence*, 1997b]. These are generally steeply to moderately dipping and cut all the deformation and it is likely that these dikes feed basaltic lavas that commonly overlie the serpentinites.

The geological history of ultramafic material exposed in tectonic windows in slow-spreading ridges commonly includes high-temperature deformation and partial melting in the mantle prior to extensive hydration and uplift into the oceanic crust. Deformed metagabbros and metaperidotites record high-temperature deformation in major shear zones under hydrous conditions. Later faulting is manifest as serpentinite shear zones. Cross-cutting diabase dikes and overlying basalts indicate that all of these processes occurred near a spreading center. Apart from a few serpentinite samples from fracture zones, the only known window into the upper mantle of fast-spread crust is at Hess Deep [*Francheteau et al.*, 1990; *Gillis et al.*, 1993a].

4. RECONSTRUCTED COLUMNAR SECTIONS

4.1 Crustal Sections from Tectonic Windows

From the preceding descriptions it is clear that seafloor escarpments do not provide simple cross sections of the oceanic lithosphere. Instead, they are the products of faulting and mass wasting that may have shuffled or buried the rock units exposed there. It is through these types of "shattered windows" that we attempt to reconstruct the original crustal structure. Constraints on the geometry and kinematics of faulting are essential for any accurate reconstruction, but these have proven very difficult to obtain. The following section summarizes current views.

The specific geometry and kinematics of faulting will determine the degree to which an original crustal structure is disrupted and modified in tectonic windows. At present both of these are very poorly constrained, so rather than attempting to make detailed reconstructions, in this paper it will be assumed that the close-spaced, stair-step faults typical of many escarpments have a more or less equal effect regardless of setting. Vertical separations across the faults and the dips of fault surfaces (not just slopes) imply modest amounts of horizontal extension (≤10%). For faults dipping

45°-60°, as commonly observed, this would result in 10-15% apparent increase in thickness of each unit. It will further be assumed that this is more or less homogeneous across the scarps that have been studied. The only exception that will be made is in the case of very broad, continuous fault zones that have been mapped along spreading centers or where other constraints make more detailed reconstructions justified.

In the following sections, crustal structures reconstructed from tectonic windows are presented for slow- and intermediate- to fast-spread lithosphere. These groups are separated to facilitate comparison of structures formed at similar magma budgets.

4.2 Tectonic Windows into Slow-Spread Lithosphere

Investigations of tectonic windows with submersibles, ROVs, and crustal drilling have provided considerable information on major rock units of the oceanic crust and the contacts that separate them. Using the assumptions outlined in the previous section, columnar sections for the tectonic windows in slow-spread can be compared (Figure 8).With the exception of the section described from the Vema Fracture Zone [*Auzende et al.*, 1989a], these reconstructed sections contrast strongly with the prevailing view of oceanic crust (Figure 1). Several notable points emerge at once: 1) Different rock unit are exposed at the seafloor. 2) There is a wide range in the thickness of various rock units. 3) There is a wide range in the internal structure and composition of the rock units. 4) In some places the anticipated ophiolite-like sequence of rock units is not present with important units, notably sheeted dikes complexes, missing. 5) Regardless of the internal structure or composition of the crustal units, they tend to be overlain locally by basaltic lavas.

The diversity of these crustal sections may reflect significant variations in the relationship between tectonic and magmatic processes along slow-spreading ridges. A more detailed look at these rock units and at site-specific relationships provides some important clues regarding these processes.

4.2.1 Vema Transform. The crustal section for the southern transform valley wall of the Vema Transform is based on two composite transects spaced about 4 km apart from 5 *Nautile* dives [*Auzende et al.*, 1989a]. Collectively these dives describe the same general sequence of rock units found in ophiolite complexes including serpentinized ultramafic rocks near the base of the slope, gabbroic rocks dominating the middle slope and passing upward into a sheeted dike complex capped by basaltic pillow lavas and lesser flows. Combining the vertical outcrop extent of each unit to make a columnar section (Figure 8a), results in a total crustal thickness of between 4 and 5 km, which might be in accord with relatively thin crust expected near transforms [*Auzende et al.*, 1989a]. The total thickness of the section and the thicknesses of the rock units exposed is open to question

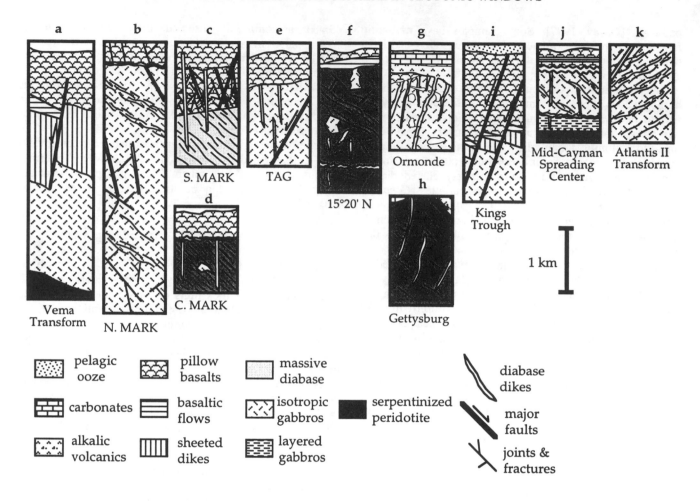

Figure 8. Columnar sections for tectonic windows in slow-spread oceanic crust. See Figure 2 for locations. Sections are constructed with the assumptions that minimal displacements have occurred across steep faults and that rock units are laterally continuous over the survey areas (few kilometers across). Columns or data from following sources: a. Vema Transform [*Auzende et al.*, 1989]; b. Northern MARK area [*Karson and Dick*, 1983; *Mével et al.*, 1991; *Karson and Lawrence* , 1997a]; c. Central MARK area [*Mével et al.*, 1991; *Karson and Lawrence*, 1997b]; d. Southern MARK area [*Karson et al.*, 1996; *J.A. Karson, unpubl. data*]; e. TAG Area [*Zonenshain et al.*, 1989]; f. MAR near 15°20'N FZ [*Cannat et al.*, 1997b]; g. & h. Ormonde and Gettysburg seamounts at Gorringe Bank [*Auzende et al.*, 1978]; i. King's Trough [*K. Klitgord, J. Casey, S. Agar, et al.*, 1990, unpubl. cruise rept.]; j. Mid-Cayman Spreading Center [*CAYTROUGH* , 1979; *Stroup and Fox*, 1981]; and k. Atlantis II Transform [*Dick et al.*, 1991; 1998].

because the contacts between them appear to be faulted. The sense and magnitude of displacements on these faults is not known.

The section exposed in this tectonic window is especially noteworthy because it is the only known place in slow-spread crust where an entire ophiolite-like assemblage is exposed. It has the most extensive and well developed sheeted dike complex known in slow-spread crust. The dikes here are subvertical and strike parallel to the regional trend of the MAR and local magnetic lineations.

4.2.2 MARK Area. The MARK Area south of the Eastern RTI of the Kane Transform is the most intensively studied area of slow-spread crust. Surface mapping with submersibles [*Karson and Dick*, 1983; *Karson et al.*, 1987; *Brown and Karson*, 1988; *Mével et al.*, 1991; *Auzende et al.*, 1993; *Karson and Lawrence*, 1997a,b] and side-looking sonar [*Kong et al.*, 1988; *Gao*, 1997; *Gao et al.*, 1998] provide dense coverage of large areas of the seafloor. Crustal drilling on ODP Legs 109 [*Detrick et al.*, 1988] and 153 [*Cannat et al.*, 1995a] provide detailed information on shallow subsurface rock units.

At least three different types of crustal structures are documented in this area (Figure 8b-d). The first columnar section represents the internal structure of the crust exposed in an extensive tectonic window that extends about 40 km from the RTI southward along the western rift valley wall [*Karson and Lawrence*, 1997a]. A different perspective comes from exposures just to the south (central MARK Area) in the same spreading segment [*Karson and Lawrence*, 1997b] and on the eastern rift valley wall of the next spreading segment (SMARK Area), several tens of kilometers to the south [*Lawrence et al.*, 1998].

The northern MARK Area, near the RTI, is dominated by variably deformed and metamorphosed gabbroic rocks which crop out from >6000 mbsl to about 2500 mbsl on the western median valley wall. This rock unit may represent an oblique cross section of the oceanic crust with a gabbroic unit several kilometers thick (Figure 8b). The gabbroic rocks in this area are highly variable including olivine gabbros, gabbros, and troctolites [*Karson and Dick*, 1983; *Cannat et al.*, 1995a]. Deformation and metamorphism are very heterogeneous. Microstructures show magmatic to crystal-plastic fabrics locally overprinted by cataclastic fabrics [*Karson and Dick*, 1983; *Cannat et al.*, 1995a]. Metamorphic effects are also highly variable in distribution and intensity. These range from low-pressure granulite facies through amphibolite facies to greenschist facies or lower grade effects [*Gillis et al.*, 1993b; *Cannat et al.*, 1997a]. Hydrothermal breccias and veins are common in this assemblage and appear to be spatially related to areas of concentrated deformation [*Delaney et al.*, 1985; *Kelley and Delaney*, 1987]. Analysis of high-temperature fluid inclusions indicate pressures of formation of about 1 kb and suggest that at least 2 km of material has been removed from above the exposed section as it was uplifted [*Kelley and Delaney*, 1987].

Material from several shallow (most <100 m) ODP holes drilled in this area yielded mainly gabbroic material [*Cannat et al.*, 1995a]. Overall the results of this study suggest that, at least at this structural level, this crustal section is composed of a collage of gabbroic plutons that have been intruded during and after tectonic extension [*Karson and Lawrence*, 1997a; *Cannat et al.*, 1995b]. The igneous activity and metamorphic relations indicate that construction and deformation of this crustal material took place beneath the median valley floor prior to uplift and faulting in the median valley wall of the spreading center.

Rare diabase dikes, less than 1 m wide, cut massive to foliated gabbroic rocks. They are generally steeply dipping and have well developed chilled margins. They have been identified both in outcrops [*Karson and Lawrence*, 1997a] and in ODP drill core [*Cannat et al.*, 1995a]. No sheeted dike complex has been found in this tectonic window.

Basaltic pillow lavas crop-out just upslope from gabbroic material. The contact between the underlying gabbroic rocks and the basalts is poorly exposed, but they are separated by <100 m along some submersible dive tracks. The vertical interval between the lower contact and the top of the slope suggests that this unit could be as much as a few hundred meters thick. Exposures of the basalts are generally highly fractured and more accurately described as a basaltic breccia. A single *Nautile* dive on the western side of the RTI massif documents a faulted sequence of pillow lavas and dikes at least 500 m thick [*Auzende et al.*, 1993].

The western median valley wall is marked by moderately to gently dipping fault surfaces that are locally continuous upslope for >100 m. These are superimposed on ductile shear zones that are generally parallel to the faults. This dip-slope is cut by fault scarps with unknown vertical separations that create bathymetric steps of a few tens of meters to about 200 m. This tectonic window is interpreted as the faulted edge of an oceanic core complex [*Karson*, 1990; *Mutter and Karson*, 1992] with an oblique crustal section exposed by slip on a family of shear zones and major detachment faults.

Exposures on the northern edge of the RTI massif along the Kane Transform include sheared and faulted diabase, gabbros, and serpentinites [*Auzende et al.*, 1993]. These rock units cannot be traced to the south and may represent an assemblage of rocks that are related specifically to the transform domain. Farther to the west along the transform wall, variably sheared serpentinites and gabbroic rocks may represent a window into another type of core complex structure [*Auzende et al.*, 1993; *Karson*, 1998].

About 40 km south of the RTI, serpentinites crop-out in a series of exposures along the western median valley wall [*Brown and Karson*, 1988; *Mével et al.*, 1991; *Karson and Lawrence*, 1997b]. Submersible studies, camera data, dredges, and ODP drilling have investigated this belt (Figure 8c; see *Karson* and *Lawrence* [1997b] for a summary). A continuous magnetic anomaly along this belt [*Schultz et al.*, 1988] suggests that the isolated areas investigated are parts of a continuous serpentinite belt that is a few kilometers wide and about 25 km long. Rather than being confined to a segment boundary [*Tucholke and Lin*, 1994; *Cannat et al.*, 1995c], the serpentinites appear to crop-out on the median valley wall for a substantial length of a discrete spreading segment. They crop-out on the walls of the median valley at the same latitude as the prominent neovolcanic ridge and the Snake Pit Hydrothermal vents. Mapping of gravity anomalies and dredging off-axis to the northwest suggests that serpentinite exposures crop-out in swath that extends for many tens of kilometers into older crust to the west [*Cannat et al.*, 1995c].

The MARK serpentinites have gently dipping high-temperature foliations that are cut by minor shear zones and a dense array of serpentine veins that dip moderately to the east. Extensive gently east-dipping serpentinite shear zones are present in surface outcrops. Locally the sheared serpentinites are overlain by angular serpentinite breccias with a carbonate matrix. The shear zones and breccias are cut and offset by steeply dipping normal faults. Paleomagnetic studies demonstrate that little if any post-serpentinization rotation has occurred [*Hurst et al.*, 1997; *Lawrence*, 1998]. In the shallow (~200 m) ODP drill holes, minor bodies of gabbro to amphibolite and veins representing melt channels were also encountered. Undeformed diabase dikes with chilled margins cut the serpentinites [*Cannat et al.*, 1995a], and probably feed nearby basaltic pillow lavas.

Basaltic pillow lavas and minor sheet flows appear to lie directly over the serpentinites. Although contacts have not been directly observed, but intact pillow basalts and lesser subhorizontal sheet flows crop-out <100 m upslope from serpentinites. The contact between these units appears to be limited to a narrow depth range, so it is roughly horizontal.

At about 22°40'N, about 100 km south of the Kane RTI, a separate spreading segment referred to as the Southern MARK (SMARK) area has a very steep asymmetrical rift valley with a relatively high, steep, eastern median valley wall. A crustal section about 2 km thick is exposed there that includes three mappable rock units (Figure 8d; Lawrence et al., 1998). The lowest unit is massive diabase cut by numerous, closely spaced, east-dipping, cataclastic shear zones. Locally undeformed diabase dikes cut the deformation fabrics. A middle unit consists of highly fractured basaltic pillow lavas and diabase dikes. The upper unit includes mainly relatively fresh basaltic pillow lavas and sparse dikes. In the middle and lower units dikes have variable orientations and paleomagnetic studies show that many have been tectonically rotated. Dikes in the less deformed upper basaltic unit have not been rotated. The upper basaltic unit appears to lie unconformably over the more deformed and metamorphosed lower basaltic and diabasic units (Figure 8d).

4.2.3 TAG Area. The TAG area at 26° N on the MAR has been the site of many detailed seafloor investigations (see *Kleinrock et al.*, [1996] for a recent summary). Submersible studies and side-scan sonar mapping have been focused on the median valley floor especially in the vicinity of the active TAG hydrothermal mound. Several studies have also focused on the median valley walls and crustal cross sections exposed there [*Zonenshain et al.*, 1989; *Eberhart et al.*, 1988; *Karson and Rona*, 1990]. Although these represent only rather widely spaced transects and significant along-strike variations are evident, a general columnar section can be constructed (Figure 8e). Although submersible observations just east of the TAG hydrothermal mound showed only variably faulted basaltic lavas and rare individual dikes [*Karson and Rona*, 1990], a tectonic window with gabbros, sheeted dikes and basaltic lavas occurs about 10 km to the south [*Zonenshain et al.*, 1989]. Here, limited gabbro exposures with locally dense dikes pass upslope into basaltic pillow lavas. Contacts between the rock units are not reported and therefore unit thicknesses are not constrained. Serpentine clasts in nearby sedimentary rocks suggests that ultramafic rocks may also crop-out nearby [*Zonenshain et al.*, 1989]. Gabbroic rocks also crop-out on fault-line scarps just off axis to the east [*Tiezzi and Scott*, 1980] and suggest large-scale faulting, a very thin volcanic cap, or both nearby.

4.2.4 MAR at 15°20'N. Both to the north and south of the 15°20'N Fracture Zone, the MAR rift valley walls have

been studied extensively by submersibles and dredging [*Cannat et al.*, 1997b; *Lagabrielle et al.*, 1998, and references therein]. The rift walls are dominated by faulted exposures of serpentinite cut by gabbroic intrusions (Figure 4f). Moderately dipping, high-temperature fabrics and serpentinite shear zones are widely developed. These extensive exposures are riddled with gabbroic dikes a few meters wide as well as larger gabbro masses on the order of a hundreds of meters across. More extensive gabbroic outcrop areas mapped to the north of the fracture zone may be relatively large, discrete intrusive bodies. Diabase dikes cut the serpentinites locally and are likely to be related to overlying lavas.

The serpentinites south of the fracture zone are capped by relatively fresh basaltic material including subhorizontal sheet flows [*Cannat et al.*, 1997b]. In addition, basaltic material covers the floor of the median valley below and appears to lap onto the deformed and altered plutonic rocks of the bounding median valley walls. Unlike some other tectonic windows along the MAR, the opposing median valley walls here both have extensive outcrops of faulted mafic to ultramafic plutonic rocks [*Cannat et al.*, 1997b; *Lagabrielle et al.*, 1998].

4.2.5 Gorringe Bank Area. Submersible studies at Gorringe Bank in the eastern Atlantic near the Azores-Gibralter plate boundary show a complex geologic structure in outcrops of oceanic crust and upper mantle rocks [*Auzende et al.*, 1978]. Oceanic lithosphere has probably been uplifted and dismembered by major thrust faults along a restraining bend in this dominantly strike-slip plate boundary.

At Ormonde Seamount (Figure 8g), a large gabbro outcrop with vertical dimensions of ~1000 m is cut by numerous mafic dikes. It is unclear how many, if any of these dikes are related to Eocene volcanism. The gabbros and dikes are overlain by alkalic lavas and carbonate sediments.

Gettysburg Seamount (Figure 8h) has extensive serpentinite outcrops. High-temperature foliations have complex orientations as do later, cross-cutting serpentinite shear zones. Gabbroic rocks appear to be faulted against the serpentinites on one edge of the massif, but the age and kinematics of the faulting are not known. Numerous dikes cut the serpentinite, but these may be at least in part related to Eocene alkaline volcanics that crop-out nearby. Basaltic pillow lavas are in close proximity to gabbro outcrops and no sheeted dike complex has been observed [*Auzende et al.*, 1978]. Overall, outcrops are heavily sedimented, fractured, and cut by relatively young alkalic intrusions and the major geologic contacts between the rock units are not well exposed.

4.2.6 Kings Trough Area. Kings Trough is located in the eastern Atlantic and crustal exposures in its rift valley walls were examined in a Russian *Mir* submersible diving program [*Klitgord et al.*, 1990]. Faulted and sedimented exposures of gabbroic rocks exposed on steep scarps up to >2000 m high are reported to pass upward into a sheeted dike complex with vertical dikes which in turn is overlain by basaltic pillow lavas (Figure 8i). Contacts separating the various units are faulted.

4.2.7 Mid-Cayman Spreading Center. The Mid-Cayman Spreading Center is a ~100 km long spreading segment that links the Swan and Oriente Transforms along the northern boundary of the Caribbean plate. The rift valley walls provide extensive tectonic windows (Figure 8j) that have been examined in two *Alvin* diving programs [*CAYTROUGH*, 1979; *Stroup and Fox*, 1981].

Serpentinized harzburgites and possible cumulate ultramafic rocks crop-out at the base of the walls of the Mid-Cayman Spreading Center. Although the contact with extensive outcrops of gabbroic rocks just upslope is not clearly delineated, it is possible that a "petrologic Moho" (contact between residual ultramafic rocks and magmatic units) is preserved at least locally here. Basaltic samples collected along transects in this area suggest that dikes cut the serpentinites.

Ultramafic and feldspathic ultramafic cumulate rocks cropping-out low on the median valley walls appear to grade upward into more continuous gabbroic material including gabbros, olivine gabbros, and troctolites. Across a poorly exposed, sediment-covered area about 2 km across and 200 m vertically, basaltic pillow lavas crop-out upslope. This narrow gap suggests that basalts may essentially lie directly over gabbroic rocks without an intervening sheeted dike complex. Relatively fresh pillow lavas are widely exposed on the floor of the median valley downslope where they may locally overlie serpentinites.

4.2.8 Atlantis II Transform. Although surface exposures have not yet been mapped, a major tectonic window into the oceanic crust has been drilled at the summit of a flat-topped massif along the Atlantis II Transform on the SW Indian Ridge (Figure 8k). This massif is ~11 m.y. old and apparently formed near an RTI before being uplifted and translated along the transform. Thus, it formed in a position comparable to the gabbroic rocks presently exposed in the northern MARK Area. Drilling during the course of two ODP legs has recovered a vertical interval ~1500 m thick through a complex assemblage of gabbroic rocks [*Dick et al.*, 1991; 1998]. Deformation, metamorphism, and intrusive relationships are heterogeneously distributed through the core. Dramatic, moderately dipping, amphibolite facies shear zones up to several tens of meters thick appear to be preferentially localized in oxide-rich gabbros. Petrographic and geochemical relations suggest that the deformation occurred at least in part before the gabbros were completely crystallized and less deformed gabbroic veins locally cut the deformation fabrics. No contacts with other rock units are known at this time.

4.2.9 Other Possible Windows. Investigations of the Atlantis Transform at 30°N on the MAR have recently shown that serpentinites, variably deformed and metamorphosed gabbroic rocks, metabasalts, and basalts that were dredged long ago [*Miyashiro et al.*, 1969] come from the edges of a dome-like massif with a strongly lineated upper surface interpreted as a detachment fault [*Cann et al.*, 1997]. Similar lineated massifs with plutonic rock exposures have been mapped farther to the south along the MAR and into older seafloor [*Tucholke et al.*, 1997]. Such lineated massifs are widely developed in the central North Atlantic and may mark sites of major detachment faulting and exposure of serpentinites [*Cann et al.*, 1997; *Tucholke et al.*, 1998].

Outcrops of serpentinites near the eastern RTI of the Oceanographer Transform occur on the oblique-trending inside-corner median valley wall adjacent to the active transform [*OTTER Team*, 1984]. They are intensely sheared, veined and weathered and no fabric orientation or kinematic information is available. Camera tows document pillow basalts upslope about 500 m, leaving little room for intervening rock units. Thus, it appears that the basalts lie directly above somewhat older, more deformed and altered serpentinites.

Recent studies of the walls of rift segments between the Oceanographer and Hayes transforms reveal that serpentinites crop-out locally [*Gràcia et al.*, 1997]. Some of these occur in faulted areas with only a few hundred meters of relief again suggesting that serpentinites are present at shallow structural levels of the crust in these areas.

Finally, several drill holes in slow-spread crust recovered coarse-grained plutonic rocks including gabbros and serpentinites within only ~100 m of the surface [*Juteau et al.*, 1988]. Basaltic material was also recovered in some of these holes and may be samples of a rubbly basaltic cap. These relations suggest that even in areas where basaltic pillow lavas dominate the surface, it is possible that they represent only a relatively thin veneer of relatively young lavas over somewhat older, variably deformed and altered mafic/ultramafic rock units. Many seafloor escarpments in slow-spread crust, especially along fracture zones and ridge-parallel escarpments, have been dredged and clearly indicate extensive exposures of serpentinites and other plutonic rocks [*Cann and Funnel*, 1967; *Aumento and Loubat*, 1971; *Tiezzi and Scott*, 1980; *Engel and Fisher*, 1975; *Dick*, 1989; *Auzende et al.*, 1993]. These areas may be important sites for future studies of tectonic windows into slow-spread lithosphere.

4.3 Implications for Processes at Slow-Spreading Ridges

Viewed collectively, the crustal structures found to date in tectonic windows into slow-spread crust raise a number of questions regarding the application of the generalized ophiolite model for oceanic crust and the current views of spatial and temporal variations of crustal accretion along slow-spreading ridges.

4.3.1. No Consistent Layered Sequence. The columnar sections assembled in Figure 8 show significant variations from the general crustal structure found in many ophiolites

(Figure 1). Even considering that there are significant uncertainties in the reconstructions of both ophiolites and these columnar sections, specific geological relations found in tectonic windows are not in accord with the ophiolite model. For example, several sections lack some of the typical rock units found in ophiolites. In particular, sheeted dike units are rarely observed in tectonic windows into slow-spread crust. In some windows all of the middle to lower crustal magmatic units are missing. In many of these locations, plutonic rock units are exposed directly on the seafloor. Several have sedimentary sequences or basaltic lavas lying unconformably over plutonic rock units which must have been exposed on the seafloor. These features clearly indicate a much broader range of relations in faulting and magmatism than that implied by the ophiolite model. The reconstructed thicknesses of rock units is not so important as their geological contacts and the implied geological histories.

The diverse geological relations observed in slow-spread crust are difficult to reconcile with marine seismic refraction data that historically have characterized the oceanic crust as a continuous layered geological structure. A number of possible explanations exist: 1) The geological variations seen in the tectonic windows may be below the resolution of seismic investigations. 2) Seismic layering may be related to fracturing and alteration rather than corresponding to major rock units. 3) The observed geological variations might not be representative of typical slow-spread crust and do not occur where seismic studies have been done. 4) "Complete" stratiform ophiolites may not have formed at slow-spreading ridges.

4.3.2. Variations Related to Segment-Scale Processes. The current view of segment-scale processes on slow-spreading ridges [*Lin et al.*, 1990; *Sempéré et al.*, 1990; *Detrick et al.*, 1995] is largely based on the morphological and geophysical expression of the MAR. Basically, individual ridge segments are considered to have magmatic construction focused near the segment mid-point with magma supply diminishing toward the segment ends marked by transform faults and non-transform discontinuities (see *Karson and Elthon* [1987] for an alternative interpretation). In addition, crust generated at the inside and outside plate corners at segment boundaries are thought to be different, with thinner crust created at inside corners [*Karson and Dick*, 1983; *Tucholke and Lin*, 1994].

The geology of crustal sections seen in tectonic windows from slow-spread crust does not consistently match this view. Some crustal sections and exposures known from dredging are consistent with relatively thick magmatic crust being generated near segment centers and very thin crust being formed at segment boundaries, but others do not. Crustal sections at the Vema Transform, the eastern Kane Transform RTI (northern MARK area) and at Atlantis II Transform (SW Indian Ridge) are all along major transform boundaries where the magmatic activity would be expected to be attenuated. In all cases gabbroic assemblages as much as several kilometers thick are present. Investigations of

several RTI's also show that extensive volcanic construction extends into these areas [*Karson and Dick*, 1983; *Auzende et al.*, 1989b; *Mameloukas-Frangoulis et al.*, 1991; *Gao et al.*, 1998; *Tivey et al.*, 1998]. Conversely, tectonic windows near segment centers in the central part of the MARK Area and the TAG Area have gabbroic or serpentinized ultramafic rocks exposed at the surface. Thus, it appears that the morphological and geophysical segmentation of slow-spreading ridges does not have any simple relationship with respect to crustal geology. These inconsistencies suggest that the distribution of magmatism on the scale of spreading segments is more complex than current models predict. Alternatively, the tectonic windows may represent anomalous crustal structures.

4.3.3. Temporal Variations in Spreading Processes. One of the most striking aspects of crustal structures found in tectonic windows in slow-spread crust is evidence of time-varying processes. Considering that each columnar section represents the crust created at a very limited part of the MAR, they commonly show that faulting and magmatic construction have varied significantly during their construction. The most obvious discontinuities are represented by lavas that unconformably overlie previously deformed, metamorphosed and uplifted mafic to ultramafic assemblages. Others are marked by dikes and plutons that crosscut somewhat older materials. These features hint at dramatic tectonic processes beneath spreading centers. For example, some tectonic windows show evidence of upper mantle material having been progressively hydrated, metamorphosed, and uplifted beneath the MAR axial valley. Others show that gabbroic rocks with magmatic, granulite, and amphibolite facies deformation fabrics were uplifted to the median valley floor prior to the intrusion of dikes and eruption of basaltic lavas. Quantifying these processes is one of the major challenges confronting mid-ocean ridge tectonics.

At present, radiometric dating techniques cannot resolve age differences between the different discordant rock units; however, the strong localization of volcanism, hydrothermal activity, and faulting in the axial valleys of slow-spreading ridges [e.g., *Macdonald*, 1982], probably limits crustal construction to this narrow region. Considering the typical widths of axial valleys and spreading rates, this would be less than a few hundred thousand years. Although at least a few kilometers of uplift can be inferred from metamorphic relations [e.g., *Kelley and Delaney*, 1987; *Vanko*, 1988], the kinematics of these processes are not established. It is noteworthy that the implied displacement and uplift rates for the movement of crust from beneath the median valley floor to the elevated rift mountains (a few kilometers per million years) are comparable to the highest known tectonic displacement rates of the Himalayas [*Hodges et al.*, 1998].

Morphological and geophysical patterns describe narrow rift valleys with thick, "geophysically defined" crust as having "robust" magma supplies and wider rift valleys with thinner crust as magmatically "starved" [*Tolstoy et al.*,

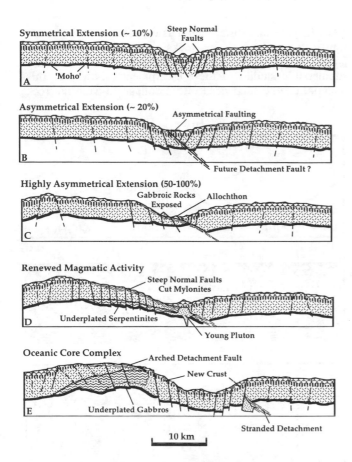

Symmetrical Extension (~ 10%)   Steep Normal Faults

A   'Moho'

Asymmetrical Extension (~ 20%)   Asymmetrical Faulting

B   Future Detachment Fault ?

Highly Asymmetrical Extension (50-100%)
Gabbroic Rocks Exposed   Allochthon

C

Renewed Magmatic Activity
Steep Normal Faults Cut Mylonites

D   Underplated Serpentinites   Young Pluton

Oceanic Core Complex   Arched Detachment Fault
New Crust

E   Underplated Gabbros   Stranded Detachment

10 km

Figure 9. Range of structures observed along the central MAR. From a-e these structures may be viewed as manifestations of spreading with decreasing magma budgets or as a sequence of stages in the evolution of an "oceanic core complex" that might develop in a single spreading center segment over time. Panels as follows: a. symmetrical spreading (similar to FAMOUS Area); b. asymmetrical spreading with major faults concentrated on one side of the rift valley (similar to TAG Area); c. highly asymmetrical spreading with a major detachment fault exposing gabbroic or ultramafic rocks (similar to central MARK Area); d. highly extended area with detachment fault truncated by younger magmatic construction (similar to N. MARK Area); and e. Oceanic core complex with dome-shaped mass of highly extended and exhumed middle- to deep-crustal and upper mantle rocks that might be isolated between crustal volumes created during intervals of spreading with higher magma budgets (similar to MARK), *Karson* [1990; 1998], N. American side of MAR at 25°-27°N- *Tucholke et al.*[1997], and E. Atlantis RTI, *Cann et al.* [1997]. Note that similar structures appear to have formed in crust with thick gabbroic units as well as crust dominated by serpentinite [*Karson,* 1998].

1993; *Detrick et al.*, 1995]. The common occurrence of major geological discontinuities suggests a "sputtering" magma supply on the scale of tens of thousands to perhaps a few million years that is superimposed on actively extending axial lithosphere. One might conclude that mechanical

extension is the "default mechanism" of plate separation, but that it is interrupted periodically by magmatism.

The range of geological expressions of the MAR median valley (Figure 9) indicate large variations in magma budget, defined as the volume of magma delivered to a spreading center per unit of plate separation. Along-strike variations in the morphology and geology of spreading centers suggests that the magma budget can change rapidly along the length of spreading centers independent of spreading rate [*Karson and Winters*, 1992; *Detrick et al.*, 1995; *Sempéré et al.*, 1993; 1997]. At one end of the spectrum, magmatism may more or less keep pace with plate separation to create an ophiolite-like crust. If magmatism ceases for varying lengths of time or is heterogeneously distributed along or across the axis, faulting will dominate, "amagmatic spreading" may occur, and under some circumstances, oceanic core complexes may develop. It is not clear at present whether core complexes form preferentially in crust with thin magmatic units or if they can develop in thick crust as well.

4.3.4. Implications for the Geometry and Kinematics of Faulting. As pointed out in early seafloor investigations, slip on closely spaced, steeply dipping faults, typical of both rift and transform valley walls, cannot expose deep crustal or upper mantle rocks in tectonic windows unless crustal rock units are initially very thin [*Francheteau et al.*, 1976]. Yet, reconstructed thicknesses of rock units and other geological observations discussed above show that relatively thick lithospheric sections have been exposed and deep crustal and upper mantle rocks have been severely thinned and dramatically uplifted. This paradox can be resolved if slip is concentrated on localized shear zones that have large displacements. Geological observations of low-angle shear zones and detachment faults in tectonic windows [*Karson and Dick*, 1983; *Karson*, 1990; *Lawrence et al.*, 1998; *Dick et al.*, 1991; *Tucholke et al.*, 1997] suggest that this is a viable mechanism. Recently acquired images of bathymetry and side-looking sonar show that the seafloor of the central North Atlantic has numerous dome-like massifs with strongly lineated upper surfaces [*Cann et al.*, 1997; *Tucholke et al.*, 1998; also see *RIDGE Multibeam Synthesis at: http://imager.ldeo.columbia.edu*]. Collectively, these observations suggest that one of the manifestations of seafloor spreading on the MAR may be the formation of oceanic core complexes analogous to core complexes formed in highly extended continental terranes [*Karson et al.*, 1987; *Karson*, 1990; *Mutter and Karson*, 1992; *Tucholke and Lin*, 1994; *Cann et al.*, 1997; *Tucholke et al.*, 1997; 1998]. Seismic reflection images of slow-spread crust have been interpreted in various ways [*McCarthy et al.*, 1988; *White et al.*, 1990; *Mutter and Karson*, 1992; *Morris et al.*, 1993]. It is likely that major crustal reflectors are related to large-scale tectonic extension; however, the geology of highly extended oceanic terranes suggests that they could correspond to a range of different geological features (Figures 10 and 11).

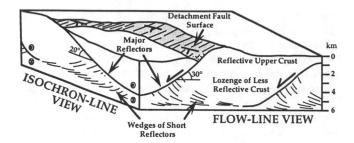

Figure 10. Interpretation of intracrustal seismic reflectors in slow-spread crust in terms of major structures observed near the MAR axis in the MARK area (24°N) [after *Mutter and Karson*, 1992]. Major reflectors on "flow-line" profiles are correlated with detachment faults and shear zones related to large-scale extension of the axial lithosphere. Reflectors on "isochron-line" profiles are interpreted as lateral ramps linking the extensional structures.

4.4 Tectonic Windows into Intermediate- to Fast-Spread Lithosphere

At present, there are only 2 areas with sufficient data for constructing columnar sections of major tectonic windows in intermediate- to fast-spread crust (Figure 12). Compared to slow-spread lithosphere, tectonic windows are rare and mainly provide views of the uppermost crustal structure including basaltic lava units and sheeted dike complexes. Outside of highly faulted exposures at the Hess Deep Rift (Figure 3), extensive exposures of the plutonic foundation of fast-spread crust are yet to be discovered and investigated in detail. The paucity of plutonic rock exposures is probably a reflection of the generally larger magma budget of faster spreading plate boundaries that may tend to produce a relatively thick, uniform, upper crustal assemblage of basaltic lavas and diabase dikes that is not dissected by major faults along spreading centers. Together, these units constitute a relatively fine-grained basaltic carapace 1.5 to 2 km thick that formed above a laterally persistent axial magma chamber [AMC; *Detrick et al.*, 1987; 1993a; *Toomey et al.*, 1990; 1994; *Caress et al.*, 1992; *Sinton and Detrick*, 1992]. Relatively coarse-grained gabbroic material is likely to be plated directly on to the hydrothermally cooled AMC roof. Especially poorly known is the nature of the middle to lower crustal units that probably crystallized from the AMC and the underlying low-velocity zone, considered to be a crystal mush [*Sinton and Detrick*, 1992].

4.4.1 Blanco Transform Wall. A major tectonic window into crust produced at the southern Juan de Fuca Ridge has been investigated in the faulted escarpments of the north wall of the Blanco Transform [*Juteau et al.*, 1995; *Naidoo*, 1998]. This exposure is several tens of kilometers long and has >2 km of relief. The lower parts of the slope are dominated by talus and debris fans. Thus, it appears that the upper 1.5- 2 km of the oceanic crust is exposed here (Figure

12a). In general, crustal materials exposed in this area are limited to basaltic volcanic rocks with sparse diabase dikes [*Juteau et al.*, 1995]. Coordinated, side-looking sonar and submersible studies of these escarpments show that this material can be subdivided into a collage of volcanic rock

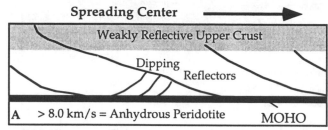

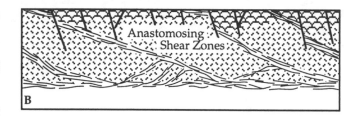

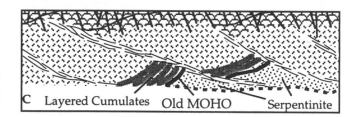

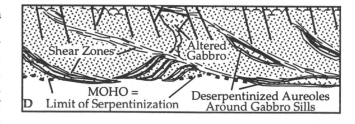

Figure 11. Possible interpretations of intracrustal seismic reflectors in slow-spread oceanic crust based on features observed in tectonic windows. a. Highly generalized pattern of reflectors commonly seen in flow-line profiles. b. Major reflectors as an anastomosing array of ductile shear zones beneath faulted upper crustal rocks. The Moho coincides with a major ductile shear zone. c. Reflections from extended, rotated, and altered lower crust and upper mantle rocks in panels separated by major shear zones. The Moho is varies spatially from a shear zone, to a serpentinization front, to an igneous mafic/ultramafic contact. d. Reflections from shear zones in heterogeneous serpentinites and minor mafic intrusions. Some intrusions may invade the shear zones. The Moho is a variably sheared serpentinization front.

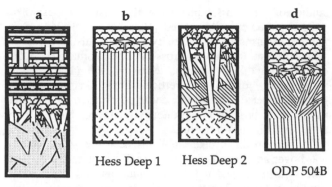

Figure 12. Columnar sections for fast-spread oceanic crust. Sections as in Figure 8; see Figure 3 for locations. Columns or data from following sources: a. Blanco Transform [*Juteau et al.*, 1995]; b. Hess Deep 1, composite of several widely spaced dives on the north rift wall [*Francheteau et al.*, 1992]; c. Hess Deep 2- composite of several dives in 2 small areas on the north and south rift walls [*Karson et al.*, 1992; *Hurst et al.*, 1994a]; and d. ODP Hole 504B [*Anderson et al.*, 1982; *Alt et al.*, 1993].

units typically a few kilometers thick and a few kilometers across in the spreading direction. Within these units, cyclic sequences of flow morphologies suggest multiple eruptive cycles [*Naidoo*, 1998]. The overall thickness of this unit is ~1500 m.

Side-scan sonar backscatter patterns and submersible sampling identify a massive diabase unit about 500 m thick that crops-out for at least several kilometers across the scarp face. This unit appears to consist of a number of thick (tens of meters) massive, tabular, generally horizontal masses. It is not clear if these are large flows, sills, or perhaps relatively the fine-grained upper portions of plutons that intruded the upper crust [*Naidoo*, 1998]. *Tivey* [1996] showed that magnetic polarity boundaries mapped across the scarp face dip toward the southern Juan de Fuca spreading center, suggesting that axial subsidence occurred as the upper crust was constructed.

Although a very substantial vertical section appears to be exposed in this area, no sheeted dikes or coarse-grained gabbroic rocks were found. Faulting is extensive and much of the basaltic material is so intensely fractured that original flow morphologies and structures are difficult to discern. Sills and upper lavas have geochemical compositions that are distinctly different from the main magmatic suite and are probably related to other magmatic systems [*Naidoo*, 1998].

4.4.2 Hess Deep Rift. Hess Deep Rift is located at the tip of a westward propagating spreading center in the Equatorial Pacific. Here, rifting appears to have cut into crust produced at the EPR about 1 million years ago. It is possible that crust created on the EPR near this microplate may not be representative of crust typically formed at fast-spreading

ridges because of its particular tectonic setting [*Zonenshain et al.*, 1980; *Searle and Francheteau*, 1986; *Lonsdale*, 1988]. On the north wall of the rift, steep fault-line scarps truncate abyssal hill lineaments nearly parallel to the EPR suggesting that the rift did not strongly influence spreading as the crust that is now exposed was created. Faulting associated with rifting and microplate rotations have clearly affected the crust in this area and disrupted this tectonic window [*Lonsdale*, 1988; *Francheteau et al.*, 1990]. The upper parts of the rift walls appear to provide windows into the uppermost 2 km of EPR crust. Exposures of mafic and ultramafic plutonic rocks lower on the slope occur in a series of fault-bounded blocks that cannot be easily reconstructed with respect to rock units in the upper parts of the scarp [*Francheteau et al.*, 1990; *Hurst and Karson*, 1991; *Wiggins et al.*, 1996]. Dredging the escarpments of this area have recovered a wide variety of plutonic rock suggesting that they are widely exposed [*Zonenshain et al.*, 1980; *Kashintsev et al.*, 1982].

The first examination of the walls of the Hess Deep Rift was a *Nautile* dive program that consisted of 4 dives spaced at a few tens of kilometers that made single transects up both the north and south rift walls [*Francheteau et al.*, 1990]. Additional dives surveyed plutonic rocks on a prominent Intra-Rift Ridge at the base of the north wall [*Francheteau et al.*, 1990; *Hekinian, et al.*, 1993], that was subsequently drilled on ODP Leg 147 [*Gillis et al.*, 1993a]. The second dive program in the area consisted of 11 *Alvin* dives concentrated in two detailed study areas at the top of the north and south walls. Each of the study areas is only about 2.5 km across and included several dives with both steep NS transects as well as nearly contour-parallel EW transects. This array of dives provides good constraints on the details of upper crustal composition and structure in these relatively small areas [*Karson et al.*, 1992; *Hurst et al.*, 1994a; *Tougas*, 1998].

Nautile dives and drilling on the Intra-Rift Ridge recovered serpentinized harzburgite, much of which is very highly altered to a pale weathering clayey material [*Francheteau et al.*, 1990; *Gillis et al.*, 1993a]. High-temperature, crystal-plastic foliation in the peridotite is expressed as prominent ledges and dips gently to the northeast, that is, oblique to the spreading direction, but generally away from the EPR axis. Whether or not the foliation has been rotated by post-spreading tectonic events is unknown. Nearby, dunites with minor segregations of chromite, plagioclase, and clinopyroxene also crop-out. Samples from these outcrops are similar to the "impregnated" dunites commonly found in cross-cutting veins in ophiolite complexes [*Boudier and Coleman*, 1981; *Nicolas*, 1989; *Girardeau and Francheteau*, 1993].

These ultramafic lithologies crop-out on steep slopes just downslope from layered, cumulate gabbroic rocks and massive gabbros, but no distinct contact was observed. These rock units may be in separate fault-bounded blocks. Less than 1000 m upslope, diabase sills and dikes and basaltic pillow lavas crop-out, raising questions regarding the internal structure of this section of fast-spread crust and the nature of rift-related faulting.

Gabbroic rocks of the Inter-Rift Ridge include layered cumulate gabbros and minor wehrlites [*Hekinian et al.*, 1993; *Natland and Dick*, 1996; *Pedersen et al.*, 1996]. ODP Leg 147 drilling penetrated gabbroic material, some of which appears on the basis of paleomagnetic data, to be in rotated blocks [*Gillis et al., 1993*a]. One hole that appears to be intact yielded 147 m of gabbroic material with a magmatic lineation in a near vertical orientation [*MacLeod et al.*, 1996a]. Uncertainties arising from faulting during the opening of the Hess Deep Rift do not permit these rock units to be linked the upper crustal rock units exposed at the top of the scarp [*Francheteau et al.*, 1990; *Hurst and Karson*, 1991; *MacLeod et al.*, 1996b; *Wiggins et al.*, 1996].

Based on the integrated results of the *Nautile* dives, *Francheteau et al.* [1992] concluded that the upper 2 km of fast-spread EPR crust is composed of three laterally continuous units (Figure 12b). Above a local exposure of gabbroic material there is an extensive sheeted dike complex composed of essentially vertical diabase dikes formed parallel to the EPR. Only minor deviations in dike attitude (dips of 85°-80°, east or west) occur locally. The observed dike unit ranges from as little as 300 to as much as 1200 m in thickness. The sheeted dikes are overlain by an assemblage of mixed extrusive units (mainly pillow basalts) and intrusive units (dikes and possible sills) ranging from 50 to 500 m thick. The uppermost unit consists of approximately 100-200 m of basaltic pillow lavas. The extent to which faulting may have influenced these thicknesses is not clearly defined in the dive areas. Hydrothermal alteration was found in the sheeted dikes and especially in the transitional or mixed intrusive/extrusive unit. Apart from the thinner volcanic units, these observations are in good agreement with the results of drilling at Hole 504 B [*Francheteau et al.*, 1992].

In the *Alvin* study areas, significant variations and additional details are documented [*Karson et al.*, 1992; *Hurst et al.*, 1994a; *Tougas*, 1998]. Just above a major talus ramp the structurally lowest unit is a mafic plutonic complex (>600 m thick) consisting of gabbroic rocks cut by diabase dikes and minor trondhjemite veins. The gabbroic rocks are intensely fractured and veined. The gabbroic unit passes abruptly (over <200 m) upward into a sheeted diabase dike complex (350 to 700 m vertical interval). The dikes are typically 0.5 to 1.5 m wide and are moderately to intensely jointed. Locally, steeply dipping cataclastic fault zones, some of which contain networks of sulfide veins, cut the dike complex. Although the dikes generally strike northwest, parallel to abyssal hills just to the north, the internal structure of the dike complex is variable and complicated. Although dikes in the lavas and upper part of the sheeted dike complex tend to be subvertical, near the base of the sheeted dikes and in the gabbroic rocks they commonly have moderate dips (60°-70°). Very low-angle dikes with dips as low as 30° occur in some places. Paleomagnetic

studies of oriented samples show that many of these dikes have been tectonically rotated [*Hurst et al.*, 1994a]. Subparallel swarms of dikes are typical and distinct swarms with different orientations cross-cut one another. The latest cross-cutting dikes tend to be relatively wide (1-2 m), only very weakly jointed, and vertical. These features indicate that the sheeted dike complex was constructed in an environment of broadly synchronous fracturing, block rotations, and igneous intrusion.

The proportion of dikes in lavas decreases gradually upward over an interval of as much as 500 m. Locally, sheeted dike swarms extend upward as much as 200 m into the overlying lavas. The upper unit of predominantly pillow basalts and basaltic breccias (<200 to 800 m vertical interval) is cut by ≤10% diabase dikes. Massive or sheeted lava flows are very rare and, except near the top of the section, the lavas are intensely fractured. Individual dikes were observed to penetrate to within <100 m of the top of the unit. Mass wasting may have removed part of the uppermost lavas in some areas.

In general, the internal crustal structure of the *Alvin* dive areas (Figure 12c) is interpreted in terms of two discrete spreading events. A first generation of dikes and lavas created at the axis were faulted, brecciated, hydrothermally altered and collapsed prior to the intrusion of later, vertical dikes and overlying less brecciated lavas. The type of brecciation and alteration found here is very similar to that found in some ophiolites [e.g., *Williams and Malpas*, 1972] but tilted and cross-cutting dikes are rarely documented [*Varga*, 1991; *MacLeod and Rothery*, 1992]. Fracturing and alteration apparently took place near the ridge axis as evidenced by the cross-cutting igneous bodies and high-temperature metamorphic features.

4.4.3 Others Windows in Intermediate- to Fast-Spread Lithosphere. Besides possible exposures along other microplate boundaries, the only other known window into the internal structure of intermediate- to fast-spread lithosphere occurs at the Mathematicians Ridge. The nature of faulting and the geologic structure are not known, but extensive outcrops of gabbroic rocks overlain by younger alkalic basalts are present. High-pressure fluid inclusions in the gabbroic rocks suggest that faulting has exhumed mid-crustal rocks [*Batiza and Vanko*, 1986].

Other plutonic rock exposures occur at some fracture zones in the Pacific and may provide access to deep-level rocks [for example, *Anderson and Nishamori*, 1979; *Hébert et al.*, 1983]. Given the paucity of these exposures and their locations along major fracture zones, it seems likely that deep drilling will be necessary to sample the plutonic foundation of fast-spread lithosphere.

4.5 Implications for Processes at Intermediate- to Fast-Spreading Ridges

Crustal structures exposed in tectonic windows are consistent with the notion that the magma budget and local magma supply are much larger and more continuous along intermediate- to fast-spreading ridges compared those of slow-spreading ridges. Faulting is very modest in fast-spread crust except perhaps at major spreading discontinuities [*Macdonald*, 1998]. The continuous internal structure of the crust reflects rather continuous processes that might create a crustal structure similar to that found in stratiform ophiolites and in accord with traditional views of the oceanic crust [*Penrose Conference Participants*, 1972].

The internal structure of the uppermost components of intermediate- to fast-spread crust is known from only a few locations so at present any generalizations must be very tentative. Structural variations at axial discontinuities can only be a matter of speculation at present. Still, the very continuous upper crustal structures that appear to be present along the walls of the Hess Deep Rift and the Blanco Transform make a strong case for relatively continuous magmatic construction during spreading. Variations that occur across these tectonic windows suggest fluctuations in magma supply that may be related to inflation and collapse of axial magmatic systems. The few exposures of the plutonic foundation of fast-spread crust cannot be easily related to upper crustal rock units and are much more dismembered than those of some tectonic windows in slow-spread crust. Thus the details of processes that create the middle to lower crust at fast-spreading ridges remain poorly constrained. The available information provides some important information regarding the nature of oceanic lithosphere and processes that produced it at intermediate- to fast-spreading ridges.

4.5.1. Seismic Layers and Rock Units. Despite the nonunique correlations between seismic velocities and oceanic rock types, there remains a tendency to equate seismic layer thicknesses with rock units in the upper oceanic crust. Several studies have interpreted seismic layer 2A as extrusive basaltic material [e.g., *Christeson et al.*, 1994]. The rapid thickening of layer 2A from ~200 to ~400 m near fast-spreading ridge axes has been used to infer the build-up of extrusive basaltic material across the neovolcanic zone [*Christeson, et al.*, 1992; 1994] and to infer variations in magma supply [*Christeson, et al.*, 1996; *Hooft et al.*, 1996; *Carbotte et al.*, 1997].

Direct observations of the thickness of extrusive material at tectonic windows and in drill holes as summarized above, show substantial variations in the thicknesses of upper crustal rock units. At Hess Deep the thickness of volcanic units may be as little as 300 m in some places, whereas at ODP Hole 504B glassy extrusives were recovered to depths of >800 m below the seafloor. So, in some places, the bottom of layer 2A might lie within the extrusive basaltic units but elsewhere be in the top of sheeted dike complexes.

The low velocity of seismic layer 2A is attributed to the relatively high porosity of basaltic lava flows resulting from original extrusive morphology, and this porosity is likely to be modified as the upper crust is constructed [*Houtz and Ewing*, 1976; *Christeson et al.*, 1992]. In tectonic windows, extrusive morphology appears to be destroyed in the upper couple hundred meters of the crust where basaltic units appear to be an intensely fractured mass

of material. It therefore seems likely that fracture porosity produced by the collapse and compaction of lavas and by faulting might also influence the velocity structure of the uppermost crust. Thus, it seems unlikely that layer 2A matches the thickness of basaltic lavas in the oceanic crust.

Seismic studies near ODP Site 504B showed that the seismic layer 2/3 boundary occurs at a depth corresponding to change in porosity and alteration within the sheeted dike complex [*Detrick et al.*, 1994]. This casts doubts on the long-standing interpretation of the seismic layer 2/3 boundary as a sheeted dike complex/gabbro contact. Uncertainties in the lateral continuity of the sheeted dike complex as well as hydrothermal alteration effects seen in drill holes and cores relative to the scale of seismic resolution make it difficult to compare seismic structure with crustal geology.

4.5.2. Building the Uppermost Crust. Despite the continuous, low-relief form of the axial summit depression of the East Pacific Rise, there is a growing body of evidence that the subsurface geology may be substantially more complex than that suggested by prevailing models. For both the upper crustal units and middle to lower crustal plutonic units very substantial vertical mass transfer is required to build the thickness of these units in the relatively narrow (few kilometers wide) locus of magmatic accretion (Figure 13). Several studies hint at an important role for axial subsidence by referring to the tiny axial summit depression as an elongate "caldera" [*Macdonald and Fox*, 1988; *Haymon et al.*, 1991; *Fornari et al.*, 1998]. However, the scale of these depressions is only tens to hundreds of meters wide and tens of meters deep and the mechanism of subsidence is not clear. In some cases it appears that collapse of volcanic plumbing systems and drainback of lavas from them may result in this scale of subsidence [*Fornari et al.*, 1998], whereas a strong case can also be made for these as the surface manifestations of dike intrusion events [*Chadwick and Embley*, 1998; *Curewitz and Karson*, 1998]. *Macdonald et al.* [1996] emphasize the role of growth faulting, at least on outward facing faults. It is likely that all of these processes play a role in near-surface vertical mass transfer. In any case, these and possibly other processes must accommodate the subsidence of axial volcanic material to depths equivalent to the vertical thickness of the basaltic volcanic layer (hundreds of meters) and this must take place within the locus of surface volcanism (a few kilometers across axis). Thus, in addition to the obvious seafloor spreading, substantial sinking of volcanic units must also occur along in the axial region. Numerous models have been proposed for this general process [*Cann*, 1974; *Dewey and Kidd*, 1977; *Rosencrantz*, 1982; *Hooft et al.*, 1996], and direct observations in tectonic windows hint at a mechanism.

At present it is not clear whether the known variations in upper crustal structure formed at fast-spreading ridges represents isolated complications, perhaps from overlapping spreading centers or other discontinuities, or if they represent part of a range of crustal structures that form as a consequence of time-varying processes typical of crustal

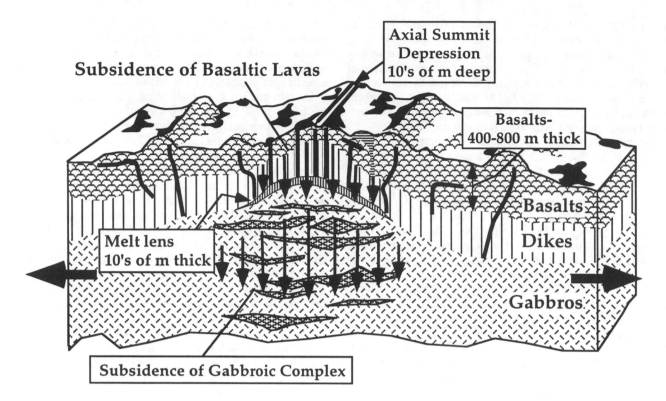

Figure 13. Crustal construction at fast-spreading ridges [after *Perfit et al., 1994*]. Despite having a rather simple morphology similar to a linear shield volcano with an axial summit depression, very significant tectonic processes must occur in the subsurface. Note that the thickness of volcanic material is approximately an order of magnitude greater than the depth of the axial summit depression, demanding that lavas erupted near the axis subside hundreds of meters as they move laterally away from the spreading center. There are many possible kinematic models that can account for this [*Cann*, 1974; *Rosencrantz*, 1982; *Karson et al.*, 1992; *Hooft et al.*, 1996]. Similarly, magmas intruded at depth either as a shallow magma lens or deeper-level sills must subside to accommodate intrusion at various levels and build the middle and lower crust [*Sleep*, 1975; *Kelemen et al.*, 1997] and to accommodate the thickening of the overlying intrusive/extrusive carapace of basaltic lavas and diabase dikes.

accretion [*Lewis*, 1983; *Lichtman and Eissen*, 1983; *Kappel and Ryan*, 1986]. Substantial temporal variations can be inferred from the present along-axis variation from broad, inflated parts of the EPR that are currently experiencing robust magmatism as opposed to narrower, rifted axial highs that may be in the process of collapsing [*Macdonald and Fox*, 1988; *Scheirer and Macdonald*, 1993; *Macdonald*, 1998].

In some areas, for example, the Hess Deep Rift, relatively simple crustal structures were observed. These conform to models of oceanic crust derived from studies of ophiolites and marine seismic refraction experiments [*Francheteau et al.*, 1992]. This type of upper crustal section might be produced during periods of robust magmatism, correlating with periods of inflated, elevated axial swells along the EPR. In contrast, areas with more complex upper crust with variably tilted blocks of sheeted dikes associated with brecciated, hydrothermal altered basaltic volcanic units, may represent

material that "collapsed" as a result of an axial deflation event. Periodic variations in upper crustal structure may correlate with periods of waxing and waning magma supply. The upper crustal units drilled at Hole 504B and 896A are generally thicker than those at Hess Deep, but despite being separated by only about 1 km, show significant variations [*Dilek*, 1998] that may also be related to axial inflation and deflation events. Lavas dipping toward the spreading center in Hole 504B suggest some degree of axial subsidence during spreading [*Pariso and Johnson*, 1989; *Dilek*, 1998]. Magnetic anomaly boundaries in the upper crust at the Blanco Transform dip gently toward the Juan de Fuca Ridge axis and also suggest progressive axial subsidence during accretion [*Tivey*, 1996].

The type of sub-axial collapse described above is well known in terrestrial basaltic calderas [e.g., *Walker*, 1993] and could represent a mechanism for vertical mass redistribution that is required to build the observed thicknesses of extrusive basaltic material [*Karson et al.*, 1992; *Hurst et al.*, 1994a]. Whether this type of collapse or other axial subsidence mechanisms occur, it is likely that they are kinematically linked with processes deeper in the crust. Somehow, processes in the underlying axial magma chamber or mush zone must make space for the subsidence of upper crustal units.

4.5.3. Middle to Lower Crustal Construction. The internal structure of middle to lower crustal assemblages of gabbroic rocks is a matter of speculation. Studies of ophiolites have yielded a wide range of observational data [*e.g., Nicolas*, 1989]. These combined with inferences from marine geological and geophysical studies suggest the processes by which the middle and lower oceanic crust are produced at spreading centers. Current thinking on the internal structure of these crustal components is largely based on an extrapolation of subaxial processes. Although it is generally believed that this part of the crust is created by the crystallization of gabbroic rocks from a more or less continuous magma supply, the details of this process are not know.

Early models of oceanic crust formed from robust magma supplies were based on features found in gabbroic assemblages of ophiolite complexes and offered a range of possibilities based on different assumptions. *Moores and Vine* [1971] suggested a collage of layered mafic/ultramafic plutons on the order of a few kilometers across based on the Troodos ophiolite. Later *Greenbaum* [1972] assumed a maximum angle of repose for cumulate layering and using the measured thickness of the cumulate pile, suggested a flat-topped, triangular magma chamber about 20 km wide. Thermal and density zoning were thought to result in ultramafic cumulates in the bottom apex of the magma chamber grading to mafic cumulates near the top. In a continuously spreading environment this would produce a consistent pattern of layers dipping toward the axis that are oblique to broader scale horizontal compositional and density layering.

Cann [1974] and *Nisbet and Fowler* [1978] speculated that magma chambers of various sizes and with varying replenishment rates could have the form of broader "onion" or narrower "leek" shaped forms, and could create concentric layering patterns in the crust. This "infinite onion" model for the crust also incorporated compositional zonation of the magma chamber and hence the resulting crust such that broad-scale compositional layering remained horizontal despite various orientations of fine-scale layering.

Using similar assumptions and thermal and mechanical arguments, *Sleep* [1975] proposed a model incorporating a small (few kilometers wide) triangular magma chamber with a pointed top and horizontal floor. Thermal zonation in the chamber was thought to result in petrological and hence density variations in cumulates across the chamber floor. Resulting differential subsidence across-axis resulted in a deeper layer of ultramafic cumulates and a shallower layer of mafic cumulates with a complex sigmoidal pattern of finer scale layering. *Dewey and Kidd* [1977] presented an expanded version of this model supported by detailed geological observations from various ophiolites. In this model, large-scale compositional layering is roughly horizontal, but fine-scale layering has a sigmoidal form, generally dipping toward the axis. Large ductile strains are predicted especially for the lower parts of this assemblage.

By mapping arcuate layering patterns in ultramafic and mafic plutonic units of the Bay of Islands ophiolite and accretion directions based on crescumulate crystal growth morphologies, *Casey and Karson* [1981] provided an observational basis for the "infinite onion" model with a relatively small magma chamber. Subsequent studies suggested that the tectonized lower parts of the arcuate forms contained medium- to high-pressure cumulate rocks that crystallized at ~10 km depth and were transported to the surface along the walls of a steep conduit before the middle to upper crust was build on top of them [*Elthon et al.*, 1982].

Field investigations in the Oman ophiolite indicated an angular relation between the mafic/ultramafic contact and cumulate layering. Assuming that the layering dipped toward the paleospreading center, the thickness of the cumulate gabbros yielded a model a Greenbaum-like model for a very wide (about 20 km) magma chamber with a flat roof [*Pallister and Hopson*, 1981]. Subsequent studies assuming the opposite dip (and age relations) in variably deformed igneous layering led later workers [*Nicolas et al.*, 1988; *Boudier et al.*, 1996] to propose a model with a relatively small, "tent-shaped" magma chamber.

In the late 1980's and early 1990's documentation of only very small magma bodies beneath the East Pacific Rise [e.g. *Detrick et al.*, 1993a] more or less eliminated the magma chamber models with large dimensions. Even models with relatively small magma chambers had to be revised for a magma lens only tens of meters thick and <1 km wide underlain by a low velocity zone of crystal mush [*Sinton and Detrick*, 1992]. For such a small magma chamber to create a magmatic crust 3-5 km thick beneath the ridge axis, two important processes seem likely. First, very substantial vertical mass transport and deformation must occur in the crystal mush region [*Sleep*, 1975; *Dewey and Kidd*, 1977; *Phipps Morgan and Chen*, 1993; *Quick and Denlinger*, 1993]. Observations of supra-solidus ("magmatic") to high-temperature crystal-plastic deformation of gabbroic rock in ophiolites may reflect this process [*Nicolas*, 1989]. Second, it is likely that numerous discrete sills and minor intrusive bodies are injected in the subaxial region. The depth of intrusion would be a function of the neutral buoyancy horizon appropriate to the melt and local crustal density structure [*Ryan*, 1993] and might range from the Moho to the upper crust [*Kelemen et al.*, 1997]–similar to the conclusions of *Moores and Vine* [1971] many years earlier.

Observations from ophiolites (Figure 6a-d) provide good evidence small-scale intrusions are significant [*Moores and*

Vine, 1971; *Harkins et al.*, 1980; *Pederson, 1986; Bédard et al.*, 1996; *Kelemen et al.*, 1997]. Following these inferences, the magmatic crust produced at fast-spreading ridges would be a composite of variably subsided and deformed intrusive bodies rather than the product of a single replenished magma body [*Henstock et al.*, 1993; *Kelemen et al.*, 1997]. The complexity of the resulting internal structure would be very substantial and may account for the highly variable character of patterns of layering and deformation reported from even the best exposed and most carefully studied ophiolites. At present, geometry and kinematics of deformation that accommodates this vertical mass transfer has not been resolved.

The evolution of these ideas is an excellent illustration of how the cross-pollination of ideas from ophiolites and constraints from marine studies have been involved in the evolution of current views of how processes at mid-ocean ridges operate. It is also noteworthy that some of the earliest models, based on basic field geological observations, have resurfaced as key elements of our current understanding.

5. CONCLUDING REMARKS

Despite the simple, rather uniform, layered seismic structure of the oceanic lithosphere derived from seismic refraction studies and ophiolite analogs (Figure 1), there is strong evidence that the internal geological structure of the oceanic crust is extremely complex and variable on scales ranging from tens of meters to tens of kilometers. As concluded by several other studies, the "seismic velocity structure" of the oceanic crust does not appear to accurately represent the arrangement of rock units in the oceanic crust, but instead primarily maps alteration and porosity effects that overlay the fundamental geological structure. The disparity between seismic velocity structure and geology is extreme in slow-spread crust or at least where the magma budget is relatively low. Even at fast spreading rates and higher magma budgets, where available geological observations do not rule out a large-scale layered structure, seismic studies probably do not map the contacts between major rock units. Seismic reflection studies have begun to reveal a complex internal structure in slow-spread crust. Whereas many of the features imaged may be fault zones related to extensional tectonics, the significance of middle to deep crustal reflections and how they relate to shallow crustal structures has not been clearly established.

Direct observations in tectonic windows into crust produced at slow-spreading rates indicate that it is generally far more heterogeneous and variable than suggested by models derived from ophiolite complexes and geophysical studies. In many places, it can be shown that the expected layered sequence of pillow lavas, sheeted dikes, and gabbroic rocks over mantle peridotites is not present. Instead, a more complex assemblage of variably deformed and metamorphosed

mafic to ultramafic rock units is documented. Coarse-grained plutonic rock units are commonly continuous for only tens to hundreds of meters. The contacts between the units are poorly documented at present, but appear to be highly variable with (both sharp and gradational) igneous and tectonic contacts that may also vary spatially. Younger, generally less deformed and metamorphosed mafic rock bodies cut the older units. A discontinuous cover of basaltic pillow lavas and flows unconformably overlies the complex plutonic assemblage in many places. Cross-cutting relationships and unconformities indicate either significant discontinuities in magma production or widely shifting (kilometers) loci of magmatism at spreading centers. Thus, slow-spread crust may resemble the ophiolite model of oceanic crust only in a very general sense. At the scale of kilometers or tens of kilometers tectonic windows reveal a heterogeneous crust that appears to have been assembled piecemeal between or during crustal deformation events.

Comparatively little is known of the internal structure of lithosphere produced at fast-spreading environments. Direct observations show that the ophiolite model may be a reasonable approximation at least for the upper crust. Studies of tectonic windows help show how a thick layer of basaltic lavas is produced in a narrow neovolcanic zone by eruption over a subsiding axial crust. At present, few observations shed light on how gabbroic units several kilometers thick are created from a relatively small magma chamber. In general, to create a layered structure similar to ophiolites, it is necessary that dramatic (kilometer-scale) and rapid (meters per year) vertical subsidence occurs beneath spreading centers in order to accommodate the vertical thickening of crustal units before the lithosphere moves laterally away from the locus of magmatism at the ridge axis. The details of these processes at different crustal levels are not yet known, but a mechanism similar to caldera collapse and growth faulting appears to be required at shallow levels, with sill intrusion and other vertical mass transport mechanisms dominating at depth. It is noteworthy that these processes are not readily apparent from the surface expression of fast-spreading ridges, nor from available seismic data.

Collectively, tectonic windows into the oceanic crust indicate that the crust formed at slow-spreading rates and/or low magma budgets is very different from that formed at intermediate and higher rates. Although the similar rock units appear to be common to all oceanic crust, the dimensions of the units, the contacts that separate them, and their internal compositions and structures are likely to be very different. One way to view these variations is in terms of how magmatic construction modifies lithospheric stretching. Stretching and related alteration of the lithospheric mantle can be thought of as the "default" response to plate separation. The result of spreading with no magmatism would be a highly deformed serpentinized peridotite crust that grades downward into essentially anhydrous mantle peridotite. This type of spreading may occur in areas of very slow plate separation or very infertile mantle [*Brown and Karson*, 1988; *Cannat*, 1993; *Sleep and Barth*, 1997]. With a small amount of decompression melting, minor dikes and/or mafic plutons may intrude the serpentinite crust and a thin or discontinuous cap of basalt may be present. With more magmatism, it is possible for igneous construction to replace stretched serpentinites in the crust. Only when sufficient magmatism occurs to fully accommodate plate separation, will stretching of the lithosphere diminish. Even in fast-spread crust, normal faults cut sheeted dikes and other magmatic units, so stretching always appears to play at least a minor role. Consequently, dike intrusion events represent only one type of plate separation or spreading event.

Direct investigations of the oceanic crust are still in their infancy, but have already hinted at the diverse nature of oceanic crust from different environments. The ophiolite model and "one-size fits all" view of oceanic crust and seafloor spreading are now outdated or too generalized to be useful. New types of investigations that bring traditional field geological techniques as well as new methods to seafloor geology and high-resolution geophysics will be required to make progress in this area.

Investigations of tectonic windows into the oceanic crust and upper mantle provide an important perspective on the internal structure and composition of rock units that constitute the oceanic lithosphere. The various types of crustal structures observed indicate that the relationship between magmatic construction and tectonic deformation varies substantially in time and space. These observations have been carried out on lateral and vertical scales that are comparable to the outcrop areas of ophiolite complexes and to the resolution of marine seismic studies. Therefore, they provide an important check on models of the oceanic crust derived from these other types of studies. A number of conclusions can be reached from the present view into tectonic windows:

1. Tectonic windows generally do not support the continuously layered model of oceanic crust and upper mantle derived from ophiolite and marine geophysics. Instead, it appears that magmatism, tectonic deformation and hydrothermal processes operate in complex patterns in time and space to create a wide range of crustal structures. The range of crustal structures appears to be very diverse at slow-spreading ridges but is probably more uniform at faster rates or higher magma budgets. Even crust produced at magmatically robust, fast-spreading ridges is very likely to have a much more complex internal structure than would be expected based on their relatively simple morphologic expression.

2. In general, slow-spread crust is significantly different from that produced at fast- to intermediate-spreading rates, but a more fundamental parameter than spreading rate in controlling oceanic crustal structure is the magma budget (magma volume per unit plate separation). Magma budget determines the extent to which magmatic construction interrupts mechanical stretching of the lithosphere during plate separation. At high magma budgets even slow-spreading settings like the Reykjanes Ridge probably produce a more or less continuous magmatic crust on the order of 5-7 km thick. At lower magma budgets, even at intermediate- or fast-spreading rates, tectonic extension could possibly result in stretching and thinning of rock units and the exhumation of deep crustal or even upper mantle rocks.

3. Magmatic construction appears to vary in a complex way in time and space with respect to spreading segments of slow-spreading ridges. These variations account for the diverse crustal structures found in tectonic windows; however, the observed variations in crustal structures are in some cases at odds with current models of segment-scale processes [*Lin et al.*, 1990; *Cannat*, 1993; *Detrick et al.*, 1995]. Specifically, in some places magma-starved crust occurs near segment centers and crust with thick magmatic assemblages occurs near segment ends.

4. The range of major fault structures along slow-spreading ridges is very similar to that of continental extensional terranes. This includes a spectrum of structures ranging from slightly faulted and fissured magmatically robust rift segments to highly faulted and extended areas including oceanic core complexes.

5. Oceanic core complexes, analogous to those of continental terranes form dome-like massifs of highly extended and uplifted deep crustal to upper mantle rocks exposed along low-angle detachment faults. Tectonic windows along rift valley walls and fracture zones provide cross sections of these terranes. Core complexes appear to be widespread in slow-spread crust of the central North Atlantic and thus they appear to be fundamentally important aspects of seafloor spreading. Understanding their architecture and kinematic evolution is a major challenge for the next generation of mid-ocean ridge studies.

6. Tectonic windows are relatively rare in lithosphere formed at intermediate- to fast-spreading rates. The available exposures reveal mainly upper crustal rock units including basaltic volcanic rocks, sheeted diabase dike complexes, and massive diabase to gabbro units. The thickness of extrusive basaltic units (commonly >500 m) cannot be accounted for simply by the build-up of lavas. Pervasive fracturing, faulting, and block rotations may reflect rapid subaxial subsidence that accommodates the growth of the extrusive units of the upper crust within a few kilometers of the small axial summit depression.

7. Very limited studies of deep crustal and upper mantle rocks from tectonic windows in fast-spread crust are not sufficient to elucidate the processes by which the plutonic portion of the crust is constructed or how mantle flow is related to crustal accretion. However, it is likely that vertical mass transfer processes including magmatic intrusion and ductile deformation in the middle to lower crust are coordinated with the sub-axial subsidence of upper crustal units at fast-spreading ridges.

Acknowledgments. I thank the numerous investigators who over the years have collaborated with me on studies of ophiolite complexes and studies of tectonic windows into the oceanic crust. These various projects have been supported by grants from the National Science Foundation Program in Marine Geology and Geophysics. J-M. Auzende provided photos from the Vema Transform and J.F. Casey provided a copy of a cruise report from King's Trough. I thanks M.H. Cormier and an anonymous reviewer whose comments helped to significantly improve this manuscript. Dan Curewitz's help was invaluable in the final preparation of the manuscript.

REFERENCES

Abbate, E., V. Bortolotti, and G. Principi, Apennine ophiolites: a peculiar oceanic crust, *Ofioliti, 1*, 59-96, 1980.

Abbots, I.L., Intrusive processes at ocean ridges: Evidence from the sheeted dyke complex of Masirah, Oman, *Tectonophys., 60*, 217-233, 1979.

Agar, S.M., Microstructural evolution of a deformation zone in the upper oceanic crust: evidence from DSDP Hole 504B, *J. Geodynam., 13*, 119-140, 1991.

Alexander, R.J. and G.D. Harper, The Josephine ophiolite: An analogue for slow- to intermediate-spreading mid-ocean ridges, in *Ophiolites and their Modern Oceanic Analogues,* edited by L.M. Parson, B.J. Murton, and P. Browning, *Geol. Soc. Lond., Spec. Publ., 60*, 3-38, 1992.

Allerton, S., B.J. Murton, R.C. Searle, and M. Jones, Extensional faulting and segmentation of the Mid-Atlantic Ridge north of the Kane fracture zone (24°00'N to 24°40'N), in *Tectonic, Magmatic, Hydrothermal, and Biological Segmentation of Mid-Ocean Ridges*, edited by C.J. MacLeod, C.J. Tyler, and C.L. Walker, 61-102, *Geol. Soc. Lond., Spec. Publ. 118*, 1996.

Alt, J.C., et al., Proc. ODP, Initial Rept., 148, 352 pp., Ocean Drilling Program, College Station, TX, 1993a.

Anderson, R.N., J. Honnorez, K. Becker, A.C. Adamson, J. C. Alt, R. Emmermann, P. Kempton, H. Kinoshita, C. Laverne, M.J. Mottl, and R.L. Newmark, DSDP Hole 504B, the first reference section over 1 km through Layer 2 of the oceanic crust, *Nature, 300*, 589-594, 1982.

Anderson, R.N. and R.K. Nishimori, Gabbro, serpentinite, and mafic breccia from the East Pacific Rise, *J. Phys. Earth, 27*, 467-480, 1979.

ARCYANA, Direct sea-floor observations in submersibles of an active rift valley and transform fault in the Atlantic near 36°50'N, *Science, 190*, 108-116, 1975.

ARCYANA, *FAMOUS: Photographic Atlas of the Mid-Atlantic Ridge: Rift and Transform Fault at 3000 m Depth*, 128 pp., Bordas, Paris, 1978.

Aumento, F. and H. Loubat, The Mid-Atlantic Ridge near 45°N. Serpentinized ultramafic intrusions, *Can. J. Earth Sci., 8*, 631-663, 1971.

Auzende, J.M., D. Bideau, E. Bonatti, M. Cannat, J. Honnorez, Y. Lagabrielle, J. Malavieille, V. Mamaloukas-Frangoulis, and C. Mével, Direct observation of a section through slow-spreading oceanic crust, *Nature, 337*, 726-729, 1989a.

Auzende, J.M., D. Bideau, E. Bonatti, M. Cannat, J. Honnorez, Y. Lagabrielle, J. Malavieille, V. Mamaloukas-Frangoulis,

and C. Mével, The MAR-Vema Fracture Zone intersection surveyed by deep submersible *Nautile, Terra Nova, 2*, 68-73, 1989b.

Auzende, J.M., M. Cannat, P. Gente, J.P. Henriet, T. Juteau, J. Karson, Y. Lagabrielle, and M.A. Tivey, Deep layers of mantle and oceanic crust exposed along the southern wall of the Kane Fracture Zone: Submersible observations, *Acad. Sci. C.R., 317*, 1641-1648, 1993.

Auzende, J.M., J.L. Olivet, J. Charvet, A. LeLann, X. Le-Pichon, J.H. Monteiro, A. Nicolas, and A. Ribeiro, Sampling and observation of oceanic mantle and crust on Gorringe Bank, *Nature, 273*, 45-59, 1978.

Ballard, R.D., W.B. Bryan, J.R. Heirtzler, G. Keller, J.G. Moore, and T. van Andel, Manned submersible observations in the FAMOUS area: Mid-Atlantic Ridge, *Science, 190*, 103-108, 1975.

Ballard, R.D., R.T. Holcomb, and T.H. van Andel, The Galapagos Rift at 86°W, 3, Sheet flows, collapse pits, and lava lakes of the rift, *J. Geophys. Res., 84*, 4507-5422, 1979.

Ballard, R.D. and J.G. Moore, *Photographic Atlas of the Mid-Atlantic Ridge Rift Valley*, 114 pp., Springer-Verlag, NY, 1977.

Barany, I. and J.A. Karson, Basaltic breccias of the Clipperton fracture zone: Sedimentation and tectonics in a fast-slipping oceanic transform, *Geol. Soc. Am. 101*, 204-220, 1989.

Batiza, R. and D.A. Vanko, Petrologic evolution of large failed rifts in the eastern Pacific: Petrology of volcanic and plutonic rocks from the Mathematician Ridge area and Guadalupe Trough, *J. Petrol., 26*, 564-602, 1986.

Bédard, J.H. and R. Hébert, The lower crust of the Bay of Islands ophiolite, Canada: Petrology, mineralogy, and the importance of syntexis in magmatic differentiation in ophiolites and at ocean ridges, *J. Geophys. Res., 101*, 25,105-25,124, 1996.

Bloomer, S.H. and J.W. Hawkins, Gabbroic and ultramafic rocks from the Mariana Trench: An island arc ophiolite, *in The Tectonic and Geologic Evolution of Southeast Asian Seas and Island, Part 2*, edited by D.E. Hayes, *Geophysical Monogr. 27*, 294-317, American Geophysical Union, Washington, D.C., 1983.

Bonatti, E., Serpentinite protrusions in the oceanic crust, *Earth Planet. Sci. Lett., 32*, 107-113, 1976.

Bonatti, E. and J. Honnorez, Sections of the earth's crust in the equatorial Atlantic, *J. Geophys. Res., 81*, 4104-4116, 1976.

Boudier, F., Microstructural study of three peridotite samples drilled at the western margin of the Mid-Atlantic Ridge, edited by W.G. Melson, et al., *Initial Reports Deep Sea Drill. Project, 45*, 603-608, 1978.

Boudier, F. and R.G. Coleman, Cross section through the peridotite in the Samail Ophiolite, southeastern Oman Mountains, *J. Geophys. Res., 86*, 2573-2592, 1981.

Boudier, F., A. Nicolas, and B. Ildefonse, Magma chambers in the Oman ophiolite: Fed from the top or bottom?, *Earth Planet. Sci. Lett., 144*, 239-250, 1996.

Bratt, S.R. and G.M. Purdy, Structure and variability of oceanic crust on the flanks of the East Pacific Rise between 11° and 13° N, *J. Geophys. Res., 89*, 6111-6125, 1984.

Brown, J.R. and J.A. Karson, Variations in axial processes on the Mid-Atlantic Ridge: The neovolcanic zone in the MARK Area, *Mar. Geophys. Res., 10*, 109-138, 1988.

Buck, W.R., Flexural rotation of normal faults, *Tectonics, 7*, 959-973, 1988.

Buck, W. R., Modes of continental lithospheric extension., *J. Geophys. Res., 96*, 20,161-20,178, 1991.

Calvert, A.J. and C.G. Potts, Seismic evidence for hydrothermally altered mantle beneath old crust in the Tydeman fracture zone, *Earth Planet. Sci. Lett., 75*, 439-449, 1985.

Cann, J.R., A model for oceanic crustal structure developed, *Geophys. J. R. Astron. Soc., 39*, 169-187, 1974.

Cann, J.R., D.K. Blackman, D.K. Smith, E. McAllister, B. Janssen, S. Mello, E Avgerinos, A.R. Pascoe, and J. Escartín, Corrugated slip surfaces formed at ridge-transform intersections on the Mid-Atlantic Ridge, *Nature, 385*, 329-332, 1997.

Cann, J.R. and B.M. Funnell, Palmer Ridge, a section through the upper part of the oceanic crust? *Nature, 213*, 661-664, 1967.

Cannat, M., Emplacement of mantle rocks in the sea floor at mid-ocean ridges, *J. Geophys. Res., 98*, 4163-4172, 1993.

Cannat, M. et al., Proc. ODP, Initial Rept., 153, 798 pp., Ocean Drilling Program, College Station, TX, 1995a.

Cannat, M., G. Ceuleneer, and J. Fletcher, Localization of ductile strain and magmatic evolution of gabbroic rocks drilled at the Mid-Atlantic Ridge (23° N), in *Proceedings of the Ocean Drilling Program, Scientific Results*, vol. 153, edited by J.A. Karson, M. Cannat, D.J. Miller, and D. Elthon, pp. 77-98, Ocean Drill. Prog., College Station, TX, 1997a.

Cannat, M., J.A. Karson, D.J. Miller, et al., Probing the plutonic foundation of the Mid-Atlantic Ridge, *Eos, Trans. AGU, 76*, 129-133. 1995b.

Cannat, M., T. Juteau, and E. Berger, Petrostructural analysis of the Leg 109 serpentinized peridotites, in *Proceedings of the Ocean Drilling Program, Scientific Results*, vol. 106/109, edited by R.S. Detrick, W.B. Bryan, J. Honnorez, and T. Juteau, pp. 47-57, Ocean Drill. Prog., College Station, TX, 1990.

Cannat, M., Y. Lagabrielle, N. de Coutures, H. Bougault, J. Casey, L. Dmitriev, and Y. Fouquet, Ultramafic and gabbroic exposures at the Mid-Atlantic Ridge: Geological mapping in the 15°N region, *Tectonophys., 279*, 193-214, 1997b.

Cannat, M., C. Mével, M. Maia, C. Deplus, C. Durand, P. Genet, P. Agrinier, A. Belarouchi, A. Dubuisson, E. Humler, and J. Reynolds, Thin crust, ultramafic exposures, and rugged faulting patterns at the Mid-Atlantic Ridge (22°-24°N), *Geology, 23*, 49-52, 1995c.

Carbotte, S.M. and K.C. Macdonald, Causes of variation in fault-facing direction on the ocean floor, *Geology, 18*, 749, 1990.

Carbotte, S.M., J.C. Mutter, and L. Xu, Contribution of volcanism and tectonism to axial and flank morphology of the southern East Pacific Rise, 17°10'-17°40'S, from a study of layer 2A geometry, *J. Geophys. Res., 102*, 10,165-10,184, 1997.

Caress, D.W., M.S. Burnett, and J.A. Orcutt, Tomographic image of the axial low-velocity zone at 12°50'N on the East Pacific Rise, *J. Geophys. Res., 97*, 9243-9263, 1992.

Carlson, R.L. and C.N. Herrick, Densities and porosities in the oceanic crust and their variations with depth and age, *J. Geophys. Res., 95*, 9153-9170, 1990.

Casey, J.F., Comparison of major- and trace-element geochemistry of abyssal peridotites and mafic plutonic rocks with basalts from the MARK region of the Mid-Atlantic Ridge,

edited by J.A. Karson et al., *Proc. Ocean Drill. Program Scientific Results*, 153, edited by J.A. Karson, M. Cannat, D.J. Miller, and D. Elthon, 181-242, 1997.

Casey, J.F., J.F. Dewey, P.J. Fox, J.A. Karson, and E. Rosencrantz, Heterogeneous nature of the oceanic crust and upper mantle: A perspective from the Bay of Islands Ophiolite, in *The Oceanic Lithosphere*, edited by C. Emiliani, The Sea, VII, pp. 305-338, Wiley, New York, 1981.

Casey, J.F., and J.A. Karson, Magma chamber profiles from the Bay of Islands ophiolite complex, *Nature, 298*, 295-301, 1981.

CAYTROUGH, Geological and geophysical investigation of the Mid-Cayman Rise spreading center: Initial results and observations, in *Deep Drilling Results in the Atlantic Ocean: Oceanic crust, Maurice Ewing Series 2*, edited by M. Talwani, et al., pp. 66-95, AGU Washington, D.C., 1979.

Ceuleneer, G. and M. Cannat, High-temperature ductile deformation of Site 920 peridotites, edited by J.A. Karson et al., *Proc. Ocean Drill. Program Scientific Results*, 153, 23-34, 1997.

Chadwick, W.W., Jr., and R.W. Embley, Graben formation associated with recent dike intrusions and volcanic eruptions on the mid-ocean ridge, *J. Geophys. Res., 103*, 9807-9825, 1998.

Christensen, N.I., The abundance of serpentinites in the oceanic crust, *J. Geol., 80*, 709-719, 1972.

Christensen, N.I., Ophiolite, seismic velocities and oceanic crust structure, *Tectonophys., 47*, 131-157, 1978.

Christeson, G.L., G.M. Kent, G.M. Purdy, and R.S. Detrick, Extrusive thickness variability at the East Pacific Rise 9°-10°N: Constraints from seismic techniques, *J. Geophys. Res., 101*, 2859-2783, 1996.

Christeson, G.L., G.M. Purdy, and G.J. Fryer, Structure of young upper crust at the East Pacific Rise near 9°30'N, *Geophys. Res. Lett., 19*, 1045-1048, 1992.

Christeson, G.L., G.M. Purdy, and G.J. Fryer, Seismic constraints on shallow crustal emplacement processes at the fast spreading East Pacific Rise, *J. Geophys. Res., 99*, 17,957-17,974, 1994.

Coleman, R.G., *Ophiolites*, 220 pp., Springer-Verlag, New York, 1977.

Cowie, P.A., C.H. Scholz, M. Edwards, and A. Malinverno, Fault strain and seismic coupling on mid-ocean ridges, *J. Geophys. Res., 98*, 17,911-17,920, 1993.

Crittenden, M.D., Jr., P.J. Coney, and G.H. Davis (Eds.), Cordilleran metamorphic core complexes, *Geol. Soc. Am. Mem., 153*, 490 pp., 1980.

Curewitz, D. and Karson, J.A., Geological consequences of dike intrusion at mid-ocean ridge spreading centers, in *Faulting and Magmatism at Mid-Ocean Ridges, Geophys. Monogr. Ser.*, edited by W.R. Buck et al., AGU, Washington, D.C., this volume, 1998.

CYAMEX Scientific Team, First manned submersible dives on the East Pacific Rise at 21°N (Project RITA): General Results, *Mar. Geophys. Res., 4*, 345-379, 1981.

Davis, G.A. and G.S. Lister, Detachment faulting in continental extension: Perspectives from the southwestern U.S. Cordillera, *Geol. Soc. Am., Spec. Paper 218*, 133-159, 1988.

Delaney, J.R., D.W. Mogk, and M.J. Mottl, Quartz-cemented, sulfide-bearing greenstone breccias from the Mid-Atlantic Ridge- samples of a high temperature hydrothermal upflow zone, *J. Geophys. Res., 92,* 9175-9192, 1985.

DeLong, S.E., J.F. Dewey, and P.J. Fox, Topographic and geologic evolution of fracture zones, *J. Geol. Soc. Lond., 136,* 303-310, 1979.

Detrick, R.S., et al., Proc. ODP, Initial Rept., 106/109, 249 pp., Ocean Drilling Program, College Station, TX, 1988.

Detrick, R. S., P. Buhl, E. Vera, J. Mutter, J. Orcutt, J. Madsen, and T. Brocher, Multi-channel seismic imaging of a crustal magma chamber along the East Pacific Rise, *Nature, 326,* 35-41, 1987.

Detrick, R.S., J. Collins, R. Stephen, and S. Swift, *In situ* evidence for the nature of the seismic layer 2/3 boundary in oceanic crust, *Nature, 370,* 288-290, 1994.

Detrick, R.S., A.J. Harding, G.M. Kent, J.A. Orcutt, J.C. Mutter and P. Buhl, Seismic structure of the southern East Pacific Rise, *Science, 259,* 499-503, 1993a.

Detrick, R.S., J.C. Mutter, P. Buhl, and I.I. Kim, No evidence from multichannel reflection data for a crustal magma chamber in the MARK Area on the Mid-Atlantic Ridge, *Nature, 347,* 61-64, 1990.

Detrick, R.S., H.D. Needham, and V. Renard, Gravity anomalies and crustal thickness variations along the Mid-Atlantic Ridge between 33° and 40°N, *J. Geophys. Res., 100,* 3767-3787, 1995.

Detrick, R.S., R.S. White, and G.M. Purdy, Crustal structure of North Atlantic fracture zones, *Rev. Geophys, 31,* 439-458, 1993b.

Dewey, J.F. and J.M. Bird, Origin and emplacement of the ophiolite suite: Appalachian ophiolites in Newfoundland, *J. Geophys. Res., 76,* 3179-3206, 1971.

Dewey, J.F. and W.S.F. Kidd, Geometry of plate accretion, *Geol. Soc. Am. Bull., 79,* 411-423, 1977.

Dick, H.J.B., Abyssal peridotites, very slow spreading ridges and ocean ridge magmatism, in *Magmatism in the Ocean Basins,* edited by A.D. Saunders and M.J. Norry, *Geol. Soc. Lond., Spec. Publ. 42,* 71-105, 1989.

Dick, H.J.B., J. Honnorez, and P.W. Kirst, Origin of abyssal basaltic sand, sandstone and gravel from DSDP Hole 396B, Leg 46, edited by L. Dmitriev et al., Initial Reports Deep Sea Drill. Project, 45, 331-340, 1978.

Dick, H.J.B., P.S. Meyer, S.H. Bloomer, S. Kirby, D. Stakes, and C. Mawer, Lithostratigraphic evolution of an in-situ section of oceanic Layer 3, edited by R.P. Von Herzen et al., *Proc. Ocean Drill. Program Scientific Results,* 118, 439-538, 1991.

Dick, H.J.B., J.H. Natland, and Leg 176 Scientific Party, A 1.5 km section of gabbroic lower crust at the SW Indian Ridge, *Eos, Trans. AGU, 79* (17), Spring Meet. Suppl., S375, 1998.

Dick, H.J.B., G. Thompson, G. and W.B. Bryan, Low angle faulting and steady state emplacement of plutonic rocks at ridge-transform intersections *Eos, Trans. AGU, 62,* 406, 1981.

Dilek, Y., Structure and tectonics of intermediate-spread oceanic crust drilled at DSDP/ODP Holes 504B and 896A, Costa Rica Rift, in Geological Evolution of Ocean Basins: Results from the Ocean Drilling Program, edited by A. Cramp, C.J. MacLeod, S.V. Lee, and E.J.W. Jones, *Geol. Soc. Lond. Spec. Publ. 131,* 179-197, 1998.

Dilek, Y., E.M. Moores, M. Delaloye, and J.A. Karson, Amagmatic Extension and Tectonic Denudation in the Kizildag Ophiolite, Southern Turkey, Implications for the Evolution of Neotethyan Oceanic Crust, in *Ophiolites Genesis*

and Evolution of Oceanic Lithosphere, edited by Tj. Peters, A. Nicolas and R.G. Coleman, pp. 485-500, Kluwer, Dordrecht, 1991.

Eberhart, G.L., P.A. Rona, and J. Honnorez, Geologic controls on hydrothermal activity in the Mid-Atlantic Ridge rift valley: Tectonics and volcanics. *Mar. Geophys. Res., 10,* 233-259, 1988.

Elthon, D., Petrology of gabbroic rocks from the Mid-Cayman Rise spreading center, *J. Geophys. Res., 92,* 658-682, 1987.

Elthon, D., J.F. Casey, and S. Komor, Mineral chemistry of ultramafic cumulates from the North Arm Mountain Massif of the Bay of Islands Ophiolite: Implications for high pressure fractionation of oceanic basalts, *J. Geophys. Res., 87,* 8717-8734, 1982.

Engel, C.G. and R.L. Fisher, Granitic to ultramafic rock complexes of the Indian Ocean ridge system, western Indian Ocean, *Geol. Soc. Am. Bull., 86,* 1553-1578, 1975.

Escartín, J. and J. Lin, Ridge offsets, normal faulting, and gravity anomalies of slow spreading ridges, *J. Geophys. Res., 100,* 6163-6177, 1995.

Fisher, R.L., and Engel, C.G., Ultramafic and basaltic rocks dredged from the nearshore flank of the Tonga Trench, *Geol. Soc. Am. Bull., 80,* 1373-1378, 1969.

Fornari, D.J., R.M. Haymon, M.R. Perfit, T.K.P. Gregg, and M. H. Edwards, Axial summit trough of the East Pacific Rise 9°-10°N: Geological characteristics an evolution of the axial zone on fast spreading mid-ocean ridges, *J. Geophys. Res., 103,* 9827-9855. 1998.

Fox, P.J., and B.C. Heezen, Sands of the Mid-Atlantic Ridge, *Science, 159,* 1367-1370, 1965.

Fox, P.J., E. Schreiber, and J.J. Peterson, The geology of the oceanic crust: compressional wave velocities of oceanic rocks, *J. Geophys. Res., 78,* 5155-5172, 1973.

Fox, P.J. and J.B. Stroup, The plutonic foundation of the oceanic crust, in *The Oceanic Lithosphere,* edited by C. Emiliani, The Sea, VII, pp. 119-218, Wiley, New York, 1981.

Francheteau, J., R. Armijo, J. L. Cheminee, R. Hekinian, P. Lonsdale, and N. Blum, 1 Ma East Pacific Rise oceanic crust and uppermost mantle exposed by rifting in Hess Deep (equatorial Pacific Ocean), *Earth Planet. Sci. Lett., 101,* 281-295, 1990.

Francheteau, J., R. Armijo, J.L. Cheminee, R. Hekinian, P. Lonsdale, and N. Blum, Dyke complex of the East Pacific Rise exposed in the walls of Hess Deep and the structure of the upper oceanic crust, *Earth Planet. Sci. Lett., 111,* 109-121, 1992.

Francheteau, J., P. Choukroune, R. Hekinian, X. LePichon, and H.D. Needham, Oceanic fracture zones do not provide deep sections in the crust, *Can. J. Earth Sci., 13,* 1223-1235, 1976.

Francheteau, J., R. Armijo, J.P. Congne, J. Girardeau, M. Consantin, R. Hekinian, D.F. Naar, R.N. Hey, and R.C. Searle, Submersible observations of the Easter microplate and its boundary, *Eos, Trans. AGU, 75* , 582, 1994.

Francheteau, J., P. Patriat, J. Segoufin, R. Armijo, M. Doucoure, A. Yelles-Chaouche, J. Zukin, S. Calmant, D.F. Naar, and R.C. Searle, Pito and Orongo fracture zones: The northern and southern boundaries of the Easter microplate (southeast Pacific), *Earth Planet. Sci. Lett., 89,* 363-374, 1988.

Francis, T.J.G., Serpentinization faults and their roles in the tectonics of slow spreading ridges, *J. Geophys. Res., 86,* 11,616-11,622, 1981.

Fryer, P., E.L. Ambos, and D.M. Hussong, Origin and emplacement of Mariana forearc seamounts, *Geology, 13,* 774-777, 1985.

Gao, D., Seafloor geology at the eastern intersection of the Mid-Atlantic Ridge and Kane Transform (23°40'N): A new perspective from side-scan sonar image analysis, Ph.D. dissertation, Duke Univ., Durham, N.C., 250 pp., 1997.

Gao, D., S.D. Hurst, J.A. Karson, J.R. Delaney, and F.N. Spiess, Computer-aided interpretation of side-looking sonar images from the eastern intersection of the Mid-Atlantic Ridge with the Kane transform, *J. Geophys. Res.,* in press, 1998.

Gillis, K., Controls on hydrothermal alteration in a section of fast-spreading oceanic crust, *Earth Planet. Sci. Lett., 134,* 473-489, 1995.

Gillis, K., et al., Proc. ODP, Initial Rept., 147, 366 pp., Ocean Drilling Program, College Station, TX, 1993a.

Gillis, K., G. Thompson and D.S. Kelley, A view of the lower crustal component of hydrothermal systems at the Mid-Atlantic Ridge, *J. Geophys., Res., 98,* 19,595-19,619, 1993b.

Girardeau, J. and J. Francheteau, Plagioclase-wehrlites and peridotites on the East Pacific Rise (Hess Deep) and Mid-Atlantic Ridge (DSDP Site 334): Evidence for magma percolation in the oceanic upper mantle, *Earth Planet. Sci. Lett., 115,* 137-149, 1993.

Goff, J.A., A global and regional analysis of near-ridge abyssal hill morphology, *J. Geophys. Res., 96,* 21,713-21,737, 1991.

Goff, J.A., A. Malinverno, D.J. Fornari, and J.R. Cochran, Abyssal hill segmentation: Quantitative analysis of the East Pacific Rise flanks 7°S-9°S, *J. Geophys. Res., 98,* 13,851-13,862, 1993.

Gràcia, E, D. Bideau, R. Hekinian, Y. Lagabrielle, and L. Parson, Along-axis magmatic oscillations and exposure of ultramafic rocks in a second-order segment of the Mid-Atlantic Ridge (33°43'N to 34°07'N), *Geology, 25,* 1059-1062, 1997.

Greenbaum, D., Magmatic processes at ocean ridges: Evidence from the Troodos Massif, Cyprus, *Nature, 238,* 18-21, 1972.

Hancock, P. L., and A.A. Barka, Kinematic indicators on active normal faults in western Turkey, *J. Struct. Geol., 9,* 573-584, 1987.

Harding, A.J., G.M. Kent, and J.A. Orcutt, A multichannel seismic investigation of upper crustal structure at 9°N on the East Pacific Rise: Implications for crustal accretion, *J. Geophys. Res.,98,* 13,925-13,944, 1993.

Harding, A.J., J.A. Orcutt, M.E. Cappus, E.E. Vera, J.C. Mutter, P. Buhl, R.S. Detrick, and T.M. Brocher, Structure of young oceanic crust at 13°N on the East Pacific Rise from expanding spread profiles, *J. Geophys. Res., 94,* 12,163-12,196, 1989.

Harkins, M.E., H.W. Green, II, and E.M. Moores, Multiple intrusive events documented from the Vourinos ophiolite complex, northern Greece, *Am. J. Sci., 280-A,* 284-295, 1980.

Haymon, R.M, D.J. Fornari, M.H. Edwards, S. Carbotte, D. Wright, and K.C. Macdonald, Hydrothermal vent distribution along the East Pacific Rise crest (9°09' - 54'N) and its

relationship to magmatic and tectonic processes on fast-spreading mid-ocean ridges, *Earth Planet. Sci. Lett., 104,* 513-534 1991.

Hébert, R., D. Bideau, and R. Hekinian, Ultramafic and mafic rocks from the Garret Transform Fault near 13°30'S on the East Pacific Rise: igneous petrology, *Earth Planet. Sci. Lett., 65,* 107-125, 1983.

Heezen, B.C., The rift in the ocean floor, *Scientific American, 203,* 98-110, 1960.

Heezen, B.C., M. Tharp, and M. Ewing, The floors of the oceans. 1. The North Atlantic, *Geol. Soc. Am., Spec. Pap. 65,* 122 pp., 1959.

Hekinian, R., D. Bideau, J. Francheteau, J.L. Cheminee, R. Armijo, Lonsdale, and N. Blum, Petrology of the East Pacific Rise crust and upper mantle exposed in Hess Deep (Eastern Equatorial Pacific), *J. Geophys. Res.,* 98, 8069-8094, 1993.

Henstock, T.J., A.W. Woods and R.S. White, The accretion of oceanic crust by episodic sill intrusion, *J. Geophys. Res.,* 98, 4143-4161, 1993.

Hess, H.H., History of ocean basins, in *Petrologic Studies: A Volume in Honor of A.F. Buddington,* edited by A.E.J. Engel *et al., Geol. Soc. Am. Mem.,* 599-620, 1962.

Hodges, K., S. Bowring, K. Davidek, D. Hawkins, and M. Krol, Evidence for rapid displacement on Himalayan normal faults and the importance of tectonic denudation in the evolution of mountain ranges, *Geology, 26,* 483-487, 1998.

Hooft, E.E., M.C. Kleinrock, and C. Ruppel, Rifting of oceanic crust at Endeavor Deep on the Juan Fernandez microplate, *Mar. Geophys. Res., 17,* 251-273, 1995.

Hooft, E.E., H. Schouten, and R.S. Detrick, Constraining crustal emplacement processes from the variation in seismic layer 2A thickness a the East Pacific Rise, *Earth Planet. Sci. Lett., 142,* 289-310, 1996.

Houtz, R. and J. Ewing, Upper crustal structure as a function of plate age, *J. Geophys. Res., 81,* 2490-2498, 1976.

Hurst, S.D., J.S. Gee, and R.M. Lawrence, Reorientation of structural features at Sites 920 to 924 using remanent magnetization and magnetic characteristics, edited by J.A. Karson et al., *Proc. Ocean Drill. Program Scientific Results,* 153, 547-559, 1997.

Hurst, S.D. and J.A. Karson, Cross sections of the Hess Deep Rift: A critical analysis, *Eos, Trans. AGU, 72,* 488, 1991.

Hurst, S.D., J.A. Karson, and K.L. Verosub, Paleomagnetic study of tilted diabase dikes in fast-spread oceanic crust exposed at Hess Deep, *Tectonics 13,* 789-802, 1994a.

Hurst, S.D., E.M. Moores, and R.J. Varga, Structure and geophysical expression of the Solea graben, Troodos ophiolite, Cyprus, *Tectonics, 13,* 139-156, 1994b.

Jaroslow, G.E., G. Hirth, and H.J.B. Dick, Abyssal peridotite mylonites: Implications for grain-size sensitive flow and strain localization in the oceanic lithosphere, *Tectonophys., 256,* 17-37, 1996.

Juteau, T., O. Bideau, G. Dauteuil, G. Manc'h, D.D. Naidoo, P. Nehlig, H. Ondreas, M.A. Tivey, K.X. Whipple, and J.R. Delaney, A submersible study in the western Blanco Fracture Zone, N.E. Pacific: Lithostratigraphy, magnetic structure and magmatic and tectonic evolution during the last 1.6 Ma, *Mar. Geophys. Res., 17,* 399-430, 1995.

Juteau, T., M. Cannat, and Y. Lagabrielle, Serpentinized peridotites in the upper oceanic crust away from transform zones: a comparison of the results of previous DSDP and

ODP legs, edited by R.S. Detrick et al., *Proc. Ocean Drill. Program Scientific Results, Part B, 106/109*, 303-308, 1988.

Kappel, E.S. and W.B.F. Ryan, Volcanic episodicity and a non-steady state rift valley along north east Pacific spreading centers: evidence from Sea MARC I, *J. Geophys. Res., 91*, 13,925-13,940, 1986.

Karson, J.A., Seafloor Spreading on the Mid-Atlantic Ridge: Implications for the Structure of Ophiolites and Oceanic Lithosphere Produced in Slow-Spreading Environments, in *Proceedings of the Symposium TROODOS 1987*, edited by J. Malpas, E.M. Moores, A. Panayiotou, and C. Xenophontos, pp. 547-555, Geol. Surv. Dept., Nicosia, Cyprus, 1990.

Karson, J.A., Geological investigation of a lineated massif a the Kane Transform: Implications for oceanic core complexes, *Phil. Trans. R. Soc. Lond. A*, in press, 1998.

Karson, J.A. and J.F. Dewey, Coastal Complex, western Newfoundland: An Early Ordovician oceanic fracture zone, *Geol. Soc. Am. Bull., 89*, 1037-1049, 1978.

Karson, J.A. and H.J.B. Dick, Tectonics of ridge-transform inter-sections at the Kane Fracture Zone, *Mar. Geophys. Res., 6*, 51-98, 1983.

Karson, J.A. and D. Elthon, Evidence for variations in magma production along spreading centers: a critical appraisal, *Geology, 15*, 127-131, 1987.

Karson, J.A. and P. J. Fox, P.J., Geological and geophysical investigation of the Mid-Cayman Spreading Center: Seismic velocity measurements and implications for the constitution of Layer 3, *Geophys. J. R. Astr. Soc, 85*, 389-412, 1986.

Karson, J.A., S.D. Hurst, S.D., R.M. Lawrence, and SMARK Cruise Participants, Upper crustal construction and faulting at a segment-scale half graben on the Mid-Atlantic Ridge at 22°30'N (SMARK Area), *Eos, Trans. AGU, 77* (17), Spring Meet. Suppl., S271, 1996.

Karson, J.A., S.D. Hurst, and P. Lonsdale, Tectonic rotations of dikes in fast-spread oceanic crust exposed near Hess Deep, *Geology, 20*, 685-688, 1992.

Karson, J.A. and P.A. Rona, Block-tilting, transfer faults and structural control of magmatic and hydrothermal processes in the TAG Area, Mid-Atlantic Ridge 26°N, *Geol. Soc. Am. Bull., 102*, 1635-1645, 1990.

Karson, J.A. and R.M. Lawrence, Tectonic setting of serpentinite exposures on the western median valley wall of the MARK Area in the vicinity of Site 920, edited by J.A. Karson et al., *Proc. Ocean Drill. Program Scientific Results, 153*, 5-22, 1997b.

Karson, J.A. and R.M. Lawrence, Tectonic window into gabbroic rocks of the middle oceanic crust in the MARK area near Sites 921-924, 61-76, 1997a.

Karson, J.A., G. Thompson, S.E. Humphris, J.M. Edmond, W.B. Bryan, J.R. Brown, A.T. Winters, R.A. Pockalny, J.F. Casey, A.C. Campbell, G. Klinkhammer, M.R. Palmer, R.J. Kinzler, and M.M. Sulanowska, , Along-axis variations in seafloor spreading in the MARK Area, *Nature, 328*, 681-685, 1987.

Karson, J.A. and A.T. Winters, Along-Axis Variations in Tectonic Extension and Accommodation Zones in the MARK Area, Mid-Atlantic Ridge 23°N Latitude, in *Ophiolites and Their Modern Oceanic Analogues*, edited by L.M. Parson,

B.J. Murton and P. Browning, *Geol. Soc. Lond. Spec. Publ. 60*, 107-116, 1992.

Kashintsev, G. L., M. I. Kuz'min, and E. N. Popolitov, Composition and structure of the oceanic crust in the vicinity of the Hess Basin (Pacific Ocean), *Geotect., 16*, 512-520, 1982.

Kelemen, P.B., K. Koga, and N. Shimizu, Geochemistry of gabbro sills in the crust/mantle transition zone of the Oman ophiolite: implications for the origin of the oceanic lower crust, *Earth Planet. Sci. Lett.*, 475-488, 1997.

Kelley, D.S. and J.R. Delaney, Two-phase separation and fracturing in mid-ocean ridge gabbros at temperatures greater than 700°C, *Earth Planet. Sci. Lett., 146, 83*, 53-66, 1987.

Kidd, R.B., R.C. Searle, A.T.S. Ramsay, H. Prichard, and J. Mitchell, The geology and formation of King's Trough, northeast Atlantic Ocean, *Mar. Geol., 48*, 1-30, 1982.

Kleinrock, M.C., et al., Detailed structure and morphology of the TAG active hydrothermal mound and its geotectonic environment, edited by S.E. Humphris, et al., Proc. Oc. Drill. Program, Initial Report, 158, 15-21, 1996.

Klitgord, K., J.F. Casey, S. Agar and Cruise Participants, Cruise Report of Russian *Mir* dives at Kings Trough, (unpublished report) 1990.

Kong, L.S., R.S. Detrick, P.J. Fox, L.A. Mayer, and W.B.F. Ryan, The morphology and tectonics of the MARK Area from Sea Beam and Sea MARK I observations (Mid-Atlantic Ridge 23° N), *Mar. Geophys. Res., 10*, 59-90, 1988.

Kong, L.S., S.C. Solomon, and G.M. Purdy, Microearthquake characteristics of a mid-ocean ridge along-axis high, *J. Geophys. Res., 97*, 1659-1685, 1992.

Lagabrielle, Y., and J-M. Auzende, Active *in situ* disaggregation of oceanic crust and mantle on Gorringe Bank: an analogy with ophiolitic massifs, *Nature, 297*, 490-493, 1982.

Lagabrielle, Y., D. Bideau, M. Cannat, J.A. Karson, and C. Mével, Ultramafic-mafic plutonic rock suites exposed along the Mid-Atlantic Ridge (10°N-30°N): Symmetrical-asymmetrical distribution and implications for seafloor spreading processes, in *Faulting and Magmatism at Mid-Ocean Ridges, Geophys. Monogr. Ser.*, edited by W.R. Buck et al., AGU, Washington, D.C., this volume, 1998.

Lagabrielle, Y. and M. Cannat, Alpine Jurassic ophiolites resemble the modern central Atlantic basement, *Geology, 18*, 319-322, 1990.

Lawrence, R.M., Paleomagnetic studies of the geometry and kinematics of faulting at the slow-spreading Mid-Atlantic Ridge (MARK Area), Ph.D. dissertation, Duke Univ., Durham, N.C., 292 pp., 1998.

Lawrence, R.M., J.A. Karson, and S.D. Hurst, Dike orientations and fault-block rotations in slow-spread oceanic crust at the SMARK Area, Mid-Atlantic Ridge at 22°40'N, *J. Geophys. Res., 103*, 663-676, 1998.

Lawson, K., R.C. Searle, J.A. Pearce, P. Browning, and P. Kempton, Detailed volcanic geology of the MARNOK area, Mid-Atlantic Ridge north of Kane transform, in *Tectonic, Magmatic, Hydrothermal, and Biological Segmentation of Mid-Ocean Ridges*, edited by C.J. MacLeod, C.J. Tyler, and C.L. Walker, *Geol. Soc. Lond., Spec. Publ. 118*, 61-102, 1996.

Lemoine, M., P. Tricart, and G. Boillot, Ultramafic and gabbroic ocean floor of the Ligurian Tethys (Alps, Corsica,

Apennines): In search of a genetic model, *Geology, 15*, 622-625, 1987.

Lewis, B.T.R., The process of formation of ocean crust, *Science, 220*, 151-157, 1983.

Lichtman, G.S. and J.-P. Eissen, Time and space constraints on the evolution of medium -rate spreading centers, *Geology, 11*, 592-595, 1983.

Lin, J., G.M. Purdy, H. Schouten, J.C. Sempéré, and C. Zervais, Evidence from gravity data for focused magmatic accretion along the Mid-Atlantic Ridge, *Nature, 334*, 627-632, 1990.

Lister, G.S. and G.A. Davis, The origin of metamorphic core complexes and detachment faults formed during Tertiary continental extension in the northern Colorado River region, U.S.A., *J. Struct. Geol., 11*, 65-94, 1989.

Lonsdale, P., Structural pattern of the Galapagos microplate and evolution of the Galapagos triple junctions, *J. Geophys. Res., 93*, 13,551-13,574, 1988.

Lotti, B., Paragone fra la roccie ofiolitiche terziare italiane e la roccie basiche pare terziarie della Scozia et dell'Irlande, *Boll. R. Comitato Geol. Italia, 76*, 1886.

Louden, K.E., Osler, J.C., Srivastava, S.P., and Keen, C.E., Formation of oceanic crust at slow spreading rates: New constraints from an extinct spreading center in the Labrador Sea, *Geology, 24*, 771-774, 1996.

Luyendyk, B.P. and K.C. Macdonald, Physiography and structure of the FAMOUS rift valley inner floor observed with a deeply-towed instrument package, *Geol. Soc. Am. Bull., 88*, 648-663, 1977.

Luyendyk, B.P. and K.C. Macdonald, A geological transect across the crest of the East Pacific Rise at 21°N latitude made from the deep submersible *Alvin, Mar. Geophys. Res., 7*, 467-488, 1985.

Macdonald, K.C., Mid-ocean ridges: fine scale tectonic, volcanic and hydrothermal processes within the plate boundary zone, *Ann. Rev. Earth Planet. Sci., 10*, 155-190, 1982.

Macdonald, K.C., Linkages between faulting, volcanism, hydrothermal activity and segmentation on fast spreading centers, in *Faulting and Magmatism at Mid-Ocean Ridges, Geophys. Monogr. Ser.*, edited by W.R. Buck et al., AGU, Washington, D.C., this volume, 1998.

Macdonald, K.C. and P.J. Fox, The axial summit graben and cross-sectional shape of the East Pacific Rise as indicators of axial magma chambers and recent volcanic eruptions, *Earth Planet. Sci. Lett., 88*, 119-131, 1988.

Macdonald, K.C., P.J. Fox, R.T. Alexander, R. Pockalny, and P. Gente, Volcanic growth faults and the origin of Pacific abyssal hills, *Nature, 380*, 125-129, 1996.

Macdonald, K.C. and B.P. Luyendyk, Deep-tow studies of the structure of the Mid-Atlantic Ridge crest near 37° N (FAMOUS), *Geol. Soc. Am. Bull., 88*, 621-636, 1977.

Macdonald, K.C. and B.P. Luyendyk, Investigation of faulting and abyssal hill formation on the flanks of the East Pacific Rise (21°N) using *Alvin, Mar. Geophys. Res., 7*, 515-535, 1985.

MacLeod, C.J., F. Boudier, G. Yaouancq, and C. Richter, Gabbro fabrics from Site 894, Hess Deep: Implications for magma chamber processes at the East Pacific Rise, edited by C. Mével et al., *Proc. Ocean Drill. Program Scientific Results, 147*, 317-329, 1996a.

MacLeod, C.J., B. Célérier, G.L. Früh-Green, and C.E. Manning, Tectonics of Hess Deep: a synthesis of drilling results from Leg 147, edited by C. Mével et al., *Proc. Ocean Drill. Program Scientific Results, 147*, 461-495, 1996b.

MacLeod, C.J. and D.A. Rothery, Ridge axial segmentation in the Oman ophiolite: evidence from along-strike variations in the sheeted dyke complex, *in Ophiolites and their Modern Oceanic Analogues*, edited by Parson, L.M., B.J. Murton, and P. Browning, 39-64, *Geol. Soc. Lond., Spec. Publ., 60*, 1992.

Malcolm, F.L., Microstructures of the Cayman Trough gabbros, *J. Geol., 89*, 675-688, 1981.

Mamaloukas-Frangoulis, V., J-M. Auzende, D. Bideau, E. Bonatti, M. Cannat, J. Honnorez, Y. Lagabrielle, J. Malavieille, C. Mével, and H.D. Needham, In-situ study of the eastern ridge-transform intersection of the Vema Fracture Zone, *Tectonophys., 190*, 55-71, 1991.

Marks, N.S., Sedimentation on new ocean crust: The Mid-Atlantic Ridge at 37°N, *Mar. Geol., 43*, 65-82, 1981.

McCarthy, J., J.C. Mutter, J.L. Morton, N.H. Sleep, and G.A. Thompson, Relict magma chamber structures preserved within the Mesozoic North Atlantic crust?, *Geol. Soc. Am. Bull., 100*, 1423-1436, 1988.

Melson, W.G., S.R. Hart, and G. Thompson, St. Paul's Rocks, Equatorial Atlantic: Petrogenesis, radiometric ages, and implications for sea-floor spreading, in *Studies in Earth and Space Sciences, The Hess Volume*, edited by R. Shagan, *Geol. Soc. Am. Mem. 132*, 241-272, 1972.

Mével, C., M. Cannat, P. Gente, E. Marion, J.M. Auzende, and J.A. Karson, Emplacement of Deep Crustal and Mantle Rocks on the West Median Valley Wall of the MARK Area (M.A.R. 23°N), *Tectonophys., 190*, 31-53, 1991.

Michael, P.J. and E. Bonatti, Peridotite compositions from the North Atlantic: Regional and tectonic variations and implications for partial melting, *Earth Planet. Sci. Lett., 73*, 91-104, 1985.

Miyashiro, A., F. Shido, and M. Ewing, Composition and origin of serpentinites from the Mid-Atlantic Ridge near 24° and 30° north latitude, *Contrib. Min. Petrol., 23*, 117-127, 1969.

Moores, E.M., Origin and emplacement of ophiolites, *Rev. Geophys. Space Physics, 20*, 735-760, 1982.

Moores, E.M. and F.J. Vine, The Troodos massif, Cyprus, and other ophiolites as oceanic crust: evaluation and implications, *Phil. Trans. R. Soc. Lond. A, 268*, 443-466, 1971.

Morris, E., R.S. Detrick, T.A. Minshull, J.C. Mutter, R.S. White, W. Su, and P. Buhl, Seismic structure of oceanic crust in the western North Atlantic, *J. Geophys, Res., 98*, 13,879-13,903, 1993.

Murton, B.J., and L.M. Parson, Segmentation, volcanism and deformation of oblique spreading centres: A quantitative study of the Reykjanes Ridge, *Tectonophys., 222*, 237-257, 1993.

Mutter, J.C. and J.A. Karson, Structural processes at slow-spreading ridges, *Science, 257*, 627-634, 1992.

Naidoo, D.D., Magmatic accretion of the upper oceanic crust, Ph.D. dissertation, Univ. of Washington, Seattle, WA, 567 pp., 1998.

Natland, J.H. and H.J.B. Dick, Melt migration through high-level gabbro cumulates of the East Pacific Rise at Hess

Deep: the origin of magma lenses and the deep crustal structure of fast-spreading ridges, edited by C. Mével et al., *Proc. Ocean Drill. Program Scientific Results, 147,* 21-58, 1996.

Nicolas, A., Structure of Ophiolites and Dynamics of Oceanic Lithosphere, 367 pp., Kluwer Academic Press, Dordrecht, The Netherlands, 1989.

Nicolas, A., and F. Boudier, Rooting of the sheeted dike complex in the Oman Ophiolite, in *Ophiolites and Their Modern Oceanic Analogues,* edited by L.M. Parson, B.J. Murton and P. Browning, *Geol. Soc. Lond., Spec. Publ. 60,* 39-54, 1992.

Nicolas, A., I. Reuber, and K. Benn, A new magma chamber model based on structural studies in the Oman ophiolite, *Tectonophys, 151,* 87-105, 1988.

Nisbet, E.G., and C.M. Fowler, The Mid-Atlantic Ridge at 37° and 45°N: Some geophysical and petrological constraints, *Geophys. J. R. Astron. Soc., 54,* 631-660, 1978.

Oceanographer Transform Tectonic Exploration and Research (OTTER) Team, Geology of the Oceanographer Fracture Zone: The Ridge-Transform Intersection, *Mar. Geophys. Res., 6,* 109-142, 1984.

Pallister, J.S., Structure of the sheeted dike complex of the Samail ophiolite near Ibra, Oman, *J. Geophys. Res., 86,* 2661-2672, 1981.

Pallister, J.S. and C.A. Hopson, Samail ophiolite plutonic suite: field relations, phase variation, cryptic variations and layering and a model of a spreading ridge magma chamber: *J. Geophys. Res., 86,* 2593-2644, 1981.

Pariso, J.E. and H.P. Johnson, Magnetic properties and oxide petrography of the sheeted dike complex in Hole 504B, edited by K. Becker et al., *Proc. Ocean Drill. Prog. Sci. Results, 111,* 159-166, 1989.

Pedersen, R.B., The nature and significance of magma chamber margins in ophiolites: examples form the Norwegian Caledonides, *Earth Planet. Sci. Lett., 77,* 100-112, 1986.

Pedersen, R.B., J. Malpas, and T. Falloon, Petrology and geochemistry of gabbroic and related rocks from Site 894, Hess Deep, editeds by C. Mével et al., *Proc. Ocean Drill. Prog. Sci. Results, 147,* 3-20, 1996.

Penrose Conference Participants, Penrose field conference on ophiolites, *Geotimes, 17,* 24-25, 1972.

Perfit, M.R., and W.W. Chadwick, Jr., Magmatism at mid-ocean ridges: Constraints from volcanological and geochemical investigations, in *Faulting and Magmatism at Mid-Ocean Ridges, Geophys. Monogr. Ser.,* edited by W.R. Buck et al., AGU, Washington, D.C., this volume, 1998.

Perfit, M.R., D.J. Fornari, M.C. Smith, J.F. Bender, C.H. Langmuir and R.M. Haymon, Small-scale spatial and temporal variations in mid-ocean ridge crest magmatic processes, *Geology, 22,* 375-379, 1994.

Peterson, J.J., P.J. Fox, and E. Schreiber, Newfoundland ophiolites and the geology of the oceanic layer, *Nature, 247,* 194-196, 1974.

Phipps Morgan, J. and Chen, Y.J., The genesis of oceanic crust: Magma injection, hydrothermal circulation, and crustal flow, *J. Geophys. Res., 98,* 6283-6298, 1993.

Purdy, G.M., New observations of the shallow seismic structure of young oceanic crust, *J. Geophys. Res., 92,* 9351-9362, 1987.

Purdy, G.M. and R.S. Detrick, The Crustal Structure of the Mid-Atlantic Ridge at 23° N from Seismic Refraction Studies. *J. Geophys. Res., 91,* 3739-3762, 1986.

Purdy, G.M. and P.D. Rabinowitz, The Kane fracture zone at the Mid-Atlantic Ridge: IPOD Sites 395 and 396, edited by W.G. Melson et al., *Initial Reports Deep Sea Drill. Project, 45* (map supplement), 1978.

Quick, J.E. and R.P. Denlinger, Ductile deformation and the origin of layered gabbro in ophiolites, *J. Geophys. Res., 98,* 14,015-14,027, 1993.

Quon, S.H. and E.G. Ehlers, Rocks of northern part of Mid-Atlantic Ridge, *Geol. Soc. Am. Bull., 74,* 1-8, 1963.

Raitt, R.W., The crustal rock, in *The Sea,* Vol. 3, edited by M.N. Hill, pp. 85-102, Wiley-Interscience, New York, 1963.

Robinson, P.T. and J. Malpas, The Troodos ophiolite of Cyprus: New perspectives on its origin and emplacement, in Oceanic Crustal Analogues, *Proceedings of the Symposium TROODOS 1987,* edited by J. Malpas, E.M. Moores, A. Panayiotou, and C. Xenophontos, 13-26, Geol. Surv. Dept., Nicosia, Cyprus, 1990.

Rosencrantz, E., Formation of uppermost oceanic crust, *Tectonics, 1,* 471-494, 1982.

Rosencrantz, E., The structure of sheeted dikes and associated rocks in North Arm massif, Bay of Islands ophiolite complex, and the intrusive process at oceanic spreading centers, *Can. J. Earth Sci., 20,* 787-801, 1983.

Rothery, D.A., The base of the sheeted dyke complex, Oman Ophiolite: implications for magma chambers at mid-ocean ridge spreading centers, *J. Geol. Soc. Lond., 140,* 287-296, 1983.

Ryan, M.P., Neutral buoyancy and the structure of mid-ocean ridge magma reservoirs, *J. Geophys. Res., 98,* 22,321-22,338, 1993.

Salisbury, M.H. and N.I. Christensen, The seismic velocity structure of a traverse through the Bay of Islands ophiolite complex, Newfoundland, an exposure of oceanic crust and upper mantle, *J. Geophys. Res., 83,* 805-817, 1978.

Scheirer, D.S. and K.C. Macdonald, Variation in cross-sectional area of the axial ridge along the East Pacific Rise, evidence for the magmatic budget of a fast spreading center, *J. Geophys. Res., 98,* 7871-7886, 1993.

Schreiber, E. and P.J. Fox, Density and P-wave velocity of rocks fromt he FAMOUS region and their implications for the structure of the oceanic crust, *Geol. Soc. Am. Bull., 88,* 600-608, 1977.

Schulz, N.J., R.S. Detrick, and S.P. Miller, Two- and three-dimensional inversions of magnetic anomalies in the MARK Area (Mid-Atlantic Ridge 23°N), *Mar. Geophys. Res., 10,* 41-57, 1988.

Searle, R. C. and J. Francheteau, Morphology and tectonics of the Galapagos triple junction, *Mar. Geophys. Res., 8,* 95-129, 1986.

Sempéré, J.-C., J.R. Cochran, and SEIR Scientific Team, The Southeast Indian Ridge between 88°E and 118°E: Variations in crustal accretion at constant spreading rate, *J. Geophys. Res., 102,* 15,489-15,506, 1997.

Sempéré, J.C., J. Lin, H.S. Brown, H. Schouten, and G.M. Purdy, Segmentation and morphotectonic variations along a slow spreading center: The Mid-Atlantic Ridge (24°00'N–30°40'N), *Mar. Geophys. Res., 15,* 153-200, 1993.

Sempéré, J.C., G.M. Purdy, and H. Schouten, Segmentation of

the Mid-Atlantic Ridge between 24° and 30°40'N, *Nature, 344*, 427-431, 1990.

Severinghaus, J.P. and K.C. Macdonald, High inside corners at ridge-transform intersections, *Mar. Geophys. Res., 9*, 353-367, 1988.

Shand, S.J., Rocks of the Mid-Atlantic Ridge, *J. Geol., 57*, 89-92, 1949.

Shaw, P.R., Ridge segmentation, faulting and crustal thickness in the Atlantic Ocean, *Nature, 358*, 490-493, 1992.

Shaw, P.R. and J. Lin, Causes and consequences of variations in faulting style at the Mid-Atlantic Ridge, *J. Geophys. Res., 98*, 21,839-21,852, 1993.

Shih, J.S-F., The nature and origin of fine-scale sea-floor relief, Ph.D. dissertation, MIT/WHOI Joint Program, Cambridge, M.A., 1980.

Sinton, J.M. and R.S. Detrick, Mid-ocean ridge magma chambers, *J. Geophys. Res., 97*, 197-216, 1992.

Sleep, N.H., Formation of oceanic crust: Some thermal constraints, *J. Geophys. Res., 80*, 4037-4042, 1975.

Sleep, N.H. and G.A. Barth, The nature of oceanic lower crust and upper mantle emplaced at low spreading rates, *Tectonophys., 279*, 181-192, 1997.

Smith, D.K. and J. Cann, Building the crust on the Mid-Atlantic Ridge, *Nature, 365*, 707-715, 1993.

Smith, D.K. et al., Mid-Atlantic Ridge volcanism from deep-towed side scan sonar images, 25°-29°N, *J. Volcan. and Geotherm. Res., 67*, 233-262, 1995.

Spencer, J., The role of tectonic denudation in warping and uplift of low-angle normal faults, *Geology, 12*, 95-98, 1984.

Spudich, P. and J. Orcutt, A new look at the oceanic crust, *Rev. Geophys. Space Phys., 18*, 627-645, 1980.

Srivastava, S.P. and C.R. Tapscott, Plate kinematics of the North Atlantic, in *The Geology of North America, The Western North Atlantic Region*, edited by P.R. Vogt and B.E. Tucholke, *Geol. Soc. Am., DNAG Ser., M*, pp. 379-404, Boulder, C.O., 1986.

Steinmann, G., Die ophiolithischen Zonen in den Mediterranen Kettengebirgen, *14th Inter. Geol. Congr., 2*, 637-668, 1927.

Stern, R.J., S.H. Bloomer, F. Martinez, T. Yamazaki, and T.M. Harrison, The composition of back-arc basin lower crust and upper mantle in the Mariana Trough: A first report, *The Island Arc, 5*, 354-372, 1996.

Stroup, J.B. and P.J. Fox, Geologic Investigations of the Cayman Trough: Evidence for thin oceanic crust along the Mid-Cayman Rise: *J. Geol., 89*, 395-420, 1981

Suess, E., *Das Antlitz der Erde*, Tempsky, Wien, 1909.

Tiezzi, L.J. and R.B. Scott, Crystal fractionation in a cumulate gabbro, Mid-Atlantic Ridge, 26°N, *J. Geophys. Res., 85*, 5438-5454, 1980.

Tivey, M.A., Vertical magnetic structure of ocean crust determined from near-bottom magnetic field measurements, *J. Geophys. Res., 101*, 20275-20296, 1996.

Tivey, M.A., A. Takeuchi, et al., A submersible study of the western intersection of the MAR and Kane FZ (WMARK), *Mar. Geophys. Res.*, in press, 1998.

Tolstoy, M., A.J. Harding, and J.A. Orcutt, Crustal thickness on the Mid-Atlantic Ridge– Bull's-eye gravity anomalies and focused accretion, *Science, 262*, 726-729, 1993.

Toomey, D.R., G.M. Purdy, S.C. Solomon, and W.S.D. Wilcock, The three-dimensional seismic velocity structure of the East Pacific Rise, near latitude 9°30'N, *Nature, 347*, 639-645, 1990.

Toomey, D.R., S.C. Solomon, and G.M. Purdy, Microearthquakes beneath the median valley of the Mid-Atlantic Ridge near 23°N: tomography and tectonics: *J. Geophys. Res., 93*, 9093-9112, 1988.

Toomey, D.R., S.C. Solomon, and G.M. Purdy, Tomographic imaging of the shallow crustal structure of the East Pacific rise at 9°30N, *J. Geophys. Res., 99*, 24,135-24,157, 1994.

Toomey, D.R., S.C. Solomon, G.M. Purdy, and M.H. Murray, Micro-earthquakes beneath the median valley of the Mid-Atlantic Ridge near 23° N: hypocenters and focal mechanisms: *J. Geophys. Res., 90*, 5443-5458, 1985.

Tougas, S., Structural analysis of East Pacific Rise Crust exposed in the Hess Deep Rift: Constraints on crustal accretion at fast-spreading ridges, M.S. thesis, Duke Univ., Durham, N.C., 70 pp., 1998.

Tucholke, B.E. and Lin, J., A geological model for the structure of ridge segments in slow-spreading ocean crust, *J. Geophys. Res., 99*, 11,937-11,958, 1994.

Tucholke, B.E., J. Lin and M.C. Kleinrock, Megamullions and mullion structure defining oceanic metamorphic core complexes on the Mid-Atlantic Ridge, *J. Geophys. Res., 103*, 9857-9866, 1998.

Tucholke, B.E., J. Lin, M.C. Kleinrock, M.A. Tivey, T.B. Reed, J. Goff, and G.E. Jaroslow, Segmentation and crustal structure of the western Mid-Atlantic Ridge flank, 25°25'-27°10'N and 0-29 m.y., *J. Geophys. Res., 102*, 10,203-10,224, 1997.

Vanko, D.A., Temperature, pressure and composition of hydrothermal fluids, with their bearing on the magnitude of tectonic uplift at mid-ocean ridges, inferred from fluid inclusions in oceanic layer 3 rocks, *J. Geophys. Res., 93*, 4595-4611, 1988.

Varga, R.J., Modes of extension at mid-ocean ridge spreading centers: evidence from the Solea graben, Troodos ophiolite, Cyprus, *J. Struct. Geol., 13*, 517-538, 1991.

Varga, R.J. and E.M. Moores, Intermittent magmatic spreading and tectonic extension in the Troodos Ophiolite: Implications for exploration for black smoker-type ore deposits, in Oceanic Crustal Analogues, *Proceedings of the Symposium "TROODOS 1987"*, edited by J. Malpas, E.M. Moores, A. Panayiotou, and C. Xenophontos, 53-74, Geol. Surv. Dept., Nicosia, Cyprus, 1990.

Vera, E.E., J.C. Mutter, P. Buhl, J.A. Orcutt, A.J. Harding, M.E. Kappus, R.S. Detrick, and T.M. Brocher, The structure of 0- to 0.2-m.y.-old oceanic crust at 9°N on the East Pacific Rise from expanded spread profiles, *J. Geophys. Res., 95*, 15,529-15,556, 1990.

Vine, F.J. and Offset Drilling Working Group, Ocean Drilling Program, Technical Note 25, College Station, TX, 1995.

Walker, G.P.L., Basaltic-volcano systems, in *Magmatic Processes and Plate Tectonics*, edited by H.M. Prichard, T. Alabaster, N.B.W. Harris, and C.R. Neary, *Geol. Soc. Lond., Spec. Publ. 76*, 3-38, 1993.

Wernicke, B. P. and G.J. Axen, On the role of isostasy in the evolution of normal fault systems, *Geology, 16*, 848-451, 1988.

Westway, R., and N. Kusznir, Fault and bed 'rotation' during continental extension: block rotation or vertical shear? *J. Struct. Geol., 15*, 753-770, 1993.

White, G.W. and J. B. Stroup, Distribution of rock types in the Mid-Cayman Rise, Caribbean Sea, as evidence for conjugate normal faulting in slowly spreading ridges, *Geology*, 7, 32-36, 1979.

White, R.S., R.S. Detrick, J.C. Mutter, P. Buhl, T.A. Minshull, and E. Morris, New seismic images of oceanic crustal structure, *Geology*, 18, 462-465, 1990.

White, R.S., R.S. Detrick, M.C. Sinha, and M.H. Cormier, Anomalous seismic crustal structure of oceanic fracture zones, *Geophys. J. R. Astr. Soc.*, 79, 779-798, 1984.

Wiggins, S.M., L.M. Dorman, B.D. Cornuelle, and J.A. Hildebrand, Hess Deep rift valley structure from seismic tomography, *J. Geophys. Res.*, 101, 22,335-22,354, 1996.

Wilkens, R.H., G.J. Fryer, and J. Karsten, Evolution of porosity and seismic structure of upper oceanic crust, *J. Geophys. Res.*, 96, 17,981-17,996, 1991.

Williams, H. and J. Malpas, Sheeted dikes and brecciated dike rocks within transported igneous complexes, Bay of Islands, western Newfoundland, *Can. J. Earth Sci.*, 9, 1216-1229, 1972.

Wolfe, C.J., G.M. Purdy, D.R. Toomey, and S.C. Solomon, Microearthquake characteristics and crustal velocity structure at 29°N on the Mid-Atlantic Ridge: The architecture of a slow spreading segment, *J. Geophys. Res.*, 100, 24, 449-24,472, 1995.

Woollard, G.P., The interrelationships of crustal and upper mantle parameter values in the Pacific, *J. Geophys. Res.*, 13, 87-137, 1975.

Zonenshain, L.P., L.L. Kogan, L.A. Savostin, A.J. Golmstock, and A.M. Gorodnitskii, Tectonics, crustal structure and evolution of the Galapagos triple junction, *Mar. Geol.*, 37, 209-230, 1980.

Zonenshain, L.P., M.I. Kuzmin, A.P. Lisitsin, Yu.A. Bogdanov, and B.V. Baranov, Tectonics of the Mid-Atlantic Rift Valley between the TAG and MARK Areas (26°-24°N), Evidence for vertical tectonism, *Tectonophys.*, 159, 1-23, 1989.

Jeffrey A. Karson, Division of Earth & Ocean Sciences, Duke University, Box 90229, Durham, NC 27708 (e-mail: jkarson@eos.duke.edu).

Structure of Modern Oceanic Crust and Ophiolites and Implications for Faulting and Magmatism at Oceanic Spreading Centers

Yildirim Dilek

Department of Geology, Miami University, Oxford, Ohio, USA

Eldridge M. Moores

Department of Geology, University of California, Davis, California, USA

Harald Furnes

Geologisk Institutt, Universitetet i Bergen, Bergen, Norway

A review of the internal structure of the upper and lower crust in modern oceanic lithosphere and in well-preserved ophiolites leads to some conclusions on the nature of interactions between magmatism and faulting during the construction of oceanic lithosphere at spreading centers. Sheeted dike complexes are made of subparallel vertical intrusions of magma parallel to the axial plane of an oceanic spreading center, and they display structures whose nature are strongly controlled by the mode of and interplay between magmatism and faulting at different seafloor spreading rates. Drilled core samples from the slow-spreading MARK area and Site 735B (Southwest Indian Ridge) record a complex history of solid-state deformation and attendant alteration of the lower crust at temperatures in excess of 700°C and continuing down to 180°C. Ductile shear zones, brittle faults, and detachment surfaces observed in the core samples and on the seafloor consistently indicate normal sense of shearing associated with tectonic extension and crustal stretching. Gabbroic rocks and serpentinized peridotites are exposed on the seafloor in the MARK area, suggesting that upper crustal units have been stripped away due to amagmatic extension. The sheeted dike complex of the intermediate-spreading oceanic crust along the Costa Rica Rift displays intense microfracturing at discrete depth intervals, as observed in ODP cores, that are possibly associated with faults or localized deformation zones. The seismic layer 2/3 boundary occurs within the sheeted dike complex and corresponds to changes in physical properties over a depth interval, rather than to the presence of a lithological change from dikes to gabbros. The dike-gabbro boundary is probably tectonic, corresponding to the fault zone drilled into in the borehole and coincides with one of the half-graben bounding and gently dipping normal faults depicted on single channel seismic reflection profiles. The fast-spreading oceanic crust drilled in the Hess Deep area in the eastern equatorial Pacific Ocean is relatively undeformed with well-preserved igneous contacts between the sheeted dike complex and the gabbros and a

Faulting and Magmatism at Mid-Ocean Ridges
Geophysical Monograph 106
Copyright 1998 by the American Geophysical Union

transition zone between the lower crustal and mantle sequences. Gabbroic core samples mainly display magmatic flow structures (subvertical foliation and associated lineation defined by anisotropic plagioclase crystals), rather than solid-state ductile deformation features. Widespread fracturing of the crust was associated with subsolidus cooling and thermal contraction after axial magma emplacement within the crustal accretion zone.

The sheeted dike complexes in the Troodos (Cyprus) and Kizildag (Turkey) ophiolites, both inferred to be of slow-spreading origin, show numerous planar to listric normal faults and structural grabens suggestive of tectonic extension, and complex intrusive relations in their plutonic units indicative of recurring and intermittent magmatic activities. Detachment surfaces within the lower crustal sequence (Troodos) or at the boundary between the crustal and mantle sequences (Kizildag) define a brittle-plastic transition zone along which the upper crustal units above was accommodated by mylonitization and denudation in lower crustal units and mantle rocks below. Sheeted dike complexes in the Semail (Oman) and Solund-Stavfjord (Norway) ophiolites, interpreted to be of fast- to intermediate-spreading origin (respectively), have steeply dipping dike intrusions and show minor dike-parallel and dike-perpendicular faults. Sheeted dikes in these ophiolites root into the underlying plutonic sequences and locally display mutually intrusive relations with differentiated plutonic rocks. These relations together with the relatively undeformed nature of the remnant oceanic crust imply robust magmatism that kept pace with seafloor spreading and associated extension during the evolution of these ophiolites. The internal structure of the sheeted dike complexes and the nature of dike-gabbro boundary in oceanic crust are affected by the vertical temperature gradient in the oceanic lithosphere and the magma supply, both of which are associated with the presence (or absence) of an active magma lens at shallow crustal depths during seafloor spreading. Limited magma supply in slow-spreading "cold" ridge systems facilitates amagmatic extension via development of planar to listric normal faults, causing block rotations in sheeted dikes and detachment surfaces and shear zones at the dike-gabbro boundaries and within the upper plutonic sequences. Extensive fracturing and faulting of upper and lower crust during periods of amagmatic extension facilitates vigorous hydrothermal circulation and hence results in alteration of the crust and serpentinization of the upper mantle. Denudation of crust and uplift of the serpentinized peridotites are a result of tectonic extension along detachment surfaces at segment ends. Subsequent replenishment of magma supply results in emplacement of dike swarms through extended crust and of shallow-level plutons, and in extrusion of lavas on unroofed plutons and upper mantle rocks. Robust magmatism at intermediate- to fast-spreading centers is a product of a melt lens perched in shallow crustal depths, creating a "hot" environment in which little magma freezing occurs, and generates relatively undeformed oceanic crust without extensive faulting and crustal thinning.

1. INTRODUCTION

Three fifths of the surface of the solid earth is made of oceanic lithosphere, all of which has been formed during the last 170 Ma at the mid-ocean ridges. Understanding the structure of the oceanic lithosphere and the mid-ocean ridges is particularly important because it provides a key to understanding the mantle and the kinematics of plate tectonics. Most models dealing with the generation and development of oceanic lithosphere have been based mainly on inferences from seafloor bathymetry, interpretations of seismic reflectors based on continental analogues, mantle flow patterns, and rock samples [e.g., Macdonald and Luyendyk, 1986; Orcutt, 1987; Dick et al., 1991; Smith and Cann, 1993]. From the earliest studies, geophysical investigations and analyses of seismic refraction profiles have suggested that the structure of oceanic crust is astoundingly uncomplicated and uniform [e.g., Hill, 1957; Raitt, 1956, 1963; Christensen and Salisbury, 1975]. Geoscientists have long equated the seismic structure of oceanic lithosphere to a simple "layer cake" sequence consisting of sediments, basaltic lavas, diabasic dikes, and a thick plutonic section overlying the mantle, with the boundary between the igneous crust and mantle corresponding to the Mohorovicic seismic discontinuity (Moho). Accretion of the lower crust was interpreted to have resulted from crystallization of magma from a steady-state magma chamber located over the upwelling mantle beneath a spreading axis. In this interpretation, dike intrusions emanating from the magma chamber form the feeders to the overlying lava extrusions on the seafloor, whereas primitive layered gabbros in the bottom of a magma chamber are overlain by more evolved isotropic gabbros in a simple stratigraphy. The upper crustal rocks composed of lava flows and dikes were equated to seismic layer 2 and lower crustal rocks consisting of isotropic and layered gabbros to seismic layer 3.

Recent geophysical studies and in-situ sampling of the oceanic lithosphere through the deep drilling projects (DSDP) and Ocean Drilling Program (ODP) have provided new information and hypotheses on its architecture and evolution [Kong et al., 1992; Mutter et al., 1985; Toomey et al., 1990; Vera et al., 1990; Sinton and Detrick, 1992; Detrick et al., 1987; Orcutt, 1987; Sempere and Macdonald, 1987; Alt et al., 1993]. The results of these studies show that the architecture of oceanic crust is highly complex both along- and across-axis of a spreading center, and that this inhomogeneous structure cannot be explained adequately with steady-state magma chamber models [Kong et al., 1988; Mutter et al., 1985; Detrick et al., 1987; Orcutt, 1987; Sempere and Macdonald, 1987; Toomey et al., 1988; Vera et al., 1990]. Modern oceanic crust is considerably thinner near large transforms [Fox et al, 1980, Dick, 1989; Mutter and Detrick, 1984] and in ridge systems with spreading rates that are lower than 2 cm/yr [Bown and White, 1994; Reid and Jackson, 1981]. Models of the internal stratigraphy of the lower crust have become more complex with the introduction of narrow accumulation zones and melt lenses, ephemeral magma chambers [e.g., Sinton and Detrick, 1992], and tectonism and hydrothermal alteration in the accretion zone at the slow-spreading rates [e.g., Dick et al., 1992]. In slow-spreading ocean ridges, magma chambers appear to be ephemeral with long intervening periods of amagmatic (tectonic) extension producing major structural disruptions of any simple "magmatic" stratigraphy. Rotated blocks of crust and high- to low-angle normal faults and detachment surfaces are present at the crest of the Mid-Atlantic Ridge [Karson, 1990; Mutter and Karson, 1992], and lower crustal rocks (undeformed to mylonitized gabbros) and serpentinized peridotites are exposed on the rift walls of various mid-ocean ridges suggesting the exhumation of deep crust and mantle sections via tectonic stretching during ocean crust generation (MARK area, Hess Deep, Site 735B at the SW Indian Ridge). Therefore, a very wide variety of structures ranging from simple graben and half-graben to oceanic core complexes with complex fault systems exist along slow-spreading axes. At fast-spreading ridge systems (i.e., East Pacific Rise), however, oceanic lithosphere is much thicker (>15 km; Vera et al., 1990] with no extensive faulting and crustal attenuation, and its internal architecture is more supportive of the seismically-defined layered structure [Phipps Morgan et al., 1994].

Certain limitations exist, however, regarding the tectonic interpretations derived from geophysical investigations, bathymetric surveys, and indirect sampling and discontinuous coring of modern oceanic lithosphere. For example, equivocal origins of crustal reflectors due to the difficulties of tracing them to the seafloor pose problems for interpreting fault geometries and fault kinematics and for better constraining the distribution of deformation in the oceanic lithosphere. Although they have been instrumental in studying the structure and petrology of lower crust and upper mantle, dredged rock samples from seafloor outcrops are problematic because the geological setting and vertical dimensions of the rock bodies from which these samples are derived are poorly constrained. Furthermore, current technology limits penetration depths of drilling and seismic resolution in young oceanic lithosphere. Ophiolites, which are structural analogues for oceanic lithosphere, provide 3-dimensional exposures and age relations to study the nature of extensional deformation and magmatic construction of crust at spreading environments. Studies of ophiolites complement significantly our knowledge of the architecture of oceanic crust derived from both seismic images and drill holes at modern spreading centers. Intact ophiolites without polyphase structural fabrics associated with obduction (emplacement) processes record significant information on the spatial and temporal interplay between structural and magmatic processes and on the role of faulting and brittle failure in development of hydrothermal systems.

However, this analogy is limited because: (1) ophiolites are fossil remnants which might have been produced by processes different than those occurring in the modern oceans, and (2) geochemical and petrological data suggest that most ophiolites might have formed in "suprasubduction zone" settings rather than in mid-oceanic settings [Pearce et al., 1984; Hawkins et al., 1984]. It is important to note, however, that even though many ophiolites may represent oceanic crust formed above subduction zones, modern back-arc basin spreading centers such as the Mariana trough and the Lau basin have recognizable or inferred symmetric magnetic stripes analogous to the mid-ocean ridge axes and those of moderate to fast-spreading rates display seismically defined axial melt bodies [Morton and Sleep, 1987; Collier and Sinha, 1990; Baker et al., 1996]. Therefore, the morphological and spreading rate-dependent characteristics of axial melt bodies and the resultant thermal structure should apply to suprasubduction as well as mid-oceanic spreading centers, although geochemical characteristics may differ.

In this paper, we review the structure and tectonics of oceanic crust as observed directly at modern spreading centers through deep drilling and associated surveys and from well-preserved Tethyan and Caledonian ophiolite complexes that we have examined in order to shed light on the relative roles of magmatism and tectonism in generation of oceanic crust, as well as on the spreading rate dependence of the internal architecture of oceanic lithosphere. The four main parts of the paper include: (1) ophiolite-oceanic crust analogy, and variations in crustal structure of modern oceanic crust; (2) internal architecture of modern oceanic crust at slow-, intermediate-, and fast-spreading centers; (3) internal structure of ancient oceanic crust as observed in ophiolites; and, (4) structural processes at oceanic spreading centers and spatial and temporal relations between faulting and magmatism.

2. OPHIOLITE-OCEAN CRUST ANALOGY

The structure and petrology of oceanic crust have been the subject of systematic studies in marine geology and geophysics during the last fifteen years, and significant advances have been made to understand better the structure and evolution of oceanic crust as a result of these studies and the investigations of ophiolites [e.g., Coleman, 1971; Moores and Vine, 1971; Gass and Smewing, 1973; Macdonald, 1982; Cann et al., 1983; Karson et al., 1987; Sempere and Macdonald, 1987; Becker et al., 1989; Nicolas, 1989; Toomey et al., 1990; Vera et al., 1990; White et al., 1990; Dick et al., 1991; Phipps Morgan et al., 1994; Tucholke and Lin, 1994; Detrick et al., 1995; Cann et al., 1997]. Ophiolites have been of particular importance in the examination of seafloor spreading structures in oceanic crust and in the reconstruction of ancient plate boundaries ever since their recognition as on-land fragments of oceanic lithosphere [e.g., Coleman 1971; Moores and Vine 1971; Moores, 1982]. The concept of an oceanic spreading center origin for ophiolites has been widely accepted based mainly on the presence of sheeted dike complexes, which characterize magmatic extension via seafloor spreading [Coleman 1971; Dewey and Bird 1971; Gass 1968; Moores and Vine 1971; Pallister and Hopson 1981]. Given the difficulty in obtaining an adequate density of spatially oriented samples from modern oceanographic studies, the advantage of studying ophiolites is that detailed, 3-D reconstructions of spreading center processes can be evaluated at a high resolution. Importantly, geologic mapping and the collection of spatially oriented samples can be used to evaluate geometric, mechanical, and kinematic models developed from the seismic and gravity studies of modern ridges [e.g.,

Henstock et al., 1993; Nicolas et al., 1993; Phipps Morgan et al., 1994].

The inferred "layer-cake" structure of oceanic lithosphere is based on the correlation of the seismically determined layering of the oceanic crust with the pseudostratigraphy found in ophiolites (Fig. 1). Recent observations and information from both modern oceanic crust and ophiolites indicate, however, that there are significant deviations from this inferred layered structure and the typical pseudostratigraphy (respectively), particularly when the slow-spreading oceanic crust and the ophiolites of an inferred slow-spreading center origin are considered. In a typical ophiolite pseudostratigraphy, the extrusive sequence and the underlying sheeted dike complex collectively make up the upper oceanic crust and correspond to seismic layers 2A and 2B, respectively (Fig. 1). Plutonic rocks, consisting of cumulate to isotropic gabbros, diorites, and differentiated trondhjemites, form the lower crust and represent the seismic layer 3 of oceanic lithosphere. Submersible observations at Hess Deep, which is a rifted portion of the fast-spreading East Pacific Rise, have shown that structural thicknesses for the upper crust are comparable to seismic layer thicknesses derived from rise-axis experiments [Francheteau et al., 1992; Phipps Morgan et al., 1994]. Drilling into the 5.9-million year-old intermediate-spreading oceanic crust at DSDP/ODP Hole 504B has penetrated 2.1 km below the seafloor (mbsf) through the extrusive sequence and sheeted dike complex and confirmed earlier models for a layered structure of the uppermost 2 km of oceanic crust [Becker et al., 1989; Alt et al., 1993].

Correlations of seismic velocities of ophiolitic rocks and their structural features with the crustal velocity structure of modern oceanic crust suggest that the ophiolite-ocean crust analogy is in general valid [Christensen and Smewing, 1981; Kempner and Gettrust, 1982], particularly for oceanic crust formed at fast-spreading ridges [Dick, 1996]. Comparison of the internal structure and stratigraphy of the Semail ophiolite with that of the 0 to 0.2 m.y. old oceanic crust at 9°N on the fast-spreading East Pacific Rise (EPR) reveals, for example, remarkable similarities between the two (Fig. 2). The velocity-depth models for expanded spread profiles across the EPR suggest the existence of an axial magma chamber with a total lateral extent of less than about 4 km but with a striking continuity of tens of kilometers along the spreading axis [Detrick et al. 1987; Vera et al. 1990]. The top of the axial magma chamber, consisting of molten material, lies 1.6 km below the seafloor and marks a low velocity zone (Fig. 2). This low velocity zone at the EPR continues for nearly 5 km at depth down to the Moho and corresponds to the 4 to 6 km thick plutonic sequence in the Semail ophiolite. The Moho at the bottom of the magma chamber constitutes a nearly 1 km thick transition zone, rather than a single discontinuity, that is composed of

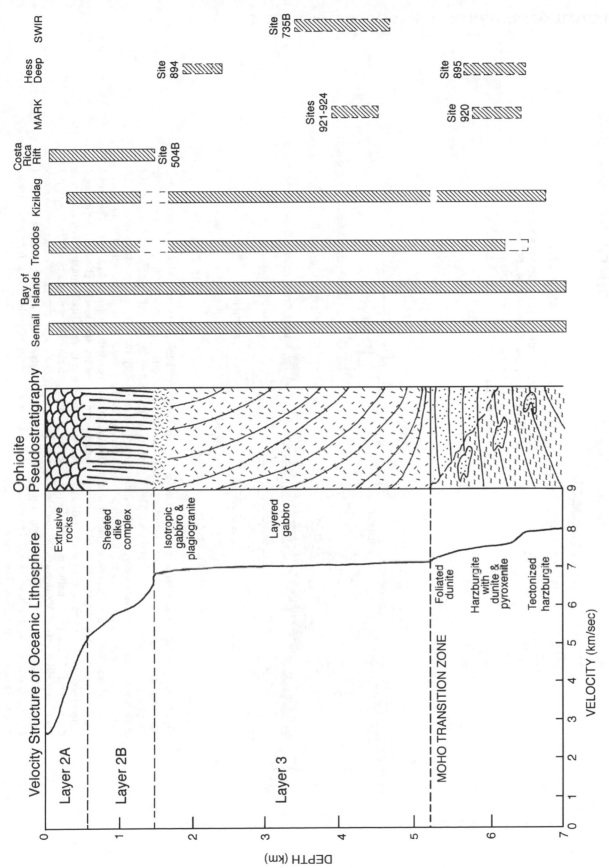

Figure 1. Ophiolite pseudostratigraphy compared to the velocity structure of oceanic lithosphere (based on seismic measurements from the East Pacific Rise; velocity structure is after Phipps Morgan et al., 1994). Columns on the right show the correlation of the exposed stratigraphy of a number of ophiolite complexes and the recovered lithostratigraphic units from several DSDP/ODP drill sites with the typical ophiolite pseudostratigraphy with no reference to crustal unit thicknesses.

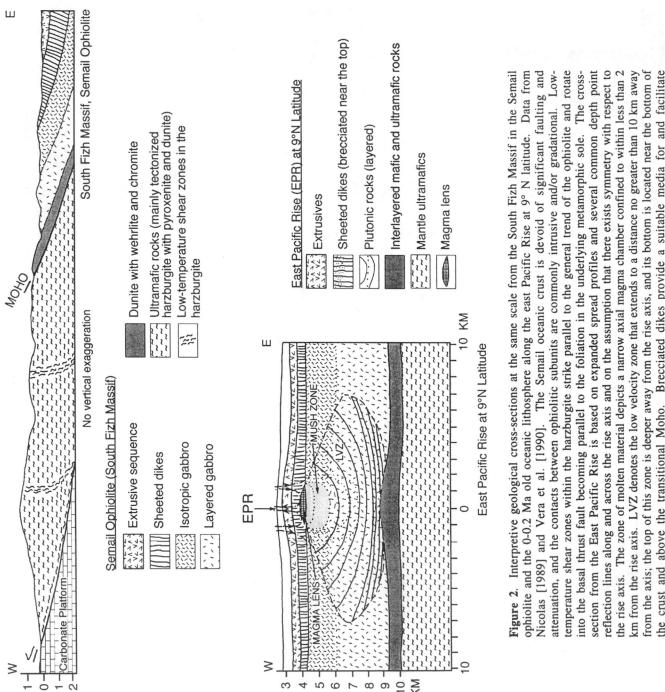

Figure 2. Interpretive geological cross-sections at the same scale from the South Fizh Massif in the Semail ophiolite and the 0-0.2 Ma old oceanic lithosphere along the east Pacific Rise at 9° N latitude. Data from Nicolas [1989] and Vera et al. [1990]. The Semail oceanic crust is devoid of significant faulting and attenuation, and the contacts between ophiolitic subunits are commonly intrusive and/or gradational. Low-temperature shear zones within the harzburgite strike parallel to the general trend of the ophiolite and rotate into the basal thrust fault becoming parallel to the foliation in the underlying metamorphic sole. The cross-section from the East Pacific Rise is based on expanded spread profiles and several common depth point reflection lines along and across the rise axis and on the assumption that there exists symmetry with respect to the rise axis. The zone of molten material depicts a narrow axial magma chamber confined to within less than 2 km from the rise axis. LVZ denotes the low velocity zone that extends to a distance no greater than 10 km away from the axis; the top of this zone is deeper away from the rise axis, and its bottom is located near the bottom of the crust and above the transitional Moho. Brecciated dikes provide a suitable media for and facilitate hydrothermal circulation, which is shown with arrows.

interlayered mafic and ultramafic rocks [Vera et al. 1990], also similar to the Moho transition zone in the Semail ophiolite [Nicolas, 1989].

3. VARIATIONS IN CRUSTAL STRUCTURE OF MODERN OCEANIC CRUST

In contrast to the relatively constant thickness (~6 km on average) and regular internal structure of intermediate- to fast-spreading oceanic crust, slow-spreading oceanic crust along the Mid-Atlantic Ridge shows variations in crustal thickness on the order of 1 to 4 km both along and across axis and is in places more discontinuous and disrupted than "normal" oceanic crust [Purdy and Detrick, 1986; Toomey et al., 1988; Blackman and Forsyth, 1991; Tolstoy et al., 1993; Detrick et al., 1993; Cannat et al., 1995b; Detrick et al., 1995; Gente et al., 1995]. Thin and discontinuous oceanic crust at slow-spreading environments is commonly associated with ridge discontinuities and segment ends as measured in seismic experiments and modeled from residual gravity anomalies. These regions with thinner than normal oceanic crust are also the sites of ultramafic and gabbroic rock outcrops on the seafloor, indicating significant deviations from the inferred layered structure of typical oceanic crust [Karson and Dick, 1983; Karson et al., 1987; Mével et al., 1991; Tucholke and Lin, 1994; Cannat et al., 1995a, 1995b]. The occurrence of these lower crustal and upper mantle rocks on the seafloor is interpreted to have resulted from denudation of upper crust by faulting and tectonic thinning [Karson, 1990; Mével et al., 1991] and/or from the absence of stable melt lens at shallow depths because of the existence of thick axial lithosphere at the ends of ridge segments [Sleep, 1975; Cannat, 1993; Sleep and Barth, 1994].

Multichannel seismic reflection and wide-angle reflection and refraction profiles along and across mid-ocean ridge systems have shown the existence of both subhorizontal and moderately to steeply dipping features in oceanic lithosphere. Subhorizontal reflectors have been interpreted to represent the transitions between the extrusive sequence and the sheeted dike complex (layer 2A/2B boundary), the sheeted dike complex and the plutonic sequence (layer 2/3 boundary), the Moho, and/or the top of magma chambers [Purdy, 1982; McCarthy et al., 1988; Harding et al., 1989; Toomey et al., 1990; Vera et al., 1990; Sinton and Detrick, 1992; Mutter, 1992; Detrick et al., 1994; Morris et al., 1993]. More steeply dipping reflectors within the upper and lower crust have been interpreted as probable faults, some of which appear to pass at depth into low-angle detachment surfaces within layer 3 or along the crust-mantle boundary [White et al., 1990; Mutter and Karson, 1992; Morris et al., 1993]. These features inferred from seismic studies and the observed seafloor bathymetry and morphology have been integrated to decipher fault geometries and kinematics within the framework of dynamic models of seafloor spreading

[Sempere and Macdonald, 1987; Macdonald, 1986; Morris et al., 1993]. The equivocal origins of crustal reflectors and problems with tracing them to the seafloor pose, however, certain limitations and problems for these interpretive models. Limitations also exist regarding the interpretations derived from the studies of dredged samples. Samples of coarse-grained mafic and ultramafic rocks, which are thought to have formed in the middle to lower crust and in the upper mantle beneath spreading axes, have been dredged from seafloor outcrops, but the geological setting and vertical dimensions of the rock bodies from which these samples are derived are poorly constrained. It is widely accepted, therefore, that direct sampling is essential to obtain unambiguous data and information pertaining to the internal structure and composition of the oceanic lithosphere.

4. INTERNAL ARCHITECTURE OF MODERN OCEANIC CRUST

For nearly 28 years the Deep Sea Drilling Project (DSDP) and the Ocean Drilling Program (ODP) have provided vital information and direct sampling opportunity in the third dimension, beneath the seafloor, in examining the modern oceanic crust. Through deep crustal and mantle drilling projects, the complexity and the heterogeneity of the internal structure of oceanic crust, both igneous and tectonic, have become apparent. Examination of core samples recovered from drill sites at and/or near mid-ocean ridge systems with various spreading rates has shown that the mode and nature of igneous accretion, faulting and deformation, and hydrothermal alteration in modern oceanic crust differ significantly as a result of variations in seafloor spreading rates. In this section, we summarize the findings of systematic studies of drilled core samples of oceanic crust from different spreading environments recovered on a number of scientific cruises of DSDP and ODP.

4.1. Slow-spreading Oceanic Crust

The Mid-Atlantic Ridge at the Kane Transform (MARK) area and its immediate environs have been the focus of numerous geophysical and bathymetric surveys and have been sampled extensively by dredging and submersible studies, and two Ocean Drilling Program (ODP) cruises (Legs 109 and 153; see Cannat et al., 1995a, for an extensive bibliography on previous studies in the MARK area). The MARK area is located just south of the Kane Transform on the Mid–Atlantic Ridge at about 23° N latitude and encompasses several segments separated by zero-offset boundary zones (Fig. 3; Purdy and Detrick, 1986]. Tectonic extension has been dominant until a very recent magmatic event in Segment-1 [Karson, 1990], and the axial rift valley here is characterized by a continuous median valley ridge, which narrows and deepens as the ridge-transform intersection (RTI) is approached. The

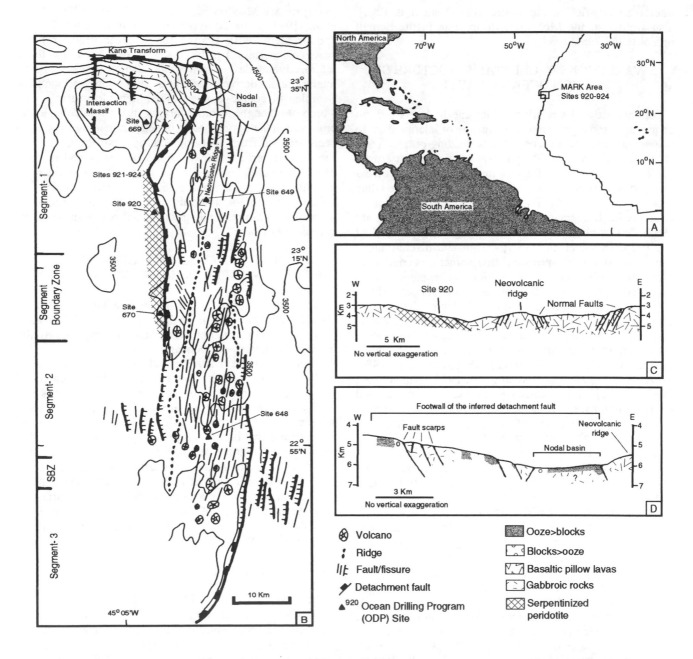

Figure 3. (A) Location of the MARK area at the Mid-Atlantic Ridge near the Kane Fracture Zone at 22°-24°N latitude. (B) Generalized tectonic map of the MARK area, showing the simplified bathymetry, ridge axis segments and boundaries, the neovolcanic ridge, fault scarps and fissures, distribution of gabbro and serpentinized peridotites in the footwall of the detachment fault, and ODP drill sites. Gabbroic outcrops are located in the Ridge-Intersection Massif south of the Kane Transform and the serpentinized peridotites are exposed farther south along the east-dipping detachment surface. Contours are in kilometers below sea level [modified from Cannat, Karson, Miller, et al., 1995a]. (C) Cross section across the MARK area near Site 920, showing the terrain and surface geology [modified from Karson and Lawrence, 1997a]. (D) Cross section across the MARK area through the Nodal Basin at 23°35'N latitude, showing the terrain and surface geology [data are from Karson and Lawrence, 1997b, and the references therein].

median valley is markedly asymmetrical; the western wall is substantially steeper and higher than the eastern wall. Photographic traverses and observations from the DSV *Alvin* and *Nautile* show that the western wall contains a diverse assemblage of variably deformed and metamorphosed basaltic, gabbroic, and ultramafic rocks [Karson and Dick, 1983; Karson et al., 1987; Brown and Karson, 1988; Mével et al.; 1991]. These lithologies occur in high-angle fault blocks in the hanging wall of an undulating low-angle detachment surface or a series of subparallel faults (Figs. 3 and 4).

Serpentinized peridotite crops out for tens of kilometers along the western rift valley south of the Kane Transform, and partially to completely serpentinized harzburgites were recovered during Legs 109 and 153 (Fig. 3; Cannat et al., 1990; Hébert et al., 1990; Juteau et al., 1990; Cannat et al., 1995a). Site 920, drilled nearly 18 km north of Site 670, is located on a gently sloping terrace at the top of a steep cliff, exposing massive to schistose serpentinized peridotite on the western wall of the rift valley. The serpentinized peridotite outcrop extends nearly continuously from about 3500 meters below the sea-level (mbsl) to about 3100 mbsl and is cut by east-facing normal (?) faults. Faults with relatively steep dips (40°-70°) cut a more pervasive, gently east-dipping (20°-40°) schistosity, which is interpreted to be related to earlier, low-angle faults and shear zones [Karson et al., 1987; Mével et al., 1991].

The predominant rock types recovered from Site 920 include serpentinized harzburgite, with lesser amounts of lherzolite, dunite, clinopyroxenite, and websterite along with variably altered olivine gabbro, gabbro, oxide-rich gabbroic rocks, amphibolitized microgabbro, amphibolite, rodingitized gabbro, and plagioclase-olivine phyric diabase [Cannat et al., 1995a]. Porphyroclastic harzburgites display heterogeneous alteration with the most intense alteration (95-100%) occurring in orthopyroxene-poor zones, and they locally contain 2- to 20-cm-wide mylonitic zones. Some mylonitic zones are spatially associated with metagabbroic intervals in the core and contain highly deformed and recrystallized ultramafic and gabbroic material, which underwent extensive grain-size reduction of primary phases.

Amphibolites and amphibolitized microgabbro intervals are recovered from both Holes 920B and 920D. The amphibolite interval in Hole 920B contains linear mafic aggregations of neoblastic brown amphibole and porphyroclasts of olivine, clinopyroxene, and magnetite forming fine-grained anastomosing bands, which are bounded by segregations of neoblastic plagioclase [Cannat et al., 1995a]. This texture defines a moderately developed foliation and a well-developed lineation and indicates formation at transitional granulite to amphibolite facies conditions.

Extensive outcrops of gabbro, metagabbro, metabasalt, and metadiabase occur in the Ridge-Transform Intersection massif (RTI) on the western wall of the Mid-Atlantic Ridge

in the MARK area (Fig. 3; Dick et al., 1981; Karson and Dick, 1983; Mével et al., 1991; Auzende et al., 1993; Cannat et al., 1995a). These rocks are exposed in a complex juxtaposition along east-facing escarpments that correspond to high-angle faults, which commonly crosscut some east-northeast dipping gentle faults along the east face of the RTI (Figs. 3 and 4; Karson and Dick, 1983; Mével et al., 1991; Cannat et al., 1995a). A large gabbro massif also makes up much of a transverse ridge and its walls occurring in the rift mountains of the slow-spreading Southwest Indian Ridge (8 mm/year half-spreading rate) and flanking the Atlantis II Transform [Dick et al., 1991]. The 435 m of nearly continuous core recovered from Hole 735B located on a wavecut platform at the crest of this transverse ridge consists mainly of primary medium- to coarse-grained gabbro, late intrusive fine- to medium-grained microgabbro, and synkinematic oxide-bearing gabbro [Dick et al., 1991].

Gabbroic samples collected from the RTI massif in the MARK area, via dredges, *Alvin* and *Nautile* dives, and drilling, include variably deformed and metamorphosed olivine gabbro, through gabbronorite and ferrogabbro to trondhjemite [Karson and Dick, 1983; Mével et al., 1991; Marion et al., 1991; Gillis et al., 1993b; Cannat et al., 1995a]. Locally, well-developed preferential orientation of clinopyroxene grains and elongation of subhedral plagioclase laths produced a magmatic fabric in some of these rocks that resulted from crystal accumulation and/or magmatic flow. Similarly, medium- to coarse-grained primary gabbroic rocks from Site 735B exhibit a magmatic foliation, defined by the planar-shaped fabric of plagioclase, olivine, and clinopyroxene grains, that is commonly parallel to the cm-scale compositional layering in these rocks [Cannat et al., 1991]. The lack of intracrystalline deformation and/or recrystallization in these minerals suggests that this texture is probably related to laminar viscous flow of the progressively cooled gabbros. This magmatic fabric in gabbroic rocks from both the MARK area and Site 735B along SWIR is locally overprinted by crystal-plastic deformation fabrics that are defined by a shape-preferred orientation of elongated olivine and pyroxene porphyroclasts with asymmetric recrystallized tails and/or by discrete shear zones (Fig. 5). Asymmetric pressure shadows and tails of recrystallized grains around porphyroclasts, drag folds, and downdip plunging lineations within these shear zones consistently give a normal sense of shearing associated with ductile deformation [Cannat et al., 1991; Dilek et al., 1997b]. Decimeter-thick cataclastic domains in the upper sections of Holes 921B and 921C in the MARK area contain closely spaced shear zones and elongated clasts of the primary minerals in a very fine-grained cataclastic matrix [Cannat et al., 1995a] and represent probable late-stage brittle fault domains.

The plutonic rocks recovered from the MARK area record a complex history of alteration and deformation

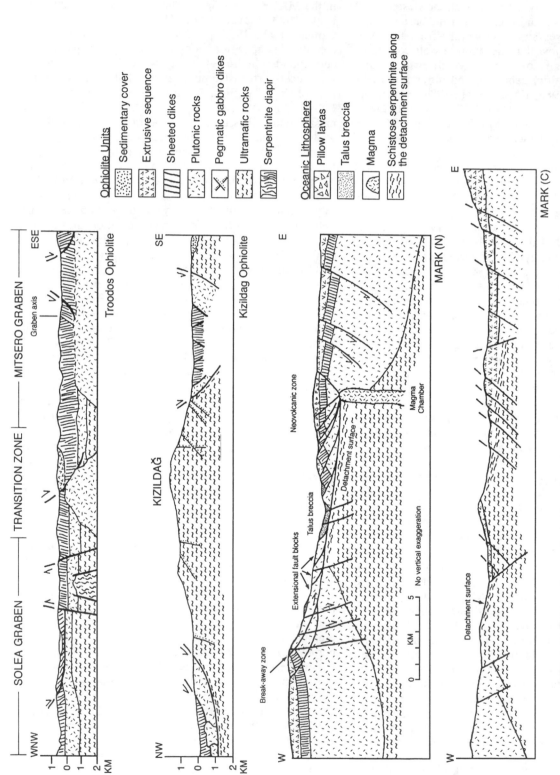

Figure 4. Interpretive geological cross-sections at the same scale from the Troodos, Kizildag, and the slow-spread oceanic lithosphere in the northern and central spreading cells, MARK (N) and MARK (C) respectively, in the MARK area along the Mid-Atlantic Ocean Ridge (no vertical exaggeration). The northern spreading cell, MARK (N), constitutes the ridge segment between the Kane transform fault and the accommodation zone south of the neovolcanic zone shown in Figure 3 and undergoes more than 80-90% tectonic extension with a higher magma budget compared to the central spreading cell, MARK (C), which occurs south of the accommodation zone. The area of MARK (C) experiences more than 100% tectonic extension with a low magma budget and is characterized by the existence of a serpentinite detachment surface. While the Troodos ophiolite resembles the early stages of the MARK (N) area, the Kizildag ophiolite is analogous in architecture to the advanced stages of the MARK (N) area and the intermediate stages of the MARK (C) area [modified from Dilek and Eddy, 1992].

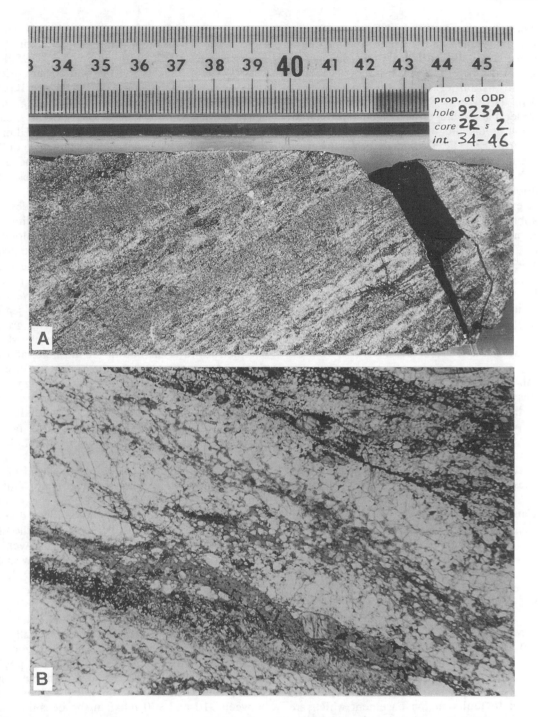

Figure 5. (A) Gneissic gabbro interval in the core from Site 923A in the MARK area (Sample 153-923A-2R-2, 34-46 cm).   Compositional banding and elongated porphyroclastic texture in the gabbro display drag features along the shear zone boundaries that indicate a normal sense of shearing.   The strong foliation in the rock is cut by actinolite veins at right angles. (B) Photomicrograph of a shear zone fabric in the olivine gabbro rock from Site 922A in the MARK area (Sample 153-922A-2R-2, 2-5 cm).   The 10-mm-thick shear zone dips moderately (36°) and has sharp boundaries with the olivine gabbro host rock. It includes lenses and bands of recrystallized plagioclase and brown hornblende and strongly recrystallized olivine and clinopyroxene porphyroclasts. Asymmetric olivine porphyroclasts contain small neoblasts with subgrains and irregular sutured boundaries and show dextral or down-dip (normal) sense of shearing [Dilek et al., 1996d].   Field of view is 7 mm.

starting at temperatures in excess of 700°C and continuing down to 330°-180°C [Kelley and Delaney, 1987; Gillis et al., 1993b; Kelley et al., 1993]. Mineral assemblages in ductile shear zones in some of the metamorphosed gabbroic rocks record high-temperature deformation and recrystallization (Fig. 5). Similarly, synkinematic metamorphic mineral assemblages in sheared gabbros from Site 735B include calcic plagioclase + clinopyroxene ± olivine ± orthopyroxene ± Fe-Ti oxide ± brown to green hornblende, covering a temperature range between granulite and upper amphibolite facies conditions [Stakes et al., 1991]. Amphibolitization and development of amphibole veins are most extensive in shear zones, whereas many undeformed gabbros are virtually unaltered, suggesting that shear zones strongly controlled hydrothermal circulation and alteration in lower oceanic crust [Dick et al., 1991; Stakes et al., 1991]. Amphibole veins, crack networks, trondhjemitic veins and intrusion breccias, and late hydrothermal breccias and veins represent successive stages of brittle failure in lower oceanic crust with its progressive cooling down to greenschist facies at Site 735B.

4.2. Intermediate-spreading Oceanic Crust

The internal architecture of intermediate-spreading oceanic crust has been studied in detail through drilling at Site 504B in the eastern equatorial Pacific Ocean. DSDP/ODP Hole 504B, situated on the southern flank of the intermediate-spreading (half rate of 3.6 cm/yr to the south and 3.0 cm/yr to the north; Hey et al., 1977] Costa Rica Rift, penetrates 2.1 km into 5.9-m.y.-old oceanic crust and provides a significant *in-situ* reference section for the physical and chemical structure of the upper oceanic lithosphere [Cann, Langseth, et al., 1983; Anderson, Honnorez, Becker, et al., 1985; Becker, Sakai, et al., 1988; Becker, 1989; Dick, Erzinger, Stokking, et al., 1992; Alt, Kinoshita, Stokking, et al., 1993]. The seafloor bathymetry in the vicinity of Site 504B is defined by east-west trending asymmetric elongated ridges and troughs with wavelengths of 5-6 km and relief of 100-150 meters (Fig. 6A; Langseth et al., 1988; Swift and Stephen, 1995; Kent et al., 1996]. Recent seismic surveys in this region reveal the existence of distinct offsets in sediment reflectors overlying the volcanic basement, and these offsets are commonly spatially associated with basement escarpments with a relief of 100 metres or more [Kent et al., 1996]. The scarps correspond to faults in the basement which are mostly active as indicated by their upward continuation through the sedimentary strata and to the seafloor. The offsets of sediment reflectors show that the downdropped side is generally to the north, towards the Costa Rica spreading axis, pointing to the existence of north-dipping normal faults in the basement (Fig. 6B). The apparent decrease in fault displacements upsection in the sedimentary strata suggests a growth fault nature of some of these normal faults and possible change in their geometry (e.g., lessening dip angles) with depth. An east-west striking,

north-dipping normal fault occurs about 1 km south of Hole 504B and passes within several hundred metres of Hole 896A (Fig. 6C; Swift and Stephen, 1995; Kent et al., 1996]. The sediments and the seafloor are offset by this fault indicating activity on it during the last 100,000 years. Hole 504B is thus located on the hanging wall of this active normal fault, whereas Hole 896A is situated on its footwall. Extrapolation of fault dip, as imaged in reflection profiles, into the basement yields a dip angle of 70°±5° for this north-dipping active fault, which can be estimated to intersect Hole 504B at a depth of ~4.8 kmbsl assuming a planar geometry (Fig. 6C).

The lithostratigraphy in Hole 504B includes [Adamson, 1985], from top to bottom (Fig. 6D), sediments (274.5 m); an upper zone of pillow lavas, pillow breccias, hyaloclastites, flows, and sills (571.5 m); a transition zone of pillow lavas, flows and dikes (209 m); and a lower zone of diabasic dike rocks (1055.6 m). Thus, the drilled section penetrates geophysically defined oceanic Layers 1, 2A, 2B, and 2C to a depth of 2.1 km beneath the seafloor [Alt, Kinoshita, Stokking, et al., 1993]. The grain size of diabasic rocks increases significantly in the sheeted dike complex towards the bottom of the hole, and combined with an apparent decrease in the occurrence of chilled dike margins in the same intervals it suggests higher temperatures and slower cooling rates near 2 km at depth during the evolution of oceanic crust [Alt, Kinoshita, Stokking, et al., 1993] and may also point to the proximity of gabbros to the bottom of the hole at 2111.0 mbsf.

Hole 504B, located on the hanging wall of the north-dipping active fault, penetrates two major fault zones at about 800 mbsf in the lower volcanic rocks and at 2111 meters below sea floor (mbsf) in the lower dikes (Fig. 6). The fault zone at 800 mbsf coincides with the lithological boundary between the volcanic sequence (layer 2B) above and the transition zone consisting of pillow and massive lava flows and dikes below [Agar, 1991], separates contrasting domains of magnetic properties [Pariso and Johnson, 1989], and marks a south-dipping (outward facing) normal fault [Dilek et al., 1996a]. The deformation zone between 840 and 958.5 mbsf is spatially associated with this fault zone and contains cataclastic zones, isoclinal microfolds, and alignment of quartz and phyllosilicate minerals as observed in the discontinuous core record [Agar, 1990, 1991]. The stockwork zone of alteration between 910 and 930 mbsf in the oceanic crust occurs within the transition zone [Honnorez et al., 1985] and is possibly related to intense hydrothermal circulation facilitated by dilation during slip on shallow faults in the deformation zone [Agar and Marton, 1995]. The fault zone at 2111 mbsf in the bottom of the hole represents a dip-slip fault defined by closely spaced east-northeast striking microfractures with steep dips and steeply plunging lineations [Alt et al., 1993; Dilek et al., 1996a] and probably marks the boundary between the dikes and the gabbros.

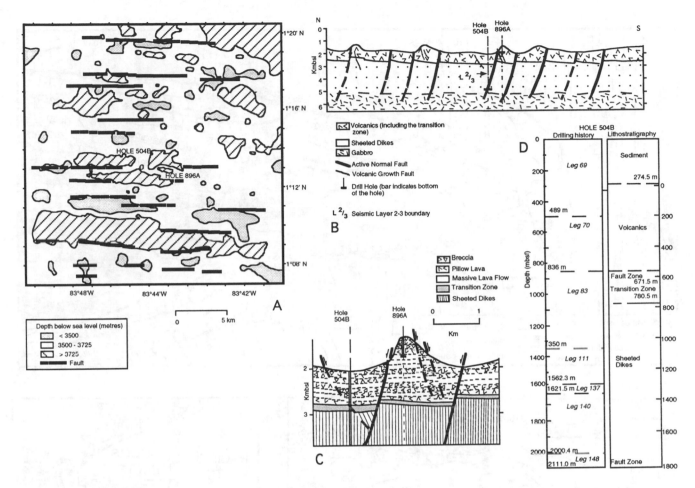

Figure 6. (A) Bathymetric map showing the fault-bounded abyssal hill fabric in the area around Sites 504 and 896 in the Panama Basin [modified from Dilek, 1998]. Location and the areal extent of the faults are based on seismic reflection data from Kent et al. [1996] and Swift and Stephen [1995]. (B) North-south trending geological cross-section across Holes 504B and 896A. The location and geometry of the active normal faults are determined based on seismic reflection data and profiles from Kent et al. [1996] and Swift and Stephen [1995]. Depth to the dike-gabbro boundary is based on average thickness of the sheeted dike complex in modern oceanic crust and is consistent with the apparent increase in grain size of diabasic rocks in the bottom of Hole 504B. (C) Geological cross-section across Holes 504B and 896A showing the volcanic stratigraphy in the extrusive sequence, transition zone between the volcanic rocks above and the sheeted dikes below, and the upper part of the sheeted dike complex. Development of the asymmetric abyssal hill fabric in this region is interpreted to have developed via synthetic (dipping towards the rift axis) and antithetic (dipping away from the rift axis) faulting and synchronous volcanism in the flanking tectonic province of the intermediate-spreading oceanic lithosphere along the Costa Rica Rift. See text for discussion. (D) Generalized lithostratigraphy and drilling history of Hole 504B. Fault zones inferred from core observations and downhole bore measurements in Holes 504B are shown [modified from Dilek, 1998].

4.3. Fast-spreading Oceanic Crust

Oceanic crust that developed along the fast-spreading (130 km/my) East Pacific Rise (EPR) at ~1 Ma is exposed in the walls of the Hess Deep rift ahead of the tip of the westward-propagating Cocos-Nazca spreading center (Fig. 7) and has been studied directly through dredging, submersible, and drilling investigations [Hey et al., 1977; Lonsdale, 1988; Francheteau et al., 1990; Karson et al.,

1992; Hekinian et al., 1993; Gillis et al., 1993a; Hurst et al., 1994; Mével et al., 1996]. The intra-rift ridge north of the Cocos-Nazca ridge propagator represents a main topographic high within the Hess Deep rift valley, corresponding to a horst structure bounded by oppositely-dipping high-angle normal faults (Fig. 7B). These faults are seismically active with average hypocentral depth located ~8 km below the summit of the Intra-rift ridge [Porras et al., 1995]. Observations made during *Nautile*

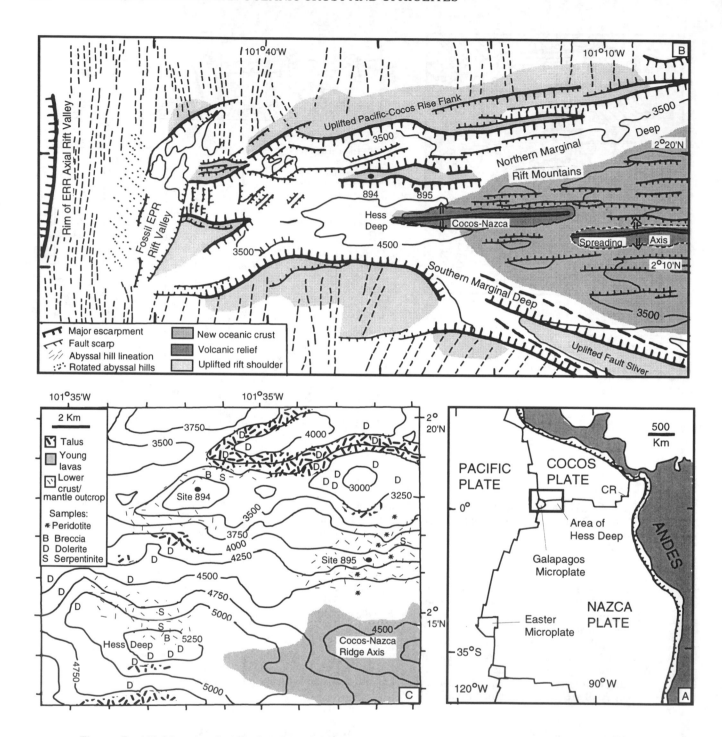

Figure 7. (A) Map showing the location of the Hess Deep area, lithospheric plates, and plate boundaries in the eastern equatorial Pacific Ocean. CR = Costa Rica Rift. (B) Simplified tectonic map of the Hess Deep area between the Cocos and Nazca plates, showing the relief, Cocos-Nazca spreading axis, fault scarps and escarpments, abyssal hill distribution, and ODP Sites 894 and 895 [modified from Londsdale, 1988]. The east-west-trending, fault-bounded horst structure north of the Cocos-Nazca ridge propagator represents the intra-rift ridge. (C) Bathymetry and distribution of rock outcrops in the vicinity of the intra-rift ridge in the Hess Deep area [modified from Francheteau et al., 1990].

and *Alvin* dives in the rift walls have revealed the occurrence of extensive outcrops of upper crustal units in a nearly 1.5-km-thick continuous section (Fig. 7C; Francheteau et al., 1990; Karson et al., 1992]. The fast-spread oceanic crust exposed along the northern rift wall contains extrusive rocks, sheeted dike complex, and mafic plutonic rocks with gradual igneous contacts.   The extrusive sequence is composed mainly of pillow basalts, basalt breccias, and less than 10% dikes in a ~600-m-thick section and is in places intensely jointed and fractured [Karson et al., 1992]. The ratio of dikes to lavas increases downsection over an interval of about 500 m into the sheeted dike complex.

The sheeted dike complex beneath the extrusive sequence in the northern rift wall of the Hess Deep rift consists of 90% or more dike intrusions with screens of gabbros and lavas in outcrops up to 700 m in thickness. Diabasic dikes are commonly 0.5 to 1.5 m wide and moderately to intensely jointed. In general, sheeted dikes have north-northwest trends and are cut by faults that are oriented both parallel (~340°) and at high angles to the local dike intrusions [Karson et al., 1992]. The dips of the dikes vary locally and range from subvertical to moderate (~40°). Some of the inclined dikes are crosscut by vertical individual dike intrusions, suggesting that they were rotated subsequent to their cooling [Hurst et al., 1994]. These rotations might have resulted from faulting and subsidence above an emptied axial East Pacific Rise magma chamber, block faulting during the opening of the Hess Deep rift, or some combination of these two processes.   The observation of multi-stage rotations of sheeted and individual dikes about both horizontal and vertical axes near the spreading axis may suggest the occurrence of complex tectonic processes at fast-spreading ridges.

A transition from sheeted dikes to dominantly plutonic rocks occurs over a vertical distance of less than 200 m on the north wall of the Hess Deep rift. The plutonic section here is more than 600-m-thick and contains gabbroic rocks that are cut by diabasic dikes and minor trondhjemite veins. Gabbroic rocks and serpentinized peridotites are also recovered by dredging and submersible dives from the fault-bounded horst (Intra-rift ridge) within the Hess Deep basin (Fig. 7C; Francheteau et al., 1990]. Hole 984G drilled during ODP Leg 147 near the summit of this intra-rift ridge penetrated 154.5 mbsf into a succession of gabbronorites and gabbros (Fig. 8A; Gillis et al., 1993a]. Gabbroic rocks recovered in Site 894G cores display a subvertical foliation and associated lineation defined by the shape-preferred orientation of anisotropic plagioclase crystals [MacLeod et al., 1996]. These fabrics, which were originally developed parallel to the axis of the East Pacific Rise, are interpreted to have formed by magmatic flow rather than solid-state ductile deformation.

Mantle rocks recovered from Site 895 located 9 km to the east of Site 894 and on the southern slopes of the intra-rift ridge are composed mainly of serpentinized tectonite

harzburgites and dunites with intercalations of troctolites, olivine gabbros, and minor basaltic dikes (Fig. 8A; Gillis et al., 1993a]. Interlayering of these mafic and ultramafic rocks at a scale of tens of centimeters to tens of meters in two stratigraphic sections is analogous to the interlayered structure of the Moho transition zone described from the Semail ophiolite in Oman (Fig. 8B; Boudier et al., 1996]. The mantle fabric in the ultramafic rocks is represented by a foliation and associated lineation that are defined by elongated spinel and orthopyroxene grains. Paleomagnetically corrected fabric measurements show that both foliation strike and lineation trend run parallel to the EPR axis and that they may thus record the mantle flow beneath the EPR during the accretion of the oceanic crust [Boudier et al., 1996]. A systematic decrease of foliation dip upsection in the ultramafic rocks implies a bend in flow trajectory towards subhorizontal at shallow depths and supports this interpretation. The pseudostratigraphy of the fast-spread oceanic crust in the Hess Deep area is thus very similar to the crustal architecture of both the Semail ophiolite and the modern oceanic crust at the East Pacific Rise (Figs. 2 and 8B).

5. INTERNAL STRUCTURE OF ANCIENT OCEANIC CRUST AS OBSERVED IN OPHIOLITES

Major differences in both the crustal thickness and the internal structure of well-preserved ophiolite complexes (Figs. 2 and 4) probably reflect variations in tectonic processes at the paleo-spreading centers along which they were developed. The variation in structural styles noted between fast and slow spreading centers may help differentiate fast from slow-spread ophiolite complexes. For example, comparison of the Semail and Troodos ophiolites bear directly on the morphology of fast and slow spreading centers, respectively.   Structural data, paleogeographic relations, and estimates of crustal thickness from Semail bear striking resemblance to those obtained from modern ocean crust forming the East Pacific Rise [Tilton et al., 1981; Nicolas and Boudier, 1995]. At the other end of the spectrum, the Troodos ophiolite shows evidence for episodic magmatism, extensive faulting, fault-controlled hydrothermal circulation, and large scale tilting of the crustal sequence [e.g., Varga and Moores, 1985; Schiffman et al., 1987; van Everdingen and Cawood, 1995], features typical of modern slow spreading centers [Karson, 1990]. The presence of low angle, syn-magmatic detachment faults and other evidence of magmatic and amagmatic extension in the Troodos ophiolite has also been interpreted as the effects of crustal accretion at a slow spreading center [Dilek and Eddy, 1992]. Thus, it may be possible to differentiate relative fast versus slow spreading ophiolites based on the style of extensional tectonics and magmatism [Casey et al., 1983; Nicolas, 1989; Dilek et al., 1990; Allerton and Vine, 1991; Dilek and Eddy, 1992; MacLeod and Rothery, 1992; Nicolas and Boudier, 1995;

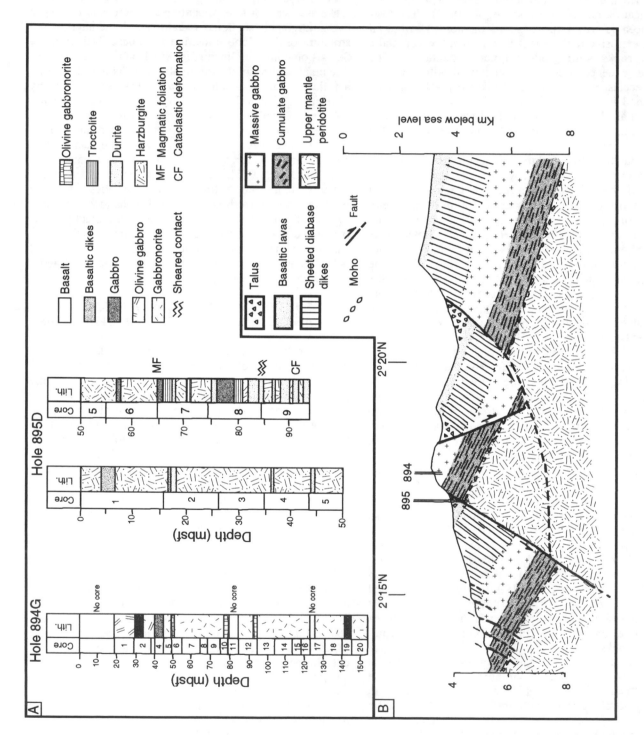

Figure 8. (A) Lithostratigraphy of Holes 894G and 895D in the Hess Deep area [data from Gillis et al., 1993a]. (B) Interpretive geological cross section across the central to northern part of the Hess Deep, at 101°32'W longitude [modified from McLeod et al., 1996]. The fault-bounded horst structure, where Hole 894 is situated, corresponds to the intra-rift ridge. See text for discussion.

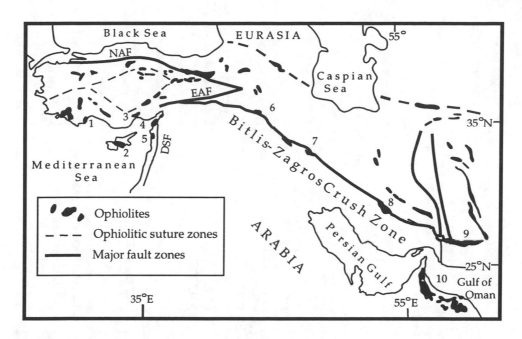

Figure 9. Distribution of the Cretaceous Neo-Tethyan ophiolites around the Arabian promontory [from Dilek and Thy, 1998, and references therein]. Key to the numbers: 1- Tekirova-Antalya (Turkey), 2- Troodos (Cyprus); 3- Mersin (Turkey), 4- Kizildag (Turkey), 5- Baër-Bassit (Syria), 6- Khoy (Iran), 7- Kermanshah (Iran), 8- Neyriz (Iran), 9- Makustan (Iran), 10- Semail (Oman).1- Troodos (Cyprus); 2- Baër-Bassit (Syria); 3- Kizildag (Turkey); 4- Guleman (Turkey); 5- Cilo (Turkey); 6- Kermanshah (Iran); 7- Neyriz (Iran); 8- Semail (Oman).

Moores et al., 1990; Dilek et al., 1997c]. The Cretaceous ophiolites occurring along the northern edge of the Arabian promontory display, for example, significant variations in terms of their crustal thickness, pseudostratigraphy, and structural architecture along strike from the Troodos massif in the west to the Semail complex in the east (Fig. 9). These ophiolites represent remnants of the Neo-Tethyan ocean, which evolved between Afro-Arabia in the south and Eurasia in the north during the Mesozoic [Dilek and Moores, 1990]. The crustal thickness of these ophiolites is estimated based on the paleo-horizontal markers, such as the lava-pelagic sediment interface and/or contacts and dike-gabbro boundaries when sheeted dikes are nearly subvertical.

5.1. Ophiolites with Inferred Slow-Spread Origin

The Troodos (Cyprus) and Kizildag (Turkey) ophiolites in the eastern Mediterranean region display a complete pseudostratigraphy and no evidence for any significant emplacement-related tectonic deformation in their subunits and/or sedimentary cover, suggesting that they did not experience any large-scale tectonic transport from their igneous environment of origin. The Troodos ophiolite forms an ESE-trending massif and consists of two parts separated by the nearly E-W striking Arakapas fault zone (Fig. 10). The northern part is composed of a central core

of peridotite and serpentinite overlain by a plutonic sequence, a sheeted dike complex, and extrusive volcanic rocks [Moores and Vine 1971; Gass and Smewing 1973] and has a N-S structural grain defined by the orientation of sheeted dikes and normal faults (Fig. 10). The second part south of the Arakapas fault zone includes the Limassol Forest complex, which consists of tectonized mantle rocks cut by plutonic and hypabyssal rocks and overlain by pillowed and massive basaltic lavas, polymictic breccias, and volcaniclastic rocks (Fig. 10; Moores and Vine, 1971; Simonian and Gass, 1978; Murton and Gass, 1986]. The structural grain in the Limassol Forest complex is defined by generally E-W striking syn-magmatic, transtensional shear zones with sinistral and/or dextral slip [MacLeod and Murton, 1993] and is interpreted to have resulted from transform-fault tectonics during the evolution of the ophiolite. Thus, the Troodos ophiolite in general preserves an excellent record of spatial and temporal variations in structural processes in a spreading center-transform fault system within the Neo-Tethys. The Kizildag ophiolite exposed in southern Turkey nearly 300 km northeast of the Troodos ophiolite forms a NE-trending massif that consists of a core of serpentinized harzburgites and dunites overlain by a plutonic sequence, sheeted dikes, and extrusive rocks [Dubertret, 1955; Dilek and Moores, 1985; Erendil, 1984; Dilek et al., 1991]. The Kizildag massif contains two main parts separated by the NW-striking high-angle

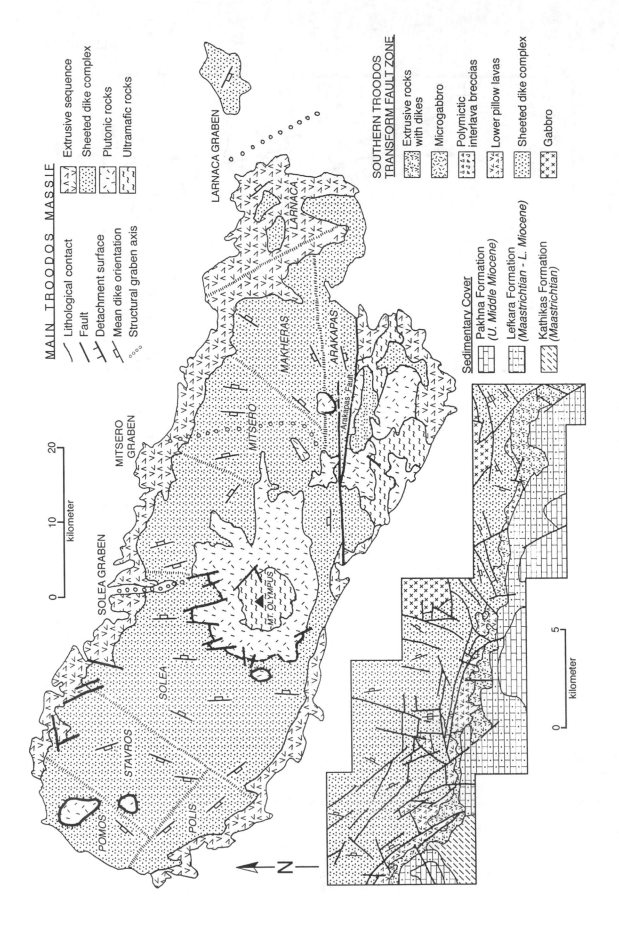

Figure 10. Simplified geological maps of the Troodos ophiolite and the Southern Troodos Transform Fault Zone. Data are from Dilek et al. [1990], Moores et al. [1990], Dilek and Eddy [1992], and MacLeod and Murton [1993].

Tahtaköprü fault (Fig. 11). In the western part of the massif, mantle rocks are exposed mainly in a laterally extensive and broad antiform, whereas the crustal units mainly occupy a structurally and topographically well-defined graben on the southeastern flank of this antiform [Dilek and Delaloye, 1992]. The eastern part of the massif consists mainly of serpentinized peridotites directly overlain by lava flows and rotated fault blocks of sheeted dike and plutonic rocks [Dilek et al., 1990].

Well-developed sheeted dike complexes in the Troodos and Kizildag ophiolites suggest extensive magmatic spreading and display numerous high- to low-angle normal faults cut by late intrusions. These crosscutting relations indicate that magmatic extension was accompanied and/or ensued by brittle deformation associated with tectonic extension, and that there was an interplay of magmatic and amagmatic extension during construction of the ophiolites. The sheeted dike complex in the main Troodos massif includes a number of structural domains characterized by concordant dike attitudes and three north-south striking structural grabens defined by blocks of dikes dipping towards the respective graben axes (Fig. 10; Verosub and Moores 1981; Varga and Moores 1985; Moores et al. 1990]. Sheeted dikes in N-S trending structural grabens, the Solea, Mitsero, and Larnaca grabens from west to east (Figs. 4 and 10), respectively, are cut by and rotated along high- to low-angle normal faults that are commonly parallel to the graben axes. Crosscutting relations between different dike intrusions and domains indicate that the structural grabens become successively younger from the Solea graben in the west to the Larnaca graben in the east [Varga and Moores, 1985; Mukasa and Ludden, 1987]. In general, the more steeply dipping sheeted dikes cut less steeply dipping dikes in these grabens (Fig. 12), and dikes become relatively less steeply dipping towards the graben axes.

Tectonic extension associated with the seafloor spreading history of the Kizildag ophiolite is manifested in different crustal architecture and structural styles in the two separate massives of the ophiolite (Fig. 11). The main outcrop of the sheeted dike complex occurring in the NE-SW oriented graben within the western massif of the ophiolite is cut by two subsets of mineralized faults [Dilek and Thy, 1998]. One subset of faults is parallel to the sheeted dikes, has generally shallower dip angles than the dikes, and displays down-dip plunging slickenside lineations. Locally, these faults form well-developed horst and graben structures, whereas in some places they are listric in geometry and are associated with rotated and tilted fault blocks of sheeted dike swarms. The slickenside data combined with the fault geometry show the general orientation of the least compressive stress (σ_3) to be ~325°, which is nearly perpendicular to the trend of the dike complex and the mean orientation of the ophiolite complex [Dilek, 1997]. The second subset of faults within the sheeted dike complex contains faults that are perpendicular

to the mean dike orientation with steep dips and oblique to subhorizontal slickenside lineations. Both the dike-parallel and dike-perpendicular faults are moderately to intensely mineralized containing epidote, chlorite, actinolite, prehnite, hematite, and quartz, suggesting that faulting was accompanied by and facilitated the hydrothermal circulation and associated mineral precipitation in the Kizildag oceanic crust.

The second massif of the ophiolite occurring east of the Tahtaköru fault lacks the coherent pseudostratigraphy observed in the main massif to the west. It consists mainly of volcanic, dike, and/or plutonic rocks directly overlying the serpentinized peridotites along low-angle faults [Dilek et al., 1991; Dilek and Thy, 1998]. The ~400-m-thick extrusive sequence near the village of Kömürçukuru, for example, overlie serpentinized peridotites along a gently SE-dipping fault and is in turn overlain stratigraphically by the Maastrichtian siliciclastic and carbonate rocks. Numerous individual dikes and sills are intrusive into the extrusive sequence and feed into different horizons of pillow and sheet lava flows. These dikes are similar in terms of their geochemistry and orientation relative to the dike swarms intruding the serpentinized peridotites in the lower plate [Dilek and Thy, 1998; Dilek, unpublished data], suggesting that dike intrusions in the peridotites and the volcanic sequence may be genetically related. This observation suggests that the volcanic rocks might have been extruded directly on the serpentinized mantle rocks, which were already exposed on the seafloor prior to this late-stage magmatic episode.

Similarly, the Ligurian ophiolites occurring in the Alpine-Apennine orogenic belt in western Europe display contact relations and structures typical of slow-spreading modern oceanic crust. The Ligurian ophiolites (Fig. 13A) represent the remnants of the Meosoic Neo-Tethys, which evolved between the Apulia and Eurasia (Ligurian Ocean), and rest tectonically on the continental margin units (Toscana Series) of the Apulian microcontinent (Fig. 13B and C). In these ophiolites, gabbros commonly occur as large intrusions in the peridotites, and serpentinized peridotites and gabbros are stratigraphically overlain by ophiolitic breccia, ophicalcite, and pelagic sedimentary rocks (Fig. 13D; Abbate et al., 1980; Lemoine et al., 1987; Lagabrielle and Cannat, 1990]. Basaltic dikes intrude both the mafic-ultramafic rocks and the overlying ophiolitic breccias. The occurrence in the breccia of detrital material derived from plagiogranite intrusions in the peridotites that are coeval with the mutually crosscutting basaltic dikes suggests that magmatism and tectonic denudation were penecontemporaneous along the Ligurian spreading center [Borsi et al., 1996]. Pillow lavas with a MORB affinity overlie the ophiolitic and sedimentary breccias and occur as isolated units within the ophiolites [Venturelli et al., 1981]. These features of the Ligurian ophiolites are analogous to those observed in young oceanic crust exposed in the MARK area and may therefore

GEOLOGICAL MAP OF THE KIZILDAĞ (TURKEY) AND BAËR-BASSIT (SYRIA) OPHIOLITES

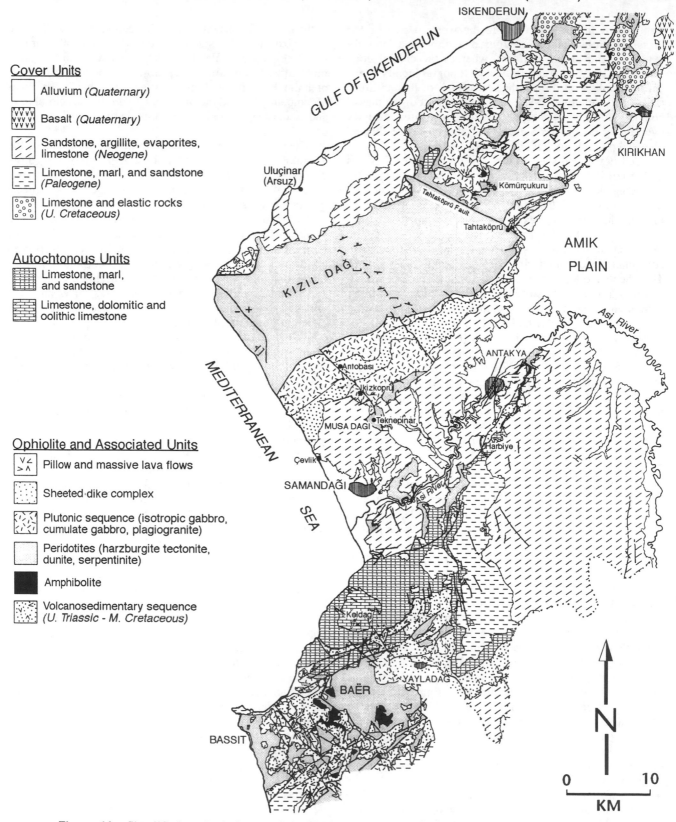

Cover Units

- ☐ Alluvium *(Quaternary)*
- Basalt *(Quaternary)*
- Sandstone, argillite, evaporites, limestone *(Neogene)*
- Limestone, marl, and sandstone *(Paleogene)*
- Limestone and elastic rocks *(U. Cretaceous)*

Autochtonous Units

- Limestone, marl, and sandstone
- Limestone, dolomitic and oolithic limestone

Ophiolite and Associated Units

- Pillow and massive lava flows
- Sheeted dike complex
- Plutonic sequence (isotropic gabbro, cumulate gabbro, plagiogranite)
- Peridotites (harzburgite tectonite, dunite, serpentinite)
- Amphibolite
- Volcanosedimentary sequence *(U. Triassic - M. Cretaceous)*

Figure 11. Simplified geological map of the Kizildag ophiolite and its environs in southern Turkey; also shown is the Baër-Bassit ophiolite in Syria farther south [modified from Dilek and Thy, 1988].

Figure 12. Crosscutting dike relations in the sheeted dike complex exposed in the eastern part of the Solea graben of the Troodos ophiolite (Kambia-Kykko Road).

suggest a similar origin via tectonic extension in a slow-spreading center with a poor magma budget [Lagabrielle and Cannat, 1990; Borsi et al., 1996].

5.2. Ophiolites with Inferred Fast- to Intermediate-spread Origin

The Late Cretaceous Semail ophiolite in Oman differs significantly from the coeval Troodos and Kizildag ophiolites in terms of its overall thickness and crustal structure and may represent an ancient analogue for fast-spreading oceanic lithosphere [Nicolas, 1989; Dilek and Delaloye, 1992; MacLeod and Rothery, 1992; Boudier and Nicolas, 1995; Nicolas and Boudier, 1995]. The Semail ophiolite occupies a NNW-trending belt over 500 km in length and 50 to 80 km in across-strike along the southeastern edge of the Arabian Peninsula (Fig. 9) and forms the uppermost tectonic nappe of a number of oceanic imbricate thrust slices emplaced southwestwards onto the autochthonous carbonate platform [Boudier and Coleman, 1981; Hopson et al., 1981; Lippard et al., 1986]. More than 8 km of the 15-km-thick pseudostratigraphy of the Semail ophiolite is composed of the mantle sequence consisting of dunitic and harzburgitic peridotites cut by pyroxenite dikes (Fig. 14). Gently dipping foliations and associated lineations in the peridotites at right angles to the average trend of the sheeted dikes represent high-temperature plastic flow structures within the upper mantle sequence [Nicolas et al., 1988]. Recent studies in the mantle peridotites have revealed the possible existence of a number of mantle diapirs, 10 x 15 km in diameter and 30 to 50 km apart, that are characterized by divergence of flow planes and flow lines and by traces of intense magmatic activity [Nicolas et al., 1988; Ceuleneer, 1991]. These diapiric structures appear to be a result of non-tectonic ridge segmentation along the strike of the Semail paleo-spreading center and to have produced via melt extraction processes crustal segments that were emplaced through and across previously accreted oceanic lithosphere [Nicolas et al., 1994]. The 8-km-wide Maqsad diapir in the Maqsad/Maharam area is interpreted to represent a mantle upwelling zone feeding into a plutonic sequence and a NW-trending sheeted dike complex penetrating into an earlier formed crustal segment with NE-striking sheeted dikes [Ceuleneer, 1991]. Thus this penetration zone preserves a boundary between crust that

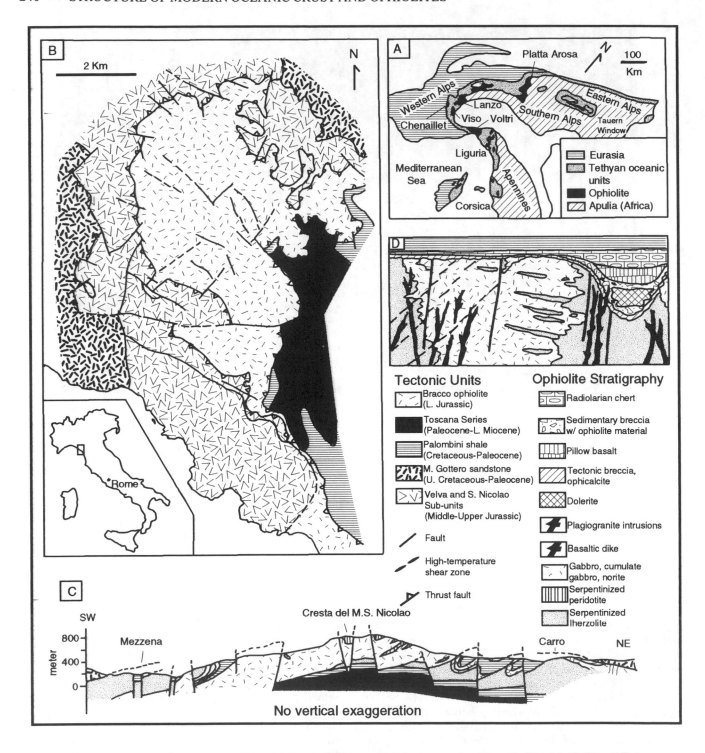

Figure 13. (A) Distribution of the Alpine-Apennine ophiolites in western Europe. (B and C) Simplified geological map and cross section of the Bracco ophiolite (in Liguria). Data are from Cortesogno et al., [1987]. (D) Reconstructed interpretive tectonostratigraphy of the Ligurian oceanic crust [data are from Abbate et al., 1980; Lemoine et al., 1987; Lagabrielle and Cannat, 1990; Borsi et al., 1996]. Tectonic breccia, ophicalcite, pillow basalt, and sedimentary breccia occur in a fault-bounded trough and are underlain by mafic and ultramafic rocks that were exposed on the seafloor as a result of tectonic extension and crustal denudation. See text for discussion.

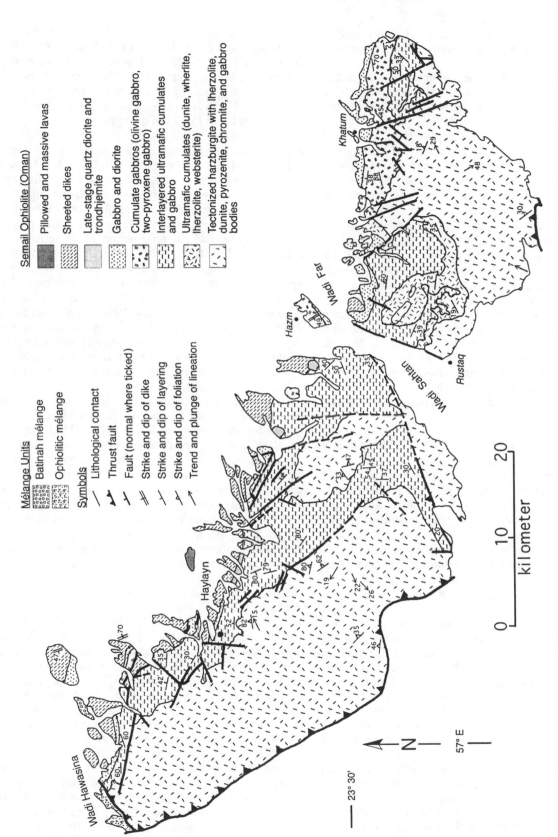

Figure 14. Simplified geological map of the Haylayn Block and its environs in the Semail ophiolite in Oman. Data are from Lippard et al. [1986], Juteau et al. [1988], and Nicolas [1989].

formed in separate axial magma chambers, and therefore coincides with "second-order" axial discontinuties as defined by Macdonald et al. [1988].

The crustal sequence of the Semail ophiolite is separated from mantle rocks by a transition zone, which is interpreted to form the flat-lying and gently undulating Moho (Fig. 2; Nicolas et al., 1988; Boudier and Nicolas, 1995]. The transition zone, locally as thick as several hundred meters, is composed dominantly of dunite with gabbro, pyroxenite, and chromitite segregations and grades upwards into layered gabbros (Figs. 2 and 14). Detailed studies of the Moho Transition Zone by Boudier and Nicolas [1995] have shown that its thickness and composition are related to the inferred geometry of the asthenospheric mantle flow. Thick (~500 m) transition zone occurs on top of mantle diapirs characterized by vertical flow and display extensive diffuse melt impregnations, dikes, and sills, all pointing to an intense magmatic activity, whereas thin (~10 m) transition zone occurs in areas of horizontal mantle flow where homogeneous mantle harzburgites are interlayered with deformed dunites over a thickness of 5 meters. In both cases, the transition zone represents a level of magmatic exchange between asthenospheric mantle below and oceanic crust above, and records active melt circulation below a spreading axis [Nicolas et al., 1994; Boudier and Nicolas, 1995].

Above the transition zone into the crustal sequence, layered gabbros are continuously roofed by foliated and isotropic gabbros (Fig. 15), and near the top of the plutonic sequence the upper isotropic gabbro and small discontinuous bodies of diorite and plagiogranite intrude the base of the overlying sheeted dike complex, suggesting recurring magmatic activities during generation of the lower crust in the Semail ophiolite (Fig. 16; Pallister and Hopson, 1981; Juteau et al., 1988]. However, some of the late-stage diabasic and granophyric dikes of the sheeted dike complex locally cut through the uppermost section of the plutonic sequence indicating that mutually intrusive relationships exist between the sheeted dike complex and the underlying plutonic sequence. In general, however, the boundary between the sheeted dike complex and the underlying plutonic sequence is marked by a root zone, in which the magmatic foliation in the high-level gabbro unit steepens and first diabasic dikes appear [Nicolas and Boudier, 1991]. Irregular-shaped proto-dikes in the root zone are made of doleritic microgabbros and display nearly vertical foliation planes marking the flow of basaltic melt within these dikes. Root zones at the base of the sheeted dike complex in the Semail ophiolite possibly represent the roof of the axial magma chamber(s) and mark a thermal boundary between the convective magma chamber system below and the hydrothermal circulation system in the sheeted dike complex above [Nicolas and Boudier, 1991].

The sheeted dike complex with an average thickness of 1.5 km consists of subvertical to vertical dikes that strike commonly parallel to subparallel to the general NNW trend of the ophiolite (Figs. 14 and 15; Pallister, 1981; Nicolas et al., 1988]. This NNW orientation of the sheeted dike complex has been interpreted as the general trend of the Semail paleo-spreading center (in the present coordinate system). However, non-tectonic deviations in dike orientations occur locally in the ophiolite, where high-level plutonic rocks in which the dikes are rooted thin out and disappear, leaving the sheeted dikes penetrating into layered gabbros and ultramafic rocks (Fig. 15; MacLeod and Rothery, 1992]. Screens of layered gabbro and/or layered ultramafic rocks between these sheeted dikes indicate that the dikes were intruded laterally into the already consolidated, pre-existing lower oceanic crust by penetration beyond the termination of an active magma chamber a few kilometers along-strike. These dike penetration zones coincide with perturbations in the spreading fabric of the sheeted dike complex and may represent propagating rift and/or overlapping spreading centers [MacLeod and Rothery, 1992; Nicolas and Boudier, 1995]. Away from these inferred propagators and local deviations in dike orientations, sheeted dikes display no evidence for significant brittle deformation via faulting and block rotation, and the dikes are concordant throughout different massifs in the ophiolite.

The Late Ordovician Solund-Stavfjord ophiolite in western Norway is a remnant of the Iapetus oceanic crust, which developed in a Caledonian marginal basin [Furnes et al., 1990], and is structurally analogous to intermediate-spreading oceanic crust (Fig. 17). The ophiolite contains three structural domains that display distinctively different crustal architecture reflecting the mode and nature of magmatic and tectonic processes operated during the multi-stage seafloor spreading evolution of this marginal basin [Dilek et al., 1997c]. Domain-I includes, from top to the bottom, an extensive extrusive sequence, a transition zone consisting of dike swarms with screens of pillow breccias, a sheeted dike complex, and plutonic rocks composed mainly of isotropic gabbro and microgabbro. Extrusive rocks include pillow lavas, pillow breccias, and massive sheet flows and are locally sheared and mineralized, containing epidosites, sulfide-sulfate deposits, Fe-oxides, and anhydrite veins, reminiscent of hydrothermal alteration zones on the seafloor along modern mid-ocean ridges. A fossil lava lake in the northern part of the ophiolite, on the island of Alden, consists of a >65-m-thick volcanic sequence composed of a number of separate massive lava units interlayered with pillow lavas and pillow breccia horizons (Figs. 17 and 18; Furnes et al., 1992]. The NE-trending sheeted dike complex contains multiple intrusions of metabasaltic dikes with one- and two-sided chilled margins and displays a network of both dike-parallel normal and dike-perpendicular oblique-slip faults of oceanic origin. There is no evidence for significant block faulting with rotation and tilting of the dikes along these faults. The dike-gabbro boundary is mutually intrusive and

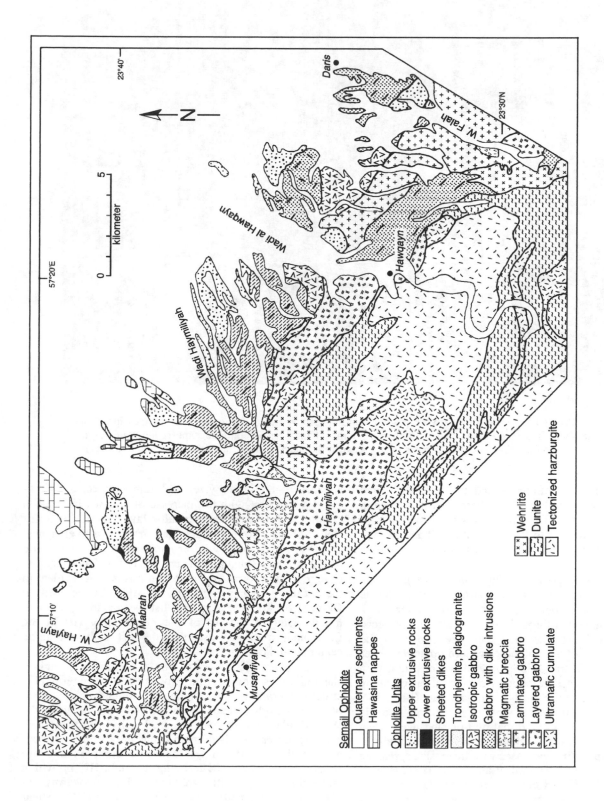

Figure 15. Simplified geological map of the southeastern part of the Haylayn massif in the Semail ophiolite, Oman [modified from McLeod and Rothery, 1992]. NW-striking numerous dike intrusions in the gabbros and ultramafic rocks near Hawqayn are interpreted to represent a NW-propagating ridge (penetration zone), whereas layered gabbro, gabbronorite, and ultramafic cumulates in the Haymiliyah area are interpreted as the southeastern terminus of an axial magma chamber [McLeod and Rothery, 1992; Juteau et al., 1988].

Figure 16. Crosscutting relations between plagiogranite intrusions and diabasic dikes in the transition zone between the sheeted dike complex and the plutonic section of the Semail ophiolite exposed in Wadi Ragmi, northern Oman. Faulted diabasic dike screens occur between the plagiogranite intrusions (light colored rock).

represents the root zone of the sheeted dike complex [Dilek et al., 1997c], similar to the dike-gabbro transition zone in the Semail ophiolite. Thus the internal architecture of Domain-I are analogous to those of intermediate- to fast-spreading oceanic crust at modern mid-ocean ridge environments.

The volcanic stratigraphy in Domain-I also suggests an intermediate-spreading rate in comparison to the volcanic stratigraphy of upper oceanic crust in modern spreading centers, as documented through deep sea drilling projects. The areal distribution of massive sheet flows in the extrusive sequence of the Solund-Stavfjord ophiolite ranges from ~3-13% on Oldra to 35% on Alden, consistent with the estimates of massive sheet flow occurrences along intermediate-spreading mid-ocean ridges (Fig. 18). Observations of eruption styles of basalt in oceanic spreading centers along the global mid-ocean ridge system show that the areal ratio of sheet versus pillow basalts in axial zones of spreading increases with increasing spreading rate (Fig. 18 A and C; Ballard et al., 1979, 1981; Bonatti and Harrison, 1988]. Higher sheet/pillow lava ratios in intermediate- to fast-spreading ridges are an artifact of initially higher total volume of erupted magma, which

appears to be of aphyric, high-temperature, and low-viscosity nature; with the increase of stresses induced in rapidly cooling submarine basaltic lava flows, lava forms tend to change from lava lakes and sheet flows to elongated and rounded pillows, broken pillow lavas, and hyaloclastites [Bonatti and Harrison, 1988]. This apparent change in lava forms and in their ratios is most evident in the extrusive sequence going from Alden in the north to Oldra in the south, suggesting that the spreading rate was much faster at the tip of the propagating rift in the Caledonian marginal basin than the rest of the spreading center behind the tip.

The ophiolitic units in Domain-II include mainly sheeted dikes and plutonic rocks with a general NW structural grain and are commonly faulted against each other, although primary intrusive relations between the sheeted dikes and the gabbros are locally well preserved. The exposures of this domain occur only in the northern and southern parts of the ophiolite complex and are separated by the ENE-trending Domain-III, in which isotropic to pegmatitic gabbros and dike swarms are plastically deformed along ENE-striking sinistral shear zones. These shear zones, which locally include fault

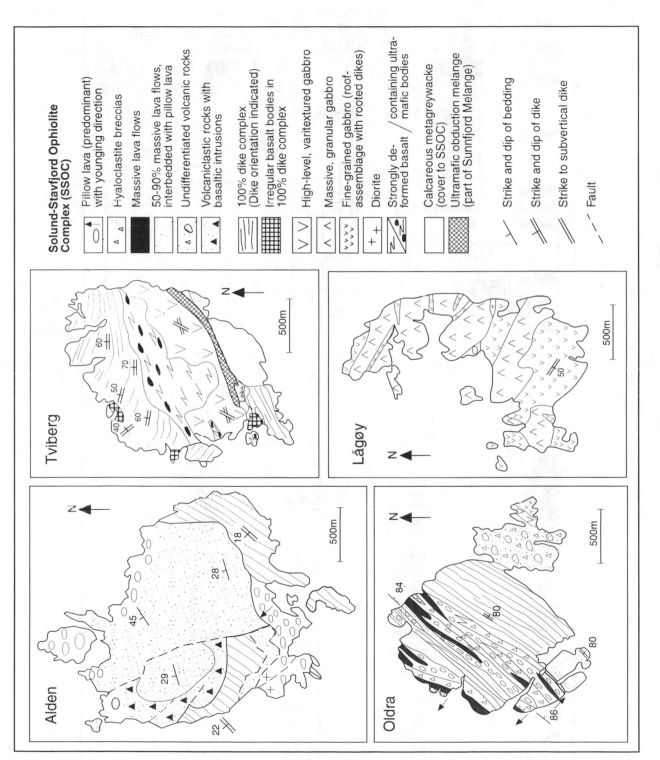

Figure 17. Simplified geological maps of different domains in the Solund-Stavfjord ophiolite in western Norway. A- Oldra; B- Alden; C- Lågøy; D- Tviberg [from Dilek et al., 1997c].

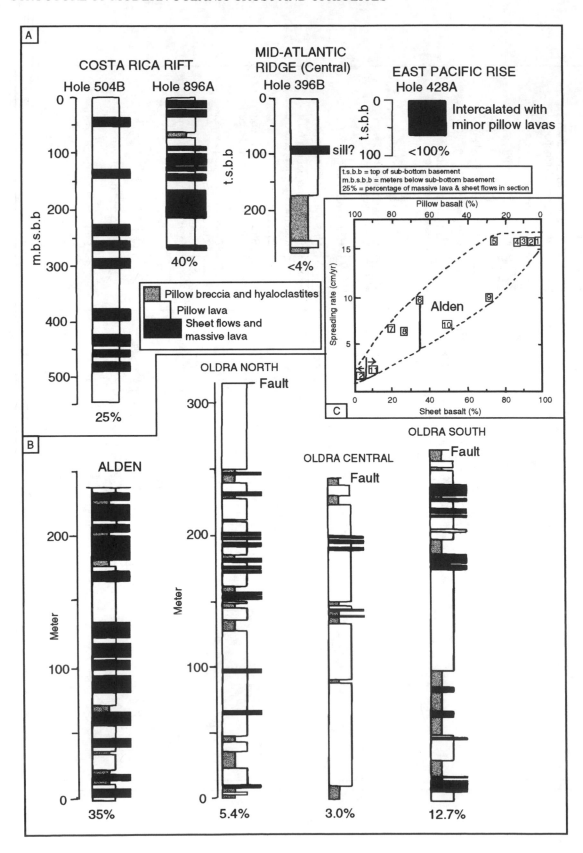

slivers of serpentinite intrusions, are crosscut by N20°E-striking undeformed basaltic dike swarms that contain xenoliths of gabbroic material [Skjerlie and Furnes, 1990]. The NW-trending sheeted dike complex in the northern part of Domain-II curves into an ENE orientation approaching Domain-III in the south. The anomalous nature of deformed crust in Domain-III is interpreted to have developed within an oceanic fracture zone or transform fault plate boundary [Skjerlie and Furnes, 1990; Dilek et al., 1997c].

The magmatic evolution of Domain-I encompasses closed-system fractional crystallization of high-Mg basaltic magmas in small ephemeral chambers, which gradually interconnected to form large chambers in which mixing of primary magmas with more evolved and fractionated magma caused resetting of magma compositions through time [Skjerlie et al., 1989; Dilek et al., 1997c]. The compositional range from high-Mg basalts to ferrobasalts within Domain-I is reminiscent of modern propagating rift basalts. The NE-trending Domain-I is therefore interpreted to represent a remnant of an intermediate- to fast-spread rift system that propagated northeastwards (in present coordinate system) into a pre-existing oceanic crust, which was developed along the NW-trending doomed rift (Domain-II) in the marginal basin [Dilek et al., 1997c]. The N20°E dikes laterally intruding into the anomalous oceanic crust in Domain-III represent the tip of the rift propagator.

6. STRUCTURAL PROCESSES AT OCEANIC SPREADING CENTERS AND SPATIAL AND TEMPORAL RELATIONS BETWEEN MAGMATISM AND FAULTING

6.1. Magmatism and Crustal Deformation at Fast- to Intermediate-Spreading Ridges

Observations on the structure and stratigraphy of modern oceanic crust and ophiolite complexes indicate that spreading rates strongly control the mode and nature of magmatism and tectonism (i.e., mechanical extension) and the interplay between these processes along oceanic spreading centers. The relatively intact pseudostratigraphy with little or no deformation of fast-spreading oceanic crust and its ancient analogs suggests that magmatism is the dominant process during the generation of oceanic crust at fast-spreading oceanic centers, and that it keeps pace with seafloor spreading and plate separation. Robust magmatic activity at fast-spreading centers is expressed externally by a continuous volcanic edifice with a minor fault-bounded summit graben and extensive recent lava flows and volcanic eruptions (Fig. 19). The gentle flanks of the axial volcanic dome are cut by numerous along-strike continuous normal faults and fault scarps that form local horst and graben structures. Direct observations of abyssal hills along the East Pacific Rise have shown that these topographic features on the seafloor, which are typically 10-20-km-long, 2-5-km-wide, and 50-300-m-high, have been produced by an interplay of faulting and magmatism near the spreading axis [Macdonald et al., 1996]. The abyssal hills appear to be bounded by normal faults dipping towards the axis and volcanic growth faults dipping away from it, creating asymmetric horst blocks. The repeated draping of fault scarps that face away from the spreading axis by syntectonic near-axis lava flows indicates that faulting and magmatism occur synchronously as the newly created oceanic lithosphere moves away and out of the active crustal accretion zone. The planar and high-angle nature of these faults shows no rotation or back tilting of the fault blocks. Similarly, the upper crust in the Semail ophiolite shows no significant faulting and seafloor block rotation, in contrast to widespread high- to low-angle normal faults associated with block rotation in the extrusive rocks and the sheeted dike complexes of the Troodos and Kizildag ophiolites.

Block faulting and tectonic rotation seem to have played a major role in the geological history of the Hess Deep area. The Intra-rift ridge as a horst structure in the northern part of the Hess Deep and the existence of gabbros and serpentinized peridotites on the seafloor here indicate that exhumation of the lower crust and upper mantle rocks was accommodated largely by extensional faulting. Francheteau et al. [1990] and McLeod et al. [1996] proposed that tectonic rotation of the fault blocks, as suggested by the

Figure 18. (A) Volcanic stratigraphy in modern upper oceanic crust at mid-ocean ridges with different spreading rates based on the DSDP-ODP drilling data. Costa Rica Rift = intermediate-spreading ridge (60-70 km/my; Alt et al., 1993); Mid-Atlantic Ridge = slow-spreading ridge (25 km/my; Dimitriev et al., 1978); East Pacific Rise = fast-spreading ridge (120 km/my; Rosendahl et al., 1980). (B) Volcanic stratigraphy in Domain-I, showing the distribution and percentage of pillow lava, pillow breccia-hyaloclastite, and sheet-massive lava flows in the Solund-Stavfjord ophiolite (unpublished data of H. Furnes and C.H. Fonneland). (C) Percentage distribution of sheet lava/pillow lava and pillow breccias in Domain-I in comparison with the areal ratios of sheet basalt/pillow basalt versus spreading rate on axial zones of various mid-ocean ridges [from Dilek et al., 1997c]. Data from Oldra reflect the average with standard deviations of three parallel profiles across-strike of the extrusive sequence. Numbers 1 through 9 mark different data locations along the East Pacific Rise: 1- 20°S, 2- 17°30'S, 3- 18°30'S, 4- 21°30'S, 5- 20°S, 6- 12°50'N (total), 7- 21°N, 8- 21°N, 9- 12°50'N (axis), while other numbers represent: 10- Galapagos Rift (86°W), 11- Mid Atlantic Ridge (37°N), 12- Red Sea (18°N).

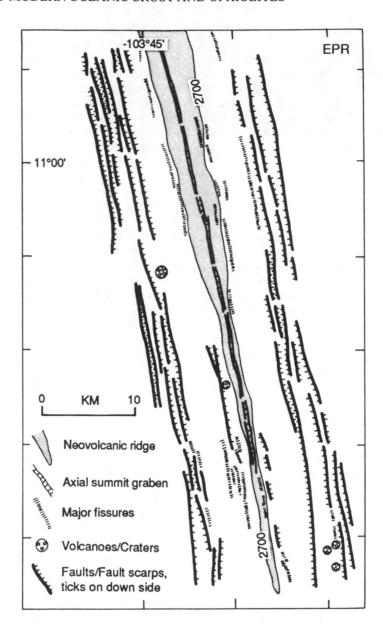

Figure 19. Simplified tectonic map of the East Pacific Rise near 11°00' N latitude [modified from Dilek and Eddy, 1992]. This segment of the east Pacific Rise near 11°00' N latitude tapers and plunges towards the Clipperton Transform (between 10°30' and 10°20' N to the south) and is characterized by a broad, dome-shaped cross-section with an axial summit graben in the center and gentle flanks with numerous along-strike continuous normal faults/fault scarps that form local horst-graben structures. The axial summit graben locally displays small en-echelon offsets and disappears before the transform fault zone. The region outlined by the 2700 m contour line generally coincides with the location of major fissures and fissure eruptions and can be taken as the neovolcanic ridge of the spreading center.

paleomagnetic data [Hurst et al., 1994; Kikawa et al., 1996], might be related to rotational normal faulting above a south-dipping low-angle detachment surface (Fig. 8B). The timing of development of the inferred detachment fault and the horst-bounding high-angle normal faults is very

late in the tectonic history of the Hess Deep area and is associated with the opening of the Hess Deep rift [McLeod et al., 1996]. Thus the extensional tectonism recorded in the oceanic crust in the Hess Deep is spatially and temporally related to amagmatic rifting in advance of the westward-propagating Coco-Nazca spreading center, and hence it does not reflect the original seafloor spreading tectonics of the fast-spread oceanic lithosphere at the north-south-trending East Pacific Rise.

Despite the absence of any evidence for solid-state deformation, the gabbroic rocks of the fast-spread oceanic lithosphere from the Hess Deep show widespread brittle fracturing and veining in their evolutionary history. The early-stage microscopic amphibole veins appear to have developed due to pervasive influx of fluids along randomly oriented microfracture networks at metamorphic temperatures of 600°-750°C [Manning and MacLeod, 1996]. Rare macroscopic amphibole veins crosscut these microscopic veins and are filled with mineral assemblages characterizing amphibolite to greenschist facies metamorphic conditions (450°-600°C). This brittle failure and associated high-temperature alteration probably occurred upon subsolidus cooling and thermal contraction of the gabbro within several tens of thousands of years after axial magma emplacement and within a few kilometers of the East Pacific Rise axis [MacLeod et al., 1996]. The late-stage, E-W oriented chlorite veins in the gabbros represent a dense array of extensional fractures that are spatially and temporally associated with the westward propagating Cocos-Nazca rift [Manning and MacLeod, 1996].

The seafloor bathymetry defined by linear hills and intervening throughs in the vicinity of Site 504B reflects a basement topography defined by fault-bounded asymmetric volcanic hills that developed within the crustal accretion zone in the intermediate-spreading environment (Fig. 6; Dilek, 1998]. Thus the physiographic expression of the seafloor structures at intermediate-spreading Costa Rica Rift are similar to those observed at the fast-spreading East Pacific Rise. However, unlike the fast-spreading oceanic crust, the intermediate-spread oceanic crust in this region shows evidence for block rotation and tilting along the faults (Fig. 6B). The fault zone occurring at the lithological boundary between the extrusive sequence and the transition zone at 800 mbsf coincides with a magnetic property boundary separating contrasting domains of natural remanent magnetization and stable magnetization inclinations [Kinoshita et al., 1989] and results in tilting of the lower pillow lavas towards the spreading axis of the Costa Rica Rift [1998]. Correlation of the lithostratigraphy in Holes 504B and 896A (1 km south of Site 504B) based on magnetic properties of the extrusive rocks show that the inclinations of the upper volcanic sequences in both holes are very close to the stable reference inclination indicating that these sequences have not been tilted [Allerton et al., 1996; Dilek et al., 1996b]. Therefore, tilting of the lower volcanic sequence in Hole 504B must have taken place prior to the extrusion of the upper volcanic sequence and possibly within and/or near the crustal accretion zone in the spreading axis environment [Dilek, 1998]. The fault zone at 2111 mbsf in the lower sheeted dike complex at Site 504B is characterized by densely populated subvertical fractures, as observed both in core samples and in the Formation MicroScanner (FMS) and Dualaterog (DLL) images of the borehole [Dilek et al., 1996a] and may correspond to one of the half-graben bounding normal faults as depicted on single channel seismic reflection profiles [Langseth et al., 1988; Dilek, 1998].

The analyses of core samples and geophysical downhole measurements indicate that the main deformation process in the sheeted dike complex of the intermediate-spread oceanic crust at Site 504B was fracturing and veining [Dilek et al., 1996a]. Veins filled with chlorite and/or actinolite represent extension fractures and form two orthogonal vein systems [Tartarotti et al., 1995; Dilek et al., 1996a]. E-SE striking and steeply dipping extensional veins are commonly parallel to the dike margins and to the orientation of the Costa Rica Rift axis; N-NE striking veins are dike-orthogonal, representing thermal contraction cracks. There is a marked increase in grain size of the diabasic dike rocks and in abundance of actinolite at the expense of clinopyroxene and plagioclase as chlorite diminishes in the core near 2 kmbsf. These changes in the core mineralogy are accompanied by a steady increase in V_p, which reaches a value of 6.8 km/s nearly 1.4-1.6 km into the basement within the sheeted dike complex [Detrick et al., 1994; Salisbury et al., 1996]. These values of the seismic wave velocities thus suggest that the boundary between seismic layers 2 and 3 corresponds to changes in physical properties of the rocks, rather than a lithological boundary, over a depth interval within the sheeted dike complex [Detrick et al., 1994; Dilek, 1998].

6.2. Crustal Deformation and the Role of Interplay between Faulting and Magmatism at Slow-Spreading Centers

The deformed nature of thin crust, exposure of lower crustal and upper mantle rocks on the seafloor, and the existence of extrusive rocks directly overlying the serpentinized peridotites in slow-spreading oceanic crust and its ophiolite analogs indicate that mechanical extension via faulting play a critical role during the in-situ evolution of slow-spreading oceanic lithosphere. Cross-cutting relations between different generations of dike intrusions and faults suggest that tectonic extension and magmatism may be episodic events that alternate through time with some overlap. The surface expression of dominantly fault-controlled rifting and seafloor spreading is characterized by a distinct, broad depression produced by inward-dipping (towards the spreading axis) normal faults creating a stairstep morphology to the valley walls, as observed on the seafloor along the Mid-Atlantic Ridge and its on-land exposure in Iceland. The Holocene to Pleistocene fissure and fault swarms occurring in recent basaltic lava flows form an active rift zone with a median valley that runs northeast through the center of Iceland. The rift zone in the northeast (Krafla) and the southwest consists of 5-10-km-wide and 40-80-km-long, linear to curvilinear zones of faults, tension fractures (mode I), and volcanic fissures (Fig. 20). Commonly, fault planes dip between 65° and 79° and become less steeply dipping with depth in the crust [Gudmundsson, 1992]. Field studies have shown that most of the normal faults in the rift zone on Iceland nucleate on large scale tension fractures in the upper part and on a set of inclined joints in the lower part of the lava flows, and that they subsequently propagate downwards into considerable crustal depths (2 to 4 km) as faults [Forslund and Gudmundsson, 1992]. In some of the Mid-Atlantic Ridge sections where the magma supply is more scarce compared to Iceland, amagmatic extension (i.e., mechanical extension) predominates, keeping pace with plate separation and seafloor spreading. In these regions, fault zones comprise interconnected normal, oblique-slip, and strike-slip faults and whole-crust detachment faults are common [Mutter and Karson, 1992]. Seismic reflection data suggest that major normal faults that are exposed on the seafloor as tectonic discontinuities bounding asymmetric half-grabens typically have an asymmetric spoon shape and are elongated parallel to the spreading axis. These detachment surfaces and rotational slip along them are responsible for the creation of large fault blocks with the three-dimensional geometry of elongate lozenges of lower crustal and upper mantle material [Mutter and Karson, 1992].

The contact between the sheeted dike complex and the underlying plutonic sequence in the Troodos and Kizildag ophiolites is locally tectonized, indicating that this rheological boundary was decoupled during the evolution of the Neo-Tethyan oceanic crust. The fault boundary in the Kizildag ophiolite is marked by several meter-thick shear zones in which both the gabbroic and diabasic dike rocks were deformed in brittle and brittle-plastic domains. The presence of hydrothermal veins, dikes, and sills along and across these shear zones suggest that deformation occurred in intra-oceanic conditions, rather than during the emplacement of the ophiolite onto the continental margin. Shear sense indicators in faulted and deformed gabbros near and within the fault zones give normal-sense of shearing compatible with the kinematics of the regional extensional deformation in the oceanic realm. Mylonitic discrete shear zones and locally well-developed foliation planes in isotropic gabbros beneath the fault zones contain textures and mineral assemblages indicating their formation under intermediate- to high temperature lithospheric conditions during crustal extension [Dilek, 1997]. In the Troodos ophiolite, gently to moderately dipping faults along E-W striking contacts between the gabbro and sheeted dikes in the western side of the Solea graben define a detachment surface (Fig. 21), along which slickensides on fault planes indicate top-to-east or northeast movement of the faulted dike blocks over the gabbro [Varga and Moores, 1985; Moores et al., 1990; Hurst et al., 1994]. The detachment surface locally contains short (~100 m) horizontal steps and steeper ramps that collectively form a gently dipping fault zone. Sheeted dikes along the detachment surface are commonly highly fractured and brecciated, whereas the gabbro structurally below the detachment surface is fractured and deformed brittlely by numerous small faults, locally creating several hundred meter-thick shear zones [Hurst et al., 1994]. The slickenside lineation orientation and shear sense indicators from the shear zones suggest the E-W extensional nature of these structures in the gabbro.

With progressive cooling of the slow-spreading oceanic crust, brittle deformation via faulting and fissuring propagates downward into the lower crust, and brittle structures merge with zones of ductile deformation producing brittle-ductile shear zones [Reynolds and Lister, 1990]. The diffuse reflectivity and associated reflecting bands in the lower crust as observed in seismic data probably represent these zones of brittle-ductile to ductile deformation. Spatial and temporal relations between discrete shear zones and hydrothermal veins, trondhjemitic intrusions, and late hydrothermal breccias in gabbroic rocks from the MARK area and Site 735B along the Southwest Indian Ridge indicate that brittle-ductile and ductile shear zones in these rocks facilitated the penetration of fluids into the lower crust, resulting in localized pervasive high-temperature alteration at temperatures above 500°C [Mével et al., 1991; Gillis et al., 1993b; Kelley et al., 1993; Dilek et al., 1997b]. Hydrothermal alteration began preferentially within and along these locally developed extensional discrete shear zones, which acted as high-permeability pathways for Na-rich fluids that caused albite enrichment in the plagioclase and Fe-enrichment in the clinopyroxene [Dilek et al., 1997b]. The concentration of Fe-Ti oxides along some of these discrete shear zones suggests that the shear zones also controlled late magmatic (evolved) and subsolidus fluid flow in the gabbroic rocks. Crack networks within and adjacent to the discrete shear zones facilitated amphibole veining with progressive cooling of the rocks below 500°C. The corresponding increase in alteration intensity with the distribution of veins indicates that the degree of alteration in the gabbroic rocks was to a large extent controlled by these structures [Gillis et al., 1993b; Dilek et al., 1997b].

The estimated high temperatures for the state of deformation of the gabbros from Site 735B at the slow-spreading Southwest Indian Ridge indicate that the solid-state ductile deformation and the associated metamorphism probably occurred near the vicinity of the crystal-mush region beneath the ridge axis [Cannat et al., 1991]. The similar paleomagnetic properties of the deformed and undeformed gabbros and the uniform inclination of the magnetic vector downhole suggest that they all cooled below their Curie temperatures during the same magnetic interval, indicating that extensional deformation occurred within the ridge axis environment [Pariso et al., 1991]. The normal sense of shearing associated with the mylonitic

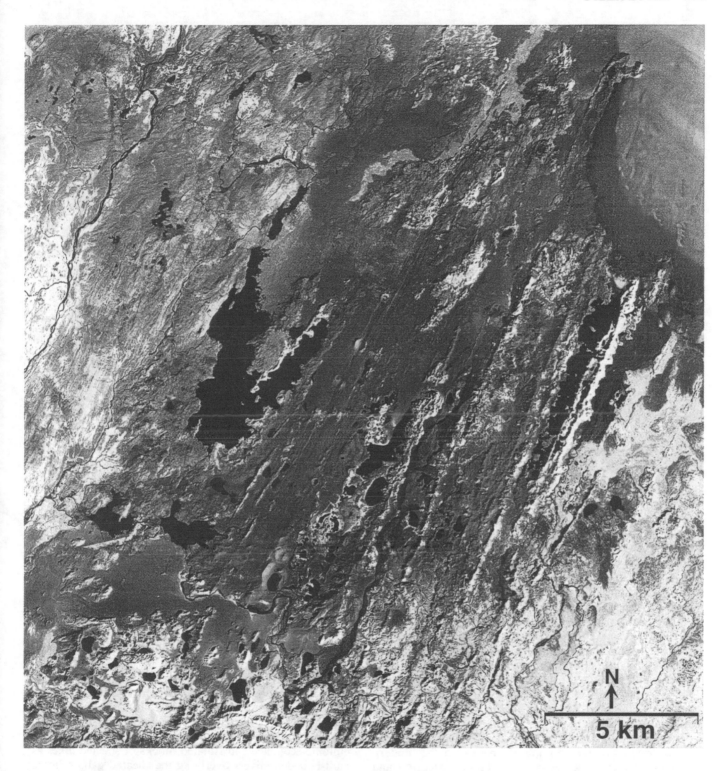

Figure 20. Landsat Thematic Mapper false color multispectral image (principal components of RGB data) of south-central Iceland, showing the distribution and spacing of faults and fault scarps and the spatial and temporal relations between faults, fissure eruptions, and volcanoes. The maximum vertical throw on the faults is 15-20 meters. The lobate feature in the northeastern corner of the image is the westernmost edge of the Vatnajökull glacier.

Figure 21. The Kakopetria detachment fault between the sheeted dike complex above and the sheared gabbro below at the town of Lemithou in the Troodos ophiolite (view is to the north-northwest). The subhorizontal detachment surface contains several faults and fault ramps within the sheared gabbro. Note the highly rotated sheeted dikes in the hanging wall of the detachment surface.

bands and shear zones suggests that ductile deformation of the gabbros near the ridge axis was related to stretching and extension of the newly crystallized crustal material. Extensional deformation of the lower crust was facilitated by normal faulting along high to listric faults, causing block rotations up to 18° as suggested by the paleomagnetic data (i.e., steep magnetic inclination of the core; Pariso et al., 1991]. The exact nature and geometry of these inferred normal faults at depth are not known; however, they are interpreted to have soled at depth into a detachment surface cutting through the entire axial lithosphere [Cannat et al., 1991; Dick et al., 1992].

The existence of dike intrusions and plutons crosscutting faults and detachment surfaces in the ophiolites and the occurrence of recent volcanic eruptions across lower crustal and mantle rocks on the modern seafloor indicate that periods of tectonic extension and amagmatic stretching are followed by renewed episode(s) of magmatism along slow-spreading ridge segments. For example, the active neovolcanic zone and isolated volcanoes in the northern cell in the MARK area point to renewed magmatism along the median valley of the ridge

axis after gabbroic rocks and serpentinized peridotites were uplifted to the seafloor in the footwall of a detachment fault dipping eastward beneath the spreading axis (Figs. 3 and 4). Local intrusions of plagiogranite dikes and stocks across the detachment surface near the town of Kalopanayiotis in the westernmost part of the Solea graben in the Troodos ophiolite [Hurst et al, 1994] also indicate that tectonic extension associated with the formation of the detachment surface was followed by renewed subsidiary magmatic activity. Similarly, the existence of late diabasic dikes crosscutting rotated fault blocks and a late lava sequence unconformably overlying tilted, earlier lavas in the Larnaca graben shows that tectonic extension was followed by a renewed igneous activity and magmatic extension. In the Kizildag ophiolite, the existence of a uniformly thin plutonic sequence and isolated blocks of sheeted dikes as fault-bounded slivers overlying the sheared gabbro in the eastern massif indicates a substantial amount of attenuation in the plutonic foundation. The pillow lavas and their feeder dikes directly overlying and crosscutting, respectively, the serpentinized peridotites also indicate that the crustal denudation and unroofing of the mantle rocks were followed by a late-stage magmatic activity during evolution of the ophiolite.   Basaltic dikes intruding the peridotites and gabbros in the Ligurian ophiolites are locally intrusive into the overlying ophiolitic breccias, indicating a renewed episode of magmatism following the exhumation of mafic and ultramafic rocks and deposition of the breccia on a tectonically active seafloor. This type of interplay between tectonic stretching and magmatic extension is a function of fluctuating magma supply to the spreading axis and appears to be a characteristic feature of oceanic crust generation at slow-spreading ridge environments [Karson, 1990; Cannat et al., 1991].

6.3. Serpentinization and Tectonism

The dominant fabric of the anastomosing foliation and the serpentine mineral phases in serpentinized peridotites from Site 920 in the MARK area indicates that much of the hydration and serpentinization in slow-spreading oceanic lithosphere occurred at temperatures around 350°-400°C [Dilek et al., 1997a]. Gabbroic veins and veinlets represent a phase of magmatic veining in the serpentinized peridotites prior to formation of the hydrothermal veins, and they are altered by Ca-enriched fluids under greenschist facies conditions at temperatures of 300°-400°C. The earliest vein generations with oblique slip-fibers indicate dilation and contemporaneous shearing that enhanced fluid circulation into the serpentinized peridotites under greenschist facies conditions (Table 1; Dilek et al., 1997a). The extensional chrysotile + magnetite veins were produced by elevated pore fluid pressures contemporaneous with stress release during exhumation of the peridotites. Reactivation of these veins during later extensional events and the generation of crack-seal veins were related to further unroofing of the serpentinites and their emplacement along the rift valley walls due to extensional tectonics [Dilek et al., 1997a].

The preservation of pseudomorphic mesh textures in serpentinized peridotites from Site 895 in the Hess Deep and of their mantle fabric elements suggests that serpentinization was a predominantly static event in this fast-spreading oceanic lithosphere [Früh-Green et al., 1996]. Petrological and stable isotope data together with microtextural relations suggest that the inception of this static metamorphism, driven by hydration reactions in the peridotites and interlayered mafic rocks, occurred at temperatures of approximately 450°-550°C and was enhanced by multiple phases of fracturing, rather than faulting. Progressive serpentinization and hydrothermal alteration from greenschist to zeolite facies were spatially and temporally related to development of multiple generations of macroscopic veins and were associated with the penetration of seawater-dominated fluids into the upper mantle rocks [Früh-Green et al., 1996].   Extensive serpentinization of the fast-spread shallow mantle at EPR was mainly facilitated by a system of steep, E-W oriented

fractures associated with the westward propagation of the Cocos-Nazca rift.

The extent and the geometry of serpentinization beneath the seafloor have significant implications for the nature of the oceanic Moho and for vertical mass transport within the spreading center system. At fast-spreading ridges with a high magma budget, serpentinization of the upper mantle is extremely limited because of the restricted nature of crustal faulting and hydrothermal circulation, and therefore the Moho is represented by a transition from gabbro to fresh ultramafic cumulates [i.e., Früh-Green et al., 1994]; at slow-spreading ridges, however, low magma supply results in extensive faulting and widespread hydrothermal alteration in the lower crust and in the upper mantle, placing the oceanic Moho between the serpentinized and fresh peridotites [Mével et al., 1991; Cannat et al., 1992]. Serpentinization of the upper mantle and its subsequent diapiric rise may also play an active role in exhumation and emplacement of the mantle material on the seafloor along slow-spreading ridges [Bonatti and Honnorez, 1976; Rona et al., 1987].

Hess [1962] originally proposed that the Moho was an alteration front in the mantle and that seismic layer 3 was largely partially serpentinized peridotite. In his model, the Moho was produced by auto-metasomatism of peridotite by deuteric water as the mantle cooled into the stability field of serpentine after being emplaced by solid-state flow to the base of the crust. Subsequent studies found, however, that it was difficult to match laboratory observations of both seismic P- and S-wave velocities for partially serpentinized peridotite to those of layer 3 [Christensen, 1972; Christensen and Salisbury, 1975], and Hess's model was rejected in favor of a gabbroic lower crust. Recent studies suggest that the ocean crust has a complex three-dimensional structure that is highly dependent on magma supply and spreading rates without large steady-state magma chambers [e.g., Whitehead et al., 1984; Detrick et al., 1990; Sinton and Detrick, 1992; Barth, 1994; Carbotte and MacDonald, 1994]. Compilations of dredge results and seismic data have indicated that a continuous gabbroic layer does not exist at slow-spreading ridges [Mutter et al., 1985; McCarthy et al., 1988; Dick, 1989; Cannat et al., 1992], and that the internal stratigraphy of oceanic crust is controlled by dynamic processes of alteration and tectonism as much as by igneous processes. The exceptional abundance of serpentinized peridotite in dredge hauls from the walls of rift valleys in fracture zones and in the rift mountains away from fracture zones [Aumento and Loubat, 1971; Cannat et al., 1992; Dick, 1989; Thompson and Melson, 1972] may suggest that serpentinite is a major component of seismic layer 3.   In these scenarios, the Moho does not correspond everywhere in the oceans to the boundary between igneous crust and the mantle, but it may be an alteration front corresponding locally to the depth of circulation of seawater down fractures and faults into the Earth's interior. Findings of the shipboard and shore-based

TABLE 1. Tectonic interpretation of low-temperature deformation and hydrothermal alteration of the upper mantle in the MARK area, Site 920, Leg-153.

Timing	Cause/Event	Effect
Early	depression of isotherms (in the absence of a magma chamber), brittle cracking down to the upper mantle, penetration of seawater along cracks (300°–400°C).	pervasive bulk-rock serpentinization
		development of mesh texture
	extensional tectonics, low-angle normal faulting accompanied by dilation (<350°C)	development of extensional V1 veins with oblique fibers
	unroofing of serpentinized peridotites (<350°C), increased fluid pressure, lower deviatoric stress, block faulting.	development of pure-extensional V2 veins
	continued unroofing of the serpentinized peridotites (<350°C), incremental opening due to stress release.	development of V3 crack-seal veins
Late	diminished fluid flow on the shoulders of the rift valley	sealing of extensional fractures by carbonate mineral + pyrite and carbonate mineral + clay veins (V4)

Vein mineralogy : V1, serpentine + actinolite ± chlorite ± talc with oblique-slip fibers; V2, chrysotile and magnetite; V3, lizardite with minor amounts of carbonate and clay minerals, pyrite, and magnetite; V4, carbonate, pyrite, and/or clay minerals. V1 = oldest, V4 = youngest vein generations [data from Dilek et al., 1997a].

studies on the core samples recovered on ODP Legs 109 and 153 indicate that some peridotites were serpentinized at over 350°C. Such temperatures correspond to depths of several kilometers below seafloor, consistent with Hess' hypothesis that the Moho could locally be a serpentinization front.

6.4. Nature of Segmentation along Spreading Centers

Along-axis variations in the processes of both accretion and deformation result in axial discontinuities at a range of scales and define ridge segmentation [Schouten et al., 1985; Macdonald et al., 1988]. Segmentation has been described on both fast- and slow-spreading ridges. On fast-spreading ridges, transform faults, large overlapping spreading centers/propagating rifts, and small overlapping spreading centers are considered first-, second-, and third-order discontinuities, respectively [Macdonald et al., 1988; Carbotte and Macdonald, 1994]. At spreading rates greater than 145 km/my, ridge offsets are taken up by propagating rifts and overlapping spreading centers, instead of transform faults [Naar and Hey, 1989], and transform-free lengths of 500 km or more are common along the fast-spreading East Pacific Rise south of the equator. At slower spreading ridges, morphological segmentation occurs on a scale of 20 to 50 km through variations in the axial depth, from highs at the segment centers to lows at the boundaries, which are accompanied by changes in the relief, width, and structural style of the axial depression [Sempere et al., 1990; Mutter and Karson, 1992]. These smaller-scale, non-transform discontinuities along slow-spreading ridges are commonly associated with oblique-slip transfer faults and/or detachment faults and thus are tectonic in origin (Fig. 3).

The nature and scale of segmentation of inferred spreading centers in the ophiolites are comparable to those observed and documented from the modern mid-ocean ridges. The fast-spread oceanic crust of the Semail ophiolite displays no evidence for transform fault segmentation along the entire 500-km-long section of the spreading ridge. However, the occurrence of mantle diapirs with dike penetration zones marks ridge segmentation on the order of 50-100 km and is interpreted to represent propagating rift and/or overlapping spreading centers [MacLeod and Rothery, 1992; Nicolas et al., 1994; Nicolas and Boudier, 1995], corresponding to second- and third-order discontinuities along the East Pacific Rise. The propagating rift tectonics of the Solund-Stavfjord ophiolite in western Norway is also a good example of second-order ridge segmentation in an intermediate-spread ancient oceanic crust. In the slow-spreading oceanic crust of the Troodos and Kizildag ophiolites, ridge segmentation is pronounced by transform faults, transfer faults, and detachment surfaces, which are analogous both in size and geometry to those documented from the Mid-Atlantic Ridge. The Arakapas fault zone in the southern part of the Troodos ophiolite represents an oceanic transform fault

boundary, which defines a first-order ridge segmentation. The detachment surface in the Solea graben in the west is a smaller-scale tectonic segmentation of a non-transform origin. In the Kizildag ophiolite, the northwest-striking Tahtaköprü fault, which separates its two massifs with distinctly different crustal architecture, is perpendicular to the general trend of the sheeted dike complex and to the inferred Kizildag spreading axis and is interpreted to represent a transfer fault juxtaposing different structural levels of the Cretaceous oceanic lithosphere [Dilek and Delaloye, 1992; Dilek and Thy, 1998]. These analogies suggest that the mode and scale of segmentation along spreading centers are closely related to focused mantle upwelling and melt migration (at fast-spreading) and mechanical extension and poor magma supply (at slow-spreading).

6.5. Effects of Thermal Structure and Magma Supply in Oceanic Crust Generation

Robust magmatism along fast-spreading ridge systems and tectonic extension alternating through time with magmatism along slow-spreading ridge systems are a result of the thermal structure of and the rate of magma supply to the ridge system [Phipps Morgan et al., 1994]. The thermal structure defined by the delicate balance between magmatic heat input and hydrothermal heat removal directly controls the axial morphology and surface expression of the spreading center. The magma chamber at a fast-spreading ridge is a narrow (~1-km-wide), thin (~50 to hundreds of meters thick) melt lens that overlies a broader (~6-km-wide), thicker (2- to 4-km) region of hot rock with a small (<3-5%) melt fraction (Fig. 22; Phipps Morgan et al., 1994). The magma injected from mantle into the crust accumulates in the melt lens at the base of the sheeted dike complex while the mush zone below subsides and forms the lower oceanic crust. This partially molten crystal mush zone is situated within a much larger, hot but largely solid "transition zone" (Fig. 23). The steep magmatic flow fabrics with vertical lineations observed in the upper-level gabbros in the Semail ophiolite and in the fast-spread modern oceanic crust (e.g., Site 894 gabbros) probably developed by magma ascent in the sub-lens mush zone and by the penetration of melt through a non-rigid matrix of crystal mush zone. Hydrothermal circulation facilitated by the brittle structures in the sheeted dike complex results in cooling of the melt lens from above, causing plating of gabbro from the lens onto the base of the sheeted dike complex. Episodic melt injection and replenishment of magma in turn result in mutually intrusive and crosscutting relations between the sheeted dike complex and the underlying gabbros through time, as observed in the Semail ophiolite. The continuous axial high along the East Pacific Rise is an artifact of the isostatic support provided by the hot zone beneath the magma lens.

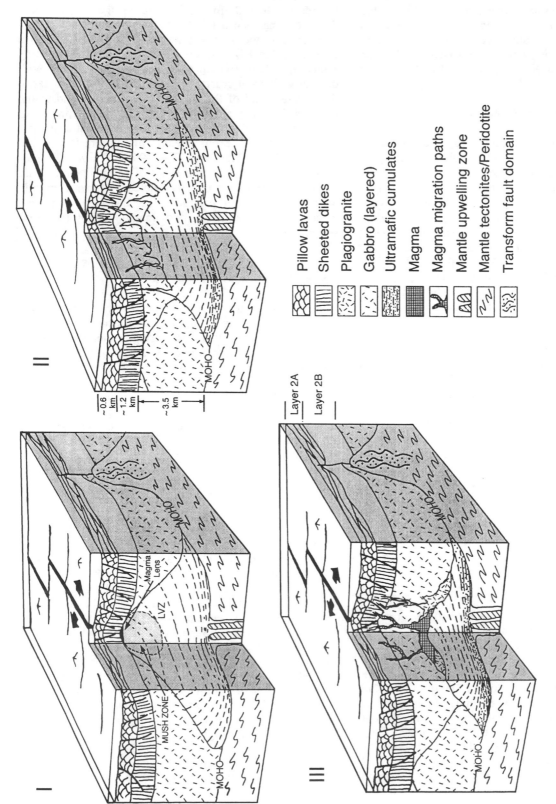

Figure 22. Schematic sequential block diagram showing the generation of oceanic crust from a magma lens at a fast-spreading ridge axis. The low-velocity zone originates at the base of the magma lens, corresponds to a region of elevated temperatures with a few percent partial melt, and extends downwards into the plutonic section. Dashed lines in the plutonic section depict layering and banding in the gabbros which is a result of crustal strain in the magma lens. This fabric becomes more pronounced downsection near the Moho. With progressive cooling of the magma lens through hydrothermal circulation, gabbro underplates the sheeted dike complex. Melt injection and replenishment of magma results in mutual intrusive relations between sheeted dikes and gabbros (stages II and III); fractionation processes near the melt lens produce highly differentiated plagiogranite and trondhjemites.

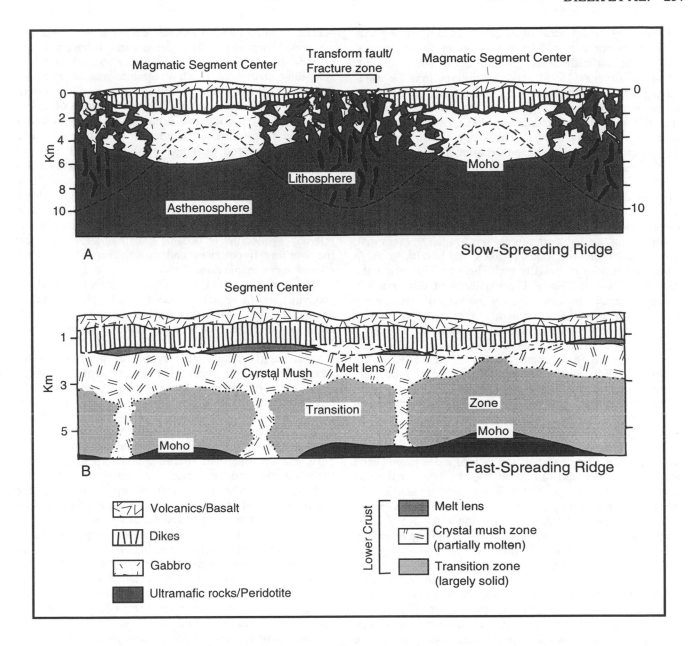

Figure 23. Along-axis profiles of the slow-spreading Mid-Atlantic Ridge (A) and the fast-spreading East Pacific Rise (B), depicting the interpretive models for the internal structure, stratigraphy, and thickness of modern oceanic crust in segment centers and segment ends at these two end-members of the global ridge system [modified from Cannat et al., 1995b; Sinton and Detrick, 1992]. The lithospheric thickness data at the Mid-Atlantic Ridge (MAR) are from Toomey et al. [1988] and Lin et al. [1990]. Segment ends in MAR represent magma-poor, thin-crust domains with significant strike-slip and oblique-slip faults (subvertical thick lines filled with dark gray shade). See text for discussion.

Results of gravity analyses have shown that magmatic accretion on the Mid-Atlantic Ridge (MAR) is highly focused at discrete centers along the ridge axis, and that large positive gravity anomalies corresponding to substantial reductions (~50%) in crustal thickness occur over spreading segments [Lin et al., 1990]. The segment centers, on the other hand, are interpreted to have a much thicker crust (Fig. 23A) as a result of large temperature gradients and a high degree of decompression melting over ascending mantle plumes. This interpretation implies a three-dimensional process of magmatic accretion along-axis of the slow-spreading MAR and suggests that the crustal and lithospheric thicknesses vary along the axis (Fig. 23A). Cannat's [1996] model suggests that the axial lithosphere is thicker at segment ends, where the thin crust is composed of a mixture of tectonically uplifted ultramafic and gabbroic rocks intruded by dikes and gabbros, and overlain by thin lava flows. This model is consistent with the observations made through submersible dives and dredging along the MAR between 22°-24° N [Cannat et al., 1995b]. In these cold regions, some of the melt extracted from the asthenosphere begin to crystallize below the crust, at depths over 15 km, resulting in relatively high pressures for the onset of fractional crystallization. The model predicts that the segment centers have much thicker crust, composed of gabbros, sheeted dikes, and basaltic lavas, in which the solidus of basaltic melts lie 1 to 3 km below the seafloor [Cannat, 1996]. The existence of thick crust in segment centers is supported by the seismic and gravity data, which suggest that the axial rift valley is replaced by an axial dome [Tolstoy et al., 1993; Detrick et al., 1995]. This configuration implies that extension of the thin lithospheric lid in segment centers at the MAR occurs mainly through diking, with only peripheral solid-state deformation in the underlying gabbroic rocks [Cannat, 1996], and suggests thermal and mechanical conditions that are analogous to those operating at the magma-rich, fast-spreading East Pacific Rise.

Episodic fashion of magmatic emplacement beneath a slow-spreading ridge signals the lack of a continuous axial magma chamber and thus points to the colder nature of accretionary processes. The axial morphology with a well-defined median valley and a structural graben system, as observed in the Mid-Atlantic Ridge and in the Troodos and Kizildag ophiolites, is a result of this "cold" environment in which crustal rocks are much colder and stronger to support large lateral variations in axial crustal structure. Thus, the brittle deformation front is pushed further downward so that faults and cracks in the upper crust can propagate and continue into the lower crust. This in turn facilitates the penetration of seawater into the lower crust and upper mantle resulting in the tectonically controlled hydrothermal alteration of slow-spreading oceanic lithosphere. The extension-induced component of the permeability structure is thus a function of axial thermal structure and plays a significant role in slow-spreading oceanic crust.

7. CONCLUSION

In this paper, we have attempted to summarize the currently available data and information on the structure and tectonics of modern oceanic crust and ophiolites, based to a considerable extent on our own work, to examine the mode and nature of interplay between magmatism and tectonism and the role of spreading rate on the internal architecture of oceanic crust and its intraoceanic deformation. Ophiolites are good structural analogs for oceanic lithosphere and provide direct access and opportunity to study seafloor spreading structures (both magmatic and amagmatic) developed during the evolution of oceanic crust. Observations and data obtained from the studies of drilled core samples from modern oceanic crust help us better constrain the nature of magmatic, metamorphic, and tectonic events and their spatial and temporal relations through time during crustal accretion, and structural studies in ophiolites benefit greatly from this vital input from direct observations of modern oceanic crust; in turn, ophiolites provide us with four dimensional (including time-age relations) exposures to examine directly the internal architecture of oceanic crust and allow us to test the working hypotheses and models on the structural evolution of oceanic crust.

The differences in the internal structure of slow-spreading and fast-spreading oceanic lithosphere and their ancient analogs result from the variations in magma supply and hence the thermal structure beneath the spreading axes. In slow-spreading environments, low and/or diminished magma supply results in freezing of magma and the absence of stable magma lens at shallow depths creating a "cold" environment, in which tectonic extension via crustal stretching and faulting predominates. This amagmatic extension is responsible for the development of detachment surfaces between the rheological/lithological boundaries, extensive fracturing and faulting which in turn facilitate pervasive hydrothermal alteration and serpentinization in the lower crust and upper mantle, denudation of lower crust and uplift of the serpentinized upper mantle, and segmentation of ridge axes. In fast-spreading centers, steady-state magma supply and the existence of a melt lens at shallow crustal depths create a "hot" environment in which there is little magma freezing and reacting with the mantle as it ascends beneath the ridge [Phipps Morgan et al., 1994]. Continuous magma injection keeps pace with seafloor spreading and plate separation, and therefore fast-spreading oceanic crust does not undergo significant amagmatic extension and thinning. This is consistent with the observations from ophiolites with inferred fast-spread origin in terms of the near absence of block fault rotation, detachment surfaces, and listric faults in them. Rift propagation and overlapping spreading centers are the main mode of segmentation in fast-spread(ing) oceanic lithosphere.

Acknowledgments. Our field work and research in ophiolites have been supported by NSF grants to Y. Dilek (EAR-9219064 and EAR-9796011) and E.M. Moores (EAR-8306959), research grants (D.41.31.147) from the Norwegian Research Council for Science and the Humanities to H. Furnes, National Geographic Society Research Committee and Miami University Faculty Research grants to Y. Dilek. Y. Dilek also acknowledges two JOI/USSAC grants in support of his post-cruise research for the ODP Legs 148 and 153 to the Costa Rica

Rift and the MARK area, respectively. Observations and interpretations presented in this paper have benefitted from close interactions and discussions with J. Karson, J. Casey, J. Alt, K. Becker, G. Harper, F. Vine, C.A. Eddy, S.D. Hurst, C. MacLeod, A. Nicolas, K. Skjerlie, R. Pedersen, R. Twiss, and R. J. Varga. Constructive reviews by A. Nicolas and J.A. Karson improved the paper.

REFERENCES

Abbate, E., V. Bortolotti, and G. Principi, Apennine ophiolite: a peculiar oceanic crust, *Ofioliti, 1,* 59-96, 1980.

Adamson, A.C., Basement lithostratigraphy, Deep Sea Drilling Project Hole 504B, *Init. Repts. DSDP, 83,* 121-128, 1985.

Agar, S.M., Fracture evolution in the upper oceanic crust: evidence from DSDP Hole 504B, in *Deformation Mechanisms, Rheology and Tectonics,* edited by R.J. Knipe and E.H. Rutter, *Geol. Soc. London Spec. Publ., 54,* 41-50, 1990.

Agar, S.M., Microstructural evolution of a deformation zone in the upper ocean crust: evidence from DSDP Hole 504B, *J. Geodynamics, 13,* 119-140, 1991.

Agar, S.M., and F.C. Marton, Microstructural controls on strain localization in ocean crust diabases: evidence from Hole 504B, *Proc. Ocean Drilling Program Sci. Results, 137/140,* 219-229, 1995.

Allerton, S., and F.J. Vine, Spreading evolution of the Troodos ophiolite, Cyprus, *Geology, 19,* 637-640, 1991.

Allerton, S., H.-U. Worm, and L.B. Stokking, Paleomagnetic and rock magnetic properties of Hole 896A, in *Proc. Ocean Drilling Program, Sci. Results, 148,* 217-226, 1996.

Alt, J.C., and H. Kinoshita, L.B. Stokking, et al., *Proc. ODP, Init. Repts., 148: College Station,* TX (Ocean Drilling Program), 1993.

Anderson, R.N., J. Honnorrez, and K. Becker, et al., *Init. Repts. DSDP, 83:* Washington (U.S. Govt. Printing Office), 1985.

Aumento, F., and H. Loubat, The Mid-Atlantic Ridge Near 45°N, XVI, Serpentinized ultramafic intrusions, *Can. J. Earth Sci., 8,* 631-663, 1971.

Auzende, J.M., M. Cannat, P. Gente, J.P. Henriet, T. Juetau, J.A. Karson, Y. Lagabrielle, and M.A. Tivey, Deep layers of mantle and oceanic crust exposed along the southern wall of the Kane Fracture Zone: Submersible observations, *C.R. Acad. Sci. Ser. 2, 317,* 1641-1648, 1993.

Baker, N., P. Fryer, and F. Martinex, Rifting history of the northern Mariana Trough: SeaMARC II and seismic reflection surveys, *J. Geophys. Res., 101,* 11,427-11,455, 1996.

Ballard R.D., R.T. Holcomb, and T.H. van Andel, The Galapagos Rift at 86°W: 3. Sheet flows, collapse pits, and lava lakes of the rift valley, *J. Geophys. Res., 84,* 5406-5422, 1979.

Ballard R.D., S. Francheteau, T. Juteau, C. Rangan, and W. Normark, The East Pacific Rise at 21°N: The volcanic, tectonic, and hydrothermal processes of the central axis, *Earth Planet. Sci. Lett., 55,* 1-10, 1981.

Barth, G.A., Ocean crust thickens approaching the Clipperton

fracture zone, *Mar. Geophys. Res., 16,* 51-64, 1994.

Becker, K., and H. Sakai, et al., Proceedings of the Ocean Drilling Program, Scientific Results, 111: College Station, TX (Ocean Drilling Program), 1988.

Becker, K., et al., Drilling deep into young oceanic crust, Hole 504B, Costa Rica Rift, *Rev. Geophys., 27,* 79-101, 1989.

Blackman, D.K., and D.W. Forsyth, Isostatic compensation of tectonic features of the Mid-Atlantic Ridge: 25°-27°30'S, *J. Geophys. Res., 96,* 11741-11758, 1991.

Bonatti, E., and J. Honnorez, Sections of the Earth's crust in the Equatorial Atlantic, *J. Geophys. Res., 81,* 4087-4103, 1976.

Bonatti, E., and C.G.A. Harrison, Eruption styles of basalt in oceanic spreading ridges and seamounts: effect of magma temperature and viscosity, *J. Geophys. Res., 93,* 2967-2980, 1988.

Borsi, L., U. Schärer, L. Gaggero, and L. Crispini, Age, origin, and geodynamic significance of plagiogranites in lherzolites and gabbros of the Piedmont-Ligurian ocean basin, *Earth Planet. Sci. Lett., 140,* 227-241, 1996.

Boudier, F., and R.G. Coleman, Cross-section through the peridotite in the Samail ophiolite, southeastern Oman Mountains, *J. Geophys. Res., 86,* 2573-2592, 1981.

Boudier, F., and A. Nicolas, Nature of the Moho transition zone in the Oman ophiolite, *J. Petrol., 36,* 777-790, 1995.

Boudier, F., C.J. MacLeod, and L. Bolou, Structures in peridotites from Site 895, Hess Deep: Implications for the geometry of mantle flow beneath the East Pacific Rise, *Proc. Ocean Drilling Program Sci. Results, 147,* 347-356, 1996.

Bown, J.W., and R.S. White, Variation with spreading rate of oceanic crustal thickness and geochemistry, *Earth Planet. Sci. Lett., 121,* 435-439, 1994.

Brown, J.R., and J.A. Karson, Variations in axial processes on the Mid-Atlantic Ridge: the median valley of the MARK area, *Mar. Geophys. Res., 10,* 109-138, 1988.

Cann, J.R., M.G. Langseth, et al., *Init. Repts. DSDP, 69:* Washington (U.S. Govt. Printing Office), 1983.

Cann, J.R., D.K. Blackman, D.K. Smith, E. McAllister, B. Janssen, S. Mello, E. Avgerinos, A.R. Pascoe, and J. Escartin, Corrugated slip surfaces formed at ridge-transform intersections on the Mid-Atlantic Ridge, *Nature, 385,* 329-332, 1997.

Cannat, M., Emplacement of mantle rocks in the seafloor at mid-ocean ridges, *J. Geophys. Res., 98,* 4163-4172, 1993.

Cannat, M., T. Juteau, and E. Berger, Petrostructural analysis of the Leg 109 serpentinized peridotites, in *Proc. ODP, Sci. Results, 106/109,* 47-56, 1990.

Cannat, M., C. Mevel, and D. Stakes, Stretching of the deep crust at the slow-spreading Southwest Indian Ridge, *Tectonophysics, 190,* 73-94, 1991.

Cannat, M., D. Bideau, and H. Bougault, Serpentinized peridotites and gabbros in the Mid-Atlantic Ridge axial valley at 57°37'N and 16°52'N, *Earth Planet. Sci. Lett., 109,* 87-106, 1992.

Cannat, M., J.A. Karson, D.J. Miller, et al., *Proc. ODP, Init. Repts., 153: College Station,* TX (Ocean Drilling Program), 798 pp, 1995a.

Cannat, M., and ten others, Thin crust, ultramafic exposures, and rugged faulting at the Mid-Atlantic Ridge (22°-24°N), *Geology, 23*, 49-52, 1995b.

Cannat, M., How thick is the magmatic crust at slow spreading oceanic ridges?, *J. Geophys. Res., 101*, 2847-2857, 1996.

Carbotte, S., and K.C. Macdonald, Comparison of seafloor tectonic fabric at intermediate, fast, and super fast spreading ridges: Influence of spreading rate, plate motions, and ridge segmentation on fault patterns, *J. Geophys. Res., 99*, 13609-13631, 1994.

Casey, J.F., J.A. Karson, D. Elthon, E. Rosencrantz, and M. Titus, Reconstruction of the geometry of accretion during formation of the Bay of Islands ophiolite complex, *Tectonics, 2*, 509-528, 1983.

Ceuleneer, G., Evidences for paleo-spreading centers in the Oman ophiolite: mantle structures in the Maqsad area, in *Ophiolite Genesis and Evolution of the Oceanic Lithosphere*, edited by Tj. Peters, A. Nicolas, and R.G. Coleman, *Kluwer Academic Publishers, The Netherlands*, 147-174, 1991.

Christensen, N.I., The abundance of serpentinites in the oceanic crust, *J. Geol., 80*, 709-719, 1972.

Christensen, N.I., and Salisbury, M.H., Structure and constitution of the lower oceanic crust, *Rev. Geophys. Space Phys., 13*, 57-86, 1975.

Christensen, N.I., and J.D. Smewing, Geology and seismic structure of the northern section of the Oman ophiolite, *J. Geophys. Res., 86*, 2545-2555, 1981.

Coleman, R.G., Plate tectonic emplacement of upper mantle peridotites along continental edges, *J. Geophys. Res., 76*, 1212-1222, 1971.

Collier, J., and M. Sinha, Seismic images of a magma chamber beneath the Lau Basin backarc spreading center, *Nature, 348*, 646-648, 1990.

Cortesogno, L., B. Galbiati, and G. Principi, Note alla "Carta geologica delle ofioliti del Bracco" e ricostruzione della paleogeografia giurassico-cretacica, *Ofioliti, 12*, 261-342, 1987.

Detrick, R., P. Buhl, J. Mutter, E. Vera, J. Orcutt, J. Madsen, and T. Brocher, Multichannel seismic imaging of magma chamber along the East Pacific Rise, *Nature, 326*, 35-41, 1987.

Detrick, R.S., J.C. Mutter, P. Buhl, and I.I. Kim, No evidence from multichannel reflection data for a crustal magma chamber in the MARK area on the Mid-Atlantic Ridge, *Nature, 347*, 61-64, 1990.

Detrick, R.S., R.S. White, and G.M. Purdy, Crustal structure of north Atlantic fracture zones, *Rev. Geophys., 31*, 439-458, 1993.

Detrick, R.S., J. Collins, R. Stephen, and S. Swift, *In situ* evidence for the nature of the seismic layer 2/3 boundary in oceanic crust, *Nature, 370*, 288-290, 1994.

Detrick, R.S., H.D. Needham, and V. Renard, Gravity anomalies and crustal thickness variations along the Mid-Atlantic Ridge between 33°N and 40°N, *J. Geophys. Res., 100*, 3767-3787, 1995.

Dewey, J.F., and J.M. Bird, Origin and emplacement of the ophiolite suite: Appalachian ophiolites in Newfoundland, *J. Geophys. Res., 76*, 3179-3206, 1971.

Dick, H.J B., Abyssal peridotites, very slow spreading ridges and ocean ridge magmatism, in *Magmatism in the Ocean Basins*, edited by A.D. Saunders and M.J. Norry, *Geol. Soc. London Spec. Pub., 42*, 71-105, 1989.

Dick, H.B.J., Lithostratigraphic constraints on oceanic crustal structure, *EOS Transactions, Am. Geophys. Union, 77*, 748, 1996.

Dick, H.B.J., G. Thompson, and W.B. Bryan, Low angle faulting and steady state emplacement of plutonic rocks at ridge-transform intersections, *EOS Transactions, Am. Geophys. Union, 62*, 406, 1981.

Dick, H.J.B., P.S. Meyer, S. Bloomer, S. Kirby, D. Stakes, and C. Mawer, Lithostratigraphic evolution of an *in-situ* section of oceanic crust, *Proc. Ocean Drilling Program, Sci. Results, 118*, 439-538, 1991.

Dick, H.J.B., P.T. Robinson, and P.S. Meyer, The plutonic foundation of a slow-spreading ridge, *Am. Geophys. Union, Geophysical Monograph 70*, 1-39, 1992.

Dick, H.J.B., J.A. Erzinger, L.B. Stokking, et al., *Proc. ODP, Init. Repts.*, 140: College Station, TX (Ocean Drilling Program), 1992.

Dilek, Y., and E.M. Moores, Structure, petrology, and origin of Kizil Dagh ophiolite, southern Turkey: Comparison with Troodos, *EOS Transactions, Am. Geophys. Union, 66*, 1129, 1985.

Dilek, Y., Seafloor spreading structure of the Kizildag ophiolite, Turkey, in *Proceedings of the International Earth Sciences Congress on Aegean Regions, IESCA 1995*, edited by Ö. Piskin, M. Ergün, M.Y. Savasçin, and G. Tarcan, Izmir-Turkey, 1, 37-52, 1997.

Dilek, Y., Structure and tectonics of intermediate-spread oceanic crust drilled at DSDP/ODP Sites 504B-896A, Costa Rica Rift, in *Geological Evolution of Ocean Basins: Results from the Ocean Drilling Program*; edited by A. Cramp, C.J. McLeod, S.V. Lee, and E.J.W. Jones, *Geological Society of London Special Publication*, 131, 179-197, 1998.

Dilek, Y., and E.M. Moores, Regional tectonics of the eastern Mediterranean ophiolites, in *Ophiolites, Oceanic Crustal Analogues*, edited by J. Malpas, E.M. Moores, A. Panayiotou, and C. Xenophontos, *Geol. Survey Dept. Cyprus*, Nicosia, 295-309, 1990.

Dilek, Y., P. Thy, E.M. Moores, and T.W.Ramsden, Tectonic evolution of the Troodos ophiolite within the Tethyan framework, *Tectonics, 9*, 811-823, 1990.

Dilek, Y., E.M. Moores, M. Deleloye, and J.A. Karson, A magmatic extension and tectonic denudation in the Kizildag ophiolite, southern Turkey: Implications for the evolution of Neotethyan oceanic crust, in *Ophiolite Genesis and Evolution of the Oceanic Lithosphere*, edited by Tj. Peters, A. Nicolas, and R.G. Coleman, *Kluwer Academic Publishers, The Netherlands*, 487-502, 1991.

Dilek, Y., and M. Delaloye, Structure of the Kizildag ophiolite, a slow-spread Cretaceous ridge segment north of the Arabian promontory, *Geology, 20*, 19-22, 1992.

Dilek, Y., and C.A. Eddy, The Troodos (Cyprus) and Kizildag (S. Turkey) ophiolites as structural models for slow-spreading ridge segments, *J. Geology, 100*, 305-322, 1992.

Dilek, Y., G.D. Harper, P. Pezard, and P. Tartarotti, Structure of the lower sheeted dike complex in Hole-504B (Leg 148), in *Proc. Ocean Drilling Program, Sci. Results, 148*, 229-243, 1996a.

Dilek, Y., G.D. Harper, J. Walker, S. Allerton, and P. Tartarotti, Structure of upper layer 2 in Hole 896A, in *Proc. Ocean Drilling Program, Sci. Results, 148,* 261-279, 1996b.

Dilek, Y., Coulton, A., and Hurst, S., Serpentinization and hydrothermal veining in peridotites at Site 920 in the MARK area (Leg 153), in *Proc. Ocean Drilling Program, Sci. Results, 153,* 35-59, 1997a.

Dilek, Y., Kempton, P.D., Thy, P., Hurst, S.D., Whitney, D.L., and Kelley, D., Structure and petrology of hydrothermal veins in gabbroic rocks from Sites 921 to 924, MARK Area (Leg 153): Alteration history of slow-spread lower oceanic crust, in *Proc. Ocean Drilling Program, Sci. Results, 153,* 155-178, 1997b.

Dilek, Y., H. Furnes, and K.P. Skjerlie, Propagating rift tectonics of a Caledonian marginal basin: Multi-stage seafloor spreading history of the Solund-Stavfjord ophiolite in western Norway, *Tectonophysics, 280,* 213-238, 1997c.

Dilek, Y., and P. Thy, Structure, petrology, and seafloor spreading tectonics of the Kizildag ophiolite (Turkey), in *Modern Ocean Floor Processes and the Geological Record,* edited by K. Harrison and R. Mills, *Geol. Soc. London Spec. Publ.,* 1998 (in press).

Dimitriev, B.R., J. Heirtzler, et al., *Initial Reports of the Deep Sea Drilling Project, 46*: Washington (U.S. Government Printing Office), 1978.

Dubertret, L., Géologie des roches vertes du nort-ouest de la Syrie et du Hatay (Turquie), *Notes Mém. Moyen-Orient, 6,* 227, 1955.

Erendil, M., Petrology and structure of the upper crustal units of the Kizildag ophiolite, in *Geology of the Taurus Belt,* edited by O. Tekeli and M.C. Göncüoglu, *Proceedings, Min. Res. Expl. Inst. Turkey, Ankara,* 269-284, 1984.

Forslund, T., and A. Gudmundsson, Structure of Tertiary and Pleistocene normal faults in Iceland, *Tectonics, 57-68,* 1992.

Fox, P.J., R. S. Detrick, and G.M. Purdy, Evidence for crustal thinning near fracture zones: Implications for ophiolites, in *Proceedings of the International Ophiolite Symposium,* edited by A. Panayiotou, *Geological Survey of Cyprus,* Nicosia, 161-168, 1980.

Francheteau, J., R. Armijo, J.L. Cheminée, R. Hekinian, P. Lonsdale, and N. Blum, 1 Ma East Pacific Rise oceanic crust and uppermost mantle exposed by rifting in Hess Deep (equatorial Pacific Ocean), *Earth Planet. Sci. Lett, 101,* 281-295, 1990.

Francheteau, J., R. Armijo, J.L. Cheminée, R. Hekinian, P. Lonsdale, and N. Blum, Dike complex of the East Pacific Rise exposed in the walls of Hess Deep and the structure of the upper oceanic crust, *Earth Planet. Sci. Lett., 111,* 109-121, 1992.

Früh-Green, G.L., A, Plas, and L.N. Dell'Angelo, and C. Lécuyer, Multi-stage hydrothermal alteration of the EPR lower crust and shallow mantle at Hess Deep: mineralogical and stable isotope constraints, *E O S Transactions, Am. Geophys. Union, 75,* 649-650, 1994.

Früh-Green, G.L., A, Plas, and L.N. Dell'Angelo, Mineralogic and stable isotope record of polyphase alteration of upper crustal gabbros of the East Pacific Rise (Hess Deep, Site 894), *Proc. Ocean Drilling Program Sci. Results, 147,* 235-291, 1996.

Furnes, H., K.P. Skjerlie, R.B. Pedersen, T.B. Anderson, C.J. Stillman, R. Suthren, M. Tysseland, and L.B. Garmann, The Solund-Stavfjord Ophiolite Complex and associated rocks, west Norwegian Caledonides: Geology, geochemistry and tectonic environment, *Geol. Mag., 127,* 209-289, 1990.

Furnes, H., R.J. Johansen, and K.P. Skjerlie, FeTi-poor and FeTi-rich basalts in the Solund-Stavfjord ophiolite complex, west Norwegian Caledonides: Relationships and genesis, *N. Jb. Miner. Mh., 4,* 153-168, 1992.

Gass, I.G., The ultramafic volcanic assemblage of the Troodos Massif, Cyprus, in *Ultramafic and Related Rocks*: John Wiley, New York, 121-134, 1967.

Gass, I.G., and J.D. Smewing, Intrusion, extrusion and metamorphism at constructive margins: evidence from the Troodos massif, Cyprus, *Nature, 242,* 26-29, 1973.

Gente, P., R. Pockalny, C. Durand, C. Deplus, M. Maia, G. Ceuleneer, C. Meevel, M. Cannat, and C. Laverne, Characteristics and evolution of the segmentation of the Mid-Atlantic Ridge between 20°N and 24°N during the last 10 million years, *Earth Planet. Sci. Lett., 129,* 55-71, 1995.

Gillis, K., C. Mével, J. Allan, et al., *Proc. ODP, Init. Repts., 147: College Station, TX (Ocean Drilling Program),* 1993a.

Gillis, K.M., G. Thompson, and D.S. Kelley, A view of the lower crustal component of hydrothermal systems at the Mid-Atlantic Ridge, *J. Geophys. Res., 98,* 19597-19619, 1993b.

Gudmundsson, A., Formation and growth of normal faults at the divergent plate boundary in Iceland, *Terra Nova, 4,* 464-471, 1992.

Harding, A.J., J.A. Orcutt, M.E. Kappus, E.E. Vera, J.C. Mutter, P. Buhl, R.S. Detrick, and T.M. Brocher, Structure of young oceanic crust at 13°N on the East Pacific Rise from expanding spread profiles, *J. Geophys. Res., 94,* 12613-12196, 1989.

Hawkins, J. W., S.H. Bloomer, C.A. Evans, and J.T. Melchior, Evolution of intra-oceanic arc-trench systems, *Tectonophysics, 102,* 175-205, 1984.

Hébert, R., A.C. Adamson, and S.C. Komor, Metamorphic petrology of ODP Leg 109, Hole 670A serpentinized peridotites: serpentinization processes at a slow spreading ridge environment, in *Proc. Ocean Drilling Program, Sci. Results, 106/109,* 103-115, 1990.

Hekinian, R., D. Bideau, J. Francheteau, J.L. Chéminée, R. Armijo, P. Lonsdale, and N. Blum, Petrology of the East Pacific Rise crust and upper mantle exposed in the Hess Deep (eastern equatorial Pacific), *J. Geophys. Res., 98,* 8069-8094, 1993.

Henstock, T.J., A.W. Woods, and R.S. White, The accretion of oceanic crust by episodic sill intrusion, *J. Geophys. Res., 98,* 4143-4161, 1993.

Hess, H.H., The history of the ocean basins, *Geol. Soc. Am. Buddington Vol.,* 599-620, 1962.

Hey, R., Tectonic evolution of the Cocos-Nazca spreading center, *Geol. Soc. Am. Bull., 88,* 1404-1420, 1977.

Hey, R., L. Johnson, and A. Lowrie, Recent plate motion in

the Galapagos ·Area, *Geol. Soc. Am. Bull., 88*, 1385-1403, 1977.

Hill, M.N., Recent geophysical exploration of the ocean floor, *Phys. Chem. Earth, 2*, 129-163, 1957.

Honnorez, J., J.C. Alt, B.-M. Honnorez-Guerstein, C. Laverne, K. Muehlenbachs, J. Ruiz, and E. Saltzaman, Stockwork-like sulfide mineralization in young oceanic crust: Deep Sea Drilling Project Hole 504B, DSDP, 83, Washington (U.S. Govt. Printing Office), 263-245, 1985.

Hopson, C.A., R.G. Coleman, R.T. Gregory, J.S. Pallister, and E.H. Bailey, Geologic section through the Samail ophiolite and associated rocks along a Muscat-Ibra transect, southeastern Oman Mountains, *J. Geophys. Res., 86*, 2527-2544, 1981.

Hurst, S.D., and J.A. Karson, Paleomagnetism of tilted dikes in fast spread oceanic crust exposed in the Hess Deep Rift: Implications for spreading and rift propagation, *Tectonics, 13*, 789-802, 1994.

Hurst, S.D., E.M. Moores, and R.J. Varga, Structure and geophysical expression of the Solea graben, Troodos ophiolite, Cyprus, *Tectonics, 13*, 139-156, 1994.

Juteau, T., M. Ernewien, I. Reuber, H. Whitechurch, and R. Dahl, Duality of magmatism in the plutonic sequence of the Semail Nappe, Oman, *Tectonophysics, 151*, 107-135, 1988.

Juteau, T., E. Berger, and M. Cannat, Serpentinized, residual mantle peridotites from the M.A.R. median valley, ODP Hole 670A (21°10'N, 45°02'W, Leg 109): Primary mineralogy and geothermometry, in *Proc. ODP, Sci. Results, 106/109*, 27-45, 1990.

Karson, J.A., Seafloor spreading on the Mid-Atlantic Ridge: implications for the structure of ophiolites and oceanic lithosphere produced in slow-spreading environments, in *Ophiolites, Oceanic Crustal Analogues*, edited by J. Malpas, E.M. Moores, A. Panayiotou, and C. Xenophontos, *Geol. Survey Dept. Cyprus, Nicosia*, 547-555, 1990.

Karson, J.A., and H.J.B. Dick, Tectonics of ridge–transform intersections at the Kane Fracture Zone, *Mar. Geophys. Res., 6*, 51-98, 1983.

Karson, J.A., G. Thompson, S.E. Humpris, J.M. Edmond, W.B. Bryan, J.B. Brown, A.T. Winters, R.A. Pockalny, J.F. Casey, A.C. Campbell, G.P. Klinkhammer, M.R. Palmer, R.J. Kinzler, and M.M. Sulanowska, Along-axis variations in seafloor spreading in the MARK area, *Nature, 328*, 681-685, 1987.

Karson, J.A., S.D. Hurst, and P. Lonsdale, Tectonic rotations of dikes in fast-spread oceanic crust exposed near Hess Deep, *Geology, 20*, 685-688, 1992.

Karson, J.A., Lawrence, R.M., Tectonic setting of serpentinite exposures on the western median valley wall of the MARK area in the vicinity of Site 920, in *Proc. Ocean Drilling Program, Sci. Results, 153*, 3-21, 1997a.

Karson, J.A., Lawrence, R.M., Tectonic window into gabbroic rocks of the middle oceanic crust in the MARK area near Sites 921-924, in *Proc. Ocean Drilling Program, Sci. Results, 153*, 61-76, 1997b.

Kelley, D.S., and J.R. Delaney, Two-phase separation and fracturing in mid-ocean ridge gabbros at temperatures greater than 700°C, *Earth Planet. Sci. Lett., 83*, 53-66, 1987.

Kelley, D.S., K.M. Gillis, and G. Thompson, Fluid evolution in submarine magma-hydrothermal systems at the Mid-Atlantic Ridge, *J. Geophys. Res., 98*, 195-79-195-96, 1993.

Kempner, W.C., and J.F. Geetrust, Ophiolites, synthetic seismograms, and ocean crustal structure. 1. Comparison of ocean bottom seismometer data and synthetic seismograms for the Bay of Islands ophiolite, *J. Geophys. Res., 87*, 8447-8462, 1982.

Kent, G.M., S.A. Swift, R.S. Detrick, J.A. Collins, and R.A. Stephen, Evidence for active normal faulting on 5.9 Ma crust near Hole 504B on the southern flank of the Costa Rica Rift, *Geology, 24*, 83-86, 1996.

Kikawa, E., P.R. Kelso, J.E. Pariso, and C. Richter, Paleomagnetism of gabbroic rocks and peridotites from Sites 894 and 895, Leg 147, Hess Deep: results of half-core and whole-core measurements, *Proc. Ocean Drilling Program ,Sci. Results, 147*, 383-391, 1996.

Kinoshita, H., T. Fruta, and J. Pariso, Downhole magnetic field measurements and paleomagnetism, Hole 504B, Costa Rica Ridge, in *Proc Ocean Drilling Program, Sci. Results, 111*, 147-156, 1989.

Kong, L.S., R.S. Detrick, P.J. Fox, L.A. Mayer, and W.B.F. Ryan, The morphology and tectonics of the MARK Area from Sea Beam and Sea MARK I observations (Mid-Atlantic Ridge 23°N), *Mar. Geophys. Res., 10*, 59-90, 1988.

Kong, L.S.L., S.C. Solomon, and G.M.J. Purdy, Microearthquake characteristics of a mid-ocean ridge along-axis high, *J. Geophys. Res., 97*, 1659-1685, 1992.

Lagabrielle, Y., and M. Cannat, Alpine Jurassic ophiolites resemble the modern Atlantic basement, *Geology, 18*, 319-322, 1990.

Langseth, M.G., M.J. Mottl, M.A. Hobart, and A. Fisher, The distribution of geothermal and geochemical gradients near Site 501/504: implications for hydrothermal circulation in the oceanic crust, *Proc. ODP, Sci. Results, 111*, 23-32, 1988.

Lemoine, M., G. Boillot, and P. Tricart, Ultramafic and gabbroic ocean floor of the Ligurian Tethys (Alps, Corsica, Apennines): In search of a genetic model, *Geology, 15*, 622-625, 1987.

Lin, J., G.M. Purdy, H. Schouten, J.-C. Sempere, and C. Zervas, Evidence from gravity data for focused magmatic accretion along the Mid-Atlantic Ridge, *Nature, 344*, 627-632, 1990.

Lippard, S.J., A.W. Shelton, and I.G. Gass, The ophiolite of Northern Oman, *Geol. Soc. London Memoir 11*, 178, 1986.

Lonsdale, P., Structural pattern of the Galapagos microplate and evolution of the Galapagos triple junctions, *J. Geophys. Res., 93*, 13551-13574, 1988.

MacCarthy, J., J.C. Mutter, J.L. Morton, N.H. Sleep, and G.A. Thompson, Relic magma chamber structures preserved within the Mesozoic North Atlantic crust, *Geol. Soc. Am. Bull., 100*, 1423-1436, 1988.

Macdonald, K.C., Mid-Ocean Ridges: Fine scale tectonics, volcanic and hydrothermal processes within the plate boundary zone, *Ann. Rev. Earth Planet. Sci., 10*, 155-190, 1982.

Macdonald, K.C., The crest of the Mid-Atlantic Ridge: Models for crustal generation processes and tectonics, in *The Geology of North America, The Western North Atlantic Region*, edited by P.R. Vogt and B.E. Tucholke, *Geological Society of America, M*, 51-68, 1986.

Macdonald, K.C., and B.P. Luyendyk, Investigation of faulting and abyssal hill formation on the flanks of the East Pacific Rise (21° N) using Alvin, *Mar. Geophys. Res., 7*, 515-535, 1986.

Macdonald, K.C., P.J. Fox, L.J. Perram, M.F. Eisen, R.M. Haymon, S.P. Miller, S.M. Carbotte, M.-H. Cormier, and A.N. Shor, A new view of the mid-ocean ridge from the behaviour of ridge-axis discontinuities, *Nature, 335*, 217-225, 1988.

Macdonald, K.C., P.J. Fox, R.T. Alexander, R. Pockalny, and P. Gente, Volcanic growth faults and the origin of Pacific abyssal hills, *Nature, 380*, 125-129, 1996.

MacLeod, C.J., and D.A. Rothery, Ridge axial segmentation in the Oman ophiolite: Evidence from along-strike variations in the sheeted dyke complex, in *Ophiolites and their Modern Oceanic Analogues*, edited by L.M. Parson, B.J. Murton, and P. Browning, *Geol. Soc. London Spec. Publ. 60*, 39-63, 1992.

MacLeod, C.J., and B.J. Murton, On the sense of slip of the Southern Troodos transform fault zone, Cyprus, *Geology, 23*, 257-260, 1993.

MacLeod, C.J., B. Célérier, G.L. Früh-Green, and C.E. Manning, Tectonics of Hess Deep: A synthesis of drilling results from Leg 147, *Proc. Ocean Drilling Program ,Sci. Results, 147*, 461-475, 1996.

Manning, C.E., and C.J. MacLeod, Fracture-controlled metamorphism of Hess Deep gabbros, Site 894: Constraints on the roots of mid-ocean ridge hydrothermal systems at fast-spreading centers, *Proc. Ocean Drilling Program, Sci. Results, 147*, 189-212, 1996.

Marion, E., C. Mével, and M. Cannat, Evolution of oceanic gabbros from the MARK/Kane fracture intersection massif. *Terra Abstr., 3*, 310, 1991.

Mével, C., M. Cannat, P. Gente, E. Marion, J.M. Auzende, and J.A. Karson, Emplacement of deep crustal and mantle rocks on the west median valley wall of the MARK area (MAR 23° N). *Tectonophysics, 190*, 31-53, 1991.

Mével, C., K. Gillis, J.F. Allan, and P.S. Meyer (editors), *Proc. Ocean Drilling Program ,Sci. Results, 147*, 1996.

Moores, E. M., Origin and Emplacement of Ophiolites, *Rev. Geophys. Space Physics, 20*, 735-59, 1982.

Moores, E.M., and F.J. Vine, Troodos massif, Cyprus and other ophiolites as oceanic crust: Evaluation and implications, *Phil. Trans. Roy. Soc. London, 268*, 443-466, 1971.

Moores, E.M., R.J. Varga, K.L. Verosub, and T.W. Ramsden, Regional structure of the Troodos dyke complex, in *Ophiolites, Oceanic Crustal Analogues*, edited by J. Malpas, E.M. Moores, A. Panayiotou, and C. Xenophontos, *Geol. Survey Dept.* Cyprus, Nicosia, 27-35, 1990.

Morris, E., R.S. Detrick, T.A. Minshull, J.C. Mutter, R.S. White, W. Su, and P. Buhl, Seismic structure of oceanic crust in the western North Atlantic, *J. Geophys. Res., 98*, 13879-13903, 1993.

Morton, J.L., and N.H. Sleep, Seismic reflections from the Lau Basin magma chamber, in *Geology and Offshore Resources of Pacific Island Arcs--Tonga Region*, edited by

D.H. Scholl and T.L. Vallier, *Circum Pacific Council for Energy and Mineral Resources, Earth Science Series*, 441-453, 1985.

Mukasa, S.B., and J.N. Ludden, Uranium-lead ages of plagiogranites from the Troodos ophiolite, Cyprus, and their tectonic significance, *Geology, 15*, 825-828, 1987.

Murton, B.J., and I.G. Gass, Western Limassol Forest Complex, Cyprus: Part of an Upper Cretaceous leaky transform fault, *Geology, 14*, 255-258, 1986.

Mutter, J.C., and R.S. Detrick, Multichannel seismic evidence for anomalously thin crust at Blake Spur fracture zone, *Geology, 12*, 534-537, 1984.

Mutter, J.C. and North Atlantic Transect Study Group, Multichannel seismic images of the oceanic crust's internal structure: Evidence for a magma chamber beneath the Mesozoic Mid-Atlantic Ridge, *Geology, 13*, 629-632, 1985.

Mutter, J.C., and J.A. Karson, Structural processes at slow-spreading ridges, *Science, 257*, 627-634, 1992.

Naar, D.F., and R.N. Hey, Speed limit for oceanic transform faults, *Geology, 17*, 420-422, 1989.

Nicolas, A., Structures of ophiolites and dynamics of oceanic lithosphere. *Kluwer Academic Publishers, The Netherlands*, 367, 1989.

Nicolas, A., G. Ceuleneer , F. Boudier , and M. Misseri, Structural mapping in the Oman ophiolites: Mantle diapirism along an oceanic ridge, *Tectonophysics, 151*, 27-56, 1988.

Nicolas, A., and F. Boudier, Rootong of the sheeted dike complex in the Oman ophiolite, in *Ophiolite Genesis and Evolution of the Oceanic Lithosphere*, edited by Tj. Peters, A. Nicolas, and R.G. Coleman, *Kluwer Academic Publishers, The Netherlands*, 39-54, 1991.

Nicolas, A., C. Freydier, M. Godard, and A. Vauchez, Magma chambers at oceanic ridges: How large?, *Geology, 21*, 53-56, 1993.

Nicolas, A., F. Boudier, and B. Ildefonse, Dike patterns in diapirs beneath oceanic ridges: The Oman ophiolite, in *Magmatic Systems*, edited by M.P. Ryan, *Academic Press*, 77-95, 1994.

Nicolas, A., and F. Boudier, Mapping oceanic ridge segments in Oman ophiolite, *J. Geophys. Res., 100*, 6179-6197, 1995.

Orcutt, J.A., Structure of the Earth: Oceanic Crust and Uppermost Mantle, *Rev. Geophys., 25*, 1177-1196, 1987.

Pallister, J.S., Structure of the sheeted dike complex of the Semail ophiolite near Ibra, Oman, *J. Geophys. Res., 86*, 2661-2672, 1981.

Pallister, J.S., and C.A. Hopson, Samail ophiolite plutonic suite: Field relations, phase variation, cryptic variation and layering, and a model of a spreading ridge magma chamber, *J. Geophys. Res., 86*, 2593-2644, 1981.

Pariso, J.E., J.H. Scott, E. Kikawa, and H.P. Johnson, A magnetic logging study of Hole 735B gabbros at the Southwest Indian Ridge, in *Proc. ODP, Sci Results, 118*, 309-321, 1991.

Pearce, J.A., S.J. Lippard, and S. Roberts, Characteristics and tectonic significance of supra-subduction zone ophiolites, in *Marginal Basin Geology*, edited by B.P. Kokelaar and M.F. Howells, *Geol. Soc. Spec. Publ. 16*, 77-96, 1984.

Phipps Morgan, J., A. Harding, J. Orcutt, G. Kent, and Y.J. Chen, An observational and theoretical synthesis of magma chamber geometry and crustal genesis along a mid-ocean ridge spreading center, in *Magmatic Systems*, edited by M.P. Ryan, *Academic Press*, 139-178, 1994.

Porras, J.J., S.M. Wiggins, and M.D. LeRoy, Microearthquake location at the Hess Deep, *EOS Transactions, Am. Geophys. Union, 76*, S273, 1995.

Purdy, G. M., The variability in seismic structure of layer 2 near the East Pacific Rise at 12°N. *J. Geophys. Res., 87*, 8403-8416, 1982.

Purdy, G.M., and R.S. Detrick, Crustal structure of the Mid-Atlantic Ridge at 23°N from seismic refraction studies, *J. Geophys. Res., 91*, 3739-3762, 1986.

Raitt, R.W., Seismic refraction studies of the Pacific Ocean basin, *Bull. Geol. Soc. Am., 67*, 1623-40, 1956.

Raitt, R.W., The crustal rock, in *The Sea*, edited by M.N. Hill, Wiley Interscience (New York), 85-102, 1963.

Reid, I., and H.R. Jackson, Oceanic spreading rate and crustal thickness, *Mar. Geophys. Res., 5*, 165-172, 1981.

Reynolds, S.J., and G.S. Lister, Folding of mylonitic zones in Cordilleran metamorphic core complexes: Evidence from near the mylonitic front, *Geology, 18*, 216-219, 1990.

Rona, P.A., L. Widenfalk, and K. Boström, Serpentinized ultramafics and hydrothermal activity at the Mid-Atlantic Ridge crest near 15°N, *J. Geophys. Res., 92*, 1417-1427, 1987.

Rosendahl, R.B., R. Hekinian, et al., *Initial Reports of the Deep Sea Drilling Project, 54*: Washington (U.S. Government Printing Office), 1980.

Salisbury, M.H., N.I. Christensen, and R.H. Wilkens, The nature of the layer 2/3 transition from a comparison of laboratory and logging velocities and petrology at the base of Hole 5094B, in *Proc. ODP, Sci. Results, 148*, 409-414, 1996.

Schiffman, P., B.M. Smith, R.J. Varga, and E.M. Moores, Geometry, conditions, and timing of off-axis hydrothermal metamorphism and ore-deposition in the Solea graben, *Nature, 325*, 423-425, 1987.

Schouten, H., K.D. Klitgord, and J.A. Whitehead, Segmentation of mid-ocean ridges, *Nature, 317*, 225-229, 1985.

Sempere, J.-C., and K.C. Macdonald, Marine tectonics: Processes at mid-ocean ridges, *Rev. Geophys., 25*, 1313-1347, 1987.

Sempere, J.-C., G.M. Purdy, and H. Schouten, Segmentation of the Mid-Atlantic Ridge between 24°N and 30°40'N, *Nature, 344*, 427-431, 1990.

Simonian, K.O., and I.G. Gass, Arakapas fault belt, Cyprus: A fossil transform fault, *Geol. Soc. Am. Bull., 89*, 1220-1230, 1978.

Sinton J.M., and R.S. Detrick, Mid-ocean ridge magma chambers, *J. Geophys. Res., 97*, 197-216, 1992.

Skjerlie, K.P., H. Furnes, and R.J. Johansen, Magmatic development and tectonomagmatic models for the Solund-Stavfjord Ophiolite Complex, West Norwegian Caledonides, *Lithos, 23*, 137-151, 1989.

Skjerlie, K.P., and H. Furnes, Evidence for a fossil transform fault in the Solund/ Stavfjord Ophiolite Complex: West Norwegian Caledonides, *Tectonics, 9*, 1631-1648, 1990.

Sleep, N.H., Formation of ocean crust: Some thermal constraints, *J. Geophys. Res., 80*, 4037-4042, 1975.

Sleep, N.H., and G.A. Barth, The nature of the oceanic lower crust and shallow mantle emplaced at low spreading rates (abstract), *EOS Transactions. Am. Geophys.Union, 75*, 626, 1994.

Smith, D., and J.R. Cann, Building the crust at the Mid-Atlantic Ridge, *Nature, 365*, 707-715, 1993.

Stakes, D.S., C. Mével, M. Cannat, and T. Chaput, Metamorphic stratigraphy of Hole 735B, in *Proc. ODP, Sci Results, 118*, 153-180, 1991.

Swift, S.A., R.A. Stephen, Hole 504B Seismic experiment: Single channel seismic survey reveals recent faulting and basement relief near crustal boreholes, *EOS Transactions. Am. Geophys.Union, 76*, F616, 1995.

Tartarotti, P., S. Allerton, and C. Laverne, Vein formation mechanisms in the sheeted dike complex from Hole 504B, in *Proc. Ocean Drilling Program, Sci. Results, 137/140*, 231-241, 1995.

Thompson, G., and W.G. Melson, The petrology of oceanic crust across fracture zones in the Atlantic Ocean: Evidence of a new kind of sea-floor spreading, *J. Geology, 80*, 526 538, 1972.

Tilton, G.R., C.A. Hopson, and J.E. Wright, U/Pb isotopic ages of the Samail opholites, Oman, with applications to Tethyan Ocean Ridge Tectonics, *J. Geophys. Res., 86*, 2763-2777, 1981.

Tolstoy, M.A., J. Harding, and J.A. Orcutt, Crustal structure on the Mid-Atlantic Ridge: Bull's eye gravity anomalies and focused accretion, *Science, 262*, 726-729, 1993.

Toomey, D.R., S.C. Solomon, G.M. Purdy, and M.H. Murray, Microearthquakes beneath the median valley of the Mid-Atlantic Ridge near 23°N: Tomography and tectonics, *J. Geophys. Res., 93*, 9093-9112, 1988.

Toomey, D.R., G. M. Purdy, S.C. Solomon, and W.S.D. Wilcock, The three-dimensional seismic velocity structure of the East Pacific Rise near latitude 9°30' N, *Nature 347*, 639-645, 1990.

Tucholke, B.E., and J. Lin, A geological model for the structure of ridge segments in slow spreading ocean crust, *J. Geophys. Res., 99*, 11937-11958, 1994.

van Everdingen, D.A., and P.A. Cawood, Dyke domains in the Mitsero graben, Troodos ophiolite, Cyprus: an off-axis model for graben formation at a spreading centre, *J. Geol. Soc. London, 152*, 923-932, 1995.

Varga, R.J., and E.M. Moores, Spreading structure of the Troodos ophiolite, Cyprus, *Geology, 13*, 846-850, 1985.

Venturelli, G., R.S. Thorpe, and P.J. Potts, Rare earth and trace element characteristics of ophiolitic metabasalts from the Alpine-Apennine belt, *Earth Planet. Sci. Lett., 53*, 109-123, 1981.

Vera, E.E., J.C. Mutter, P. Buhl, J.A. Orcutt, A.J. Harding, M.E. Kappus, R.S. Detrick, and T.M. Brocher, The structure of 0- to 0.2 m.y.-old oceanic crust at 9°N on the East Pacific Rise from expanded spread profiles, *J. Geophys. Res., 95*, 15529-15556, 1990.

Verosub, K.L., and E.M. Moores, Tectonic rotations in extensional regimes and their paleomagnetic consequences for oceanic basalts, *J. Geophys. Res., 86*, 6335-6349, 1981.

White, R.S., R.S. Detrick, J.C. Mutter. P. Buhl, Minshhull, and E. Morris, New seismic images of oceanic crustal structure, *Geology, 18*, 462-465, 1990.

Whitehead, J.A., Jr., H.J.B. Dick, and H. Schouten, Λ mechanism for magmatic accretion under spreading centres, *Nature, 312*, 146-148, 1984.

Y. Dilek, Department of Geology, Miami University, Shideler Hall, Oxford, OH 45056, U.S.A.

H. Furnes, Geologisk Institutt, Universitetet i Bergen, Allégt. 41, 5007 Bergen, Norway

E.M. Moores, Department of Geology, University of California, Davis, CA 95616, U.S.A.

Periodic Formation of Magma Fractures and Generation of Layered Gabbros in the Lower Crust Beneath Oceanic Spreading Ridges

Peter B. Kelemen

Dept. of Geology and Geophysics, Woods Hole Oceanographic Institution;

Einat Aharonov

Lamont-Doherty Earth Observatory

There must be a transition from continuous porous flow to transient formation of melt-filled fractures in the region of magma transport below oceanic spreading centers. This transition may occur at permeability barriers that impede porous flow of melt, at the base of the oceanic crust and within the lower crust. We summarize evidence for formation of melt-filled lenses at the base of the crust in the Oman ophiolite, and evidence indicating that melt is very efficiently extracted from these lenses, probably in fractures. We also discuss the possible formation of melt lenses elsewhere in the oceanic lower crust. We then present a simple physical model for the periodic formation of melt-filled fractures originating in a melt lens beneath a permeability barrier. Finally, quantitative models show how modal layering in lower crustal gabbros can form as a result of the periodic pressure changes associated with fracture formation. An ancillary result of chemical modeling is the quantification of a thermal gradient in the lower crust of the Oman ophiolite during igneous accretion beneath a spreading center, from 1165 to 1195°C near the dike/gabbro transition to ~1240°C near the crust/mantle transition. This and other data for the Oman gabbros support models in which much of the lower crust forms by crystallization in sills at a variety of depths, from the dike/gabbro transition to the base of the crust.

1. INTRODUCTION

Many models for melt extraction from the mantle and igneous accretion of the oceanic crust at spreading ridges emphasize continuous porous flow processes (e.g., *McKenzie* [*1984*]; *Spiegelman* [*1993a,b*]; *Aharonov et al.* [*1995, 1997*]) and, although this is debated, there is evidence in support of the hypothesis that melt transport in the melting region is dominantly by focused porous flow (e.g., *Kelemen et al.* [*1995a,b, 1997a*]). However, it is clear that the formation of sheeted dikes is periodic, comprising the "quantum event of upper crustal accretion"

[*Delaney et al. 1994*]. Thus, somewhere in the region of melt migration there must be a transition from continuous to punctuated flow. In this paper, we propose that the two periodic phenomena in oceanic crust, sheeted dikes and layered gabbros, may have a common physical basis. The transition from continuous to punctuated melt transport may occur mainly at the base of the lower crustal, layered gabbros, where a permeability barrier blocks porous ascent of melt, giving rise to gradually increasing melt pressure, in excess of lithostatic pressure. We present a physical model in which excess melt pressure is periodically relieved by formation of melt-filled fractures. This model can quantitatively explain the genesis of melt-filled fractures and layering in gabbro. We conclude with a discussion of various kinematic models for the formation of the lower oceanic crust, with an emphasis on constraints from igneous petrology and ophiolite gabbro compositions.

Faulting and Magmatism at Mid-Ocean Ridges
Geophysical Monograph 106
Copyright 1998 by the American Geophysical Union

1.1. Transition from Continuous to Periodic Melt Flow

Nicolas (e.g., [*1986, 1990*]) and *Maaloe* (e.g., [*1981*]) have proposed that the transition from porous flow to flow in periodically formed fractures occurs in the mantle melting region. However, some recent work reaffirms the idea that, though melt-filled fractures do form beneath ridges, melt extraction from the adiabatically upwelling mantle may occur mainly by porous flow throughout the melting region [*Kelemen et al. 1995a,b 1997a*]. In this view, melt-filled fractures in the asthenosphere are not required, given available field, geochemical and physical data, to explain the genesis of mid-ocean ridge basalt (MORB) or residual mantle peridotites. However, none of the data can be used to rule melt-filled fractures in the melting region [*Kelemen et al. 1997a*], and the presence or absence of such fractures is a matter of debate.

In any case, it is certain that porous flow is an important melt transport process at very shallow depths in the mantle beneath ridges; Boudier and co-workers (e.g., *Boudier and Nicolas* [*1995*]) and Ceuleneer and co-workers (e.g., *Ceuleneer* [*1990*]; *Ceuleneer and Rabinowicz* [*1992*]) have emphasized the prevalence of "impregnated peridotite", formed by crystallization of melt moving by porous flow, in the crust/mantle transition zone of the Oman ophiolite (Moho transition zone, or MTZ). Similar features are also found in the shallow mantle (MTZ?) at the East Pacific Rise, as drilled at Hess Deep [*Gillis, Mével, Allan et al. 1993; Dick and Natland 1996; Boudier et al. 1996a*].

Gabbroic sills are abundant in parts of the MTZ in the Oman ophiolite [*Benn et al. 1988; Ceuleneer 1990; Boudier and Nicolas 1995; Boudier et al. 1996b*]. Geochemical study of these sills, as well as "impregnations", in the Oman MTZ shows that they are very pure "cumulates"; i.e., they are magmatic rocks but they do not have plausible liquid compositions. Instead, they formed by partial crystallization of silicate liquids, from which most of the melt was extracted [*Benoit et al 1996; Kelemen et al. 1997b; Korenaga and Kelemen 1997*]. Their compositions are determined by the stoichiometry of minerals that crystallized from a melt. These characteristics are shared by lower crustal, layered gabbros in the Oman ophiolite and in gabbros dredged from the mid-ocean ridges [*Smewing 1981; Pallister and Hopson 1981; Pallister and Knight 1981; Browning 1982, 1984; Juteau et al 1988*; Meyer *et al.* 1989]. Thus, the MTZ and the lower crust commonly include regions of high melt fraction, which undergo partial crystallization, from which melt has later been extracted to form other parts of the crust.

Theoretically, melt-filled fractures originating in the mantle might propagate all the way through the crust to the sea floor (e.g., *Nicolas* [*1986, 1990*]). However, the rarity of primitive, mantle derived magmas forming dikes and lavas in ophiolites and along the mid-ocean ridges (summaries in *Sinton and Detrick* [*1992*]; *Langmuir et al.* [*1992*]; *Kelemen et al.* [*1997b*]), and the abundance of gabbro sills and "impregnated" peridotites in the MTZ, suggests that propagation of melt-filled fractures from the mantle source to the sea floor has not occurred.

Instead, the evolved compositions of dikes and lavas suggest that they are derived from crustal "magma chambers" where primitive, mantle-derived melts have been modified by crystal fractionation. This hypothesis is supported by geochemical studies showing that gabbro sills in the MTZ, and lower crustal gabbros, are in major and trace element exchange equilibrium with the liquids that formed sheeted dikes and lavas in the upper crust [*Pallister and Knight 1981; Kelemen et al. 1997b*]. Even if magma transport in fractures below the MTZ is an important melt migration mechanism, the presence of anelastic, melt-rich regions at the MTZ would generally stop propagation of these fractures. Instead any melt ascending in melt-filled fractures in the mantle would be temporarily trapped in sills. In this paper, we concentrate on melt extraction at and above the MTZ.

Kelemen et al. [*1997b*] and *Korenaga and Kelemen* [*1997, 1998*] proposed that diffuse porous flow over vertical distances larger than meters to tens of meters was not the main process of melt extraction from gabbros in the MTZ and lower crust. Instead, most melt extraction must have been in fractures or channels of focused porous flow. There is little evidence for either melt-filled fractures or porous flow channels in the lower crustal section of the Oman ophiolite, though there are some centimeter-scale dikes. The general paucity of vertical melt migration features may be a consequence of transposition of initially steep features, such as dikes, due to ductile extension of the crust beneath a spreading ridge (e.g., *Quick and Denlinger* [*1993*]).

It may also be that melt-filled fractures in the lower oceanic crust rarely form dikes at all. Melt extraction in cracks involves an initial pressure decrease followed by near adiabatic ascent. In the lower crust, rocks surrounding melt-filled fractures will be hot, close to the magmatic temperature. Under these circumstances, conductive heat loss to the walls will be slow, and ascending silicate liquids will be undersaturated in solid phases. Only in the latest stage of melt transport in fractures, as fracture width and melt velocity decrease, will crystallization occur in lower crustal conditions. In some cases, this may form cm-scale dikes. If fracture width decreases to the crystal size in surrounding rocks, crystallization of melt may "heal" the fracture without a trace. In addition, if rocks surrounding fractures are permeable, porous flow of melt into the wall rock may also heal fractures with no trace.

1.2. Physical Models for Periodic Formation of Melt-Filled Fractures From Magma Chambers

In this paper, we adopt the hypothesis that periodic or intermittent melt extraction arises by formation of melt-filled fractures. Physical models for periodic melt extraction from magma chambers at the MTZ and within the crust have been previously proposed by *Gudmundsson* (e.g., [*1986, 1990*]) and *Ida* [*1996*]. *Gudmundsson* emphasized the role of tensile stresses due to lithospheric spreading, and did not incorporate the effects of magmatic influx on melt pressure, whereas in this paper we emphasize the role of continuous influx, with plate-scale tension due to spreading entering the problem only in the sense that it decreases the tensile stress necessary to fracture rocks overlying a magma chamber. The model of *Ida* [*1996*] is similar to that adopted in this paper, except that *Ida* used a viscous rheology for the rocks hosting melt-filled fractures, whereas we use an elastic rheology for rocks surrounding fractures. In *Ida's* formulation, melt-filled fractures close by viscous flow of the wall rocks. By contrast, in our model fractures become very narrow due to elasticity of the wall rocks. When crack width and melt velocity become very small, melt in fractures solidifies to form dikes or, if the fractures are very narrow, the two walls simply anneal together.

An analogous physical problem is the development and episodic fracture of overpressured fluid "compartments" in sedimentary basins (e.g., *Hunt* [*1990*]; *Dewers and Ortoleva* [*1988; 1994*]). *Nur and Walder* [*1992*] proposed a simple analytical model to account for this process which resembles the one in this paper in some respects. However, the rheology of rocks in sedimentary basins is considerably different from that adopted for oceanic gabbros here.

A mechanism involving periodic "plugging" and "unplugging" of magmatic conduits has also been proposed by *Whitehead and Helfrich* [*1991*], who emphasize the role of temperature dependent magmatic viscosity. Our model can be viewed as a simplified, limiting version of the *Whitehead and Helfrich* approach. They included continuous changes in melt viscosity, whereas we assume a step function in viscosity; melt viscosity is constant up to the solidification temperature, at which point the viscosity becomes "infinite". Also, we assume the fracture opens and closes elastically, in response to changes in melt pressure, and may heal completely by crystallization, whereas *Whitehead and Helfrich* assumed a constant conduit width.

Our specific hypothesis is that permeability barriers at the MTZ are created by crystallization of multiply saturated basaltic melts within a porous medium. Rapid crystallization reduces porosity and permeability [*Korenaga and Kelemen 1997; Kelemen et al. 1997b*]. Other permeability barriers may be produced within the oceanic lower crust via a variety of crystallization and melt extraction mechanisms. Magma ascending by porous flow beneath these permeability barriers will pond and form melt-filled sills. In these sills, continuous influx of magma leads to increasing magmatic overpressure that periodically exceeds the tensile strength of the overlying rock, giving rise to melt-filled fracture. We present some simple physical models of this proposed mechanism, and show how it could give rise to the kinds of igneous modal layering that are observed in the Oman ophiolite in sills in the MTZ (e.g., *Boudier et al.* [*1996b*]) and in lower crustal gabbros [*Pallister and*

Hopson 1981; Smewing 1981; Browning 1982; Juteau et al. 1988]. If magma supply to the MTZ is mainly by continuous porous flow, then sills at this level record the transition from porous flow to periodic fracture. In any case, once sills are present, they are unlikely to be traversed by melt-filled fractures, and will instead be fed by porous flow and/or melt-filled fractures from below.

1.3. Origin of Modal Layering in Gabbros

In the light of our proposed model, the presence of layered gabbros in the MTZ and in the lower crust of the Oman ophiolite can be viewed as complementary to the presence of sheeted dikes in the upper crust, in that both arise from periodic formation of melt-filled fractures. Periodic pressure change has been previously proposed as a mechanism for the formation of igneous modal layering (e.g., *Cameron* [*1977*]; *Lipin* [*1993*]; also please see Section 5.5, "Pressure fluctuations" in *Naslund and MacBirney* [*1996*]). However, this is by no means a unique explanation for the origin of layering in gabbroic plutons. Other proposed mechanisms include (a) infusions of new magma into a pre-existing magma chamber, (b) periodic gravitational crystal settling due to "avalanches" or to intermittent magmatic convection, (c) sidewall crystallization from vertically stratified, chemically independent, "double diffusive" convection cells in a large magma chamber, (d) kinetic mechanisms for spatially and/or temporally periodic crystal nucleation and growth, and (e) intrusion and crystallization of sills within previously crystallized gabbroic rocks (e.g., review in *Naslund and McBirney* [*1996*]). We believe that there is ample evidence that each of these mechanisms has operated to produce igneous modal layering at some times and places, and we will not attempt an exhaustive evaluation of each of them in this paper.

Many of these explanations are most applicable to magma chambers of considerable vertical extent, emplaced within relatively cold wall rocks. *Korenaga and Kelemen* [*1997*] conclude that periodic pressure change, together with subsequent infusions of new magma, is the most likely mechanism for formation of layering in the Oman MTZ sills. In this paper, we present liquid line of descent models that quantify this process.

In extending this idea to formation of "normal" oceanic crust, it is necessary to add the caveat that obvious subhorizontal, modal layering is rare or absent in available samples of gabbros dredged and drilled from the mid-ocean ridges. However, layering is rarely visible in small samples of Oman gabbros, and this is also true of our observations of diamond drill core from layered gabbros in East Greenland such as the Skaergaard and Kap Edvard Holm intrusions. Generally, outcrops at least 10's of centimeters in diameter are required to discern modal layering, which is evident because of its lateral continuity along strike, and complementary lack of continuity across strike. Also, drilling recovered samples from upper gabbros and from shallow mantle, formed along the East Pacific Rise (e.g., *Gillis, Mével, Allan et al.* [1993]), and some mafic gabbros have been dredged from Hess Deep and from fracture zones in the Pacific (e.g., *Hekinian et al.* [*1992, 1993*]; *Constantin et al.* [*1996*]), very few samples of lower crustal gabbros from the middle of fast- to medium-spreading ridge segments have been obtained to date. Due to the compositional similarity of gabbro samples from the oceans (e.g., *Meyer et al.* [*1989*]; *Hekinian et al.* [*1993*]; *Dick and Natland* [*1996*]) and gabbros in Oman, we infer that layered gabbros will, in fact, prove to be common beneath mid-ocean ridges.

1.4. Sills in the MTZ in the Oman Ophiolite

In this paper, we rely on evidence from layered gabbros in the lower crust [*Smewing 1981; Pallister and Hopson 1981; Pallister and Knight 1981; Browning 1982, 1984; Juteau et al. 1988*] and in gabbroic sills in the MTZ in the Oman ophiolite [*Benn et al. 1988; Ceuleneer 1990; Boudier and Nicolas 1995; Boudier et al. 1996b; Kelemen et al. 1997b; Korenaga and Kelemen 1997*] to constrain processes in the lower crust and at the MTZ on-axis beneath fast- to medium-spreading oceanic spreading ridges.

The MTZ sills play a particularly large role in our reasoning. An important question is whether some of the gabbros in the MTZ might be huge inclusions of lower crustal gabbro enclosed within intrusive dunites. We find this implausible on the following grounds. (1) Interdigitated contacts between gabbro lenses and surrounding dunite preclude the possibility that the gabbro lenses are imbricate fault slices of lower crustal gabbro (e.g., Figure 2 in *Korenaga and Kelemen* [*1997*]). (2) Modal layering within the gabbroic lenses, and the external contacts of the lenses, are always parallel to the regional trend of the crust/mantle transition, and never show rotation with respect to other lenses or layering in lower crustal gabbros. If the lenses were xenoliths that had been intruded by a magma, some of them would be rotated. (3) *Benoit et al.* [*1996*] report compositionally similar gabbro lenses within residual, shallow mantle harzburgites (within about 1 vertical km below the crust/mantle transition). These cannot be xenoliths. (4) While small dunite intrusive bodies are associated with some wehrlite intrusions into gabbros in Oman, in general the formation of true dunites by intrusion of a mixture of melt plus olivine, followed by extraction of all but a few percent of the melt, is a rare event that is unlikely to form the great thicknesses of dunite observed in the Oman MTZ. Thus, we remain convinced that most or all of the gabbro lenses in the Oman MTZ were emplaced as sills intrusive into surrounding peridotite.

Sills in the Oman MTZ are generally 0.1 to 10 meters thick and 10's to 1000's of meters long, with aspect ratios (height/length) less than 0.01. They show evidence for flattening by magmatic compaction (pure shear) and/or magmatic deformation (simple shear), so that their current shapes are not the same as their shapes during magmatic

crystallization. However, it seems likely that they have not undergone more than 10:1 flattening. Also, it is important to realize that sills may have been built up by successive crystallization from several melt lenses. Each lens would probably have had a similar horizontal extent, compared to the sills, but individual lenses my have had a smaller vertical dimension. Thus, during magmatic crystallization the melt lenses were sub-horizontal, tabular bodies with heights less than 10 meters and aspect ratios (height/length) less than 0.1 [*Korenaga and Kelemen 1997*].

A somewhat independent estimate of "magma chamber height" can be derived from geochemical data. The magnitude of compositional variation of layered gabbro sections within MTZ sills and lower crustal gabbros in Oman permits estimation of the relative mass of initial liquid to crystals. Given the current thickness of such gabbro sections, and estimates of the amount of flattening, one can estimate "magma chamber heights". Using this technique, *Browning* [*1982*] and *Korenaga and Kelemen* [*1997*] estimated that "magma chambers" within both MTZ sills and lower crustal gabbros were on the order of 0.1 meters to 10's of meters. Since igneous layers in these sections commonly extend for more than 100 meters, this again indicates that the "magma chambers" were sub-horizontal sills, both in the MTZ and in the sites of crystallization of lower crustal gabbros.

1.5. Sills in the MTZ Beneath Fast- to Medium-Spreading Ridges

We infer that the sills in the Oman MTZ are similar to sills in the MTZ beneath fast- to medium-spreading ridges on the following grounds. The Oman ophiolite contains pillow lavas and sheeted dikes with major and trace element characteristics similar to MORB. The presence of a continuous layer of sheeted dikes underlying pillow lavas indicates that the igneous crust of the ophiolite formed at a submarine spreading ridge. In detail, the rare earth elements (REE) and other incompatible trace elements have lower concentrations at a given Cr concentration than most MORB (e.g., *Alabaster et al.* [*1982*]; *Pearce et al.* [*1981*]). Also, in some of the northern massifs, andesitic lavas comprise part of the extrusive section. These characteristics suggest that the ophiolite may have formed in a super-subduction zone setting. However, strong similarities between the compositions of Oman lavas in general and MORB are compelling evidence that petrogenetic processes in formation of the ophiolite were similar to processes operating at a normal mid-ocean ridge. On the basis of radiometric age data, subdued crustal thickness variations, a general lack of paleo-fracture zones, and other geological observations, it is probable that the ophiolite formed at a fast- to medium-spreading ridge (e.g., *Tilton et al.* [*1981*]; *Nicolas* [*1989*]).

An essential caveat is that ophiolites intrinsically preserve features formed "off-axis", in the oceanic litho-sphere, and features formed during emplacement of the ophiolite along a compressional plate margin, as well as features formed "on-axis" beneath a submarine spreading ridge. In the case of the Oman ophiolite, igneous ages and emplacement ages are virtually the same, within error [Boudier et al. 1985; Michard et al. 1991]; Hacker et al. 1996], so that on-axis structures may have been affected by the emplacement process. For these reasons, it is possible that the sills in the MTZ in Oman are not characteristic of the MTZ beneath a normal mid-ocean ridge.

However, we hypothesize that sills along the MTZ are, in fact, an on-axis characteristic of normal fast- to medium-spreading ridges on the following grounds:

(1) Major and trace element geochemistry of the gabbroic sills indicates that their constituent minerals are in exchange equilibrium with the liquids that formed the sheeted dikes and lavas of the ophiolite [Kelemen et al. 1997b]. This is also true of minerals in lower crustal gabbros [Pallister and Knight 1981; Kelemen et al. 1997b]. Mass balance requires that the liquid extracted from gabbroic sills in the MTZ and/or from lower crustal gabbros must comprise a substantial proportion of the crust (e.g., Browning [1982]; Pallister [1984]; Kelemen et al. [1997b]). Thus, the liquids extracted from the sills and/or lower gabbros must be represented by the sheeted dikes and overlying lavas; most or all of the gabbros must have formed beneath a spreading ridge.

(2) Nearly solid gabbros are present immediately above the Moho within a few km of the ridge axis at fast-spreading ridges such as the East Pacific Rise (e.g., Vera et al. [1990]). There is likely to be substantial intergranular porous flow of melt near the MTZ. Intergranular flow of melt leads to solid-liquid equilibrium over distances larger than the solid grain size [Spiegelman and Kenyon 1992]. Thus, the MTZ must be the level of plagioclase (± clinopyroxene) saturation in ascending magmas at fast- to medium-spreading ridges (see next section). Thus, gabbroic rocks should become abundant within the MTZ on-axis at fast- to medium-spreading ridges.

(3) Seismic data indicate high melt proportions near the MTZ beneath active spreading ridges ([Dunn and Toomey 1997]; and submitted manuscript, Crawford et al. 1998). Other workers suggest that a highly reflective Moho, both on-axis and up to 30 km off-axis, may represent the emplacement of plutons at this level [Garmany 1989, 1994]. Thus, there is evidence for accumulation of melt near the MTZ at fast-spreading, mid-ocean ridges.

On this basis, we proceed with our hypothesis that sills form on-axis in the MTZ at fast- to medium-spreading ridges, and play an important role in crustal accretion.

2. CRYSTALLIZATION OF ASCENDING BASALT NEAR THE MTZ

For the purposes of this paper, we define the oceanic crust as that portion of the oceanic plates composed of >95% plagioclase-rich rocks (troctolite, olivine gabbro, gabbro, gabbronorite, etc.) plus fine-grained basalt over vertical

distances greater than 10 meters. The MTZ represents the region in which plagioclase-bearing rocks are mixed with mantle peridotite, and neither rock type comprises more than 95% of the rock over vertical distances greater than 10 meters. By this definition, beneath slow spreading ridges the MTZ may be present over a considerable depth interval (e.g., Cannat [1996]). However, at fast spreading ridges, the geothermal gradient is likely to be nearly adiabatic within the ascending mantle all the way up to the MTZ (e.g., Sleep [1975]). In such a thermal structure, the MTZ will be a narrow depth interval immediately beneath the crust, perhaps a few hundred meters thick, in which any basalt ascending by porous flow becomes saturated in plagioclase (± clinopyroxene). If this were not the case, and melts saturated only in olivine came into contact with gabbroic rocks above the MTZ, the liquids would dissolve plagioclase (plag) ± clinopyroxene (cpx), moving the Moho upward, until the liquids themselves became saturated in plag ± cpx. Conversely, if melts became saturated with plag ± cpx below the MTZ, they would form gabbroic rocks and the MTZ would extend downward.

When mantle-derived basaltic liquids similar to MORB become saturated in plag ± cpx, their mass drops rapidly with continued cooling. By comparison, olivine ± spinel fractionation at higher temperatures consumes relatively little liquid mass per °C of cooling. In Figure 1, we illustrate results of calculations using several different petrological models for the crystallization of MORB. In each, we have taken a melt composition proposed to be a mantle-derived liquid parental to MORB, and modeled fractional crystallization during cooling from above its liquidus through the temperature of plag and cpx saturation. As can be seen, crystallization rates on the order of 1 to 3%/°C are predicted over a 20 to 40°C interval below the temperature of olivine + plag (± cpx) saturation. The results of our calculations are consistent with experimental results summarized by Sinton and Detrick [1992] in their Figure 7.

The high rate of crystallization below olivine + plag (± cpx) saturation, in % liquid/°C, is due to the pseudo-eutectic nature of multiply-saturated basalt at crustal pressures, analogous to the isobarically invariant eutectic composition in the simple system forsterite (olivine) - anorthite (plag) - diopside (cpx). Cooling of olivine-saturated liquids similar to parental MORB in this system leads to minor olivine crystallization, followed by co-saturation in olivine + plag, and then olivine + plag + cpx. At the eutectic in the simple system, all of the magma is consumed by crystallization at constant temperature before cooling can continue. In the natural system of course, olivine + plag + cpx saturation is not truly eutectic, but the analogy with the simple system explains the large decrease in magma mass over a small cooling interval below the temperature of multiple saturation.

Cooling, crystallizing liquids at the MTZ will become increasingly evolved, with higher Fe/Mg for example. In this way, they will diverge from the composition of magmas in equilibrium with mantle peridotite. Where they

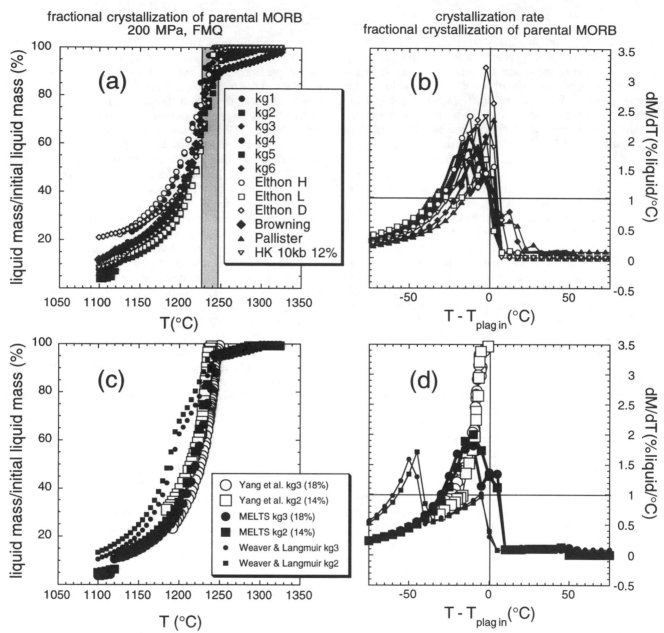

Figure 1. Modeling results for fractional crystallization of proposed, parental mid-ocean ridge basalt (MORB) liquids, H_2O-free at 200 MPa along the Fayalite-Magnetite-Quartz oxygen fugacity buffer. The main result is that the crystallization rate, in % crystallized/°C (labelled dM/dT, %/°C), increases dramatically to more than 1%/°C at the temperature of plagioclase saturation in all of these liquid compositions. The range of plagioclase saturation temperatures is indicated by a vertical grey bar in Figure 1a. Figures 1a and 1b illustrate results for a variety of proposed parental MORB compositions, calculated using the MELTS silicate liquid solution model of *Ghiorso and Sack* [*1995*]. Compositions labeled "kg" are calculated liquid compositions for polybaric fractional melting of the mantle, based on parameterization of experimental results [*Kinzler and Grove 1991*]. kg2 is calculated for an average of 14% partial melting, and kg3 is for an average of 18% partial melting. The compositions labeled "Elthon" are proposed parental MORB liquids based on study of MORB glasses, oceanic gabbros and peridotites, and experimental and theoretical studies [*Elthon et al. 1992*]. The compositions labeled Browning and Pallister are based on mass balance calculations and average rock compositions from the Oman ophiolite [*Browning 1982; Pallister 1984*]. The composition labeled HK 10kb 12% is an experimental liquid composition from *Hirose and Kushiro* [*1993*], produced by 12% batch melting of mantle peridotite. Figures 1c and 1d illustrate results for liquid compositions kg2 (14% melting) and kg3 (18% melting), using several different models [*Weaver and Langmuir 1991; Ghiorso and Sack 1995; Yang et al. 1996*]. Although the results of the various models differ, particularly in the temperature of calcic pyroxene (cpx) saturation, for our purposes they are similar in predicting a large increase in the rate of crystallization after cooling below the temperature of plagioclase saturation.

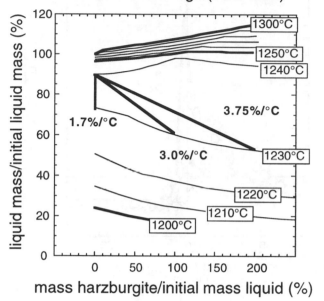

assimilation of harzburgite
200 MPa, FMQ, equilibrium xln
Kinzler & Grove kg2 (14% melt)

Figure 2. Modeling results for equilibrium crystallization of parental MORB liquid composition kg2 plus varying proportions depleted mantle harzburgite, calculated using the MELTS model, H2O-free at 200 MPa along the Fayalite-Magnetite-Quartz oxygen fugacity buffer (FMQ). The kg2 liquid composition is from *Kinzler and Grove [1991]*. The harzburgite reactant is an average composition for shallow, residual mantle peridotite samples dredged from the mid-ocean ridges, from *Dick [1989]*. Curves show the results of isothermal melt/rock reaction at a variety of temperatures. As previously predicted by *Kelemen [1990]*, and experimentally verified by *Daines and Kohlstedt [1994]*, reaction between residual mantle harzburgite and olivine-saturated basaltic liquid at constant temperature dissolves pyroxene and precipitates a smaller mass of olivine, producing an increase in the liquid mass. Just below the temperature of plagioclase saturation, at 1240°C, reaction produces a liquid saturated only in olivine [*Kelemen 1990*], and liquid mass increases during isothermal reaction. By contrast, where liquids are well into the olivine + plagioclase (± clinopyroxene) saturated region, isothermal reaction produces a decrease in melt mass [*Kelemen 1986, 1990*]. Thus, below the temperature of plagioclase saturation, combined cooling and reaction generally produces a larger crystallization rate than cooling alone.

come into contact with mantle peridotite, the combined effect of cooling and melt/rock reaction will decrease liquid mass still further compared to the effect of cooling alone. This process has been extensively discussed in previous papers (e.g., *Kelemen [1986, 1990]*), and is illustrated quantitatively in this paper in Figure 2. Reaction between crystallizing liquid and mantle harzburgite nearly doubles the rate of crystallization with cooling.

3. PERMEABILITY BARRIERS AT THE MTZ

As discussed in the previous section, multiply saturated liquids crystallize tens of percent of their mass over a 20 to 40°C temperature interval below plagioclase saturation. We infer that this temperature is attained by melts migrating by porous flow at the depth of the MTZ. *Sparks and Parmentier [1992]* and *Spiegelman [1993c]* proposed that crystallization of cooling magma entering the conductively cooled lithosphere forms a permeability barrier in the mantle. In a variant on these ideas, *Kelemen et al. [1997b]* proposed that the depth of multiple saturation immediately beneath a ridge axis may become a permeability barrier, restricting the ascent of magmas rising by porous flow and leading to the formation of sills in the MTZ.

Korenaga and Kelemen [1997] quantified the conditions under which this can occur. If there is a forced flux of porous melt flow across an interval in which the melt is crystallizing within its pore space, there are two possible end-member consequences: (1) the solid matrix will deform to maintain constant porosity at constant melt pressure, or (2) the porosity will decrease and the melt pressure will rise. The outcome depends upon the relative rates of viscous deformation required to keep the pore space open and of crystallization in the pore space, which in turn depends upon the flux of melt and the rate of cooling. Given that oceanic crust at fast- to medium-spreading ridges attains nearly 100% of its thickness within about 2 km of the ridge axis (e.g., *Vera et al. [1990]*), it is relatively straightforward to estimate the flux of melt that must pass through the MTZ beneath a spreading ridge. The cooling rate is far more difficult to determine.

One way to constrain the cooling rate is via seismic data indicating that lower crustal gabbro immediately above the Moho is essentially solid (porosity < 5%) within about 5 km of the East Pacific Rise at 9°N (e.g., *Vera et al. [1990]*; *Toomey et al. [1990]*). Given a half spreading rate of about 0.05 m/yr, an initial magmatic temperature of 1300 to 1250°C, and a solidus temperature (porosity < 5%) for lower crustal gabbros of ~ 1200°C (Figure 3), this yields cooling rates ~ 10^{-3} °C/yr, and lateral temperature gradients ~ 0.01 to 0.02°C/m. On the basis of this estimate, it is clear that thermal energy is efficiently removed from the oceanic lower crust near spreading ridges, and that this can drive igneous crystallization.

Another way to constrain the cooling of the crust beneath a spreading ridge is via comparison of near-solidus temperatures for MTZ sills, lower gabbros, upper gabbros, and lavas from the Oman ophiolite (Figure 3). For an on-axis porosity of ~10%, calculated temperatures for MTZ sills are ~ 1240°C and temperatures for lower crustal gabbros range from ~ 1205 to 1225°C. For an on-axis porosity of 30 to 10%, calculated temperatures for the shallowest gabbros are ~ 1190 to 1170°C. These estimates are discussed in more detail in section 8.2 of this paper.

Use of these values to constrain the on-axis geotherm requires the assumption that all of the rocks in the Oman

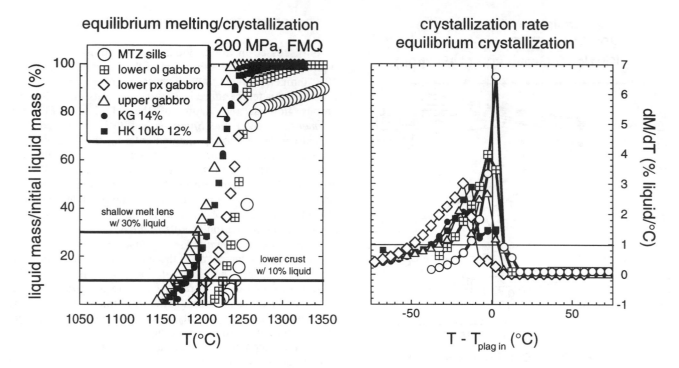

Figure 3. Modeling results for equilibrium crystallization of gabbro compositions from the Oman ophiolite, calculated using the MELTS model, H₂O-free at 200 MPa along FMQ. The composition for sills from the Moho Transition Zone (MTZ) is an average of 26 samples from *Korenaga and Kelemen* [*1997*]. The other gabbro compositions are averages from *Browning* [*1982*]. Equilibrium crystallization results for two parental MORB liquid compositions are shown for comparison. Because equilibrium crystallization consumes liquid more rapidly than fractional crystallization, all of the calculated crystallization curves are steeper than for fractional crystallization (compare Figure 1). The upper gabbro composition shows crystallization behavior similar to the liquid compositions. Averages from the layered, lower gabbros, and gabbro sills in the MTZ, have nearly vertical melting curves over the range from 10 to >80% liquid. The melting curves are at progressively higher temperatures for samples from the upper gabbros, the lower gabbros, and the MTZ, which probably indicates a temperature gradient of 45 to 75°C beneath the spreading ridge during formation of the lower crust in Oman.

crustal section formed on-axis. If they did, then the temperatures imply an average vertical temperature gradient of ~ 0.01°C/m. While it is unclear how this average vertical temperature gradient is distributed within the crust, it is apparent that the base of the crust is hotter than the average lower crust, and this in turn is hotter than the depth at which the uppermost gabbros crystallize, suggesting a relatively smooth temperature gradient. Thus, heat removal beneath the ridge axis may be sufficient to drive crystallization of melt near the MTZ.

It is difficult to be more precise about the magnitude of heat flux away from the lower crust beneath spreading ridges. Y. J. Chen and J. Phipps Morgan (personal communication 1998) have suggested that the rate of heat removal is small, and therefore can drive only a limited amount of basalt crystallization near the base of the crust, ~ 10% of the total crustal mass. However, this is based on assumptions concerning the amount and spatial distribution of hydrothermal circulation that are difficult to justify in detail, as Chen and Phipps Morgan would agree (e.g.,

Phipps Morgan and Chen [*1993*], p. 6287). If hydrothermal circulation in the lower crust 2 km off-axis beneath fast-spreading ridges is more vigorous than they have assumed, then more crystallization could occur near the base of the crust.

Using the oceanic crustal flux, cooling rates of 0.01 to 0.02°C/m, and rates of crystallization of multiply saturated basalt of 1% to 3%/°C), *Korenaga and Kelemen* [*1997*] found that crystallization can fill porosity more rapidly than viscous flow of the matrix maintains constant porosity, so that the result is a low porosity, permeability barrier.

Implicit in this discussion is the inference that the permeability barrier at the MTZ must be subhorizontal or concave downward beneath a spreading ridge, essentially parallel to sub-horizontal isotherms. While this inference is in accord with generalized thermal models for the oceanic crust (e.g., Sleep [1975]), the vigor and geometry of hydrothermal convection could lead to locally steep isotherms, and this is a subject which needs to be evaluated in detail. If isotherms near the MTZ are indeed subhorizontal, magma ascending due to buoyancy by porous flow will accumulate beneath permeability barriers. This may be responsible for formation of both gabbroic sills and more diffuse, "impregnated peridotites", in the MTZ.

Korenaga and Kelemen found that pressure in melt ascending due to buoyancy beneath a sub-horizontal permeability barrier may increase for two reasons: (1) due to forced flux of melt combined with elastic deformation of the solid matrix, and/or (2) by localizing the pre-existing vertical pressure gradient in the fluid within the region immediately surrounding the barrier. The second type of pressure effect has been numerically modeled in a simplified, rigid system by *Aharonov et al.* [*1997*].

Although the initiation of the permeability barrier and an underlying layer rich in accumulated melt requires porous flow, porous flow need not be the only mechanism of melt transport beneath the MTZ. If melt ascends in fractures , the presence of a weak layer or melt-filled lens will tend to create an anelastic barrier to fracture propagation. Thus, sills in the MTZ, once formed, will effectively trap all ascending melt, regardless of the mechanism of melt transport in the underlying mantle.

4. PERMEABILITY BARRIERS WITHIN CRYSTALLIZING GABBROS

Permeability barriers may form within crystallizing, lower crustal gabbros as well as along the MTZ. For example, it has been proposed that anorthosite bands, composed almost entirely of plag, form permeability barriers in layered intrusions (e.g., *Boudreau and McCallum* [*1986*]; *Boudreau* [*1988*]). Anorthosite bands are observed in lower crustal gabbros in Oman (e.g., *Pallister and Hopson* [*1981*]) and in the Bay of Islands ophiolite (e.g., *Bédard* [*1991*]). Such bands may be permeability barriers within the crystallizing oceanic lower crust.

Another point where permeability barriers might arise within crystallizing gabbros is at the point of low calcium pyroxene saturation. (In this paper, low calcium pyroxene is symbolized as "opx", since it is often orthopyroxene). At this point liquid compositions are analogous to an isobarically invariant reaction point in a simple system, in this case the point forsterite-cpx-opx in the system forsterite - diopside - silica, and forsterite-plag-opx in the system forsterite-anorthite-silica. Thus, as for ol-plag-cpx co-saturation, crystallization rates (%/°C) may increase dramatically at the level of plag-cpx-opx saturation, reducing porosity where liquids ascend by porous flow.

Finally, permeability barriers may arise within "cumulate" gabbros simply as a result of melt extraction. Most lower crustal gabbros in the Oman ophiolite are composed of unzoned, refractory minerals crystallized from basaltic magma at high temperature; the remaining liquid was later extracted [*Pallister and Hopson 1981; Browning 1982, 1984; Juteau et al. 1988; Kelemen et al. 1997b; Korenaga and Kelemen 1997*]. After melt extraction, these cumulate gabbros had a very small intergranular porosity, even at temperatures where basaltic liquid compositions are more than 60% liquid (Figure 3). Thus, after melt extraction, cumulate gabbros could constitute low porosity barriers to the ascent of magma by porous flow.

In all of these cases, accumulation of large liquid fractions beneath permeability barriers will create low viscosity, anelastic layers - in the limit of 100% porosity, these will be sub-horizontal, melt-filled lenses - that cannot be traversed by propagating fractures from below. Thus, once one or more lenses are present in the crust, they will act as barriers or traps for liquid ascending by melt-filled fracture. Instead, liquid from any fractures propagating from below will be trapped in these lenses, as previously suggested by *Gudmundsson* [*1986*].

5. MODEL FOR BEHAVIOR OF A CONTINUOUSLY FILLED MELT LENS

In this section we investigate a simplified model of the behavior of a viscoelastic melt lens below a sub-horizontal layer of low permeability rock. Stresses in the lens may rise due to continuous melt influx from below and accumulation of liquid. The rising stress may finally result in hydrofracture. If the stress state is isotropic, fracturing may preferentially occur in the horizontal direction (forming or extending melt lenses) due to the presence of a strong, sub-horizontal layer above. However, beneath mid-ocean ridges the tectonic environment is expected to produce a horizontal extensional component, such that $\sigma_H < \sigma_V$, where σ_H and σ_V are the horizontal and vertical stress components, respectively. In addition to tectonic stresses, the inflation of a melt-filled lens may generally produce a horizontal extensional component on the walls of the lens, as demonstrated by calculations for lacoliths (see *Turcotte & Schubert* [*1982*], p. 120).

As the lens inflates, the horizontal stress in the overlying layer will have an increasing extensional component, until a stress state is reached where vertical fractures will form. Magma flowing through vertical fractures exits the lens and relieves the elevated stress. As melt flow rates in the fracture drop, the fracture closes elastically. As the fracture becomes very narrow and melt flow velocities decrease, melt may crystallize to form a dike, or may "leak" out by porous flow. In either case, under many circumstances the fracture may be considered to close, or heal.

We loosely describe the melt-filled body beneath a permeability barrier as a lens because permeability barriers in igneous systems may often be parallel to isotherms, and in many geological systems isotherms affected by conductive cooling to the Earth's surface are sub-horizontal. Also, evidence from sills in the Oman ophiolite crust/mantle transition zone indicates that they crystallized from melt-filled lenses with vertical dimensions more than 100 times smaller than their horizontal dimensions. We describe the melt lens as having some initial (deflated) volume V_0, and an additional volume contribution derived from elastic inflation of the lens by incoming melt, V_e. A 2D melt lens is drawn schematically in Figure 4.

To simulate the dynamic behavior of the system, we need to know the relationship of melt flux to the melt pressure in the lens. We use a relationship between the elastic expansion and the magma pressure, appropriate for elongate bodies such as cracks and lacoliths [*Rubin 1995a*],

$$V_e \propto \frac{l^2}{G} p, \tag{1}$$

where l is the horizontal length of the lens, G is the elastic stiffness of the host rock, and $p = P - \rho g h$ is the melt pressure in excess of lithostatic pressure (ρ is the density of the overlying crustal column, g is the gravitational constant, h is the depth beneath the Earth's surface). The proportionality coefficient depends on geometry, and has a value of one for cracks [*Rubin 1995b*].

We use equation (1) and the fact that the volume of the melt lens changes due to input and output melt fluxes to describe how the pressure in the chamber evolves with time. We make a simplifying assumption that as the melt lens inflates or deflates, its lateral extent remains constant. Given these constraints

$$\frac{l^2}{G} \frac{\partial p}{\partial t} = Q_0 - Q_1 - \frac{1}{\nu} p, \tag{2}$$

where Q_0 is a continuous volumetric influx of melt and Q_1 is the volumetric flux out of the lens through fractures that form at the top of the lens. The last term in equation (2) is a pressure relaxation term, with ν being a relaxation parameter, allowing the chamber to exhibit Maxwell viscoelasticity (e.g., *Wilkinson* [*1960*]). The relaxation parameter allows for irreversible volume changes in the melt lens that arise from porous flow of melt outward into the surrounding rock, viscous deformation of the walls of the melt lens, and crystallization of melt.

In what follows, we introduce three postulates for our coupled melt lens-fracture model. The first is an assumption regarding when the surrounding rock will fracture and what width of fracture will form. The second postulates laminar melt flow in fractures, and the final assumption approximates the conditions under which a fracture will heal.

(1) We assume that the low permeability layer overlying the magma chamber will fracture when $p > p_{max}$, where p_{max} is a critical melt pressure in excess of lithostatic stress that is needed to fracture the rock. In addition, we assume that a fracture instantaneously reaches its final length, h, and neglect effects due to fracture propagation. Once the fracture is open, we assume that low permeability rock surrounding the fracture behaves elastically (simple Hookeian spring

model) with respect to the open fracture of width a, with elastic stiffness G (e.g., *Rubin* [*1995a*]), giving

$$a = \frac{h}{G} \left(p + \Delta\sigma_H \right). \tag{3}$$

The deviation of the horizontal stress from lithostatic stress is $\Delta\sigma_H = \rho g h - \sigma_H$, and may be calculated given a specific geometry of the melt lens and the tectonic stresses.

(2) We assume that once a fracture is open, melt flow through the fracture is laminar and constant with height, so it can be modeled with an average melt velocity

$$\omega = -\frac{a^2}{12\mu} \left(\frac{\partial p}{\partial z} + \Delta\rho g \right) \approx \frac{a^2}{12\mu h} \left(p + \Delta\rho g h \right) \tag{4}$$

where ω is melt velocity, μ is the melt viscosity, $\Delta\rho g$ is the buoyancy term arising from different densities of melt and solid, and z is depth taken to be positive upward. The pressure difference between melt and surrounding rock at the top of the crack is taken to be zero.

(3) We next approximate the conditions under which a fracture of length h, with magma flowing with velocity ω, will close, "heal", or solidify. Two competing processes occur during melt flow through a fracture: heating of the wall due to advection of hot melt, and cooling of the melt

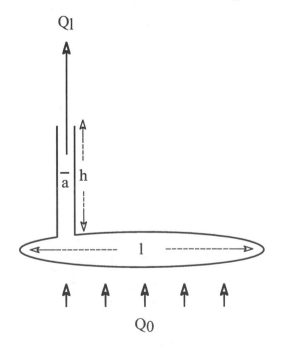

Figure 4. Schematic illustration of a melt lens below a low permeability layer, with an open crack allowing for melt flux out the top of the lens.

due to contact with colder, surrounding rock. The latter process leads to crystallization along the walls and decreasing fracture width. In our dynamic problem, pressure and flow rates in the fracture are both varying with time, due to decreasing pressure and volume in the melt lens as melt is removed. An analytical solution for solidification of a dike when both the pressure and the flux are varying is not yet available. However, solutions for solidification of melt driven by a constant pressure (e.g., [*Bruce and Huppert, 1989; Lister and Dellar, 1996*]) can be used here with some simplifying assumptions. *Lister and Dellar* [*1996*] find that if melt is driven by a constant pressure in the source region, there exists a critical Peclet number, Pe_c, below which solidification will occur, and above which melt-back will occur. The Peclet number is the ratio of the time scale for melt transport through the fracture, $t_t \sim h/\omega$, and the time scale for solidification to form a dike, $t_s \sim a^2/\kappa$, where κ is the diffusivity of heat, and is thus defined by equation (2.6) of *Lister and Dellar* [*1996*] as

$$Pe = \frac{(p + \Delta\rho gh)a^4}{h^2 \mu \kappa} \quad (5)$$

When flow rate is slower than solidification rate, the Peclet number of the system is smaller than the critical value and the fracture will solidify. Although in reality solidification occurs over some time interval, we will approximate it here as occurring instantaneously if the Peclet number is below Pe_c. The value of Pe_c depends on the melt temperature in the lens, T_0, the temperature of the rock surrounding the fracture, T_∞, the solidification temperature, T_L, the latent heat of crystallization, L, the specific heat capacity, c, and diffusivity of heat, κ. Reasonable parameters for our problem may be $L = 8 \times 10^5$ J/kg, $\kappa = 10^{-6} m^2/s$, $c = 730$ J/kg°C, $T_0 = 1230$°C, 1230°C$>T_\infty>1220$°C, and $T_L \sim 1225$ to 1230°C, of which the most uncertain are the temperatures of the melt and the surrounding rock. Using these values and Figure 5 of *Lister and Dellar* [*1996*], we determine that Pe_c is of order 1 for our system.

In order to use Pe_c as a criterion for solidification, we note that our system has a continuously changing Peclet number, since both driving pressure and fracture width change with time due to depletion of the source reservoir. When a fracture is formed, the pressure is high and the fracture is wide, so that the value of the instantaneous Peclet number as described in equation (5) is large. As the pressure drops, the value of the instantaneous Peclet number is also reduced. Thus, our approximate fracture healing criterion is that the fracture will heal when and if

$$Pe = Pe_c \sim 1 \quad (6)$$

In mathematical terms, the governing equations for the

fracture width, the pressure in the chamber, and the flow in the fracture (equations (3)-(6)) exhibit hysteresis:

for $p \geq p_{max}$

$$\rightarrow a = \frac{h}{G}(p + \Delta\sigma_H), \omega = \frac{a^2}{12\mu h}(p + \Delta\rho gh); \quad (7a)$$

for $p < p_{max}$ & $\omega a^2 > \omega_c a_c^2$

$$\rightarrow a = \frac{h}{G}(p + \Delta\sigma_H), \omega = \frac{a^2}{12\mu h}(p + \Delta\rho gh); \quad (7b)$$

and for $p < p_{max}$ & $\omega a^2 \leq \omega_c a_c^2$

$$\rightarrow a = 0, \omega = 0, \quad (7c)$$

where a_c and ω_c are fracture width and average melt velocity when $Pe=Pe_c$, i.e., when melt in the fracture solidifies to form a dike. From equations (4), (5) and (6),

$$\omega_c a_c^2 \sim \frac{\kappa h}{12}. \quad (8)$$

Note that when $a=0$, the width of the melt-filled fracture is 0, but a residual dike of finite width, formed by earlier crystallization along the fracture walls, may remain in the rock record.

5.1. Non-dimensionalization

We non-dimensionalize the problem for ease of solution. The primed non-dimensional variables are: $p' = p/p^*$, $p'_{max} = p_{max}/p^*$, $e' = \Delta\sigma_H/p^*$, $a' = a/a^*$, $t' = t/t^*$, $Q'_0 = Q_0 t^*/l^2$, $\omega' = \omega/\omega^*$, where the starred variables are characteristic dimensional quantities: $p^* = \Delta\rho gh$, $\omega^* = a^{*2}p^*/(12\mu h)$, $t^* = h/\omega^*$, $a^* = hp^*/G$. Finally we define a characteristic Peclet number, $Pe^* = p^*a^{*4}/(\kappa h^2 \mu)$. The non-dimensional equations are then

$$\gamma\dot{p}' = Q'_0 - \delta\omega'a' - Rp'; \quad (9)$$

for $p' \geq p'_{max}$

$$\rightarrow a' = p' + e', \omega' = a'^2(p' + 1); \quad (10a)$$

for $p' < p'_{max}$ & $Pe^*\omega'a'^2 > 1$

$$\rightarrow a' = p' + e', \omega' = a'^2(p' + 1); \quad (10b)$$

and for $p' < p'_{max}$ & $Pe^*\omega'a'^2 \leq 1$

$$\rightarrow a' = 0, \omega' = 0, \quad (10c)$$

where

$$\gamma = \frac{p^*}{G} \quad (11)$$

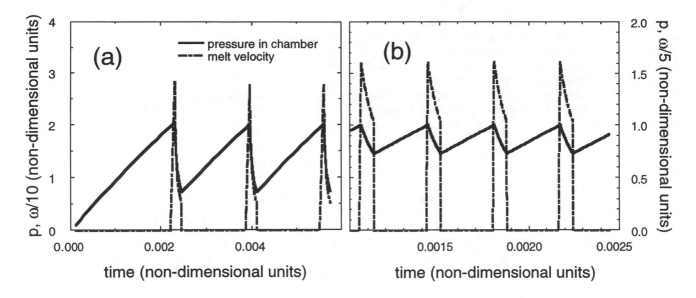

Figure 5. Melt pressure and flow velocity numerically calculated for $\gamma = 5 \times 10^{-5}$, $R = 0.005$, $Q'_0 = 0.05$, and $\delta = 0.02$. (a) has $p'_{max} = 2$ and (b) has $p'_{max} = 1$. For both cases, oscillatory behavior is established, in which pressure (solid line) rises in the melt lens until it reaches $p' = p'_{max}$. At that point a fracture is opened, and melt flow (dashed line) is maximal. As time progresses and pressure is released, flux through the fracture is reduced, until finally the healing criteria is reached and flow in the dike ceases due to crystallization or annealing of the fracture walls. At this time, pressure in the melt lens begins to rise again. In (a), the cycles are more asymmetric than in (b), because the pressure rise time in (a) is longer due to higher p'_{max}, as explained in the text.

is a coefficient expressing the inflation and deflation ability of the melt lens,

$$R = \frac{p^* t^*}{l^2 v} \qquad (12)$$

measures the relative amount of relaxation of pressure on the time scale of melt flow, and

$$\delta = \frac{a^* h}{l^2}. \qquad (13)$$

5.2. Solution:

We use the initial conditions of $p'(0)=0$ and $a'(0), \omega'(0)=0$, and numerically calculate solutions for equations (9) and (10). We find that there are three regimes for solutions:

(1) If $Q'_0/R > p'_{max}$ (see equation (9)), the rock surrounding the melt lens will fracture. There are then two possibilities: 1a. oscillating fracture and healing, or 1b. a continuously open fracture and steady state melt flow.

(1a) In the case of oscillatory fracture, initial melt velocity in the fracture is high, but as the pressure is released the velocity decreases, and the crack heals before the melt velocity reaches a steady value. Then the pressure increases

again, and the process repeats itself in a cyclic manner (Figure 5).

The time $\Delta t'$ for the rise of pressure between cycles is obtained by solving equation (9) with a closed fracture, ($a'=0$), and an initial condition of $p'=0$,

$$\Delta t'_r = \frac{\gamma}{R} \ln \left(\frac{Q'_0}{Q'_0 - Rp'_{max}} \right). \qquad (14)$$

Where "viscous" pressure dissipation is negligible, $R \to 0$, and the rise in pressure between cycles is linear:

$$\Delta t'_r = \gamma \frac{p'_{max}}{Q'_0}. \qquad (15)$$

Similarly, the decay time $\Delta t'_d$ of pressure due to melt expulsion through the fracture can be calculated from equation (9). Although the rise time, $\Delta t'_r$ increases without bound with increasing p'_{max}, the decay time, $\Delta t'_d$, approaches a constant value as p'_{max} increases, with increasingly longer rise times as can be seen in Figure 5.

(1b) The second possible scenario following fracture of the rock surrounding the melt lens is an open fracture, with steady state width and melt velocity, in which the output flux is equal to the input flux (Figure 6). In our model we have not included melt-back, but this is possible for case

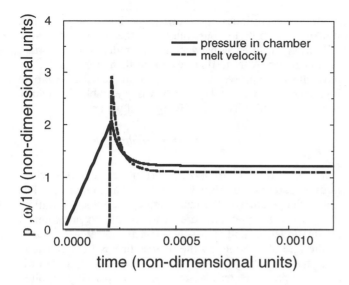

Figure 6. Pressure and flow velocity calculated numerically with flux into the melt lens, Q'_0, ten times greater than in Figure 5. Otherwise, all values are the same as in Figure 5a. Initially, melt pressure (solid line) rises in the melt lens until it reaches p'_{max}. Then a fracture is opened and melt flow (dashed line) is maximal. As time progresses, pressure drops. Outflux through the fracture is reduced and reaches a steady value equal to Q'_0. As a consequence, the crack width and melt velocity are never small enough to allow for the fracture to heal.

(1b) when input flux is large enough and/or hot enough, as shown by *Bruce and Huppert* [1989]. Neglecting "viscous" pressure dissipation during the time that the fracture is open, and assuming e' ~ 1, this steady state will occur if the input flux is high, or the solidification time is long. The criterion for continuous, steady outflux is

$$Q'_0 > \frac{\delta}{Pe^{*4/5}}   \quad (16)$$

(2) If $Q'_0/R < p'_{max}$ then the system reaches a steady state (see equation (9)) where the pressure in the melt lens is elevated relative to the initial pressure, but has not reached the point where it can fracture the surrounding rock. The larger the relaxation parameter, or the smaller the melt input, the more likely it is that this steady state will be reached before fracture occurs (Figure 7).

6. SCALED CALCULATIONS, AND COMPARISON TO OBSERVED PHENOMENA

Field observations indicate that the case of oscillatory release of melt does occur beneath oceanic spreading ridges. It is possible to obtain some quantitative estimates of the physical process from the simple model we have developed

in the previous section. First, an estimate for p_{max} can be made using laboratory measurements of the strength of rocks. Using this estimate, and ignoring viscous dissipation, we can obtain the time between consecutive fracture events, Δt_r, using equation (15), and translating back into dimensional variables

$$\Delta t_r = \frac{l^2}{GQ_0} p_{max}.   \quad (17)$$

The melt flux through the Moho beneath a fast-spreading mid-ocean ridge must be ~ 600 m^2/yr per unit length of the ridge. This is distributed over a cross-sectional region about 4 km wide, perpendicular to the ridge axis. If we take a typical MTZ melt lens to be 0.5 m high by 100 m long (volume of 10 m^2), there would be about 40 sills across 4 km, and the flux into each would be 15 m^2/yr.

The elastic stiffness of rocks, G, may be ~ 10^{10} Pa (10 GPa) for gabbroic rocks at temperatures near their solidi (e.g., *Turcotte and Schubert* [1982]), though the actual value is uncertain. Like the elastic stiffness, the melt pressure required to form tensile fractures in gabbroic rocks, particularly partially molten gabbroic rocks, is poorly known. We will use 5 10^7 Pa (50 MPa), which is the estimated the tensile strength of partially molten peridotite peridotites (e.g., *Kelemen et al.* [1997a]; *Nicolas* [1986,

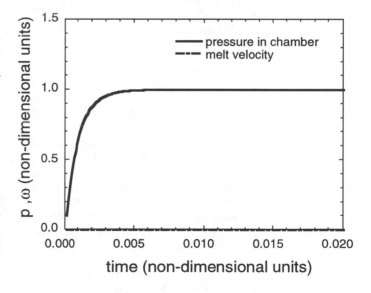

Figure 7. In this Figure, R, the "viscous dissipation coefficient", is ten times greater than in Figure 5. Otherwise, all parameters are as in Figure 5b. In this case, a steady pressure (solid line) is established when influx is compensated by viscous pressure dissipation (due to porous flow out of the melt lens and/or viscous expansion of the lens). This pressure is less than required to fracture the overlying rock, so there is no melt flow in fractures.

1990]), assuming that externally imposed tensile stresses are negligible.

With these approximations, melt-filled fractures originating at the MTZ may be estimated to occur once every three years. If viscous dissipation of melt pressure were included in the estimation, the frequency of fracture formation would be smaller. Also note that this calculation only includes the time to reduce melt flux to the critical value for "healing". Since the healing process itself may take considerable time, the recurrence intervals estimated using equation (17) should be considered minimal.

A similar estimate can be made for the "shallow melt lens" seismically imaged near the dike/gabbro transition along the East Pacific Rise (e.g., *Harding et al.* [*1989*]; *Sinton and Detrick* [*1992*]). If igneous accretion is assumed to take place continuously over the full height of the crustal section, at this level in the crust the melt flux would be about one third of the flux through the Moho, cross-sectional magma chamber width may be ~ 10^3 m, G may be closer to 10^{11} Pa [*Turcotte and Schubert 1982*], and melt lens volume may be ~ 10^4. These values give a frequency of about one fracture per 2.5 years, similar to the rate inferred from the width of sheeted dikes (~ 1 m wide dikes at a spreading rate of 0.1 m/yr yields 1 dike/10 years).

Finally, we can estimate the volume of melt expelled during each cycle, Q_{tot}, by noting that in each cycle the flux into the lens is equal to the flux out of the lens (neglecting crystallization), since at the end of each cycle the lens returns to the same deflated state. Thus, the total melt flux expelled during each diking event will be approximately

$$V_{tot} = Q_0 \Delta t_r = \frac{l^2}{G} p_{max}. \qquad (18)$$

Individual melt removal events from MTZ sills might thus have cross-sectional volumes ~ 50 m^2. This is of the same order as the cross-sectional volume estimated for melt lenses near the MTZ. Thus, individual fracture events may completely empty melt lenses near the base of the crust. Eruptions generated from the shallow melt lens would have cross-sectional volumes of ~ 500 m^2, much smaller than the shallow melt lens volume. This is consistent with the hypothesis that a shallow melt lens is present at steady-state beneath fast-spreading ridges.

7. MODAL LAYERING FORMED BY PERIODIC PRESSURE CHANGE

In this section of the paper, we present quantitative models showing how modal banding may be formed by small changes in magmatic pressure. Several different techniques can be used to calculate "liquid lines of descent" as a result of changing pressure and/or temperature in natural silicate liquid systems (e.g., *Weaver and Langmuir* [*1991*]; *Ghiorso and Sack* [*1995*]; *Yang et al.* [*1996*]). Each

of these techniques its own special strengths and weaknesses, but for our purposes the results are very similar; the differences between these models not important in this paper. After testing several of them (Figure 1), we have chosen to present results only from the MELTS model [*Ghiorso and Sack 1995*] which was preferable for mainly because of flexibility in specifying input parameters.

7.1. Modally Graded Layering in Gabbro Lenses

In section 7.1, proposed primitive, mantle-derived liquids parental to MORB are used in modeling. The initial temperature is the temperature of olivine saturation. It is assumed that pressure changes due to inflation of magmatic lenses and melt-filled fracture are rapid compared to the rate of cooling, so the temperature is held constant. The pressure change used in modeling is 50 Mpa. This is an estimate for tensile strength of partially molten peridotites (e.g., *Kelemen et al.* [*1997a*]; *Nicolas* [*1986, 1990*]) and - by analogy, in the absence of other estimates - we use it here for partially molten gabbros. The magnitude of the pressure change used in the model is quite arbitrary. If the initial, olivine-saturated liquid were slightly cooler, and thereby closer to plag saturation at the initial pressure, then a much smaller pressure change could be used in modeling, with almost exactly the same quantitative result.

Since the liquidus for anhydrous silicate liquids has a positive P/T slope, the effect of increasing pressure at constant temperature is to cause crystallization. In this case, increasing pressure leads to olivine crystallization, followed by saturation in olivine + plag (Figure 8). The predicted crystallization of plag due to increasing pressure at constant temperature may seem counterintuitive to some petrologists, since increasing pressure decreases the stability of plag relative to olivine. In considering only the liquidus surface for primitive basalts, it is true that increasing pressure moves the plag saturation point further from the composition of mantle-derived liquids. However, it must be kept in mind that, because of the positive P/T slope of the liquidus surface, the plag saturation point at a relatively high pressure is at a higher temperature than the plag saturation point at a lower pressure. In the course of increasing pressure at constant temperature, mantle-derived liquids are driven below their liquidus surface and onto multiply saturated cotectics, as illustrated schematically in Figure 9. Thus, increasing pressure drives crystallization and changes the saturated phases from olivine only, to olivine + plag.

One graded layer could be produced by fracture and rapid decompression, followed by fracture healing, gradual pressure increase, and crystallization. During a second isothermal decompression/compression cycle, any melt from the first cycle remaining in the melt lens would be above its liquidus and no crystallization would occur. Slow cooling, and/or influx of additional, olivine-saturated liquid, is required to return to olivine crystallization.

If crystallization occurs at the base of a magma lens, pressure increase is gradual, and pressure decrease is sudden, this process can give rise to "normal", modally graded layers, with only olivine at the base, grading into gabbro at the top, and then a sharp upper contact with the next olivine layer. "Reversely" graded layers could form simultaneously at the top of a magma lens, but might be expected to be less common since olivine will settle in basaltic liquids at MTZ pressures. Both normally and reversely graded layers are observed within gabbroic sills in the Oman MTZ, and within lower crustal layered gabbros in Oman, but normally graded layers are much more common (e.g., *Pallister and Hopson* [*1981*]).

7.2. Phase Layering in Lower Crustal Gabbros

In addition to layers grading from dunite to gabbro, a variety of other types of modal layering are observed in gabbroic sills and lower crustal gabbros in Oman. Particularly obvious in outcrop are pure plagioclase ("anorthosite") bands, from a few cm to > 50 cm in thickness. These layers are difficult to understand as the result of crystal fractionation during cooling of a mantle-derived basalt, since in liquids produced by such a process, plag only crystallizes along cotectics with olivine and/or cpx. However, anorthosite layers are a likely result of crystallization along an accreting surface, combined with rapid decompression of cotectic liquids followed by cooling. Because the plag primary phase volume increases in size with decreasing pressure, liquids on plag - olivine and/or plag - cpx cotectics at a high pressure will be in the plag-only stability field at a lower pressure, as illustrated in Figures 9 and 10. This process could produce anorthosite layers grading into gabbroic rocks within a magma lens, or sills composed only of plagioclase, either by subsequent removal of liquid from the crystallizing lens before it returns to olivine and/or cpx saturation, or by injection of plag-saturated liquid into previously crystallized gabbro, crystallization of an anorthosite sill, and extraction of residual liquid.

More subtle types of modal layering could also be produced by periodic pressure changes, changing the proportions of minerals crystallizing from the same liquid where cotectics are curved in PT space. In offering these examples, we do not mean to suggest that this is an exhaustive review of the possible processes in layer formation due to pressure change. On the contrary, we simply wish to provide a few examples from the rich inventory of possible processes that could form igneous modal layering in a regime of periodically changing pressure. However, where magmatic temperatures are so low that liquids are well within the domain of plag + cpx + opx + oxide saturation, formation of obvious modal layering as a result of periodic pressure change is relatively unlikely. This may help to account for the paucity of modal layering observed in "upper gabbros" near the base of the sheeted dikes in the Oman ophiolite (e.g., *Nicolas et al.* [*1988*]) and in oceanic

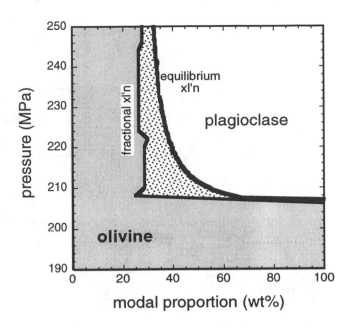

Figure 8. Results of a quantitative model for formation of an igneous layer with an olivine-rich base and a plagioclase-rich top, in terms of the modal proportion of minerals produced over a given increment of pressure increase. Crystallization is assumed to take place on the floor of a melt lens. The initial liquid composition is the calculated product of fractional crystallization of parental MORB liquid kg2 [*Kinzler and Grove 1991*] at 200 MPa from 1300 to 1244°C. At 200 MPa and 1244°C, this liquid composition is just above the temperature for plagioclase saturation. With increasing pressure, plagioclase saturation is reached. Schematic phase relations for this process are illustrated in Figure 9. Fractional crystallization produces a discontinuous change in the proportions of minerals crystallized, whereas equilibrium crystallization gradually changes the proportion of minerals in all of the products of crystallization. Production of a graded layer probably requires a process intermediate between perfect fractional and perfect equilibrium crystallization. All MELTS calculations are for H_2O-free liquids along the Fayalite-Magnetite-quartz oxygen fugacity buffer (FMQ).

crust formed at the East Pacific Rise (e.g., *Gillis, Mével, Allan et al.* [*1993*]).

8. DISCUSSION

8.1. More Layers Than Dikes: Several Melt Lenses Within the Crust?

One might ask if each modal layer records the formation of one sheeted dike? Apparently not. If individual modal layers average about 100 meters in horizontal extent and 0.1 m in height (e.g., *Pallister and Hopson* [*1981*]; *Smewing* [*1981*]; *Browning* [*1982, 1984*]; *Benn et al.* [*1988*]; *Boudier et al.* [*1996b*]), over a vertical extent of layered

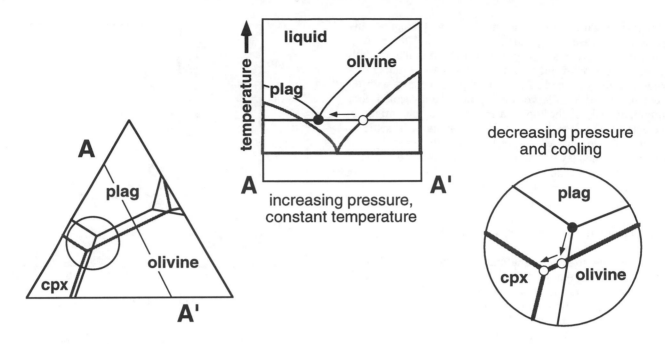

Figure 9. Schematic illustration of the system Forsterite (olivine) - Diopside (cpx) - Anorthite (plag), with cotectics at two different pressures in the range of 100 to 300 MPa, based on phase equilibrium data summarized by *Morse [1980]*. Solid lines and filled symbols indicate phase relations at relatively high pressure; grey lines and open symbols are for relatively low pressure. The temperature/composition section A-A' elucidates the process quantitatively illustrated in Figure 8. Increasing pressure at constant temperature leads to olivine crystallization, which in turn drives the liquid to the olivine-plag cotectic. Crystallization on the floor of a melt lens as a result of this type of process may form graded layers with olivine-rich bases and plag-rich tops. The lower inset shows phase relations for the process quantitatively illustrated in Figure 10. Cooling liquid saturated in olivine, plag and cpx undergoes rapid decompression. Initially, at lower pressure the liquid is not saturated in any solid phase. Because decreasing pressure increases the size of the plag primary phase volume relative to olivine and cpx, after cooling liquid at low pressure the liquid initially saturates in plag. Furthermore, because decreasing pressure has also expanded the olivine primary phase volume relative to cpx, the second phase to crystallize is olivine, followed by cpx after continued cooling. This type of process may form pure plagioclase layers ("anorthosite") and olivine + plagioclase layers ("troctolite") within layered gabbro intrusions.

gabbro of the order of 3 km, there will be ~ 10^4 layers in a 100 m wide section of crust perpendicular to the spreading direction. In the same section, if sheeted dikes average about 1 m in width (e.g., *Pallister [1981]*), there will be only 10^2 dikes. If modal layers had an average horizontal extent of 1 km, there would be 10^4 layers and 10^3 dikes in section 1 km wide. Only if layers were more than 1 m thick and 1 km long, or if dikes were less than 0.1 m wide, could the number of modal layers be equal to the number of dikes. While the average dimensions of modal layers are uncertain, an average of 1 meter thick by 1 km long is far too large.

Therefore, if each layer within gabbros uniquely records a melt-filled fracture event, there must be ~ 10 to 100 more fracture events in the lower crust than there are sheeted dikes. This result is in accord with the estimates for fracture frequency derived using equation (17) in section 6. Such a result implies that melt-filled fractures do not all reach the level of sheeted dike formation; many may terminate at overlying magma lenses or within other mechanical boundaries within the crystallizing crust. Thus, our preferred model for the accretion of oceanic layered gabbros at fast spreading ridges is that there are several magma lenses extant at any given time, at levels ranging from the MTZ to the base of the sheeted dikes. The lenses at the MTZ may be fed primarily by continuous porous flow, but the lenses higher in the crust are likely to be fed mainly by injection from melt-filled fractures originating from other lenses at greater depth.

The notion that there are several different depths at which gabbros are formed at mid-ocean ridges is supported by data from the Oman ophiolite that show compositional differences between "upper gabbros", "lower gabbros", and gabbroic sills in the MTZ (e.g., *Pallister and Hopson [1981]; Browning [1982]; Juteau et al. [1988]; Kelemen et al. [1997]; Korenaga and Kelemen [1997]*). In general, Mg/(Mg+Fe) and Ca/(Ca+Na) is higher in MTZ sills than

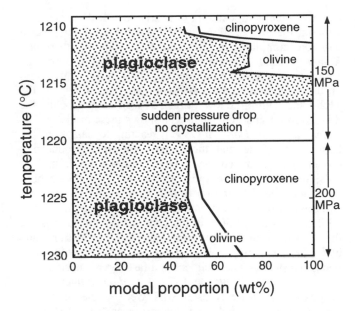

Figure 10. Results of a quantitative model for formation of an igneous anorthosite layer within gabbros. The initial liquid composition is the calculated product of fractional crystallization of parental MORB liquid kg2 [*Kinzler and Grove 1991*] at 200 MPa from 1300 to 1220°C. At 200 MPa, from 1230 to 1220°C, the liquid is saturated in olivine, plag and cpx. Rapid decompression to 150 MPa at 1220°C, followed by continued cooling leads to an interval of no crystallization, followed by plagioclase-in at 1216.5°C, olivine-in at 1214.5°C, and cpx-in at 1211.5°C. All MELTS calculations are for H_2O-free liquids along FMQ.

in lower crustal, layered gabbros within 2 km above the MTZ. These layered gabbros, in turn, have generally higher values than in upper gabbros, within 1 km of the base of the sheeted dikes. These data are consistent with a vertical zonation of temperature in the oceanic crust, with liquids that crystallize at higher levels attaining a larger degree of compositional fractionation as compared to primitive magmas entering the MTZ.

8.2. Crystallization Temperatures for Gabbros and Thermal Structure Beneath Ridges

Because many Oman gabbros are cumulates, i.e., products of small amounts of crystallization from a larger melt mass, after which the residual liquid was efficiently extracted, their calculated melting behavior is different from that of a basaltic liquid. Specifically, because they retain nearly cotectic proportions of olivine, cpx and plag, with nearly constant composition, they resemble eutectic melt compositions in some respects. Thus, as illustrated in Figure 3, the calculated % liquid in gabbro compositions changes dramatically from about 10% to >50% over a very narrow temperature interval. This crystallization behavior

can be used to estimate the temperature at which the cumulates formed, as already discussed in section 3.

Here, we consider the temperature/depth information in more detail. The average composition of 26 samples of sills from the MTZ [*Korenaga and Kelemen 1997*] gives has calculated liquid fraction of 10% at a temperature of ~ 1240°C. We compare this to temperatures for average crustal rock compositions compiled by *Browning* [*1982*]. Olivine gabbros (number of samples, n=6), typical of the lower, layered gabbro section, give a temperature of ~ 1225°C at 10% liquid. Pyroxene gabbros (n=3), with little or no olivine, are also found in the layered gabbro section. Their average composition gives ~ 1205°C at 10% liquid.

Browning's average for non-layered, upper gabbros (n=12) closely resembles a liquid composition (e.g., *Kelemen et al.* [*1997b*]). Such upper gabbros crystallized at lower temperatures than any of the layered gabbros. The calculated temperature at 10% liquid is ~ 1165°C. As noted in previous sections of this paper, seismic evidence indicates that there may be a "shallow melt lens", with more than 10% liquid, near the dike/gabbro transition directly beneath the ridge axis at some fast-spreading ridges (e.g., *Harding et al.* [*1989*]; *Sinton and Detrick* [*1992*]). Since many of these melt lenses support shear waves, *Hussenoeder et al.* [*1996*] estimate that they generally contain <30% liquid. The calculated temperature for the upper gabbro composition of *Browning* [*1982*] at 30% liquid is ~ 1195°C.

It should be noted that the calculated temperature values are dependent on the choice of modeling technique, pressure, oxygen fugacity, and so on. However, the temperature difference from one rock to another is more robust and better constrained by these calculations than the absolute temperatures.

Another important issue is whether all of the gabbroic rocks in Oman are coeval, or whether some formed well off-axis, so that the apparent vertical gradient in crystallization temperature is an artifact of some kind. There is increasing evidence that magmatism in the northern part of the ophiolite may have continued for more than 1 million years (e.g., *Perrin et al.* [*1994*]). This protracted magmatism is temporally related to large tectonic rotations of the northern part of the ophiolite relative to a hotspot reference frame, and relative to the southern part of the ophiolite. Unfortunately, lavas in the southern part of the ophiolite are not well exposed and there are few geochemical data for them. However, the chemical differences that distinguish the early and later volcanics ought to be discernible as two distinct compositional types in gabbros as well. In fact, this is the case in the northern part of the ophiolite (e.g., *Smewing* [*1981*]; *Juteau et al.* [*1988*]). However, geochemical data for gabbros from the southern ophiolite massifs, where we have concentrated our research, suggest that >90% of the igneous rocks form a single liquid line of descent. [*Pallister and Hopson 1981; Pallister and Knight 1981; Browning 1982; Benoit et al. 1996; Kelemen et al. 1997b; Korenaga* and Kelemen 1997]. Thus, we infer that most or all of the gabbros in the southern massifs form a single suite crystallized beneath a spreading center.

One additional point that should be considered is that, even if all of the Oman gabbros are coeval and formed beneath a spreading ridge, their igneous temperatures do not necessarily reflect temperatures along a steady-state geotherm beneath that ridge. This is most true for upper parts of the crustal section, where wall rocks would have been relatively cold, and magma transport was dominated by brittle intrusion of dikes and sills. Thus, in the upper crust directly beneath a ridge, there probably is no steady state geotherm, and instead every depth may be characterized by a variety of temperatures over time. In contrast, for the MTZ and lower gabbros, magmatic temperatures and wall rock temperatures were probably quite similar. For example, Korenaga and Kelemen [1997] calculated thermal Peclet numbers for diffuse, porous, intergranular melt flow near the Moho at fast-spreading ridges, and concluded that liquid and solid would maintain the same temperature during such a process.

In summary, we believe that vertical reconstructions of composition and temperature based on average data for many samples from Oman are likely to be representative of a coeval suite of rocks formed beneath a spreading ridge. Thus, the data from the Oman ophiolite are consistent with a lower crust below the ridge axis that had systematic vertical variation in both composition and temperature, with a temperature gradient of ~ 45 to 75°C from the base of the crust to the "shallow melt lens" below the base of the sheeted dikes.

8.3. Igneous Accretion of the Lower Oceanic Crust

On the basis of field data from Oman and seismic data from the East Pacific Rise, it has been proposed that the lower oceanic crust is formed by partial crystallization of gabbros in a single, shallow melt lens near the dike/gabbro transition, followed by ductile flow downward and outward from this melt lens [Nicolas et al. 1988, 1993; Quick and Denlinger 1993; Phipps Morgan and Chen 1993; Henstock et al. 1993], as originally proposed on theoretical grounds by Sleep [1975] and Dewey and Kidd [1977]. We shall refer to these as "conveyor belt" models. Because the gabbros do not preserve structures formed by crystal plastic deformation, it was proposed deformation occurred by intracrystalline glide coupled with pressure solution, in partially molten gabbros with porosities greater than 10% (e.g., Nicolas [1992]; Nicolas and Ildefonse [1996]).

Boudier et al. [1996] and Kelemen et al. [1997b] have questioned this end-member hypothesis on the grounds that some structures in the lower, layered gabbros such as modally graded layering could not survive the large shear strains required by "conveyor belt" models. In addition, Kelemen et al. [1997b] showed that the lower, layered

gabbros were compositionally nearly identical to gabbroic sills in the MTZ; thus, it is possible to form layered gabbroic cumulates in sills near the Moho. Furthermore, Kelemen et al. showed that lower crustal gabbros are compositionally different from upper gabbros. Thus, both groups proposed emplacement of gabbroic sills near the Moho and within the lower crust, as well as just below the dike/gabbro transition. Figure 11 shows these models.

Schouten and Denham [1995], reacting to data from the East Pacific Rise and the Oman ophiolite, produced a model with two "conveyor belts" originating from two sills, one at the Moho and the other near the dike/gabbro transition. Our illustration of this hypothesis, Figure 11a, is thermally quite an extreme case, because it requires crystallization of about 50% of the crustal mass at the base of the crust, where heat removal by conduction and hydrothermal convection may not be efficient, especially beneath a steady-state melt lens at a fast-spreading ridge.

Boudier et al. [1996] and Quick and Denlinger [1993] proposed that lower crustal sills were emplaced into a gabbro "conveyor belt" originating near the dike/gabbro transition. Kelemen et al. [1997b] proposed an alternative hypothesis in which the entire lower crust is constructed of "sheeted sills", each emplaced at its current depth in the crustal section, without vertical transport by ductile flow.

Whereas Kelemen et al. [1997b] proposed the "sheeted sill" model as one of several, equally plausible alternatives, we now tentatively prefer it, for several reasons.

(1) It seems unlikely on thermal grounds that 50% of the crust crystallizes along the Moho, as in Figure 11a.

(2) As noted in an earlier section of this paper, Korenaga and Kelemen [1998] argued that correlation between vertical variation of mineral compositions, coupled with the lack of a systematic, vertical compositional trend, precludes large amounts of melt transport by porous flow through lower crustal gabbros in Oman. If the main mechanism of melt transport through the lower crust were by diffuse porous flow, then the observed correlations, for example between forsterite content in olivine and anorthite content in plagioclase, would have been obliterated by solid/liquid exchange reactions and diffusion.

Therefore, although small amounts of diffuse porous flow must have played a role in melt extraction from cumulate gabbros, the main mechanism of melt transport through the lower crust must have been in zones of focused flow, such as melt-filled fractures. This seems to rule out the presence of a large (ca. 10%), interconnected melt porosity throughout the lower oceanic crust, such as is required in the "conveyor belt" hypotheses to produce low bulk viscosities. It remains possible that formation of vertical chemical variation postdates "conveyor belt" deformation.

(3) Analysis in sections 3 and 8.2 suggests that there was a gradient of 45 to 75°C from the dike/gabbro transition to the MTZ beneath the ridge axis during formation of the Oman ophiolite, as in Figure 11. As shown by Phipps

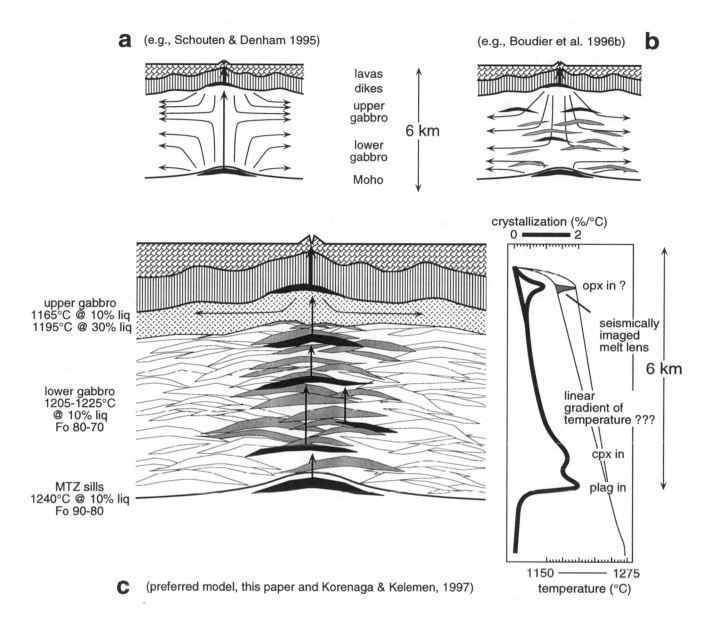

Figure 11. Recent models for the formation of the oceanic lower crust, based on seismic observations from the East Pacific Rise and field constraints from the Oman ophiolite. (a) is redrawn from *Schouten and Denham* [*1995*], who illustrated a scenario in which gabbros crystallize mainly in two melt lenses, one in the MTZ and one near the dike/gabbro transition. (b) is redrawn from *Boudier et al.* [*1996*]. They proposed that most gabbros crystallize in a melt lens near the dike/gabbro transition, but as these move downward and away from the melt lens by melt-lubricated ductile deformation they are intruded by lower crustal, gabbroic sills. A similar scenario was suggested by *Quick and Denlinger* [*1993*]. (c) is redrawn from *Kelemen et al.* [*1997b*], who proposed an end-member scenario in which there is no vertical ductile transport of gabbro in the oceanic lower crust, which is instead constructed of sheeted sills, emplaced at approximately their final depth below the sea-floor. Similar proposals have also been made by J. Bédard and co-workers (e.g., *Bédard et al.* [*1988*]; *Bédard* [*1991*]). We tentatively conclude that (c) is most consistent with observations from Oman, including the thermal gradient we infer from crystallization calculations for gabbro compositions (Figure 3). On the lower right, we schematically illustrate the inferred thermal gradient and vertical distribution of crystallization rates in the crust forming beneath a spreading ridge axis. A linear temperature gradient is adopted for simplicity. Along such a gradient, there could be three local maxima in the crystallization rate, corresponding to the appearance of plag, cpx and opx on the liquidus of cooling, crystallizing basaltic magma. Each of these might form a permeability barrier, with melt lenses ponding beneath it. Formation of melt lenses below these barriers, and elsewhere in the lower crust, followed by replenishment of these lenses with relatively hot magma from below, could dramatically perturb the linear thermal gradient illustrated here. This, in turn, would give rise to still more variations in crystallization rate and permeability in a vertical section below the axis of an active spreading ridge.

Morgan and Chen [*1993*], "conveyor belt" processes may result in small, nearly adiabatic temperature gradients in the lower crust beneath a ridge axis.

(4) In section 8.1, we suggested that - if both modal layers in gabbros and sheeted dikes record melt fracture events - the fact that the number of layers is much larger than the number of dikes suggests that there are many melt lenses at different depths within the crust at any given time.

In light of these observations and inferences, we suggest that the "sheeted sill" model accounts better for the composition of lower crustal gabbros in the Oman ophiolite than any "conveyor belt" model proposed to date.

9. CONCLUSION

Periodic formation of melt-filled fractures can result from magma accumulation beneath a permeability barrier in a visco-elastic solid medium. Periodic pressure change, caused by continuous influx and periodic extraction of liquids, can potentially explain a variety of modal layering phenomena in gabbroic rocks, not only in the oceanic crust but perhaps also in some layered intrusions. This hypothesis seems particularly likely to apply to the generation of the oceanic crust, where sheeted dikes preserve a clear record of periodic melt-filled fracture and, at least in Oman, have the compositional characteristics of liquids extracted from lower crustal, layered gabbros. Thus, the two most evident features indicative of periodic crystallization in igneous crust formed at oceanic spreading centers, modal layering in lower crustal gabbros and sheeted dikes, may be complementary features formed by the same physical process.

Acknowledgments. Generous assistance with field work was provided by Hilal Al Azri, Benoit Ildefonse, Françoise Boudier, Adolphe Nicolas, Greg Hirth, Emilie Hooft and Rachel Cox. John Delaney's insistent aphorism, "sheeted dikes are the quantum event in oceanic crust formation", inspired this paper. Jack Whitehead provided crucial guidance. We benefited from discussions with Alan Rubin, Greg Hirth, Hans Schouten, Alexander MacBirney, Stu McCallum, Alan Boudreau, Marc Spiegelman, Jun Korenaga, Ken Koga, Adolphe Nicolas, Françoise Boudier, Benoit Ildefonse, Dan McKenzie, and others, and reviews by Dave Sparks, Tom Parsons, and Ro Kinzler. This work was supported by NSF grants OCE-9314013, OCE-9416616 and OCE-9711170 (Kelemen) and a Lamont-Doherty Postdoctoral Fellowship (Aharonov).

REFERENCES

Aharonov, E., J. A. Whitehead, P. B. Kelemen, and M. Spiegelman, Channeling instability of upwelling melt in the mantle, *J. Geophys. Res., 100*, 20,433-20,450, 1995.

Aharonov, E., M. Spiegelman and P. B. Kelemen, 3D flow and reaction in porous media with implications for the Earth's mantle and for sedimentary basins, *J. Geophys. Res., 102*, 14,821-14,833, 1997.

Alabaster, T., J. A. Pearce and J. Malpas, The volcanic stratigraphy and petrogenesis of the Oman ophiolite complex, *Contrib. Mineral. Petrol., 81*, 168-183, 1982.

Bédard, J. H., Cumulate recycling and crustal evolution in the Bay of Islands ophiolite, *J. Geol., 99*, 225-249, 1991.

Bédard, J.H., R.S.J. Sparks, R. Renner, M.J. Cheadle, and M.A. Hallworth, Peridotite sills and metasomatic gabbros in the eastern layered series of the Rhum complex, *J. Geol. Soc. Lond., 145*, 207-224, 1988.

Benn, K., A. Nicolas and I. Reuber, Mantle-crust transition zone and origin of wehrlitic magmas: Evidence from the Oman Ophiolite, *Tectonophysics, 151*, 75-85, 1988.

Benoit, M., M. Polvé, and G. Ceuleneer, Trace element and isotopic characterization of mafic cumulates in a fossil mantle diapir (Oman ophiolite), *Chem. Geol., 134*, 199-214, 1996.

Boudier, F. and A. Nicolas, Nature of the Moho Transition Zone in the Oman ophiolite, *J. Petrol., 36*, 777-796, 1995.

Boudier, F., J. L. Bouchez, A. Nicolas, M. Cannat, G. Ceuleneer, M. Misseri, and A. Montigny, Kinematics of oceanic thrusting in the Oman ophiolite: Model of plate convergence, *Earth Planet. Sci. Lett., 75*, 215-222, 1985.

Boudier, F., C. J. MacLeod and L. Bolou, Structures in peridotites from Site 895, Hess Deep: Implications for the geometry of mantle flow beneath the East Pacific Rise, in *Proc. Ocean Drill . Prog. Sci. Res., 147*, edited by C. Mevel, K. M., Gillis, J. F. Allan, and P. S. Meyer, pp. 347-356, 1996a.

Boudier, F., A. Nicolas and B. Ildefonse, Magma chambers in the Oman ophiolite: Fed from the top or the bottom?, *Earth Planet. Sci. Lett., 144*, 239-250, 1996b.

Boudreau, A. E. and I. S. McCallum, Investigations of the Stillwater Complex: III. The Picket Pin Pt/Pd Deposit, *Econ. Geol., 81*, 1953-1975, 1986.

Boudreau, A. E., Investigations of the Stillwater Complex. IV. The role of volatiles in the petrogenesis of the J-M Reef, Minneapolis adit section, *Can. Miner., 26*, 193-208, 1988.

Browning, P., *The Petrology, Geochemistry and Structure of the Plutonic Rocks of the Oman Ophiolite*, Ph.D. thesis, 405 pp., The Open University, United Kingdom, 1982.

Browning, P., Cryptic variations within the cumulate sequence of the Oman ophiolite: Magma chamber depth and petrological implications, *Geol. Soc. London Spec. Pub., 13*, 71-82, 1984.

Bruce, P. M. and H. E. Huppert, Thermal control of basaltic fissure eruptions, *Nature, 342*, 665-667, 1989.

Bruce, P. M. and H. E. Huppert, Solidification and melting along dykes by the laminar flow of basaltic magma, in *Magma Transport and Storage*, edited by M. P. Ryan, pp. 87-101, J. Wiley and Sons, New York, 1990.

Cameron, E. N., Chromite in the central sector, eastern Bushveld Complex, South Africa, *Amer. Miner., 62*, 1082-1096, 1977.

Cannat, M., How thick is the magmatic crust at slow spreading oceanic ridges?, *J. Geophys. Res., 101*, 2847-2857, 1996.

Ceuleneer, G. , Evidence for a paleo-spreading center in the Oman ophiolite: Mantle structures in the Maqsad area, in *Ophiolite Genesis and Evolution of the Oceanic Lithosphere*, edited by Tj. Peters, A. Nicolas, and R. Coleman, pp. 147-173, Kluwer, Dordrecht, 1991.

Ceuleneer, G. and M. Rabinowicz, Mantle flow and melt migration beneath oceanic ridges: Models derived from observations in ophiolites, in *Geophysical Monograph 71, Mantle Flow and Melt Generation*, edited by J. Phipps Morgan, D. K. Blackman and J. M. Sinton, pp. 123-154, American Geophysical Union, Washington DC, 1992.

Constantin, M., R. Hekinian, D. Bideau, and R. Hebert, Construction of the oceanic lithosphere by magmatic intrusions: Petrological evidence from plutonic rocks formed along the fast-spreading east Pacific Rise, *Geology, 24*, 731-734, 1996.

Daines, M. J. and D. L. Kohlstedt, A laboratory study of melt migration, *Phil Trans. Roy. Soc. London A, 342*, 43-52, 1993.

Daines, M.J., and D.L. Kohlstedt, The transition from porous to channelized flow due to melt/rock reaction during melt migration, *Geophys. Res. Lett., 21*, 145-148, 1994.

Delaney, J. R., J. A. Baross, M. D. Lilley and D. S. Kelley, Is the quantum event of crustal accretion a window into a deep hot biosphere?, *EOS, 75*, 617, 1994.

Dewers, T. and P. Ortoleva, The role of geochemical self-organization in the migration and trapping of hydrocarbons, *Applied Geochemistry, 3*, 287-316, 1988.

Dewers, T. and P. Ortoleva, Nonlinear dynamical aspects of deep basin hydrology: Fluid compartment formation and episodic fluid release, *Am. J. Sci., 294*, 713-755, 1994.

Dewey J. F. and S. F. Kidd, Geometry of plate accretion, *Geol. Soc. Amer. Bull., 88*, 960-968, 1977.

Dick, H. J. B., Abyssal peridotites, very slow spreading ridges and ocean ridge magmatism, in *Magmatism in the Ocean Basins*, edited by A. D. Saunders and M. J. Norry, pp. 71-105, Geological Society Special Publications, 1989.

Dick, H. J. B. and J. H. Natland, Late stage melt evolution and transport in the shallow mantle beneath the East Pacific Rise, *Sci. Res. Ocean Drill. Prog., 147*, 103-134, 1996.

Dunn, R. A., and D. R. Toomey, Seismological evidence for three-dimensional melt migration beneath the East Pacific Rise, *Nature, 388*, 259-262, 1997.

Elthon, D., M. Stewart and D. K. Ross, Compositional trends of minerals in oceanic cumulates, *J. Geophys Res, 97*, 15,189-15,199, 1992.

Garmany, J., Accumulations of melt at the base of the oceanic crust, *Nature, 340*, 628-632, 1989.

Garmany, J., Mapping the occurrence of melt intrusions at the base of young oceanic crust near 9 30'N on the EPR, *EOS, 75*, 602, 1994.

Ghiorso, M. S. and R. O. Sack, Chemical mass transfer in magmatic processes, IV: A revised and internally consistent thermodynamic model for the interpolation and extrapolation of liquid-solid equilibria in magmatic systems at elevated temperatures and pressures, *Contrib. Mineral. Petrol., 119*, 197-212, 1995.

Gillis, K., C. Mével, and J. Allan, et al., *Proceedings of the Ocean Drilling Program, Initial Reports, 147*, 366 pp., 1993.

Gudmundsson, Å., Formation of crustal magma chambers in Iceland, *Geology, 14*, 164-166, 1986.

Gudmundsson, Å., Emplacement of dikes, sills and crustal magma chambers at divergent plate boundaries, *Tectonophysics, 176*, 257-275, 1990.

Hacker, B. R., J. L. Mosenfelder, and E. Gnos, Rapid emplacement of the Oman Ophiolite; thermal and geochronologic constraints, *Tectonics, 15*, 1230-1247, 1996.

Harding, A. J., J. Orcutt, M. Kappus, E. Vera, J. Mutter, P. Buhl, R. Detrick and T. Brocher, The structure of young oceanic crust at 13°N on the East Pacific Rise from expanding spread profiles, *J. Geophys. Res., 94*, 12,163-12,196, 1989.

Hekinian, R., D. Bideau, M. Cannat, J. Francheteau, and R. Hébert, Volcanic activity and crust-mantle exposure in the ultrafast Garrett transform fault near 13°28'S in the Pacific, *Earth Planet. Sci. Lett., 108*, 259-275, 1992.

Hekinian, R., D. Bideau, J. Francheteau, J.L. Cheminee, R. Armijo, P. Lonsdale, and N. Blum, Petrology of the East Pacific Rise crust and upper mantle exposed in Hess Deep (eastern equatorial Pacific), *J. Geophys. Res., 98*, 8069-8094, 1993.

Henstock, T. J., A. W. Woods and R. S. White, The accretion of oceanic crust by episodic sill intrusion, *J. Geophys. Res., 98*, 4143-4161, 1993.

Hirose, K. and I. Kushiro, Partial melting of dry peridotites at high pressures: Determination of compositions of melts segregated from peridotite using aggregates of diamond, *Earth Planet. Sci. Lett., 114*, 477-489, 1993.

Hunt, J. M., Generation and migration of petroleum from abnormally pressured fluid compartments, *Am. Assoc. Petrol. Geol. Bull., 74*, 1-12, 1990.

Hussenoeder, S. A., J. A. Collins, G. M. Kent, R. S. Detrick, A. J. Harding and J. A. Orcutt, Seismic analysis of the axial magma chamber reflector along the southern East Pacific Rise from conventional reflection profiling, *J. Geophys. Res., 101*, 22,087-22,105, 1996.

Ida, Y., Cyclic fluid effusion accompanied by pressure change: Implications for volcanic eruptions and tremor, *Geophys. Res. Lett., 23*, 1457-1460, 1996.

Juteau, T., M. Beurrier, R. Dahl, and P. Nehlig, Segmentation at a fossil spreading axis: The plutonic sequence of the Wadi Haymiliyah area (Haylayn Block, Sumail Nappe, Oman), *Tectonophys., 151*, 167-197, 1988.

Kelemen, P. B., Assimilation of ultramafic rock in subduction-related magmatic arcs, *J. Geol., 94*, 829-843, 1986.

Kelemen, P. B., Reaction between ultramafic rock and fractionating basaltic magma I. Phase relations, the origin of calc-alkaline magma series, and the formation of discordant dunite, *J. Petrol., 31*, 51-98, 1990.

Kelemen, P. B., J. A., Whitehead, E. Aharonov, and K. A. Jordahl, Experiments on flow focusing in soluble porous media, with applications to melt extraction from the mantle, *J. Geophys. Res., 100*, 475-496, 1995a.

Kelemen, P. B., N. Shimizu and V. J. M. Salters, Extraction of mid-ocean-ridge basalt from the upwelling mantle by focused flow of melt in dunite channels, *Nature, 375*, 747-753, 1995b.

Kelemen, P. B., G. Hirth, N. Shimizu, M. Spiegelman and H. J. B. Dick, A review of melt migration processes in the asthenospheric mantle beneath oceanic spreading centers, *Phil. Trans. Roy. Soc. London A, 355*, 283-318, 1997a.

Kelemen, P.B., K. Koga and N. Shimizu, Geochemistry of gabbro sills in the crust/mantle transition zone of the Oman ophiolite: Implications for the origin of the oceanic lower crust, *Earth Planet. Sci. Lett., 146*, 475-488, 1997b.

Kinzler, R. J. and T. L. Grove, Primary magmas of mid-ocean ridge basalts 2. Applications, *J. Geophys. Res.*, *97*, 6907-6926, 1992.

Korenaga, J. and P. B. Kelemen, The origin of gabbro sills in the Moho transition zone of the Oman ophiolite: Implications for magma transport in the oceanic lower crust, *J. Geophys. Res.*, *102*, 27,729-27,749, 1997.

Korenaga, J. and P. B. Kelemen, Melt migration through the oceanic lower crust: A constraint from melt percolation modeling with finite solid diffusion, Earth Planet. Sci. Lett., *in press*, 1998.

Langmuir, C. H., E. M. Klein and T. Plank, Petrological systematics of mid-ocean ridge basalts: Constraints on melt generation beneath ocean ridges, *in American Geophysical Union Monograph 71, Mantle Flow and Melt Generation*, edited by J. Phipps Morgan, D. K. Blackman and J. M. Sinton, pp. 183-280, American Geophysical Union, Washington DC, 1992.

Lipin, B. R., Pressure increases, the formation of chromite seams, and the development of the ultramafic series in the Stillwater Complex, Montana, *J. Petrol.*, *34*, 955-976, 1993.

Lippard, S. J., A. W. Shelton, and I. G. Gass, *The Ophiolite of Northern Oman.* (178 pp.) Oxford: Blackwell, 1986.

Lister, J. R., The solidification of buoyancy-driven flow in flexible-walled channels, *J. Fluid Mech.*, *272*, 21-65, 1994.

Lister, J.R. and P. J. Dellar, Solidification of pressure-driven flow in a finite rigid channel with application to volcanic eruptions, *J. Fluid Mech.*, *323*, 267-283, 1996.

Maaloe, S., Magma accumulation in the ascending mantle, *J. Geol. Soc. London*, *138*, 223-236, 1981.

McKenzie, D., The generation and compaction of partially molten rock, *J. Petrol.*, *25*, 713-765 1984.

Meyer, P. S., H. J. B. Dick and G. Thompson, Cumulate gabbros from the Southwest Indian Ridge, 54°S-7°16'E: Implications for magmatic processes at a slow spreading ridge, *Contrib. Mineral. Petrol.*, *103*, 44-63, 1989.

Michard, A., F. Boudier and B. Goffé, Obduction versus subduction and collision in the Oman case and other Tethyan settings, in *Ophiolite Genesis and Evolution of the Oceanic Lithosphere*, edited by Tj. Peters, A. Nicolas, and R. Coleman, pp. 447-467, Kluwer, Dordrecht, 1991.

Morse, S. A., *Basalts and Phase Diagrams*, 493 pp., Springer-Verlag, Berlin, 1980.

Naslund, H. R. and A. R. MacBirney, Mechanisms of formation of igneous layering, in *Layered Intrusions*, edited by R. G. Cawthorn, pp. 1-43, Elsevier, 1996.

Nicolas, A., A melt extraction model based on structural studies in mantle peridotites, *J. Petrol.*, *27*, 999-1022, 1986.

Nicolas, A., *Structures of Ophiolites and Dynamics of Oceanic Lithosphere*, 367 pp., Kluwer Academic, Norwell, Mass., 1989.

Nicolas, A. Melt extraction from mantle peridotites: Hydrofracturing and porous flow, with consequences for oceanic ridge activity, *in Magma Transport and Storage*, edited by M. P. Ryan, pp. 159-174, J. Wiley and Sons, New York, 1990.

Nicolas, A., Kinematics in magmatic rocks with special reference to gabbros, *J. Petrol.*, *33*, 891-915, 1992.

Nicolas, A., I. Reuber and K. Benn, A new magma chamber model based on structural studies in the Oman ophiolite, *Tectonophysics*, *151*, 87-105, 1988.

Nicolas, A., C. Freydier, M. Godard and A. Vauchez, Magma chambers at oceanic ridges: How large?, *Geology*, *21*, 53-56, 1993.

Nicolas, A., B. Ildefonse and F. Boudier, Flow mechanism and viscosity in basaltic magma chambers, *Geophys. Res. Lett.*, *23*, 2013-2016, 1996.

Nur, A. and J. Walder, Hydraulic pulses in the Earth's crust, *in Fault Mechanics and Transport Properties of Rocks*, pp. 461-473, Academic Press, 1992.

Pallister, J. S., Structure of the sheeted dike complex of the Samail ophiolite near Ibra, Oman, *J. Geophys. Res.*, *86*, 2661-2672, 1981.

Pallister, J. S., Parent magma of the Semail ophiolite, Oman, *Geol. Soc. London Spec. Pub. 13* , 63-70, 1984.

Pallister, J. S. and C. A. Hopson, Samail ophiolite plutonic suite: Field relations, phase variation, cryptic variation and layering, and a model of a spreading ridge magma chamber, *J. Geophys. Res.*, *86*, 2593-2644, 1981.

Pallister J. S. and R. J. Knight, Rare-earth element geochemistry of the Samail ophiolite near Ibra, Oman, *J. Geophys. Res.*, *86*, 2673-2697, 1981.

Pearce, J. A., T. Alabaster, A. W. Shelton and M. P. Searle, The Oman ophiolite as a Cretaceous arc-basin complex: Evidence and implications, *Phil. Trans. R. Soc. Lond. A*, *300*, 299-317, 1981.

Perrin, M., M. Prevot, and F. Bruere, Rotation of the Oman ophiolite and initial location of the ridge in the hotspot reference frame, *Tectonophysics*, *229*, 31-42, 1994.

Phipps Morgan, J. and Y. J. Chen, The genesis of oceanic crust: Magma injection, hydrothermal circulation, and crustal flow, *J. Geophys. Res.*, *98*, 6283-6297, 1993.

Rubin, A. M., Propagation of magma-filled cracks, *Ann. Rev. Earth Planet. Sci.*, *23*, 287-336, 1995a.

Rubin, A. M., Getting granite dikes out of the source region, *J. Geophys. Res.*, *100*, 5911-5929, 1995b.

Schouten H. and C. Denham, Virtual ocean crust, *EOS, 76*, S48, 1995.

Sinton J. M. and R. S. Detrick, Mid-ocean ridge magma chambers, *J. Geophys., Res.*, *97*, 197-216, 1992.

Sleep, N. H., Formation of oceanic crust: Some thermal constraints, *J. Geophys. Res.*, *80*, 4037-4042, 1975.

Smewing, J. D., Mixing characteristics and compositional differences in mantle derived melts beneath spreading axes: Evidence from cyclically layered rocks in the ophiolite of northern Oman, *J. Geophys. Res.*, *86*, 2645-2659, 1981.

Sparks, D. W. and E. M. Parmentier, Melt extraction from the mantle beneath mid-ocean ridges, *Earth Planet. Sci. Lett.*, *105*, 368-377, 1991.

Spiegelman, M. Flow in deformable porous media. Part 1: Simple analysis, *J. Fluid Mech.*, *247*, 17-38, 1993a.

Spiegelman, M., Flow in deformable porous media. Part 2: Numerical analysis - the relationship between shock waves and solitary waves, *J. Fluid Mech.*, *247*, 39-63, 1993b.

Spiegelman, M. Physics of melt extraction: Theory, implications and applications, *Phil. Trans. R. Soc. London A, 342*, 23-41, 1993c.

Spiegelman, M., and P. Kenyon, The requirements for chemical disequilibrium during magma migration, *Earth Planet. Sci. Lett.*, *109*, 611-620, 1992.

Tilton, G. R., C. A. Hopson and J. E. Wright, Uranium-lead ages of the Samail ophiolite, Oman, with applications to Tethyan ocean ridge tectonics, *J. Geophys. Res.*, *86*, 2763-2775, 1981.

Toomey, D. R., G. M. Purdy, S. Solomon and W. Wilcock, The three dimensional seismic velocity structure of the East Pacific Rise near latitude 9°30'N, Nature, *347*, 639-644, 1990.

Turcotte, D. and G. Schubert, *Geodynamics: Application of Continuum Physics to Geological Problems*, 450 pp., John Wiley and Sons, New York, 1982.

Vera, E. E., P. Buhl, J. C. Mutter, A. J. Harding, J. A. Orcutt and R. S. Detrick, The structure of 0-0.2 My old oceanic crust at 9°N on the East Pacific Rise from expanding spread profiles, J. Geophys. Res., *95*, 15,529-15,556, 1990.

Weaver, J. and C. H. Langmuir, Calculation of phase equilibrium in mineral-melt systems, *Computers and Geosciences*, *16*, 1-19, 1990.

Whitehead, J. A. and K. R. Helfrich, Instability of flow with temperature-dependent viscosity: A model of magma dynamics, *J. Geophys. Res.*, *96*, 4145-4155, 1991.

Wilkinson, W., editor, *Non-Newtonian Fluids: Fluid Mechanics, Mixing and Heat Transfer*, Pergamon Press, 1960.

Yang, H. J, R. J. Kinzler, and T. L. Grove, Experiments and models of anhydrous, basaltic olivine-plagioclase-augite saturated melts from 0.001 to 10 kbar, *Contrib. Mineral. Petrol.*, *124*, 1-18, 1996.

Peter B. Kelemen, Dept. of Geology & Geophysics, McLean Laboratory, MS #8, Woods Hole Oceanographic Institution, Woods Hole, MA 02543, USA; peterk@cliff.whoi.edu.

Einat Aharonov, Lamont-Doherty Earth Observatory, Palisades, NY 10964, USA; einat@ldeo.columbia.edu.

The Rheology of the Lower Oceanic Crust: Implications for Lithospheric Deformation at Mid-Ocean Ridges

Greg Hirth[1], Javier Escartín[2,3], and Jian Lin[1]

[1]Dept. of Geology and Geophysics, Woods Hole Oceanographic Institution, Woods Hole, MA 02543, U.S.A.

[2]Dept. of Geological Sciences, University of Durham, Durham DH1 3LE, England

[3]Dept. of Geology and Geophysics, University of Edinburgh, Edinburgh EH9 3JW, Scotland

An analysis of experimentally determined flow laws for gabbroic rocks indicates that the lower oceanic crust is considerably stronger than commonly assumed in many modeling studies. A combination of experimental, geochemical and petrological constraints suggest that the majority of the lower oceanic crust is almost dry prior to hydrothermal alteration. Several geological and geophysical observations also indicate that the strength of the lower ocean crust approaches that of dry gabbroic rock. These observations are inconsistent with commonly used rheological models for the oceanic lithosphere that incorporate a weak lower crust that decouples a strong upper mantle from a brittle upper crust.

INTRODUCTION

Our current understanding of mid-ocean ridge dynamics, including the processes that control axial topography and the formation of oceanic crust, has been dramatically aided by the formulation of thermo-mechanical models for the oceanic lithosphere [e.g., *Sleep and Rosendahl*, 1979; *Phipps Morgan et al.*, 1987; *Lin and Parmentier*, 1989; *Chen and Morgan*, 1990; *Phipps-Morgan and Chen*, 1993; *Neumann and Forsyth*, 1993, *Shaw and Lin*, 1996]. The incorporation of experimentally determined flow laws has proven to be instrumental in enhancing the ability of these models to explain geophysical observations.

In most rheological models for the oceanic lithosphere the crust is assumed to be substantially weaker than the underlying mantle [e.g., *Chen and Morgan*, 1990]. A critical aspect of such models is the existence of a ductile lower crust that decouples a thin brittle crustal layer from a significantly stronger and potentially brittle lithospheric mantle [e.g., *Chen and Molnar*, 1983; *Kirby*, 1980]. However, recent experimental deformation studies conducted on "dry" gabbroic rocks [*Mackwell et al.*, 1994] indicate that such layered rheological models may significantly underestimate the strength of the lower crust.

In this paper we first review experimentally determined flow laws for gabbroic rocks. We then discuss the application of these flow laws to the dynamics of the oceanic crust using "strength versus depth" diagrams. In this part of the paper we also address experimental constraints on the depth of the brittle-to-ductile and brittle-to-plastic transitions at mid-ocean ridges. Next we review geochemical and petrological constraints that indicate that much of the lower oceanic crust has a very low water content prior to hydrothermal alteration and describe geological and geophysical observations that support this conclusion. Finally we discuss the implications of the different crustal rheologies for the dynamics of lithospheric extension at mid-ocean ridges.

Faulting and Magmatism at Mid-Ocean Ridges
Geophysical Monograph 106
Copyright 1998 by the American Geophysical Union

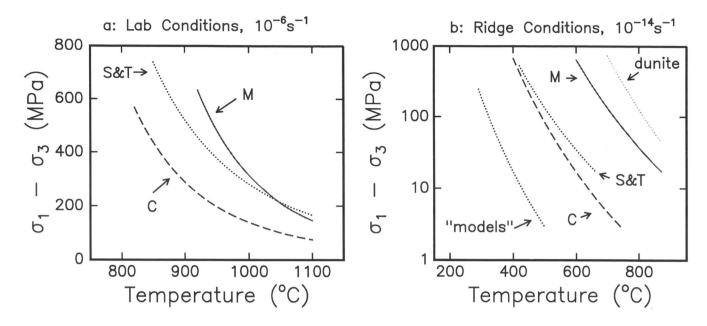

Figure 1. Plots of differential stress versus temperature. (a) Stresses calculated for a strain rate of 10^{-6} s^{-1} using a flow law for dry diabase [*Mackwell et al.*, 1994 (M)], and flow laws for untreated (i.e., "as-is") diabase [*Caristan*, 1982 (C); *Shelton and Tullis*, 1981 (S&T)]. (b) Stresses (note log scale) calculated for a strain rate of 10^{-14} s^{-1} using the same flow laws. For comparison, the stresses predicted for the crustal flow law used in numerous modeling studies ("models") [e.g., *Chen and Morgan*, 1990; *Neumann and Forsyth*, 1993; *Shaw and Lin*, 1996], and for dry dunite [*Chopra and Paterson*, 1984] are also shown.

EXPERIMENTAL CONSTRAINTS ON THE RHEOLOGY OF THE LOWER OCEANIC CRUST

Numerous studies have been conducted to determine the rheological properties of peridotite, due to its importance during mantle flow. In contrast, only a few studies have been conducted on the rheological properties of gabbroic rocks [*Shelton and Tullis*, 1981; *Kronenberg and Shelton*, 1980; *Caristan*, 1982; *Fredrich et al.*, 1990; *Mackwell et al.*, 1994; *Mackwell et al.*, 1998]. All of these studies were conducted using diabase, as opposed to coarse-grained gabbro, because diabase has a more optimal grain size/sample size ratio. The flow stress of diabase at laboratory conditions, as well as that predicted for geological strain rates in the mid-ocean ridge environment, are illustrated using plots of differential stress versus temperature in Figure 1. The flow stress was calculated using the flow laws of *Shelton et al.* [1981], *Caristan* [1982], and *Mackwell et al.* [1994] using the relationship:

$$\sigma = \left(\frac{\dot{\varepsilon}}{A \exp\left(-Q/RT\right)} \right)^{\frac{1}{n}}$$

where σ is differential stress, n is the power-law exponent, ε is strain rate, A is a material parameter, Q is the activation energy for creep, R is the gas constant, and T is absolute temperature. For comparison, the flow stress calculated using the "crustal" flow law used in the modeling studies of *Chen and Morgan* [1990], *Neumann and Forsyth* [1993] and *Shaw and Lin* [1996] and that calculated using the dry dunite flow law of *Chopra and Paterson* [1984] are also shown in Figure 1b.

A large difference in lithospheric strength is predicted by the different flow laws. For example at a temperature of 600°C, the "weak crust" flow law used in the modeling studies predicts a stress three orders of magnitude less than that of *Mackwell et al.*'s diabase flow law (Figure 1b). As illustrated in Figure 2, represented in terms of effective viscosity (i.e., $\eta_{eff} = \sigma/\varepsilon$), the flow laws show more than three orders of magnitude variation in viscosity when extrapolated to geologic strain rates. The fundamental difference in the experimental studies is that extensive effort was made by *Mackwell et al.* [1994] to analyze the rheology of dry diabase. In contrast, the experiments of *Caristan* [1982] and *Shelton and Tullis* [1981] were conducted on samples with small amounts of alteration phases that dehydrated at

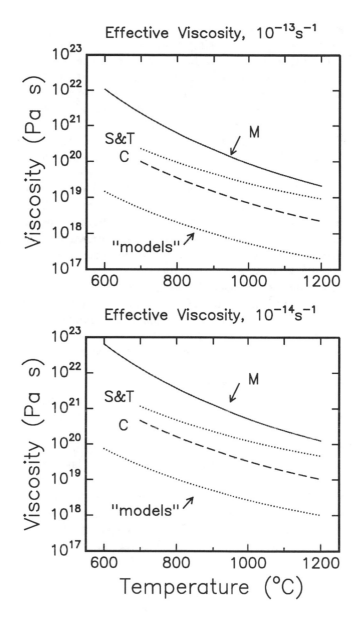

Figure 2. Plot of effective viscosity ($\eta_{eff} = \sigma/\dot\varepsilon$) versus temperature calculated using the same flow laws illustrated in Figure 1 and strain rates of 10^{-13} s^{-1} and 10^{-14} s^{-1}.

high temperature liberating water (for further discussion see *Mackwell et al.*[1994]). It is clear that the distinction between a "wet" and "dry" rheology is important for modeling the strength and dynamics of the oceanic lithosphere.

Before discussing the application of experimentally determined flow laws to mid-ocean ridges, several caveats must be considered. (1) The flow law of *Shelton and Tullis* [1981] was only published in an abstract. However, due to

its inclusion in several review papers, this flow law has been used extensively in modeling studies. The pre-exponential factor in the flow law has been misquoted (or changed) in several papers, which has lead to some confusion in the literature. In the last 15 years, many improvements have been made in the sample assemblies used to measure rheological properties in the pressure range of 1.0 to 2.0 GPa [*Borch and Green*, 1989; *Gleason and Tullis*, 1993]; comparison of rheological data collected using the improved assemblies with older data indicates significant errors may exist in the flow laws reported by *Shelton and Tullis* [1981]. (2) The flow laws of *Caristan* [1982] were determined under conditions at which samples deformed by a combination of brittle and plastic processes. The brittle processes were presumably enhanced by the existence of low effective confining pressures owing to the presence of melt [*Fredrich et al.*, 1990]. (3) Because diabase is polymineralic, its rheological properties may be controlled by different minerals at laboratory and natural conditions [*Kronenberg and Shelton*, 1980; *Tullis et al.*, 1991].

While the "dry" flow law for diabase has only recently been published, the importance of water on the rheology of silicates has been appreciated for many years. In fact, experimental studies have previously demonstrated a large effect of water on the strength of feldspathic rocks [*Tullis and Yund*, 1980]. While the resolution of the rheological data obtained during these experiments may not have been optimal, the influence of water was significantly greater than the errors in the estimation of differential stress. Such "water-weakening" effects are observed to be greater at higher pressures [e.g., *Tullis et al.*, 1979]. Subsequent studies on other rock types (quartzite, dunite) indicate that this "pressure effect" can be incorporated in the flow law with a pre-exponential water fugacity term [e.g., *Kohlstedt et al.*, 1995; *Post et al.*, 1996].

The rheological structure of the lithosphere is often modeled using strength versus depth profiles in which the strength of an upper "brittle" layer is limited by frictional resistance (i.e., Byerlee's law, *Byerlee* [1978]), and the strength of a lower "ductile" layer is limited by power law creep [e.g., *Goetze and Evans*, 1979; *Brace and Kohlstedt*, 1980]. While this two-mechanism model provides some insight into the rheological behavior of the lithosphere [e.g., *Sibson*, 1982; *Chen and Molnar*, 1983], its application to geodynamics is limited in some regards, including its simplistic "description" of the brittle-plastic transition (BPT).

Both field and experimental observations demonstrate that the BPT occurs over a wide range of conditions and involves several different deformation mechanisms [e.g., *Carter and Kirby*, 1978; *Scholz*, 1988; *Fredrich et al.*,

1989, *Tullis and Yund*, 1977, 1992; *Hirth and Tullis*, 1994]. These observations indicate that two-mechanism strength envelopes significantly overestimate the strength of rocks near the brittle-plastic transition [e.g., *Kirby*, 1980; *Carter and Tsenn*, 1987]. We emphasize that the strength versus depth curves provide only maximum stress estimates. For example, grain size reduction associated with the faulting process or possibly dynamic recrystallization, can result in significant weakening in fault zones due to an enhancement of diffusion creep [e.g., *Kirby*, 1985; *Rutter and Brodie*, 1988; *Jaroslow et al.*, 1996]. A combination of rheological and microstructural observations demonstrate that significant weakening of experimentally deformed feldspathic rocks can occur associated with grain-boundary migration recrystallization [*Tullis and Yund*, 1985] or a transition to diffusion creep [*Tullis and Yund*, 1991].

While a constitutive law for the brittle-plastic transition has not been formulated, experimental studies have lead to important advances in our understanding of the processes responsible for its occurrence. Two empirical observations appear to be relatively robust: (1) the transition from localized to distributed deformation (i.e. the transition from faulting to semi-brittle flow) occurs when the strength of the rock is lower than the frictional stress predicted by Byerlee's law, and (2) the transition to fully crystal plastic deformation (i.e. with no contribution from microcracking) occurs when the strength is lower than the confining pressure [see reviews by *Evans et al.*, 1990; *Kohlstedt et al.*, 1995].

We illustrate the possible importance of different crustal rheologies using a series of schematic strength versus depth profiles in Figure 3. The profiles were calculated using temperatures predicted for a long-segment of the northern Mid-Atlantic Ridge (model NMARL, *Shaw and Lin*, 1996) for a ~80 ky old lithosphere (i.e., 1 km off axis with a spreading rate of 12 mm/yr). Stresses were calculated using the thermal profiles calculated for the mid-point and 7 km along-axis from the mid-point of the segment (Figure 3a). The differential stress in the mantle lithosphere was calculated using a dislocation creep flow law for dry dunite [*Chopra and Paterson*, 1984]. The stress in the brittle deformation regimes was calculated using Byerlee's law with the assumption that σ_3 is the horizontal stress and $\sigma_1 = \Delta\rho gh$ with a hydrostatic pore-fluid pressure, where $\Delta\rho = \rho_{rock} - \rho_{water}$, and h is the depth below the seafloor.

The strength versus depth curves shown in Figure 3b illustrate the thick rheological "decoupling" zone in the lower crust that is predicted by the "weak" lower crustal flow laws used in the modeling studies [*Chen and Morgan*, 1990; *Neumann and Forsyth*, 1993; *Shaw and Lin*, 1996]. In contrast, there is almost no discontinuity in the strength

at the Moho (crust-mantle boundary) when differential stresses are calculated using *Mackwell et al.*'s [1994] dry diabase flow law (Figure 3c). However, notice that large changes in stress are predicted for the lower crust and upper mantle in Figure 3c for relatively small changes in the geotherm (Figure 3a). An intermediate scenario is illustrated in Figure 3d when differential stress is calculated using the "wet" diabase flow law of *Caristan* [1982].

As stated above, a major limitation of the two-mechanism strength versus depth models is in the description of the BPT. This limitation is potentially mitigated at mid-ocean ridges owing to discontinuities in the effective confining pressure with depth related to the transition from hydrostatic to lithostatic pore-fluid pressure. The fractured rocks at depths above the BPT are likely to be under hydrostatic pore-fluid pressure, $\sigma_{vertical} = \Delta\rho gh$, while uncracked rocks below the BPT are under lithostatic pressure. Furthermore, the pressure beneath the BPT also includes a component due to the weight of the water column between the sea-surface to the seafloor, $\sigma_{vertical} = \rho_{rock}gh + \rho_{water}gh_{wc}$, where h_{wc} is the depth of the water column. The influence of these constraints can be analyzed using the strength envelopes by calculating the depth at which Goetze's criterion is satisfied, that is where $(\sigma_1 - \sigma_3)_{plastic\ flow} = \sigma_3$. Because the effective pressures beneath the BPT are lithostatic and include a component due to the depth of the water column, the depth at which Goetze's criterion is satisfied corresponds closely to the depth of the crossover between Byerlee's law and the plastic flow law (Figure 4). This analysis suggests that in the mid-ocean ridge environment the simple "two-mechanism" strength profiles provide a good estimate for the depth of the BPT. The crustal rheologies illustrated in Figures 3b-d thus predict large differences in the depth and the temperature of the rocks at the BPT.

Owing to the pressure dependence of brittle deformation, the temperature at the BPT depends on the geotherm. Specifically, the BPT will occur at higher temperatures in regions with high geothermal gradients because the crossover in strength between the frictional strength and the plastic strength of the rocks occurs at lower stress. Such a distinction is important in the mid-ocean ridge environment because the geotherm along the ridge axis varies as a function of spreading rate as well as the thermal effects associated with fracture zones and focused magma emplacement [e.g., *Sleep*, 1975; *Phipps Morgan and Forsyth*, 1988].

The thermal effect on the depth of the BPT is illustrated for the different crustal rheologies in Figure 5. Based on our analysis of Goetze's criterion in the mid-ocean ridge environment, we assumed that the depth to the BPT is well constrained by the crossover between the frictional strength

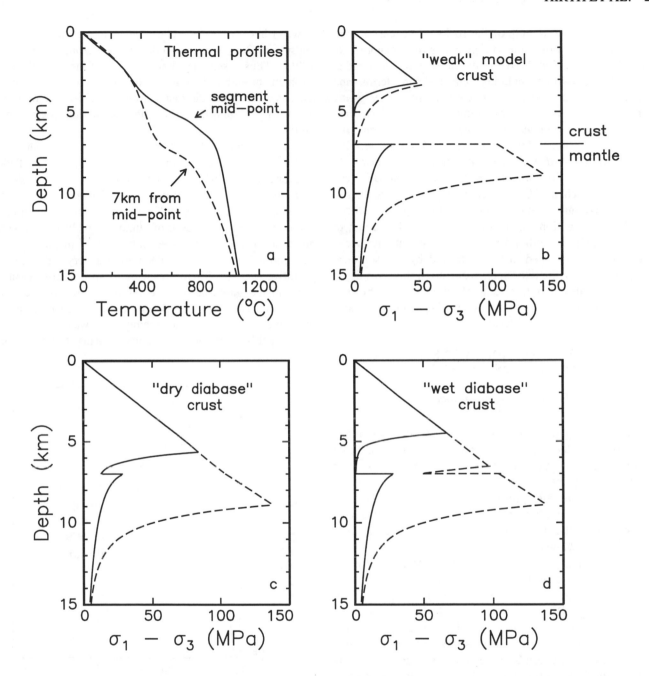

Figure 3. Strength versus depth profiles for the oceanic lithosphere near slow-spreading ridges. A strain rate of 10^{-14} s^{-1} was used in all calculations. (a) Thermal profiles used in calculations. The solid line represents the thermal profile for ~80 ky old lithosphere (i.e., 1 km off-axis with a spreading rate of 12 mm/yr) at the mid-point of a ridge segment (NMARL model of *Shaw and Lin*, 1996). The dashed line is a profile 7 km along-axis from the mid-point of the same segment. (b) Strength in plastic flow regime of the crust calculated using the flow law used in modeling studies [e.g., *Chen and Morgan*, 1990; *Neumann and Forsyth*, 1993; *Shaw and Lin*, 1996]. (c) Strength in plastic flow regime of the crust calculated using the flow law for dry diabase [*Mackwell et al.*, 1994]. (d) Strength in plastic flow regime of the crust calculated using the flow law for "wet" diabase [*Caristan,* 1982]. The dashed and solid curves in Figures 3b-3d are the stresses calculated for the dashed and solid line geotherms in Figure 3a. The strength in the lithospheric mantle is calculated using *Chopra and Paterson's* [1984] flow law for dry dunite.

and the plastic flow strength of the rocks. The curves were calculated by equating the differential stress required for plastic flow (equation 1) with the stress required for frictional sliding (which includes ρgh pressure terms and therefore depth), and solving for temperature. The "weak crust" flow law used in modeling studies [*Chen and Morgan*, 1990; *Neumann and Forsyth*, 1993; *Shaw and Lin*, 1996] predicts a ~450°C lower temperature at the BPT than *Mackwell et al.*'s [1994] dry diabase flow law. In addition, for each flow law, the temperature at the BPT can vary by as much as 300°C, depending on the geotherm. We emphasize that the curves illustrated in Figure 5 do not depend on assumptions about the geotherm. The depth to the BPT and the depth limit of seismicity at mid-ocean ridges are often assumed to be controlled by an isotherm. The application of the two-mechanism strength versus depth models illustrates that this generalization is probably not correct, especially at lower pressures (i.e., young oceanic crust or at fast-spreading ridges).

THE WATER CONTENT AND STRENGTH OF THE LOWER CRUST

Geochemical and Petrologic Constraints

A number of geochemical and petrological observations on mid-ocean ridge basalts (MORB) and gabbros indicate that the water content of the lower oceanic crust is low prior to hydrothermal alteration. The first constraint to consider is the amount of water present in primitive MORB (i.e., MORB that has not undergone fractional crystallization). Geochemical analyses indicate that primitive MORB contains 0.1-0.2 wt% water [e.g., *Michael*, 1988]. At a pressure of ~50 MPa, this amount of water is significantly below the solubility of water in basalt, indicating that MORB is unlikely to have lost water by degassing in the ridge environment [e.g., *Dixon et al.*, 1988]. In addition, 0.1-0.2 wt% is significantly less than the solubility of water in basalt at 100-200 MPa (i.e., the pressure range of the lower oceanic crust). The solubility of water in MORB in this pressure range is ~3.0-6.0 wt% [e.g., *Sisson and Grove*, 1993], indicating that at pressures appropriate for the lower oceanic crust, MORB is only ~5% saturated with respect to water.

The solubility of water in MORB is significantly greater than that in a gabbroic rock. Thus, during the initial stages of crystallization of gabbro, the water that is present will reside dominantly in the melt phase. For example, the solubility of water in diopside at 200 MPa is ~0.01 wt% [*Bai et al.*, 1994]. While we are not aware of data for the solubility

of water in Ca-rich plagioclase, the solubility of water in K-feldspar at 200 MPa is ~0.004 wt% [*Kronenberg et al.*, 1996]. Thus, assuming that the ratio of the solubilities of water in the gabbroic minerals to that in the melt can be used as a proxy for the distribution coefficient for water between MORB and the solid, the distribution coefficient is ~0.001.

A comparison of the composition of MORB to that of the plutonic lower crust from the Southwest Indian Ridge [*Meyer et al.*, 1989] and the Oman ophiolite [e.g., *Pallister and Knight*, 1981; *Browning*, 1982; *Kelemen et al.*, 1996; *Korenaga and Kelemen*, 1997] indicates that the lower crust is predominately composed of cumulate gabbro. The compositions of these cumulates indicate that the lower crust was formed by ~10 to 50% fractional crystallization of primitive MORB. Therefore, with a partition coefficient for water in the gabbroic rocks of ~0.001, most of the water present in primitive MORB will be extracted to form the lavas and dikes. A significant amount of water could reside in the gabbros if melt is trapped during crystallization.

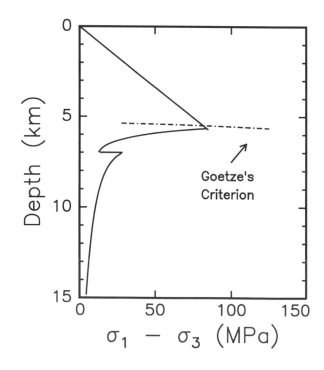

Figure 4. Strength versus depth plot illustrating that the crossover between Goetze's criterion (i.e., the brittle-plastic transition) and the dry diabase flow law corresponds closely to the crossover in strength between Byerlee's Law and the diabase flow law. The strength profile is the same as that illustrated by the solid line in Figure 3c.

However, geochemical analyses of cummulate gabbros indicate that very little melt is trapped during the crystallization process [e.g., *Natland et al.*, 1991]. In summary, while small amounts of water can result in significant weakening in rocks and minerals, most of the water present in primitive MORB will reside in melts extracted to form the lavas and dikes, and the majority of the lower crust is likely to be almost dry prior to hydrothermal alteration.

There are some important caveats to this point. (1) Highly fractionated basaltic melt will be rich in water, so even small amounts of trapped (and subsequently crystallized) melt can lead to the presence of water in more fractionated crustal rocks. (2) Chemical analyses of fluid inclusions in oceanic gabbros indicates the presence of magmatic volatiles [e.g., *Kelley*, 1997]. However, even where water is present, the water fugacity will be relatively low owing to the presence of other volatile phases (i.e., CO_2) and the relatively low pressure -compared to experimental conditions- of the lower oceanic crust. These observations suggest that even when water is present during deformation, the crust would be stronger than predicted by the flow laws for "wet" diabase determined in the laboratory.

Geological and Geophysical Constraints

Geologic observations suggest that the strength of the lower ocean crust approaches that of dry diabase in the mid-ocean ridge environment. For example, gabbros from Hess Deep (Ocean Drilling Program Site 894) show evidence for hydrothermal alteration at temperatures of 700-750°C [*Manning et al.*, 1996]. The lack of evidence for dislocation creep microstructures in these gabbros indicates that plastic deformation required a larger stress than the brittle deformation, and thus these alteration temperatures provide a minimum estimate for the temperature at the BPT. Analyses of oceanic gabbros from slow-spreading ridges demonstrate that semi-brittle deformation in the temperature range of ~500-720°C is associated with hydrothermal alteration [e.g., *Cannat et al.*, 1991; *Mevel and Cannat*, 1991]. The initiation of brittle deformation required for hydrothermal alteration at these conditions is inconsistent with the temperature at the BPT predicted by either the wet diabase flow laws or the "weak crust" flow laws used in modeling studies (Figure 5). In contrast, peak alteration temperatures of 720-750°C [*Vanko and Stakes*, 1991; *Manning et al.*, 1996] in oceanic gabbros approach the temperature predicted for the BPT (~700-800°C) using *Mackwell et al.*'s [1994] dry diabase flow law (Figure 5).

Plastic deformation of gabbroic rocks at granulite grade conditions is observed in samples from slow-spreading

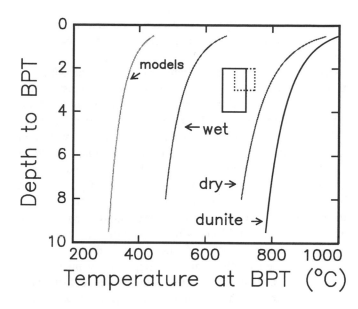

Figure 5. Depth to the brittle-plastic transition (BPT) versus temperature at the BPT calculated for the same flow laws used in Figure 1b. The curves were calculating by equating the differential stress required for plastic flow (equation 1) with the stress required for frictional sliding (which includes $\rho g h$ pressure terms and therefore depth), and solving for temperature. At shallow depths, the BPT occurs at higher temperatures (and therefore lower stresses) owing to the pressure dependence of brittle deformation. The effect is less pronounced at greater depths due to the exponential temperature dependence of plastic flow. Curves are illustrated for the "weak crust" flow law (models), *Caristan's* [1982] diabase flow law (wet), *Mackwell et al.'s* [1994] diabase flow law (dry) and *Chopra and Paterson's* [1984] dunite flow law (dunite). The peak hydrothermal alteration temperatures determined for oceanic gabbros from Hess Deep (dashed box) [*Manning et al.*, 1996] and the Southwest Indian Ridge (solid box) [*Vanko and Stakes*, 1991] are also shown.

ridges (e.g., Ocean Drilling Program Site 735b), but only in localized shear zones (e.g., *Cannat*, 1991; *Dick et al.*, 1991). The strongly deformed gabbros comprise only ~1-2% of the 735b drill core, while "undeformed" to "weakly foliated" rocks make up ~85% of the core [*Dick et al.*, 1991]. An analysis of recrystallized grain size in the granulite-grade (~850°C) gabbroic shear zones from site 735b suggests that plastic deformation occurred at stresses (~75 MPa) on the order of those predicted by Byerlee's law for conditions near the base of the oceanic crust [*Newman et al.*, 1994]; these rocks apparently preserve deformation conditions near the BPT. These observations are consistent with the strength versus depth curve illustrated in Figure 3c calculated for the "segment mid-point" geotherm of a slow-spreading ridge using *Mackwell et al.*'s dry diabase flow law.

Geophysical observations are also consistent with the stronger crustal rheology predicted for oceanic crust with a low water content. For example, at ~29°N along the Mid-Atlantic Ridge, microearthquakes are observed at "crustal" and "mantle" depths (i.e., above and below the Moho) in both the center and the end of the ridge segment [*Wolfe et al.*, 1995]; this particular segment exhibits a well developed "bull's-eye" gravity anomaly [*Lin et al.*, 1990]. Thus, at this location, no evidence was found for an aseismic (ductile) lower crust between a brittle mantle and a brittle upper crust. In addition, while it is not clear that these earthquakes are all "tectonic" (i.e, normal faults), they do not support the hypothesis that the depth of seismicity shallows towards the center of ridge segments due to thicker and weaker crust and/or higher temperatures [e.g., *Kong et al.*, 1992]. However, a strong temperature dependence for the depth to the BPT is indicated by comparing the depth of microseismicity at 29°N and 35°N on the Mid-Atlantic Ridge [*Barclay et al.*, 1995]. At 35°N the lower crust is apparently aseismic, however, the upper mantle in this location is aseismic as well.

Models for the viscosity of the lower crust of Iceland, constrained by GPS measurements, indicate a viscosity of ~3x10^{19} Pa s [*Pollitz and Sacks*, 1996]. This value is "bracketed" by the effective viscosities predicted by the diabase flow laws for a strain rate of 10^{-14}s^{-1} and a temperature between 1100°C and 1200°C (Figure 2). At a strain rate of 10^{-13}s^{-1}, the dry diabase flow law [*Mackwell et al.*, 1994] predicts a viscosity of ~10^{19} Pa s for this temperature range. In contrast, a strain rate of less than 10^{-15}s^{-1} is required for the "weak crust" model to be appropriate.

IMPLICATIONS

In order to illustrate further the effect of crustal rheology on the dynamics of the lithosphere at slow-spreading ridges, we show strength versus depth cross-sections predicted for three end-member rheological models (Figure 7). The thermal structure and along-axis crustal thickness variations in all three models, shown in Figure 6, were again constrained using the NMARL model of *Shaw and Lin* [1996]. Model 1 was calculated using the "weak-crust" flow law (Fig 7a), while in Model 2 the strength of the ductile crust was calculated using the flow law for dry diabase (Figure 7b). In both of these models the strength of the brittle regions of the lithosphere was calculated using Byerlee's law with a hydrostatic pore-fluid pressure, and the strength in

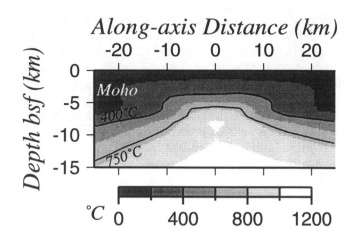

Figure 6. Temperature structure of slow-spreading oceanic lithosphere. The cross-section is parallel to the axis of a 50 km long ridge segment and located 1 km off-axis. The Moho (bold line) indicates crustal thickening from the end of segments towards the center. The temperature contours were taken from the NMARL model of *Shaw and Lin* [1996].

ductile regions of the mantle was determined using *Chopra and Paterson's* [1984] dry dunite flow law. For comparison, the influence of serpentinization on the strength of the lithosphere is illustrated in Model 3 (Figure 7c). Serpentinization is restricted to the vicinity of discontinuities and to temperatures <400 °C. Serpentinite is assumed to deform in the brittle regime with a strength limited by its own friction law; we have assumed that the serpentinite is dominantly lizardite with minor amounts of chrysotile, consistent with geologic observations [e.g., *Aumento and Loubat*, 1971; *O'Hanley*, 1996]. In this case, the coefficient of friction for the serpentinite is ~0.3 [*Reinen et al.*, 1994; *Moore et al.*, 1997; *Escartín et al.*, 1997a].

The fundamental differences in lithospheric structure and deformation mechanisms are emphasized in Figures 7d-7f by normalizing the strength predicted for each model in Figures 7a-7c by the strength predicted by Byerlee's friction law. The ductile lower crust that "decouples" the brittle crust from the mantle at the center of the segment in Model 1 (Figure 7a, 7d) is absent in Model 2 (Figures 7b,7e). Serpentinization restricted to the end of ridge segments may locally reduce the strength of the lithosphere by ~45% relative to that predicted by Byerlee's law (Figure 7f). When integrated over the total lithospheric thickness, serpentinization may reduce the overall lithospheric strength near ridge-axis offsets by ~30% [*Escartín et al.*, 1997b].

At the center of slow-spreading ridge segments faults have smaller throws and are more closely spaced than at the end

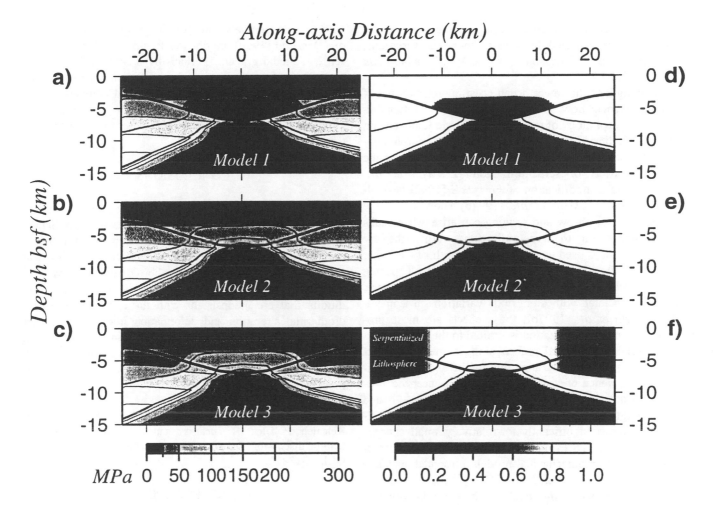

Figure 7. Maximum strength predicted for three different rheological models of the oceanic lithosphere. (a) The strength of the plastic mantle and crust are given by the flow laws for dry olivine and the "weak" crustal flow laws used in previous modeling studies, respectively. In this model a ductile lower crust decouples the brittle crust from the brittle mantle at the center of the segment. (b) In Model 2 the strength of the crust in the plastic flow regime is calculated using the flow law for dry diabase. (c) In Model 3 the strength of the lithosphere is decreased owing to the presence of serpentinized peridotite. Serpentinites are included at temperatures <400 °C, and restricted to the ends of the segment, tapering to unserpentinized lithosphere at 15 km from the discontinuity. The strength of serpentinized lithosphere is calculated using a friction coefficient of 0.3. (d-f) Cross-sections corresponding to Figures 7a-7c with strengths normalized by Byerlee's Law using densities of 2700 and 3300 kg/m^3 for the crust and mantle respectively, and assuming lithostatic loading with hydrostatic pore-fluid pressure. The normalized stress plots emphasize the differences in lithospheric strength predicted by the different crustal flow laws. The locations of the 400°C, and 750°C isotherms, corresponding to the thermal structure illustrated in Figure 6, are shown in each panel. In Figure 7f, brittle- unserpentinized lithosphere is illustrated by light gray (normalized strength = 1), while serpentinized lithosphere in shown by a darker gray.

of segments [e.g., *Shaw*, 1992]. This change in fault morphology has been interpreted to be a consequence of systematic variations in lithospheric thickness associated with along-axis variations in crustal thickness and temperature [e.g., *Shaw and Lin*, 1993; *Shaw and Lin*, 1996]. Similarly, *Chen and Morgan* [1990] use a weak lower crustal

layer to explain the absence of axial grabens at fast-spreading ridges. Such thermomechanical models were subsequently modified to explain the absence of axial grabens along some portions of the Mid-Atlantic Ridge [*Shaw and Lin*, 1996]. Dynamic models including the effects of a weak lower crust have also been used to explain the "axial-valley paradox" (i.e., along-axis bathymetric variations are apparently isostatically compensated by crustal-thickness variations, while across-axis topography is not) [*Neumann and Forsyth*, 1993].

If the strength of the lower ocean crust approaches that of dry gabbroic rock, how do we explain the observations related to the morphology of the axial valley and fault spacing at slow spreading ridges? Changes in fault morphology along ridge segments may occur as a manifestation of the rheological effects of serpentinization [*Escartin et al.*, 1997b]. Using a modification of *Forsyth's* [1992] flexural-faulting model, *Escartin et al.* [1997b] show that large variations in fault throw and spacing can arise when faults are weakened owing to serpentinization. When the dry diabase flow law is used to model the strength of the lower oceanic crust, the effects of serpentinization on fault spacing are shown to be more important that changes in lithospheric thickness associated with along axis variations in temperature [*Escartin et al.*, 1997b]; these effects are most pronounced when serpentinization is limited to the fault zones.

The absence of an axial graben along some portions of slow-spreading ridges may arise due to the significant temperature dependence of strength at conditions near the brittle-plastic transition. As illustrated in Figures 3c, 7b and 7e, the strength of both the lower crust and mantle are predicted to decrease significantly for relatively small changes in the geotherm. To evaluate this hypothesis we reassess the results of thermomechanical models presented by *Shaw and Lin* [1996]. *Shaw and Lin* [1996] assumed that the lithosphere exhibits a "pita-pocket" rheology in which a weak crustal layer is present at the crust-mantle boundary, but only in the center of the ridge segments. They used such a rheological structure to analyze the processes responsible for the transition in fault spacing and axial valley morphology along slow-spreading ridge segments. However, it is important to emphasize that most of their model predictions depend on the presence of a weak lower crust *and* upper mantle, not the presence of a decoupling zone between the crust and mantle. Thus, in regions where the mantle *and* crust beneath a spreading ridge are hot (i.e., fast spreading ridges or during active melt migration at slow spreading ridges), the axial graben may not form.

While *Chen and Morgan* [1990] show that the spreading rate dependence of axial valley morphology is well predicted by including a weak lower crust, their "olivine-only" models may actually be more applicable to the mid-ocean ridge environment because the rheological properties of dry gabbroic rocks are similar to peridotite (e.g., Figures 1 and 3c). In this case, the axial high observed at faster spreading rates may arise owing to changes in the center of moment during extension of the lithosphere [e.g., *Eberle and Forsyth*,

1996]. Alternatively, the "olivine-only" models of *Chen and Morgan* [1990] are not too far from predicting the absence of a axial graben. Perhaps by including the temperature effects of the latent heat of crystallization, or by analyzing the "volume" of the axial graben and comparing it to the volume of volcanic constructions [e.g., *Shaw and Lin*, 1996] such models would provide a better fit to the observations.

CONCLUSIONS

A comparison of constraints provided by geological, geochemical, and geophysical observations with predictions imposed by experimentally determined flow laws for diabase indicates that the rheology of the lower crust at mid-ocean ridges approaches that of dry gabbroic rock. Flow laws used in previous modeling studies predict deformation conditions that are inconsistent with the temperatures of hydrothermal alteration and deformation microstructures observed for oceanic gabbros recovered from slow-spreading ridges. These observations suggests that the lower oceanic crust is considerably stronger than commonly assumed in most modeling studies.

Acknowledgments. We would like to thank B. Evans, P. Kelemen, H. Dick, B. Tucholke and P. Meyer for numerous helpful discussions. This work was supported by NSF grant OCE-9313812.

REFERENCES

Aumento, F. and H. Loubat, The mid-Atlantic ridge near 45°N. Serpentinized ultramafic intrusions, *Can. J. Earth Sci., 8*, 631-663, 1971.

Bai, Q., Z.-C. Wang, G. Dresen, S. Mei and D.L. Kohlstedt, Solubility and stability of hydrogen in diopside single crystals, *EOS, Trans. AGU, 75*, 652, 1994.

Barclay, A.H., D.R. Toomey, G.M. Purdy, and S.C. Solomon, Crustal anisotropy and tomographic imaging of Vp/Vs structure beneath the axial high of a slow-spreading ridge segment, *EOS, Trans. AGU, 76*, F608, 1995.

Borch, R.S., and J.W. Green II, Deformation of peridotite at high pressure in a new molten salt cell: Comparison of traditional and homologous temperature treatments, *Phys. Earth Planet. Inter., 55*, 269-276, 1989.

Brace, W. F., and D. L. Kohlstedt, Limits on lithospheric stress imposed by laboratory experiments, *J. Geophys. Res., 85*, 6248-6252, 1980.

Browning, P., *The petrology, geochemistry, and structure of the plutonic rocks of the Oman ophiolite*, Ph.D. thesis, The Open University, Milton Keynes, England, 1982.

Byerlee, J. D., Friction of rocks, *Pure Appl. Geophys., 116*, 615-626, 1978.

Cannat, M., Plastic deformation at an oceanic spreading ridge: a microstructural study of site 735 gabbros (Southwest Indian Ocean), in *Proceeding of the Ocean Drilling Program, Scientific Results*, edited by R.P. Von Herzen, P.T. Robinson, et al., *Vol. 118*, 399-408, 1991.

Cannat, M., C. Mevel, and D. Stakes, Normal ductile shear zones at an oceanic spreading ridge: Tectonic evolution of site 735 gabbros (Southwest Indian Ocean), in *Proceeding of the Ocean Drilling Program, Scientific Results*, edited by R.P. Von Herzen, P.T. Robinson, et al., *Vol. 118*, 415-429, 1991.

Caristan, Y., The transition from high temperature creep to fracture in Maryland diabase, *J. Geophys. Res.*, *87*, 6781-6790, 1982.

Carter, N. L., and S. H. Kirby, Transient creep and semi-brittle behavior of crystalline rocks, *Pure Appl. Geophys.*, *116*, 807-839, 1978.

Carter, N. L., and M. C. Tsenn, Flow properties of continental lithosphere, *Tectonophys.*, *136*, 27-63, 1987.

Chen, Y. and W. J. Morgan, A nonlinear rheology for mid-ocean ridge axis topography, *J. Geophys. Res.*, *95*, 17583-17604, 1990.

Chen, W. P., and P. Molnar, Focal depths of intracontinental and intraplate earthquakes and their implications for the thermal structure and mechanical properties of the lithosphere, *J. Geophys. Res.*, *88*, 4183-4214, 1983.

Chopra, P.N., and M.S. Paterson, The role of water in the deformation of dunite, *J. Geophys. Res.*, *89*, 7861-7876, 1984.

Dick, H.J.B., P.S. Meyer, S. Bloomer, S. Kirby, D. Stakes, and C. Mawer, Lithostratigraphic evolution of an in-situ section of oceanic layer 3, in *Proceeding of the Ocean Drilling Program, Scientific Results*, edited by R.P. Von Herzen, P.T. Robinson, et al., *Vol. 118*, 439-538, 1991.

Dixon, J.E., E. Stolper and J. R. Delaney, Infrared spectroscopic measurements of CO_2 and H_2O in Juan de Fuca Ridge basaltic glasses, *Earth Planet. Sci. Lett.*, *90*, 87-104, 1988.

Eberle, M.A., and D.W. Forsyth, A dynamic mechanism that can explain the origin of the axial high on fast spreading ridges, *EOS, Trans. AGU,*, 77, F664, 1996.

Escartín, J., G. Hirth and B. Evans, Non-dilatant brittle deformation of serpentinites: Implications for Mohr-Coulomb theory and the strength of faults, *J. Geophys. Res.*, *102*, 2897-2913, 1997a.

Escartín, J., G. Hirth and B. Evans, Effects of serpentinization on the lithospheric strength and the style of normal faulting at slow-spreading ridges, *Earth Planet. Sci. Lett.*, *151*, 181-189, 1997b.

Evans, B., Fredrich, J.T., and Wong, T.-f., The brittle to ductile transition in rocks: Recent experimental and theoretical progress, in *The Brittle-Ductile Transition - The Heard Volume, Geophys. Monogr. Ser.* vol 56, edited by A.G. Duba, W.B. Durham, J.W. Handin, H.F. Wang, pp. 1-20, AGU, Washington, D.C., 1990.

Forsyth, D. W., Finite extension and low-angle normal faulting, *Geology*, *20*, 27-30, 1992.

Fredrich, J.T. and B. Evans, High temperature fracture and flow of Maryland Diabase, *EOS, Trans. AGU, 71*, 1750, 1990.

Fredrich, J. T., B. Evans, and T.-F. Wong, Micromechanics of the brittle to plastic transition in Carrara Marble, *J. Geophys. Res.*, *94*, 4129-4145, 1989.

Früh-Green, G. L., A. Plas and C. Lécuyer, 1996, Petrologic and stable isotope constraints on hydrothermal alteration and serpentinization of the EPR shallow mantle at Hess Deep (Site 895), *in Proc. Ocean Drilling Program Sci. Res*, C. Mével, K. M. Gillis, J. F. Allan and P. S. Meyer eds., *147*, 255-291, College Station, TX, 1996.

Gleason, G.C., and J. Tullis, Improving flow laws and piezometers for quartz and feldspar aggregates, *Geophys. Res. Lett.*, *20*, 2111-2114, 1993.

Goetze, C., and B. Evans, Stress and temperature in the bending lithosphere as constrained by experimental rock mechanics, *Geophys. J. R. ast. Soc.*, *59*, 463-478, 1979.

Hirth, G., and J. Tullis, The brittle-plastic transition in experimentally deformed quartz aggregates, *J. Geophys. Res.*, *99*, 11,731-11,747, 1994.

Jaroslow, G. E., H. J. B. Dick, and G. Hirth, Abyssal peridotite mylonites: implications for grain-size sensitive flow and strain partitioning in the oceanic lithosphere, *Tectonophys.*, *256*, 17-37, 1995.

Kelemen, P.B., K. Koga, and N. Shimizu, Geochemistry of gabbro sills in the crust-mantle transition zone of the Oman ophiolite: implications for the origin of the oceanic lower crust, *Earth Planet Sci. Lett.*, *146*, 475-488, 1997.

Kelley, D.S., Fluid evolution in slow-spreading environments, *Proc. Ocean Drill. Program Sci. Res.*, 153, 399-418, 1997.

Kirby, S. H., Tectonic stresses in the lithosphere: constraints provided by experimental deformation of rocks, *J. Geophys. Res.*, *89*, 6353-6363, 1980.

Kirby, S. H., Rock mechanics observations pertinent to the rheology of the continental lithosphere and the localization of strain along shear zones, *Tectonophys.*, *119*, 1-27, 1985.

Kohlstedt, D.L., B. Evans and S.J. Mackwell, Strength of the lithosphere: Constraints imposed by laboratory experiments, *J. Geophys. Res.*, *100*, 17,587-17,602, 1995.

Kong, L. S., S. C. Solomon and G. M. Purdy, Microearthquake Characteristics of a mid-ocean ridge along-axis high, *J. Geophys. Res.*, *97*, 1659-1685, 1992.

Korenaga, J. and P.B. Kelemen, The origin of gabbro sills in the Moho transition zone of the Oman ophiolite: Implications for magma transport in the oceanic lower crust, *J. Geophys. Res.*, *102*, 27729-27749, 1997.

Kronenberg, A.K., and G.L. Shelton, Deformation microstructures in experimentally deformed Maryland diabase, *J. Struct. Geol.*, *2*, 341-353, 1980.

Kronenberg, A.K., Stationary and mobile hydrogen defects in potassium feldspar, *Geochim. Cosmochim. Acta.*, *60*, 4075-4094, 1996.

Lin, J., and E.M. Parmentier, Mechanisms of lithospheric extension at mid-ocean ridges, *Geophys. J.*, *96*, 1-22, 1989.

Lin, J., G. M. Purdy, H. Schouten, J.-C. Sempéré and C Zervas, Evidence from gravity data for focused magmatic accretion along the Mid-Atlantic Ridge, *Nature, 344,* 627-632, 1990.

Mackwell, S. J., M. E. Zimmerman, D. L. Kohlstedt, and D. S. Scherber, Dry deformation of diabase: Implications for tectonics of Venus, *Lunar Planet. Sci.,* XXV, 817-818, 1994.

Mackwell, S.J., M.E. Zimmerman, and D.L. Kohlstedt, High-temperature deformation of dry diabase with application to tectonics on Venus, *J. Geophys. Res., 103,* 975-984, 1998.

Manning, C. E., P. Weston, and K. I. Mahon, Rapid high-temperature metamorphism of East Pacific Rise gabbros from Hess Deep, *Earth Planet. Sci. Lett., 144,* 123-132, 1996.

Mevel, C., and M. Cannat, Lithospheric stretching and hydrothermal processes in oceanic gabbros from slow-spreading ridges, in *Ophiolite Genesis and Evolution of the Oceanic Lithosphere,* edited by T.J. Peters, A. Nicolas and R.G. Coleman, pp. 295-314, Kluwer Academic, Netherlands, 1991.

Meyer, P. S., H. J. B. Dick, and G. Thompson, Cumulate gabbros from the Southwest Indian Ridge, 54°S-7°16'E: Implications for magmatic processes at a slow spreading ridge, *Contrib. Miner. Petrol., 103,* 44-63, 1989.

Michael, P., The concentration, behavior and storage of water in the suboceanic upper mantle: Implications for mantle metasomatism, *Geochim. Cosmochim. Acta, 52,* 555-566, 1988.

Moore, D., D. A. Lockner, M. Shengli, R. Summers and J. D. Byerlee, Strength of serpentinite gouges at elevated temperatures, *J. Geophys. Res, 102,* 14,787-14,802, 1997.

Natland, J.H., P.S. Meyer, H.J.B. Dick, and S.H. Bloomer, Magmatic oxides and sulfides in gabbroic rocks from Hole 735B and the later development of the liquid line of descent, in *Proceeding of the Ocean Drilling Program, Scientific Results,* edited by R.P. Von Herzen, P.T. Robinson, et al., *Vol. 118,* 75-112, 1991.

Neumann, G. A. and D. W. Forsyth, The paradox of the axial profile: Isostatic compensation along the axis of the Mid-Atlantic Ridge?, *J. Geophys. Res., 98,* 17891-17910, 1993.

Newman, J., A. K. Kronenberg, and W. M. Lamb, Strength of the young oceanic crust, *EOS, Trans. AGU, 75,* 329, 1994.

O'Hanley, D. S., *Serpentinites. Records of tectonic and petrological history,* Oxford Monographs on Geology and Geophysics No. 34, 277 p., Oxford University Press, New York, 1996.

Pallister, J. S., and R. J. Knight, Rare-Earth element geochemistry of the Semail Ophiolite near Ibra, Oman, *J. Geophys. Res., 86,* 2673-2697, 1981.

Phipps Morgan, J., E. Parmentier, and J. Lin, Mechanisms for the origin of mid-ocean ridge axial topography: implications for the thermal and mechanical structure of accreting plate boundaries, *J. Geophys. Res., 92,* 12823-12836, 1987.

Phipps Morgan, J. and D.W. Forsyth, 3-D Flow and temperature perturbations due to a transform offset: Effects on oceanic crustal and upper mantle structure, *J. Geophys. Res., 93,* 2955-2966, 1988.

Phipps Morgan, J., and Y.J. Chen, Dependence of ridge-axis morphology on magma supply and spreading rate, *Nature, 364,* 706-708, 1993

Pollitz, F.F., and I.S. Sacks, Viscosity structure beneath northeast Iceland, *J. Geophys. Res., 101,* 17,771-17,793, 1996.

Post, A.D., J. Tullis, and R.A. Yund, Effects of chemical environment on dislocation creep of quartzite, *J. Geophys. Res., 101,* 22,143-22,155, 1996.

Reinen, L. A., J. D. Weeks and T. E. Tullis, The frictional behavior of lizardite and antigorite serpentinites: experiments, constitutive models, and implications for natural faults, *Pure Appl. Geophys., 143,* 318-358, 1994.

Rutter, E. H., and K. H. Brodie, The role of tectonic grain size reduction in the rheological stratification of the lithosphere, *Geol. Rundschau, 77,* 295-308, 1988.

Scholz, C. H., The brittle-plastic transition and the depth of seismic faulting, *Geol. Rundschau, 77,* 319-328, 1988.

Shaw, P. R., Ridge segmentation, faulting and crustal thickness in the Atlantic Ocean, *Nature, 358,* 490-493, 1992.

Shaw, P. R. and J. Lin, Causes and consequences of variations in faulting style at the Mid-Atlantic Ridge, *J. Geophys. Res., 98,* 21839-21851, 1993.

Shaw, W. J. and J. Lin, Models of ocean ridge lithospheric deformation: Dependence on crustal thickness, spreading rate, and segmentation, *J. Geophys. Res., 101,* 17977-17993, 1996.

Shelton, G. and J. Tullis, Experimental flow laws for crustal rocks, *EOS, Trans. AGU, 62,* 396, 1981.

Sibson, R. H., Fault zone models, heat flow, and the depth distribution of earthquakes in the continental crust of the United States, *Bull. Seismol. Soc. Am., 72,* 151-163, 1982.

Sisson, T.W., and T.L. Grove, Experimental investigations of the role of H_2O in calc-alkaline differentiation and subduction zone magmatism, *Contrib. Mineral. Petrol., 113,* 143-166, 1993.

Sleep, N.H., Formation of the oceanic crust: Some thermal constraints, *J. Geophys. Res., 80,* 4037-4042, 1975.

Sleep, N.H., and B.R. Rosendahl, Topography and tectonics of mid-ocean ridge axes, *J. Geophys. Res., 84,* 6831-6839, 1979.

Tullis, J. and R.A. Yund, Experimental deformation of dry Westerly granite, *J. Geophys. Res., 82,* 5705-5718, 1977.

Tullis, J., and R.A. Yund, Hydrolytic weakening of experimentally deformed Westerly granite and Hale albite rock, *J. Struct. Geology, 2,* 439-451, 1980.

Tullis, J., and R.A. Yund, Dynamic recrystallization of feldspar: A mechanism for ductile shear zone formation, *Geology, 13,* 238-241, 1985.

Tullis, J., and R.A. Yund, Diffusion creep in feldspar aggregates: experimental evidence, *J. Struct. Geol., 13,* 987-1000, 1991.

Tullis, J., and R.A. Yund, The brittle-ductile transition in feldspar aggregates: An experimental study, in *Fault Mechanics and Transport Properties of Rocks: A Festchrift for W. F. Brace,* edited by B. Evans and T.-f. Wong, 89-119, Academic Press, London, U.K., 1992.

Tullis, J., G.L. Shelton, and R.A. Yund, Pressure dependence of rock strength: implications for hydrolytic weakening, *Bull. Mineral.*, *102*, 110-114, 1979.

Tullis, T.E., F. Horowitz, and J. Tullis, Flow laws of polyphase aggregates from end member flow laws, *J. Geophys. Res., 96*, 8081-8096, 1991.

Vanko, D.A. and D.S. Stakes, Fluids in oceanic layer 3: evidence from veined rocks, Hole 735B, Southwest Indian Ridge, *Proc. Ocean Drill. Program Sci. Results*, *118*, 181-215, 1991.

Wolfe, C., G. M. Purdy, D. R. Toomey and S. C. Solomon, Microearthquake characteristics and crustal velocity structure at 29°N of the Mid-Atlantic Ridge: The architecture of a slow-spreading segment, *J. Geophys. Res.*, *100*, 24449-24472, 1995.

J. Escartín, Department of Geology and Geophysics, University of Edinburgh, Edinburgh EH9 3JW, Scotland

G. Hirth and J. Lin, Department of Geology and Geophysics, Woods Hole Oceanographic Institution, Woods Hole, MA 02543, U.S.A.

Mechanics of Stretching Elastic-Plastic-Viscous Layers: Applications to Slow-Spreading Mid-Ocean Ridges

Alexei N. B. Poliakov

Laboratoire de Géophysique et Tectonique, Université de Montpellier II, Montpellier, France

W. Roger Buck

Lamont-Doherty Earth Observatory of Columbia University, Palisades, NY

The pattern of faulting and axial valley topographic relief are among the best observed features of mid-ocean ridges, but little work has been done showing how faulting and topographic relief might be linked. We approach this problem with numerical simulations of plastic shear zones (analogs of faults) and of the generation of topography within a stretching elastic-plastic-viscous layer. The layer is defined by contours of constant viscosity that deepen with distance from a center of symmetry. The plastic yield strength is described by a depth-dependent friction-controlled yield stress and a cohesion. An "axial valley" flanked by higher areas forms after finite extension of the layer. The maximum axial topographic relief scales with the near axis thickness of the extending part of brittle layer. The wavelength of the axial valley "shoulders" is longer than that of previous studies which did not include lithospheric elasticity. If cohesion is reduced with strain, we see finite offset on discrete normal faults. We find that the location of faults changes with time, and that resulting topographic relief resembles observed abyssal hills both visually and statistically. The relief of the hills scales with the strain-dependent cohesion loss. In a second set of calculations, we include a crustal layer that accretes by dike intrusion on top of a stretching mantle layer. The presence of a magma-filled dike is assumed to allow shear displacement at negligible levels of shear stress. We have not been able to model unlimited stretching as we did with the calculations with no crustal accretion. Our preliminary results suggest that magmatism is likely to reduce, but may not totally suppress, the offset of normal faults at a ridge. The results of both sets of calculations suggest that temporal variability in the pattern of faulting and irregularity in abyssal hill topography may be an intrinsic effect of stretching a lithospheric layer and may not require time variations in the magma supply as suggested in some recent models.

INTRODUCTION

Axial valleys were one of the first features recognized at mid-ocean ridges but the mechanics of their formation has

Faulting and Magmatism at Mid-Ocean Ridges
Geophysical Monograph 106
Copyright 1998 by the American Geophysical Union

been disputed since their discovery. These topographic lows, typically 1 to 2 kilometers deep and 20 to 30 kilometers wide [*Macdonald*, 1982], are seen at ridges where the plate-separation rate is less than 8 cm/yr [*Small*, 1994; *Malinverno*, 1993] (Figure 1). Gravity data shows that axial relief is not related to local isostatic support of thinner crust under the valleys. Rather, the mass deficit in the valley is regionally compensated by the flanking topographic highs [*Cochran*, 1979]. Slip on normal faults is thought to produce the topographic relief of axial valleys. However there is considerable controversy about how faults form.

Many questions about faulting and axial relief are unresolved. Among the most disputed is the link between faulting and magmatism at ridges. Is faulting suppressed completely during periods of magmatic activity at a ridge? According to many workers, the answer to this question is yes [e.g. *Karson et al.*, 1987; *Shaw and Lin*, 1993; *Thatcher and Hill*, 1995; *Freed et al.*, 1995], but in some preliminary calculations we consider the possibility that faulting at a ridge occurs mainly during periods of magmatism.

Another key question is what controls the spacing and offset of faults at ridges? There is some indication that there are two distinct types of normal faults at slow-spreading ridges. In most areas, normal faults are small, with about 100 m of offset, and they form and slip within a few kilometers of the neovolcanic zone at a ridge [*Macdonald and Lyendyke*, 1977; *Laughton and Searle*, 1979]. Along a few parts of ridges where the crust may be anomalously thin, faults with offsets of many kilometers tectonically expose gabbroic lower crust and the peridotitic mantle [*Karson et al.*, 1987; *Kong et al.*, 1988; *Cann et.al.*, 1997]. Many models of faulting at ridges are based on the assumption that the lithospheric thickness controls the fault spacing, with thicker lithosphere producing more widely spaced faults [e.g. *Shaw and Lin*, 1996]. It is not yet clear if such models can explain the formation of typical faults so close to the center of accretion, or explain the occurrence of two modes of faulting.

One of the most studied questions about ridge tectonics concerns the effect of axial strength structure on topography. Most models for axial topographic relief indicate that great strength, resulting from low temperatures at depth, are needed to support the topography of an axial valley. This is consistent with the observation that the depth and magnitude range of earthquake activity on slow spreading ridges is greater than for fast spreading ridges which lack axial valleys [*Toomey et al.*, 1988; *Huang and Solomon*, 1988; *Solomon et al.*, 1988]. However, the relative importance of crustal strength and mantle strength is not clear. For example, the models of *Parmentier* [1987], *Phipps-Morgan et al.* [1987], *Lin and Parmentier* [1989], and *Phipps-Morgan and Chen* [1993], predict that the total strength of the crust and mantle lithosphere control the axial relief. *Chen and Morgan* [1990a,b] infer that axial valleys can form only when the mantle is strong and the lower crust is strong enough to transmit the stresses related to mantle stretching to the upper crust.

We investigate some of these questions using a combination of numerical methods and mechanical analysis. Our numerical approach allows us to do several things that were not possible in previous studies. We can consistently treat elastic, plastic and viscous deformation of a model lithospheric structure undergoing stretching. Zones of localized plastic shearing, analogous to faults, develop spontaneously wherever a yield criteria is met. The surface is displaced and the effect of topographic stresses are included in the model.

Before describing the details of our model we review previous models for axial valleys and abyssal hills. In the main section of this paper we consider the case of steady-state stretching of a single lithospheric layer. This might be thought of as representing the end-member case of stretching a mantle layer with no crust accreting on top. This also allows us to understand some of the physics of stretching and faulting without the added complexity of crustal accretion. Next, we consider a few cases of the effect of crustal accretion on faulting and axial relief. Finally, we discuss possible implications as well as limitations of these models.

PREVIOUS MODELS FOR AXIAL VALLEYS

Axial valleys are generally explained as a result of plate separation, but there has been much discussion about how to link observed topographic and fault patterns to the material properties and geometry of the deforming ridge. Models have emphasized either viscous, plastic, or elastic properties of the ridge. There is no consensus about how or whether the crust deforms in a different way than the underlying mantle.

The first widely discussed dynamic mechanism for axial valley formation focused on the viscous behavior of the deforming region at a ridge [*Sleep*, 1969; *Lauchenbruch*, 1973]. Asthenosphere is assumed to upwell at the site of plate separation. Viscous shear traction on the surrounding lithosphere lifts the plate edge and decreases the level of viscous rise under the axial valley. This is sometimes termed the "viscous head-loss' model since the pressure head is reduced by viscous flow stresses. This simple model gives quantitative predictions for the relation of axial relief

Southeast Indian Ridge 49.9S,115E   Mid-Atlantic Ridge 22.6N,44.9W

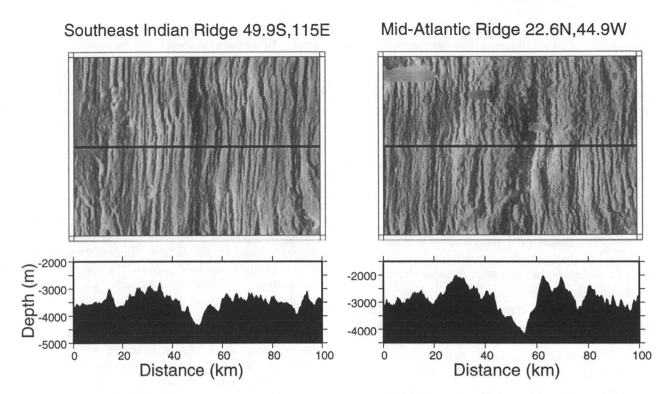

Figure 1. Ridge parallel abyssal hill topography. These shaded relief bathymetric maps show segments of the intermediate-spreading-rate Southeast Indian Ridge [data from *Cochran et al.,*1997] and the slow-spreading-rate Mid-Atlantic Ridge [data from *Gente et al.,* 1995]. The maps are oblique Mercator projections, centered on the locations given at the top of the map and rotated so that the axial valley and abyssal hill relief trend in the local north-south direction. Solid lines show the location of profiles shown below. The abyssal hills show along-strike continuity for tens of kilometers (i.e. greater than the lithospheric thickness) suggesting that two-dimensional models may be relevant to the processes forming these features.

to the effective viscosity and geometry of the deforming region at a ridge.

The viscous head-loss model has been criticized on two grounds. First, if the behavior of a ridge is viscous then axial relief should relax when it stops moving. Such relaxation at abandoned ridges is not seen [*Freed et al.,* 1995]. Second, if the viscosity of the asthenosphere is as small as many argue, then the model may require an unrealistic geometry [*Phipps-Morgan et al.,* 1987]. For the head-loss model to give the required amplitude of axial relief with a low viscosity asthenosphere requires a deep and narrow conduit. In this view, the conduit is a hot region surrounded by cold, strong mantle lithosphere. Even thermal models in which the advection is forced to mimic flow in a conduit result in a fairly smooth gradation in temperature and lithospheric thickness under a ridge [e.g. *Phipps-Morgan et al.,* 1987]. For spreading-center-models where the temperature, rheology and flow are linked, such as *Chen and Morgan* [1990b], a conduit-like flow structure does not develop.

By definition, the lithosphere is stronger than the asthenosphere, but it must also deform at a ridge axis. It is plausible that the deformation of this strong material controls the formation of axial valley relief. This is the basis of the lithospheric necking model first applied to continental rifts by *Artemyev and Artyuskov* [1971] and related to the relief at ridges by *Tapponier and Francheteau* [1978]. Necking refers to extensional deformation concentrated at the thinnest part of the lithosphere. *Tapponier and Francheteau* [1978] suggested that necking would pull down the surface of the axial lithosphere, and regional compensation would raise flanking highs. Several numerical and analytic studies of lithospheric necking have confirmed this suggestion, and have found that axial relief should scale with lithospheric thickness [*Zuber and Parmentier,* 1986; *Phipps-Morgan et al.,* 1987; *Lin and Parmentier,* 1990]. These models assumed that a region at the ridge deforms by slip on closely spaced faults which can be approximated as plastic flow. These studies are important because they showed that reasonable plastic (faulting) stresses and lithospheric geometries could result in a deep axial valley. However, they did not consistently show which regions of a ridge would in fact yield and slip on faults. Also, no distinction is made in these models between crust and mantle, either in terms of rheology or in the way crust is accreted.

The scaling of the amplitude of topographic relief with layer geometry and strength in these simple stretching models was explained in terms of a "moment-balance" argument [*Parmentier*, 1987; *Phipps-Morgan et al.* 1987]. According to this view, the stresses at the ridge axis are just at failure, or yield, while a short distance from the axis the stresses are less than the yield stress and are distributed fairly uniformly through the thickness of the lithosphere. The integral of the horizontal stress with depth equals the tectonic force, and is the same at all distances from the axis. Since the stresses are distributed over a greater depth off axis than on axis, the center of moment of the applied stresses is deeper off axis. This difference in the first moment of the stress distribution has to be balanced by stresses associated with topographic relief. However, as we show below, the assumed stress distribution is not simple, and so this argument may not yield quantitative estimates of axial relief.

Several studies aimed at continental rifting have looked at the link between faulting and topography. *Vening-Meinez* [1950] suggested that normal faults might form the boundaries of a graben in areas of extension. The regional isostatic response to the normal fault offset should produce both an axial valley and flanking rift shoulders. One of the first models that explicitly linked the thermal structure of stretched lithosphere with its rheology was done by *Braun and Beaumont* [1987]. These workers treated the lithosphere as an elastic-plastic-viscous material, and produced rift valleys with moderate relief. However, their grid was not fine enough for localized faults to develop. To explain how rifting leads to topographic relief, *Weissel and Karner* [1989] and *Braun and Beaumont* [1989] introduced the concept of the "level of necking". They assume the strain in a rift zone is distributed in such a way as to minimize the deformation of the strongest part of the lithosphere. Material moves vertically toward the depth where the yield stress is greatest, (the level of necking). Since that level is deep in the lithosphere the surface is pulled down. A regional isostatic response to the downwarped surface produces flanking topographic highs. Our results show that this may not be a good description of the necking process.

Chen and Morgan [1990a,b] and *Neumann and Forsyth* [1993] coupled the thermal and rheologic field in a model of lithospheric stretching to consider the controls on axial

relief at mid-ocean ridges. These numerical models treat the strong lithosphere as being visco-plastic. Lithospheric deformation is modeled as the flow of a viscous fluid. For the warm areas, the viscosity depends on temperature and stress, while in the coldest areas, the effective viscosity equals the ratio of the stress needed for faulting divided by the local model strain rate. *Chen and Morgan* [1990a,b] also included a crustal layer that accreted over the mantle. This study was the first to confirm the idea that the spreading rate could control whether an axial valley is seen or not, since axial valleys develop only for low model spreading rates. *Chen and Morgan* [1990a] claimed that an axial valley only forms when the crust is cool, and therefore strong enough that significant stresses can be transmitted from the mantle to the surface. Though these models give us great insight into how axial valleys may form, it is hard to analyze the relative importance of the plastically deforming regions and the viscous flowing areas below. Also, there is neither deflection of the surface nor elastic plate bending; the topography is computed a posterori from the stresses. All the material deforms in a diffuse, effectively viscous manner, so that faults do not form.

It is generally thought that much of the topographic relief of slow spreading ridges results from normal fault offset. Recent models have looked at how faults might slip at a ridge. *Shaw and Lin* [1993] argue that the spacing and throw of should be proportional to the thickness of the lithosphere at a ridge, based on an earlier analysis of stresses associated with normal fault offset [*Forsyth*, 1992]. *Thatcher and Hill* [1995] and *Freed et al.* [1995] look at the topography produced by offset of prescribed faults in elastic lithosphere. The plastic deformation away from prescribed faults and the interaction of faults with the viscous material below have not been included in such models. These workers follow a common assumption that faulting can only occur when magmatism is not active at a ridge [e.g. *Karson et al.*, 1987]. According to this view, plate separation is accommodated by thousand-to-million-year-long intervals of magmatism followed by similarly long periods of faulting. *Malinverno and Pockalny* [1990] suggest that such temporal variation in faulting could explain the genesis of abyssal hills, the basement relief seen throughout the ocean floor.

Unfortunately, there is no hard evidence for long amagmatic periods on ridges. *Kappel and Ryan* [1986] show that volcanic activity on some intermediate spreading rate ridges may be episodic on a long time scale, but not that intrusive activity is similarly episodic. Along the ridge in Iceland we know that magmatic events occur on a short O (100 yr.) time scale [e.g. *Tryggvason*, 1984].

In some sense there is a paradox within the modeling studies. *Chen and Morgan* [1990b] predict an axial valley even with steady accretion of crust. The models of faulting such as *Shaw and Lin* [1993], *Thatcher and Hill* [1995] and *Freed et al.* [1995] require long periods of no magmatic accretion of crust to get stresses high enough for faulting. Our models are a first attempt to consistently model faulting and axial valley formation.

MODEL FORMULATION

We wish to understand ways that lithospheric rheologic properties may influence the pattern of faults and axial valley relief seen at ridges. We approach this problem using a numerical method that allows mechanically consistent interactions among elastic, plastic, and viscous deformations at a two-dimensional spreading center. Faults, or regions of concentrated plastic deformation, develop spontaneously wherever stresses reach an assumed yield criteria. Warm regions can behave in a viscous manner with no localization of strain. We do not prescribe where the transition between brittle faulting and viscous flow takes place. Instead, we prescribe material properties such as the temperature dependence of viscosity, and brittle properties such as cohesion (see Figure 2).

Rheology

We assume that deformation occurs either as elastic strain, fault slip or viscous flow, depending on which deformation mechanism occurs at the lowest stress. Plastic behavior is meant to approximate seismic faulting and cataclastic flow, with a yield stress that is pressure-dependent and strain-rate independent. The viscous deformation approximates intracrystalline plasticity and diffusive creep that are pressure-independent, strain-rate dependent. Formally, these rheologies are implemented by allowing either elasto-plastic or visco-elastic deformation. The word 'plasticity' as used here describes the Mohr-Coulomb plasticity theory which is described in detail elsewhere [e.g. *Vermeer and de Borst*; 1984, *Vermeer*, 1990; *Rudnicki and Rice*, 1975; *Poliakov and Herrmann*, 1994]. Briefly, this theory is based on two assumptions: one is that a yield stress governs when a material fails (here it is assumed to be linearly proportional to the pressure), and the second is a flow rule for material at the yield point (we assume that it is incompressible, i.e. material does not change volume during plastic deformation).

The shallow, cold part of a ridge is treated as an elastic material as long as stresses are below those required to cause faulting. The elastic behavior is described in terms of Lame's Parameters which are both set equal to 3×10^{10} Pa. We assume that faulting occurs when the shear stress is large enough to allow slip on a surface with friction and cohesion. The shear stress difference required for faulting is [*Jaeger & Cook*, 1979]:

$$\tau = \mu\sigma_n + S \qquad (1)$$

where S is the cohesion, and μ is the coefficient of friction. Making the usual assumption that the minimum principal stress is horizontal [eg. *Anderson*, 1951], it can be shown that the stress difference for slip on optimally oriented faults increases linearly with depth. In a discussion of results below we designate the slope of the brittle yield stress with depth as C. For a friction coefficient of 0.8 *Brace and Kohlstedt* [1980] showed that C=22 MPa/km if the pore pressure in the rock is zero, and C=13 MPa/km if the pore pressure is hydrostatic. We assume the same friction coefficient [*Byerlee,* 1978] and that hydrostatic pore pressures.

Faults only form in brittle materials which become weaker when they break. In our model, brittle deformation is localized in "fault zones" due to our assumption of a strain and strain rate dependent reduction in shear strength, or cohesion. The reduction of cohesion during deformation is represented as:

$$S = S_0 \exp[-\varepsilon^*/\varepsilon_r], \qquad (2)$$

where S_0 is the cohesive strength, ε^* is the local effective plastic strain, and ε_r is an assumed constant. If we set the value of ε_r to 0.3, the cohesion is reduced to 1/e of its original value at 30% effective local strain. Making ε_r extremely small can lead to numerical difficulties since elastic strain energy would be quickly liberated as cohesion is lost.

Another numerical difficulty, related to artificial diffusion of strain, led us to introduced a kind of healing of inactive faults. Thus, the time rate of change of the effective plastic strain is given by:

$$d\varepsilon^*/dt = d\varepsilon/dt - \varepsilon^*/\tau, \qquad (3)$$

where τ is a time constant, and $d\varepsilon/dt$ is the total strain rate.

We assume that the stress needed for viscous (dislocation creep) deformation depends on both temperature and stress. Both laboratory measurements and solid state theory indicate that the form of this dependence is:

$$\sigma_d = \left(\frac{\dot{\varepsilon}}{A}\right)^{1/n} \exp\left(\frac{E}{nRT}\right) \qquad (4)$$

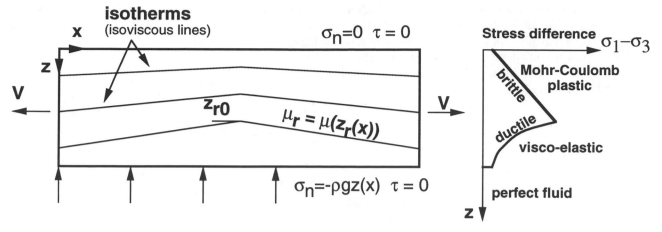

Figure 2. Model setup for stretching a single lithospheric layer. Boundary conditions and lines of constant temperature and viscosity are shown on the left. A representative lithospheric yield stress profile is shown the right. The brittle area is approximated by Mohr-Coulomb elasto-plasticity, while the ductile area is treated as visco-elastic. The transition between brittle and ductile behavior is not preset, but depends on the model strain rate distribution.

where E is activation energy, R is a gas constant and T is absolute temperature, and A is a constant. For this study we take n=1 so the viscosity is Newtonian.

We can approximate the spatial variation of viscosity in a simple way when the viscosity is strongly temperature dependent. Assuming a linear dependence of temperature with depth [$T(z) = T_r + dT/dz*(z-z_r(x))$], we can approximate equation (4) as:

$$\mu(x,z) = \mu_r \exp\left[\frac{-(z - z_r(x))}{z_e}\right] \quad (5)$$

where μ_r is the viscosity at $z=z_r$, and

$$z_e = \frac{RT_r^2}{E\left(\partial T/\partial z\right)} \quad (6)$$

is the vertical length scale over which the viscosity decreases by a factor of e.

Olivine is likely to be the mineral which controls the rheology of the mantle. For olivine the activation energy, E in Equation (4), is ~500 kJ/mole [*Kirby and Kronberg*, 1983]. We consider representative values of z_e between 100m and 1000m, corresponding to temperature gradients of roughly 200°K/km and 20°K/km if the reference temperature at $z_r = 1000°K$.

Temperature Field

The precise temperature structure of a ridge is not well known. However, there is one feature about the ridge

thermal structure that is fairly certain: the depth to a given isotherm increases with distance from the axis of spreading [see models by *Morton and Sleep*, 1985; *Phipps-Morgan and Chen*, 1993]. In our mechanical model, we assume that temperatures increase linearly with depth and the depth, to a given isotherm increases linearly with distance from the spreading center (Figure 2). Temperatures are fixed and do not evolve during our numerical calculations. The temperature structure controls the stresses needed for viscous flow at a given rate, so with increasing depth, the stresses required for flow decrease. Thus, far field extensional displacements are expected to result in concentrated deformation at the spreading center where cold strong material is the thinnest.

Numerical Method

We have used an explicit Lagrangian finite-element method similar to the FLAC (Fast Lagrangian Analysis of Continua) technique as developed by *Poliakov et al.*, [1993; 1994; 1996] and based on the approach of *Cundall* [1989]. This method is a direct solution of Newton's second law for every grid point including the effects of inertia. However, inertial stresses are kept low compared to characteristic tectonic stresses in order to approximate quasi-static processes. FLAC is a very powerful technique for simulating non-linear rheological behavior at very high resolution because the explicit time-marching scheme does not require storage of large matrices which are needed for implicit methods. This method is Lagrangian (i.e. numerical grid follows the deformation), and this method has successfully simulated localized deformation (approximating faults) in elasto-plastic materials [*Hobbs and Ord*, 1989; *Cundall*, 1990; *Poliakov and Herrmann*, 1994; *Poliakov et al.*, 1994; *Hassani and Chery*, in press], visco-elastic flow [*Poliakov et al*, 1993] and visco-elasto-plastic diapirism [*Poliakov et al.*, 1996].

The simulation of very large deformation leads to severe distortion of the numerical grid and/or overlapping the finite elements. Strongly distorted elements decrease the accuracy of the results, and overlapping elements produce erroneous results. This numerical difficulty led us to introduce re-meshing (re-gridding) during the calculations as following. The quality of mesh is checked periodically, and if the smallest angle of any triangular element is below a critical value (generally set at 10 degrees), re-meshing is triggered. The procedure consists of three parts: defining the new domain of calculations, creating the 'new' undistorted grid in the new domain, and interpolating the information from the 'old' distorted grid to the 'new' one.

The upper boundary of the new domain corresponds to that of the old domain, thus preserving the surface topographic relief. Due to stretching, the domain of calculations becomes thinner and longer. After the remeshing, we add material at the bottom and remove it from the sides. The new bottom boundary is set at a constant depth. Each column of grid elements is uniformly stretched or compressed so that correct layer thickness is preserved.

The new quasi-regular grid is only slightly distorted to adjust to the topography. To interpolate the information from the old grid to the new one, we find which 'old' elements are closest to the centers of 'new' triangles. We then assign the values of the stresses and strains from the old to the new element. The temperature field is set according to the location of the new element center.

Due to errors of interpolation, stresses in elements are not completely equilibrated with stresses in the neighboring elements. This disequilibrium of stresses (and thus forces) produces artificial accelerations and oscillations of the nodes. However, after few tens of numerical steps, these undesired effects are damped and stresses come into equilibrium. Because this transition period lasts only a very short time, compared to the total time of stretching, the deformation caused by remeshing and damping are small. We have tested different criteria to trigger remeshing and find that the essential results presented here do not depend on the amount of strain between remeshing.

Boundary Conditions

At the surface, shear and normal stresses are assumed to be zero. At the side boundaries we apply constant horizontal velocities and zero shear stress (as shown in figure 2). Due to the decrease of the viscosity with depth, a perfect fluid is a good approximation of mantle with very low viscosity. Thus, at the bottom of our model region we apply normal stress equal to the hydrostatic pressure in the mantle and zero shear stress (a Winkler foundation). The temperature is fixed at 0°C at the top surface, the conductive heat flux is zero at the side boundaries. On the bottom boundary, the temperature is maintained at the initial value set there.

RESULTS FOR STRETCHING A SINGLE LAYER

We now show the results of stretching a layer of elastic-plastic-viscous material with a fixed temperature field. The parameters that describe the individual cases are given in the figure captions or in Table 1. In the first section of results, we discuss cases with no cohesion reduction. These lead to axial valley structures and interesting stress patterns, but we do not develop anything resembling abyssal hills. In the next section of results we consider cases where the loss of cohesion with strain leads to faults that have finite offset and result in abyssal-hill-like relief.

We found that resolution of evolving sets of faults requires at least 20 mesh points in the vertical dimension in the thinnest part of the brittle zone. Model times for the calculations discussed below range from a CPU week to more than a CPU month. Therefore, the range of parameters tested was necessarily limited.

We have varied several parameters that control the model evolution in the absence of crustal accretion. These include the thickness of the brittle layer at the axis, the cohesion, and strain dependence of cohesion (ε_r and τ). The spreading rate and viscous length scale (z_e) were changed in concert with the brittle layer thickness, since they should not be independent. Since faster spreading rates should produce larger thermal gradients at depth, the brittle layer and the length scale for viscosity reduction are thinner.

Results for model runs with no cohesion loss are shown in Figure 3. Regions of high strain rate are referred to as faults or fault zones. At the earliest stage of stretching, the fault pattern is dominated by inward dipping normal faults. Offsets of such faults leads to axial valley relief. Note that the velocity field shows a triangular region moving down at the axis. With continued extension, the strain field changes and the fault pattern becomes dominated by outward dipping faults. Offset of the outward dipping faults reduces the topographic relief at the axis. Eventually steady state axial topography develops, with no topographic roughness carried off axis.

The results for the initial stage of stretching, when faults are inward dipping and the strongest part of the lithosphere does not move vertically, are superficially consistent with

Table 1. Model Parameters

Parameters defining the 'long' model runs described in the text in which remeshing allows development of multiple generations of faults. For these cases the initial cohesion, S_0, is 20 MPa and the slope of the reference isoviscous contour, dz/dx, is 0.1. The brittle layer thickness, z_b, is the thickness at the axis over which deformation is elasto-plastic rather than visco-elastic. The plate velocity, u_p, is the half-spreading rate. z_e is the depth interval that viscosity changes by a factor of $1/e$. ε_r is the plastic strain required for loss of all cohesion. The annealing time, τ, controls the rate of 'healing' of plastic strain, and so restoration of cohesion, (see text).

Case	brittle thickness (km) z_b	plate velocity (cm/yr) u_p	viscous depth scale (m) z_e	reference strain ε_r	annealing time (s) τ
g.10	10	5.0	1000	0.03	∞
g.10-12	10	5.0	1000	0.3	10^{12}
g.3-10	3	1.5	300	0.3	10^{10}
g.3-12	3	1.5	300	0.3	10^{12}

the "level of necking" concept of *Braun and Beaumont* [1989] discussed above. There are, however, two problems with this interpretation. First, all depths of the axial lithosphere strain, the base of the brittle part just strains in a more concentrated region. Second, the strain pattern changes as the lithosphere is deformed and the depth where vertical velocities are zero, migrates to the surface .

Finite slip on inward dipping faults produces axial topographic relief. This topography apparently changes the stress field in such a way to inhibit further movement on those inward dipping faults. The mode of faulting and therefore the strain pattern evolves as the topography changes. For steady state topography both inward and outward dipping fault planes within the axial valley are at a state of incipient failure. Apparently, these faults, at a range of distances from the axis, are active at the same time. In the following section we present a simple scaling argument for the critical topographic relief at which slip on inward dipping faults would be replaced by slip on outward dipping faults.

Simple Scaling of Axial Valley Relief and Brittle Layer Geometry

Estimates of how fault offset changes the force needed for continued plate separation may allow us to gain insight into how axial valley relief depends on the geometry and strength of the brittle layer and on the layer density. Here we ignore viscous stresses in the region below the brittle layer. Consider that the brittle layer thickness (z_b) increases

linearly with distance from the axis of spreading. We assume that either of two sets of faults can accommodate extension, as illustrated in Figure 4. The inward dipping fault set (A in Figure 4) is initially easier to slip on because it cuts a thinner section of the layer than the outward dipping set (B) consistent with the hypothesis of *Carbotte and Macdonald* [1990]. The difference in thickness cut by the two fault sets (Δz) is roughly equal to $z_{b0}\, dz_b/dx$, where z_{b0} is the brittle layer thickness of the axis. The yield stress at the base of the layer is $C\, z_{b0}$, where C is a constant defined above . The difference in tectonic force for initial slip on set A compared to set B is $C\,(z_{b0})^2\, dz_b/dx$.

We assume that slip on fault set A builds up topographic relief and changes the lithospheric stresses such that continued slip on those faults is more difficult than on fault set B. *Forsyth* [1992] showed that the tectonic force change inhibiting fault slip increases with the vertical component of fault throw, w, as $w\alpha\Delta\rho g$, where α is the flexural parameter for the layer, $\Delta\rho$ is the density contrast between the layer and the overlying water or air, and g is the acceleration of gravity. We assume that this force change describes only the inhibition of slip on fault set A and that the force needed for slip on fault set B is not affected. When this increase in tectonic force for continued slip on set A equals the extra force initially needed for slip on set B then fault set A is replaced by fault set B. When that happens the topographic relief reaches maximum value w_{max}. We thus expect that $w_{max} \sim C\,(z_{b0})^2\,(dz_b/dx)/\alpha\Delta\rho$.

This scaling relation is similar to that based on the moment balance argument discussed by *Phipps-Morgan et*

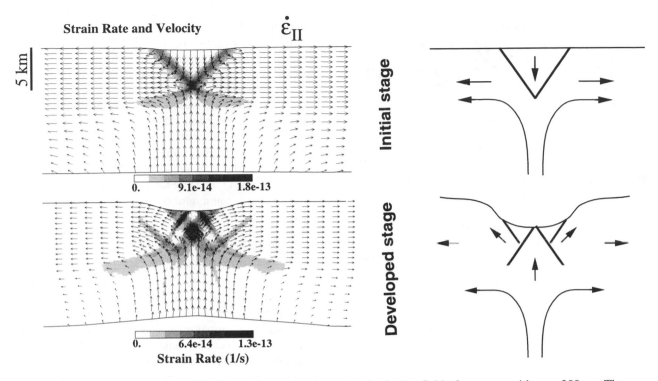

Figure 3. Cross-sections of the instantaneous strain rate and velocity fields for a case with $z_e = 300m$. The depth z_r to a reference viscosity of 10^{20} Pas was set so that the brittle ductile transition occurred at approximately 5 km depth at the axis. The slope of this reference isotherm (i. e. dz_r/dx) was set to 0.2. The upper frame shows the initial stage of rifting after approximately 200 m of extension while the lower frame shows the changes after about 2 km of extension. Cartoons to the right of the model results illustrate how the distribution of faults changes with the amount of extension.

al. [1987] but the meaning is different. First, the arguments given here clearly show that the fault distribution should depend on the topography. As a layer is stretched, topography builds up and the geometry of faulting changes. Second, we do not specify how the horizontal normal stress distribution varies with distance from the center of necking. The key to the moment balance argument is that the stresses away from the axis are distributed over a deeper extent than the stresses at the ridge. This stress pattern is not seen in our numerical results. For example, Figure 5 shows that the region of large stress differences is not uniformly distributed through the lithosphere as defined by an isotherm, but it is concentrated over a depth range similar to the depth range of faulting at the axis.

Geometry of the Brittle-Ductile Transition

At shallow depths the viscosity is so great that large strain deformation occurs by localized plastic flow (or faulting) at stresses lower than required for viscous flow. However, at greater depths the viscosity is low enough that

viscous flow occurs at a lower stresses than for faulting. The horizontal normal stresses at the same depth are relatively uniform at different distances from the axis. (Figure 5). At the axis, the stresses are at the yield values, while off axis stresses are below yield. The depth at which the stresses for viscous flow and plastic faulting are equal is called the depth of the brittle-ductile transition.

One interesting result is that the depth of the brittle-ductile transition occurs at a nearly constant depth across the region of stretching even though there are considerable lateral temperature variations across that region. Because viscous stresses depend so strongly on temperature one might assume that the brittle-ductile transition follows an isotherm. However, the viscous stresses also depend on strain rate, so that lateral variations in strain rate can and apparently do affect the depth of the brittle ductile transition.

Effect of Cohesion Reduction with Strain

The calculations with zero cohesion developed axial topographic relief that remained completely steady in time

Fault Set A Active

Fault Set B Active

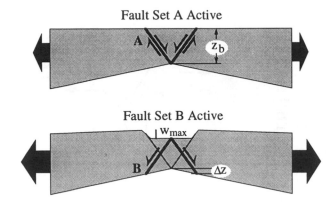

Figure 4. Geometry assumed in deriving an approximate scaling between the geometry of an extending brittle layer and the topographic relief caused by fault offset. (A) shows that inward dipping normal faults should initially cut the thinnest part of the lithosphere. When a limiting amount of topographic relied w_{max} is developed then the changes stress field would favor slip on an outward dipping set of faults (B) even though they cut a portion of the lithosphere that is thicker by Δz.

after an initial transient phase. The cases with finite cohesion drop never develop completely steady state topography, but the long-wavelength axial valley does not vary in width or depth by more than a factor of 2. For cases with faults that strain weaken, the distribution of faulting and evolution of topography depend upon two factors: (1) the geometry of the temperature field (which controls the distribution of viscosity and therefore the thickness of the brittle layer); and (2) how much weakening occurs on a given fault. We discuss only results of considering two values of the axial layer thickness (3 and 10 km). All the cases had the same slope of the isotherm marking the 10^{20}Pa-s viscosity (i.e. $dz_i/dx = 0.1$) and the same value of cohesion (20 MPa). We also considered two values of the characteristic strain for cohesion reduction (ε_r) and a range of values for the strain healing time (τ). We illustrate the results of varying the strength geometry and the cohesion reduction parameters with the results of four long model calculations.

The first case, shown in Figure 6 (case g10, Table 1), demonstrates how faulting and axial relief develop. For this example the brittle lithosphere at the axis is ~ 10 km thick (thus the number 10 in the case designation). The characteristic strain for cohesion reduction, ε_r, is 0.03, and there is no annealing of effective strain (i.e. $\tau = \infty$).

The first thing we see during a run is the development of two faults, one on either side of the axis of symmetry, that dip towards the axis. Offset of these faults forms an axial valley with flanking topographic highs. After the initial

faults have slipped over a kilometer, new faults form, and the initial faults cease slipping. By the second or third fault generation the pattern of faulting becomes more complex, with both inward and outward dipping faults being active.

There was some creation of new fault-like relief away from the spreading axis. We interpret this to be small magnitude faulting that accommodates the strain of plate bending associated with the flanking topography. The plates unbend as they move away from the rift shoulder mountains (the topographic highs flanking the axial valley). Unbending causes stresses to exceed yield and also implies some straining of the layer. Faulting occurs in this case because the strain required for a large reduction in cohesion is so small ($\varepsilon_r = 0.03$, so most cohesion is lost after 3% strain).

After 5 to 10 km of stretching faults to cease to develop finite offset (Figure 6b). For greater amounts of separation a smooth axial valley forms with no new abyssal hills formed in the valley and rafted out to the flanks. This smoothing of topography indicates a numerical problem that led us to introduce annealing of effective strain in our formulation of fault weakening. Within the model fault zones the model elements become so strained that numerical accuracy is degraded. Thus, we overlay a new more regular grid on the old deformed one. The regridding causes artificial diffusion of all quantities passed to the new grid.

The effective strain numerically diffuses through the region of active faulting in the axial valley. Eventually all elements in the region, even though they have not plastically deformed, are "tagged" with an effective strain value that is large (compared to ε_r). Then the entire region numerically loses cohesion. Within this non-cohesive area

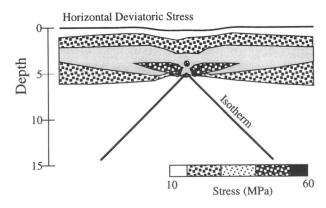

Figure 5. The horizontal deviatoric stress in the central part of stretched lithosphere for a case with $z_e = 300m$ and $dz_i/dx = 1.0$. The solid line shows the position of a reference isoviscous line where $z = z_r$. Note that the contours of constant stress are much closer to being horizontal than is the isoviscous line.

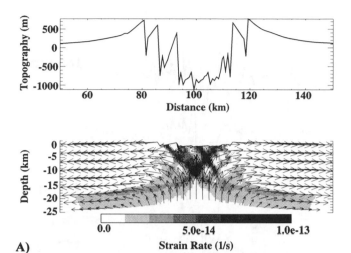

A)

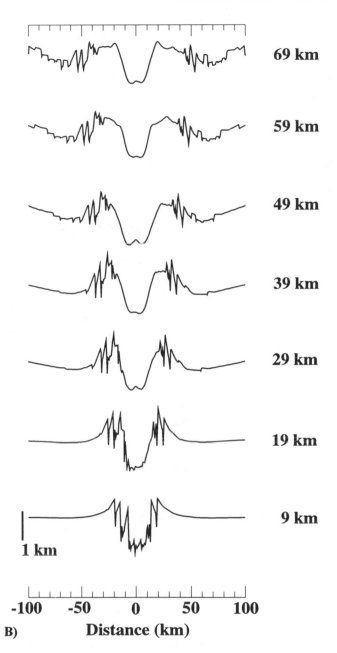

69 km

59 km

49 km

39 km

29 km

19 km

9 km

1 km

-100   -50   0   50   100

B)   **Distance (km)**

Figure 6(a). Cross-section of strain rate, velocity and topography for a calculation g.10 with rapid loss of cohesion with strain ($\varepsilon_r = 0.03$) and no strain annealing. The time shown is after 9 km of extension. We show only the central, 100 km wide, part of the calculation domain which had an initial width of 200 km and depth of 50 km. The brittle deformation at the axis extends down to about 12 km in at the axis. This is the depth at which the viscous stresses and the brittle stresses are the same.
(b). Sequence of model topographic profiles for the case g.10. Note that after 9 km, offset no sizable abyssal hills are formed, and the axial relief is smooth at wavelengths less than about 10km (see text for further explanation).

faults do not form and accommodate finite slip. The plastic deformation (strain rate) is distributed fairly uniformly in much the way it would be in a layer of dry sand. This is the kind of behavior we found for cases with no cohesion reduction. We conclude that artificial diffusion of effective strain is responsible for the progressive change in the offset of faults in the model axial valley. Cases described below included annealing of strain so that areas which were not straining would heal fast enough that numerically diffused strain would disappear.

In the next three model runs we increased the strain needed for cohesion loss by a factor of 10 (now $\varepsilon_r = 0.3$) and considered different values of the annealing time. These changes got rid of the problem of widespread cohesion loss due to artificial strain diffusion.

Figure 7 shows results when the brittle layer at the axis was about 10 km thick for case g.10-12 (here the second number is the annealing time, τ, in $\log_{10}$ seconds). The relief on individual faults is somewhat smaller than for case g.10 and this is a result of the greater strain needed for cohesion loss. Essentially, faults do not become very weak when they first form due to the larger value of ε_r and they

do not become as weak because of annealing. Figure 7(b) shows that the first feature formed is a valley, generated by one and then a pair of faults dipping in toward the axis. The next faults dipped outward, away from the axis. Eventually, faults formed at a variety of distances from the axis in what appears to be a chaotic fashion. Note that the amount of stretching shown in Figures 6(b), 7(b), 8(b), and 9(b) are all different.

An axial valley is a consistent feature of this model case, but the shape of the valley changes in time. Smaller scale and more random topographic relief is formed within and at

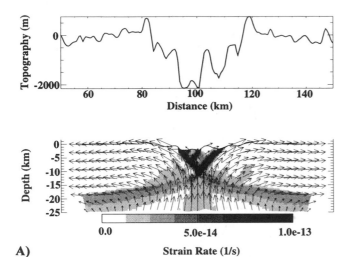

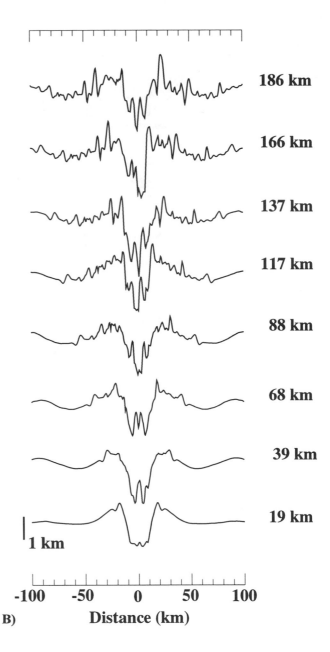

A)

Figure 7(a,b). Same as 6(a,b) except for $\varepsilon_t = 0.3$ and annealing is included with $\tau = 10^{12}$ seconds (case g.10-12). The pattern of topography, both what is axial valley-like and abyssal hill-like, varies in an irregular way after a start-up period when topography is symmetric and regular.

the edge of this axial valley. This abyssal hill-like relief is rafted away from the axis without further significant deformation. Below, we analyze the abyssal-hill-like relief for this cases in some detail and compare it to data from ridges.

The last two cases shown have thinner brittle lithosphere at the axis (3 km) and only differ in the annealing time. For case g.3-10 the annealing time was 10^{10} seconds. The effect of annealing depends on the spreading rate and the grid size. Here the effective annealing time is about one tenth that as in case g.10-12. Even when only one fault is actively straining, the cohesion drop there is not the full 20 MPa. The average throw on faults is much smaller for this case than for case g.10-12 due to the smaller cohesion reduction (Figure 8). Also, the axial valley is much narrower .

For the last case (g.3-12) the annealing time is 100 times longer than for g.3-10. The average relief on faults is much greater for this case, since annealing has almost no effect (Figure 9). The axial valley is now harder to distinguish from the abyssal-hill-like relief.

Analysis of Model Abyssal Hills

Perhaps the most interesting result is that fault offset scales with cohesion drop. Localized zones of high strain rate develop in all the models, including cases with zero cohesion. For cases with zero cohesion reduction, the region of localized deformation builds up negligible offset

before the zone of plastic deformation shifts to another position. Once steady-state axial valley relief is achieved, deformation is smoothly distributed for the cases with zero cohesion.

The effect of finite cohesion that drops in the area of a "fault" is that deformation is always concentrated in narrow plastic zones cutting the brittle layer. Finite offset across these "faults" produces topographic relief that is formed in the near axis region and then laterally rafted away. The pattern of plastic strain, including the location of initiation and cessation of "faulting", changes through time in an irregular way. This is what makes the frozen, off axis

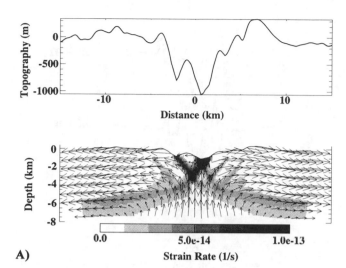

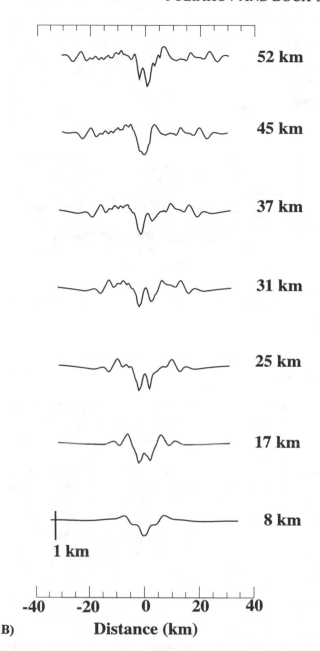

Figure 8(a,b). Same as 6(a,b) except for $z_{r\,0} = 3$ km and $\varepsilon_r = 0.3$ and annealing is included with $\tau = 10^{10}$ seconds (case g.3-10).

topography resemble the complex pattern of relief which characterizes abyssal hills formed at many ridges.

Since model case g.10-12 produced relief that looked similar to that seen around some slow spreading ridges (see Figure 1), we analyze the model topography of this case in some detail. Statistical properties of the model topography compare well with similar properties estimated for seafloor topography. To analyze abyssal hill topography, we considered only the region more than 30 km from the model axis, using the last profile shown in Figure 7 (labeled 208 km). Long wavelength topography was removed by subtracting a second-order polynomial fit through the profile on each side of the axis. The "roughness" of the abyssal hills was estimated in terms of the square root of the mean of the squared topography. We found values of 176 m and 210 m for roughness of the left and right side profiles, respectively. An autocorrelation analysis was used to find a characteristic width of the topographic relief of 7.5 km for the left side and 10 km for the right side profiles. Both the roughness and characteristic width values fall in the middle of the range estimated for abyssal hills close to the Mid-Atlantic Ridge (e.g. *Goff*, 1991).

CRUSTAL ACCRETION AND MANTLE STRETCHING

Model Formulation Including Crust

For a given brittle layer thickness at the axis we find that fault offset depends on the cohesion loss on an active fault. In recent work for a slightly different lithospheric geometry, *Lavier et al.* [1998] shows that fault offset could be unlimited in a brittle layer thinner than ~10km. In those calculations the initial cohesion had to be fairly large (~40 km) and the strain for cohesion loss (ε_r) had to be larger than assumed here (~1) to get fault offsets larger than the brittle layer thickness. This result may offer an explanation for the very large offset, low-angle normal faults seen at many ridge ridge-transform intersections at slow-spreading ridges [e.g. *Karson et al.*, 1987; *Tucholke et al.*, 1998; *Cann et al.*, 1997] For such large fault offsets the exposed footwall of the fault is rotated to a very low dip and may even become flat [*Buck*, 1988; 1993] Naturally, other

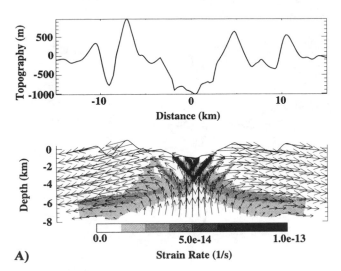

Figure 9(a,b). Same as 6(a,b) except for $z_{r\,0} = 3$ km and $\varepsilon_r = 0.3$ and annealing is included with $\tau = 10^{12}$ seconds (case g.3-12).

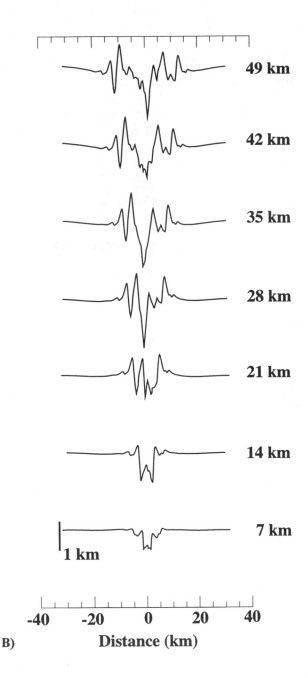

mechanisms that allow weakening of active faults, such as serpentization [e.g. *Escartin et al.*, 1997], may also promote larger fault offsets.

Since faults along many sections of slow-spreading ridges do not have extremely large offsets, something must act to limit the offsets there. The material properties could vary between different sections of the axis, but this is hard to justify. Alternatively, there is evidence that the amount of magma delivered to different sections of a ridge may be different [e.g. *Lin et al.*, 1989; *Tolstoy et al.*, 1993].

A major portion of the plate separation at a ridge may be accommodated by magmatic accretion which may in tern affect the topography and faulting at the ridge. We expect that crustal accretion should reduce the offset of normal faults at ridges. On average, the crust is about 6 kilometers thick at mid-ocean ridges [*Chen*, 1993]. Figure 10 shows the approximate pattern of long-wavelength topography, crustal structure and faulting that we envision for a slow spreading ridge. The process of crustal accretion may be very complex, involving intrusion and extrusion of magma and flow of hot, ductile crust. The common denominator of these processes is that they may accommodate plate separation at much lower stress levels than typical estimates of the stress required for faulting as shown in Figure 11. This has led a number of authors, as noted earlier, to contend that the observed crustal faults at a ridge imply long time intervals in which there is no magmatism.

Here, we consider how the position of dike intrusion may affect the pattern of mantle faulting, crustal faulting, deformation and associated topographic relief of a ridge even if magmatic accretion is steady in time. The key difference

with our single layer calculations is that a layer of crust is assumed to accrete by dike intrusion on top of a stretching mantle layer (Figure 11). Dikes should intrude the elastic part of the crust where it is under tension [e.g. *Lister and Kerr*, 1991]. The exact position of dike intrusion is difficult to specify since magma has to be delivered to an area that is under tension. The stress state of the upper crust will depend on the thermal and thus strength structure of the crust and on the pattern of mantle deformation.

When magma is present, with magma pressure close to the lithostatic pressure, then both tensile failure (diking)

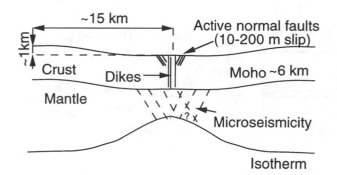

Figure 10. Schematic of the inferred structure of a slow-spreading mid-ocean ridge including features such as crustal faults that are active close to the ridge axis, and microseismicity indicating brittle behavior of the mantle.

and shear failure (faulting) may occur at very low levels of tectonic stress [see *Price and Cosgrove*, Chapter 2, 1990]. Vertical slip may then occur easily when a dike forms. Such magma-assisted normal faulting is seen in eroded sections on Iceland [*Gudmundsson*, 1992]. A slightly different mechanism for slip on normal faults in association with magmatic activity has been studied by *Rubin and Pollard* [1987]. They show that if magma in a dike does not reach the surface, then a normal fault may form to accommodate the shallow strain. We have not yet developed a method of simulating dikes that do not reach the surface, nor can we model the long term accretion of crust. (This would require a scheme to regrid and then intrude new material.) Thus in these preliminary models we can only look at how a modest amount of extension (~1 km) is affected by intrusion of dikes in different positions.

We approximate the intrusion of a dike of low viscosity magma by specifying the stress on a prescribed, initially vertical interface embedded in the top part of the model lithosphere. Normal stresses on the interfaces are set equal to the lithostatic pressure in the crust, and the shear stress is set to zero. A dike is placed either in the center of the assumed symmetrical lithospheric thermal structure or a set distance form that center. The boundary conditions on the bottom, sides and top are the same as for the single layer stretching cases. With these boundary conditions, the crust would not be deflected if it were not attached to strong mantle that was being stretched.

The crust is treated in the same way as the mantle, in that it can deform in either an elastic, plastic or viscous manner depending on the stress conditions. Laboratory measurements indicate that crustal material such as diabase may be weaker than olivine over a range of temperatures [e.g. *Shelton and Tullis*, 1981; *Kirby and Kronenberg*, 1983]. Thus, in some model cases, the crust is assigned a

lower viscosity than the mantle. We implement this rheology by prescribing a different reference isotherm in the crust than in the mantle. Both isotherms correspond to the same reference viscosity (see Figure 11). The viscosity-depth relation (i.e. the e-folding depth scale z_e) is assumed to be the same in the crust as in the mantle. The depth of the reference isotherm is varied so that the crust can be either brittle throughout, or have a weak viscous region at its base. We also consider different crustal thicknesses since the thickness of crust appears to vary by as much as a factor of 2 along some slow spreading ridge segments [e.g. *Kuo and Forsyth*, 1988; *Lin et al.*, 1990; *Tolstoy et al.*, 1993; *Detrick et al.*, 1995].

There is no reason why the dikes cutting and forming the crust at a ridges must be in the center of the ridges structure. One might expect that the position of diking might vary in time as the ridge stress field evolves. In the single layer calculations, we saw time variations in the position of faults. If the position of faults cutting the mantle lithosphere at a ridge affect the position of dikes, then the position of intrusion could move with the faults. Our present numerical approach does not allow us to place a dike interface wherever find the stresses most favorable for dike formation. However, we can see if different dike positions significantly affect the deformation of the crust.

Results for Models with Intruded Crust

For the cases where the dike is placed at the center of symmetry of the ridge thermal structure, the presence of the crust strongly suppresses faulting that deflects the Moho (the interface between the crustal and mantle). Just as in the single layer calculations, it is slip on inward dipping faults that produces deflection of the top of the stretching layer. With the ridge centered, dike slip on these faults is strongly reduced in favor of slip on outward dipping faults. This is most clearly shown in Figure 12a for a 6 km brittle crustal layer overlying mantle lithosphere. For that case the lithosphere deformed in a brittle manner down to ~3 km below the Moho. Almost no slip occurs on inward-dipping faults. The crust is not faulted and a small (~50 m high) axial high results from stretching.

When the thickness of the crust is much less than the mantle lithospheric thickness, crust-cutting normal faults develop, as shown in Figure 12b. In that example, the 3 km thick crust is entirely brittle, and overlies a ~6 km thick region of mantle deforming in a brittle manner. The crust-cutting faults for this case only account for a minor amount of the separation of the plates. In the limit that the crustal layer is negligibly thin, the results approach those of the single layer case.

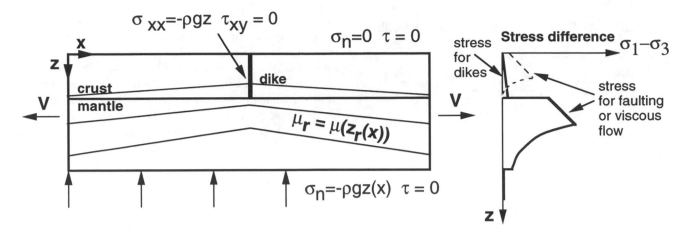

Figure 11. Model setup that includes an accreted crustal layer. The mantle part of the model is treated the same as in the one-layer model in Figure 2. The crustal layer is intruded by a vertical "dike" where the normal stress is set equal to the lithostatic stress and the shear stress is zero. The crustal layer may have a different viscosity structure than the mantle. On the right two stress profiles are shown for the crust: in a dike (solid) and the yield stress away from a dike (dashed).

The effect of weaker lower crust on the deformation pattern is shown in Figure 12c. The crust is 6 km thick (as for the case in Figure 12a), but it behaves in a brittle manner only down to ~3 km. Some deformation of Moho occurs for this situation, though no crust-cutting faults develop. A broad, shallow (~100 m deep) axial valley develops before the mantle begins to stretch mainly by outward dipping faults and topographic relief stops increasing.

Using scaling arguments similar to those described above for axial relief, we can understand why this dike geometry suppresses slip on inward dipping faults. As for the single layer case, the inward-dipping fault geometry is initially the favored geometry of deformation. The inward-dipping faults must cut the brittle part of the crust, requiring more force than that needed to fault the mantle layer alone. The dike accommodates crustal extension at a negligible stress level so that when there is slip on outward-dipping faults which intersect the dike, no extra work is done deforming the crust . Thus, only when the brittle crust is very thin can faults cut through the crust and mantle as easily as faults that link up with the dike.

Results for a Dike Offset from the Model Symmetry Axis

We have placed a dike interface about 3 km to one side of the center of symmetry in a model whose rheologic structure is equivalent to the case shown in Figure 12c. This dike geometry favors slip on a mantle-cutting faults that link the dike with the thinnest part of the mantle thermal lithosphere (Figure 12c). For this case, several

hundred meters of plate separation produces a large downward drop of the surface, as well as continued slip on the original outward dipping normal faults in the mantle (Figure 12d). Were we to put a dike interface on the other side of the center of thermal symmetry, further topographic relief could be produced by mantle faults intersecting the bottom of that dike.

It is clear that the position of dikes strongly affects the deformation and surface topography produced by these models. Dikes might form within a region a few kilometers wide at a ridge, with the position at a given time partly determined by underlying fault activity. If this can happen, then our results raise the possibility that crustal accretion accommodated by dikes may not completely suppress the topography produced by mantle stretching. Our single layer models may thus help us understand how mantle thermal structure may affect the topographic relief at a ridge. Also, magma-assisted faulting that we associate with diking is consistent with the observation that faults at slow spreading ridges are primarily active within a few kilometers of the area of most recent diking.

CONCLUSIONS

Perhaps the most important result of these idealized models of lithospheric stretching is that they show several factors that effect the distribution and offset of faults accommodating extension at a ridge. First, changes in the pattern of faulting occur spontaneously during stretching of a brittle layer, and can change the pattern of topography. This suggest the possibility that the kind of fault pattern

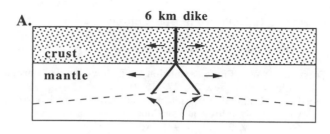

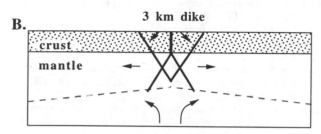

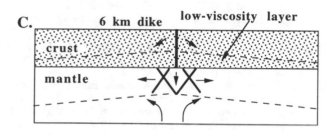

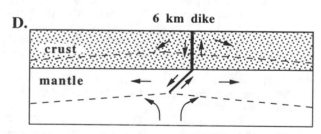

Figure 12. Cartoon representations of results of stretching a model with a crustal "dike" in different positions offset form the center of the thinnest part of a stretching mantle layer. The distance of dike emplacement from the center of symmetry is about equal to the brittle thickness of the mantle layer at the axis.

changes that may produce abyssal hill topography need not depend on magmatic supply variations. Temporal variations in the topography may be an intrinsic effect of extending a brittle layer. Second, little understood parameters describing fault weakening with strain, control the amount of offset on individual faults. For a range of parameters, fault slip may even be unlimited.

We developed an argument that topographic relief at a ridge should scale with the axial lithospheric thickness, and the slope of the lithospheric thickness with distance from the axis. The pattern of stress we map out for a single

stretching layer suggests that the "moment balance" scaling argument [*Phipps-Morgan et al.*, 1987] may be very difficult to apply to make predictions. Certainly, moments must be balanced for mechanical equilibrium in a system, but predictions from this scaling argument can only be made if the stress pattern at a ridge is known. Several authors suggest that the horizontal stress pattern a finite distance from the ridge will be close to uniform with depth over the cold, strong part of the lithosphere. This is definitely not the case in our models.

Our model strain patterns change with time in a way suggesting that the concept of a "depth of necking" controlled by the strength of the lithosphere [*Braun and Beaumont*, 1989] is an incomplete description of topographic relief caused by lithospheric extension. What might be called the "depth of necking" in our models can be defined as the depth where there is little vertical motion (though there is still shear strain deformation). With this definition, the depth of necking shallows with the amount of extension in our single layer calculation. If there is steady state topography, then this depth has to be at the surface at the axis of stretching.

We have not resolved the problem of how crust accretes and faults on top of stretching mantle. However, we have made progress in simulating brittle faulting and magmatic accretion. Our preliminary work shows the unsurprising result that accretion can suppress faulting. The lateral position of a model dike, relative to the thinnest lithosphere, can affect the pattern of crustal faulting. Since stretching a brittle layer can produce time-variable fault patterns, topography and stresses, it seems possible that tectonics controls whether magmatism is intrusive or extrusive. We do not expect that steady magmatic accretion will necessarily damp out the kind of tectonic variability we have simulated.

Future models that include regridding and so unlimited amounts of plate separation may allow us to demonstrate whether mantle stretching with steady crustal accretion can produce the observed patterns of topography and faulting. We expect to be able to model dikes that do not reach the surface, as well as simulate long term accretion plus faulting in the near future. In these models we hope to add in simple thermal models in the manner that *Chen and Morgan* [1990b] did in their steady-state visco-plastic models of stretching. One key goal is to see whether abyssal hill morphology and roughness patterns could be matched without invoking time variable magmatism. It is also possible that crustal faulting can only occur during long amagmatic periods. Perhaps careful dating of magmatic and tectonic events at ridges, using methods such as those described by *Perfit and Chadwick* (this volume), may allow discrimination between these models.

Acknowledgements.   Reviews by Norman Sleep, Paul Delaney, Jian Lin, and Anjana Shah greatly improved this manuscript. This work was supported by NSF grant OCE 94-02995 and INT 96-03350. Lamont contribution number 5831.

REFERENCES

Anderson, E.M., *The Dynamics of Faulting*, 206 pp., Oliver and Boyd, Edinburgh, 1951.

Artemmjev, M. E., and E. V. Artyushkov, Structure of the Baikal rift and the mechanism of rifting, *J. Geophys Res.,* 76, 1197-1211, 1971.

Brace, W.F. and D.L. Kolstedt, Limits on lithospheric stress imposed by laboratory experiments, *J. Geophys. Res.,* 85, 6248-6252, 1980.

Braun, J., and C. Beaumont, A physical explanation of the relation between flank uplifts and the breakup unconformity at rifted continental margins, *Geology, 17,* 760-764, 1989.

Braun, J., and C. Beaumont, Styles of continental rifting: Results from dynamic models of lithospheric extension, Canadian Society of Petroleum Geologists, Memoir 12, 241-258, 1987.

Buck, W. R., Effect of Lithospheric thickness on the formation of high- and low-angle normal faults, *Geology, 21,* 933-936, 1993.

Buck, W. R., Flexural Rotation of Normal Faults, *Tectonics, 7,* 959-973, 1988.

Byerlee, J. D., Friction of rocks, *Pure Appl. Geophys., 116,* 615-626, 1978.

Cann, J. R., D.K. Blackman, D.K. Smith, E, McAllister, B. Janssen, S. Mello, E. Avgerinos, A.R. Pascoe, J. Escartin, Corrugated slip surfaces formed at ridge-transform intersections on the Mid-Atlantic Ridge, *Nature,* 385, 329, 1997.

Carbotte, S., and K. C. Macdonald, Causes of variations in fault-facing direction on the ocean floor, *Geology, 18,* 749-752, 1990.

Chen, Y.J., Oceanic crusted thickness versus spreading rate, *Geophys. Res. Lett., 29,* 177-203, 1993.

Chen, Y. and W.J. Morgan, Rift valley/no rift valley transition at mid-ocean ridges, *J. Geophys. Res., 95,* 17,571-17,582, 1990a.

Chen, Y., and W.J. Morgan, A nonlinear rheology model for mid-ocean ridge axis topography, *J. Geophys. Res., 95,* 9275-9282, 1990b.

Cochran, J. R. An analysis of isostacy in the worlds oceans, 2; Mid-ocean ridge crests, *J. Geophys. Res., 84,* 4713-4729, 1979.

Cundall P.A., Numerical experiments on localization in frictional materials. Ingenieur Archiv. v. 58: 148-159, 1989.

Cundall, P.A., Numerical modeling of jointed and faulted rock. In: A. Rossmanith (Editor) *Mechanics of Jointed and Faulted Rocks.* A.A. Balkema. Rotterdam. pp.11-18, 1990.

Detrick, R. S., H. D. Needham, and V. Renard, Gravity anomaly and crustal thickness variations along the mid-atlantic ridge between 33° and 40°N, *J. Geophys. Res.,* 100, 3767-3787, 1995.

Escartin, J., G. Hirth, and B. Evans, Effects of serpentinization on the lithospheric strength and the style of normal faulting at slow-spreading ridges, *Earth and Planetary Science Letters,* 151, 181-189, 1997.

Forsyth, D.W., Finite extension and low- angle normal faulting, *Geology,* 20, 27-31, 1992.

Freed A., J. Lin, and H. J. Melosh, Long term survival of axial valley morphology at abandoned slow-spreading centers, *Geology,* v. 23, 11, p. 971, Nov. 1, 1995.

Goff, J. A. Near-ridge abyssal hill morphology. *J. Geophys. Res.* 98, 21713-21737, 1991.

Gudmundsson, A., Formation and growth of normal faults at the divergent plate boundary in Iceland, *Tera Nova,* 4, 464-471, 1992.

Hassani R. and Chery J. Control of extensional tectonics by crustal rheology: numerical experiments, *J. Geophys. Res.,* (submitted, 1998).

Hobbs, B.E. and A. Ord., Numerical simulation of shear band formation in a frictional-dilatational materials. Ingenieur-Archiv, 59, 209-220, 1989.

Huang, P. Y., and S. C. Solomon, Centroid depth of mid-ocean ridge earthquakes: Dependence on spreading rate, *J. Geophys. Res., 93,* 13,445-13,477, 1988.

Jaeger, J.C. and N.G.W. Cook, *Fundamentals of Rock Mechanics, Third Edition,* Fletcher and Sons Ltd. , Norwich, Great Britian, pp. 593, 1979.

Kappel, E.S. and Ryan, W.B.F., Volcanic Episodicity and a Non-Steady State Rift Valley along Northeast Pacific Spreading Centers: Evidence from SeaMARC I, *J. Geophys. Res., 91,* 13,925-13,940, 1986.

Karson, J.A., G. Thompson, S.E. Humphris, J.M. Edmond, W.B. Bryan, J.R. Brown, A.T. Winters, R.A. Pockalny, J.F. Casey, A.C. Campbell, G. Klinkhammer, M. R. Palmer, R.J. Kinzler, and M.M. Sulanowska, Along-axis Variations in Seafloor Spreading in the MARK area, *Nature, 328,* 681-685, 1987.

Kirby, S.H. and A. K. Kronenberg, Rheology of the lithosphere: Selected topics, *Rev. Geophys., 25,* 1219-1244, 1983 .

Kong, L.S.L., R.S. Detrick, P.J. Fox, L.A. Mayer, W.B.F. Ryan, The Morphology and Tectonics of the Mark Area from Sea Beam and SeaMARC I Observations (Mid-Atlantic Ridge 23°N), *Marine Geophysical Researches, 10,* 59-90, 1988.

Kuo, B.Y. and D.W. Forsyth, Gravity anomalies of the ridge-transform system in the South Atlantic between 31 and 34.5°S: Upwelling centers and variations in crustal thickness, *Mar. Geophys. Res., 10,* 205-232, 1988.

Lachenbruch, A. H., A simple mechanical model for oceanic spreading centers, *J. Geophys. Res., 78,* 3395-3417, 1973.

Laughton, A. S. and R. C. Searle, Tectonic processes on slow spreading ridges, in Deep Drilling Results in the Atlantic Ocean: Ocean Crust, Maurice Ser., Vol. 2, edited by M. Talwani, C. G. Harrison and D. E. Hayes, pp. 15-32, AGU, Washington D.C., 1979.

Lavier, L., W.R. Buck, and A. Poliakov, Models of Low-angle Normal Faulting and Ductile Shear Zone Development During Continental Rifting, *Trans. Am. Geophys. Union. (EOS),* 79, p. 5336, 1998.

Lin J. and Parmentier E.M., A Finite Amplitude Necking Model of Rifting in Brittle Lithosphere, *J. Geophys. Res., 95,* 4909-4923, 1990.

Lin, J., and E.M. Parmentier, Mechanisms of lithosphere extension at mid-ocean ridges. *Geophys. J., 96*, 1-22, 1989

Lin, J., G.M. Purdy, H. Schouten, J.C. Sempere, and C. Zervas, Evidence from gravity data for focused magmatic accretion along the Mid-Atlantic Ridge. *Nature, 344*, 627-632, 1990.

Lister, J.R., and R.C. Kerr, Fluid-Mechanical Models of Crack Propagation and Their Application to Magma Transport in Dykes, *J. Geophys. Res., 96*, 10,049-10,077, 1991.

Macdonald, K.C., Mid-ocean ridges: Fine scale tectonic, volcanic and hydrothermal processes within the plate boundary zone, *Annu. Rev. Earth Planet. Sci., 10*, 155-190, 1982.

Macdonald, K.C. and Luyenduk B., Deep-Tow studies of the structure of the Mid-Atlantic Ridge near 37N (FAMOUS), *Geol. Soc. Am. Bull., 88*, 621-636, 1977.

Malinverno, A., Transition between a valley and a high at the axis of the mid-ocean ridges, *Geology, 21*, 639-642, 1993.

Malinverno, A. and R.A. Pockalny, Abyssal hill topography as an indicator of episodicity in crustal accretion, *Earth Planet. Sci. Lett., 99*, 154-169, 1990.

Morton, J. L and N. H. Sleep, A mid-ocean ridge thermal model: Constraints on the volume of axial hydrothermal flux, *J. Geophys. Res., 90*, 11,345-11,353, 1985.

Neumann, G. A., and D. W. Forsyth, The paradox of the axial profile: Isostatic compensation along the axis of the Mid-Atlantic ridge?, *J. Geophys. Res., 98*, 17,891-17,910, 1993.

Parmentier, E. M., Dynamic topography in rift zones: Implications for lithospheric heating, *Philos. Trans. R. Soc. London, Ser. A, 321*, 23-25 1987.

Perfit, M. and Chadwick, Constraints on magmatism from field and geochemical studies of volcanic rocks at mid-ocean ridges, AGU Monograph, Magmatism and Rifting at Mid-Ocean Ridges (in prep 1998).

Phipps Morgan, J., and Y. Chen, The genesis of oceanic crust: Magma injection, hydrothermal circulation, and crustal flow, *J. Geophys. Res., 98*, 6283-6298, 1993.

Phipps Morgan, J., E.M. Parmentier, and J. Lin, Mechanisms for the Origin of Mid-Ocean Ridge Topography: Implications for the thermal and mechanical structure of accreting plate boundaries, *J. Geophys. Res., 92*, 12823-12836, 1987.

Poliakov A.N.B., and H.J. Herrmann, Self-organized criticality in plastic shear bands, *Geophys. Res. Lett.*, 21,2143-2146, 1994.

Poliakov A.N.B., H.G. Hermann, Y. Podladchikov, and S. Roux, Fractal Plastic shear bands. *Fractals, 2*, 567-581, 1994.

Poliakov A.N.B., Y. Podladchikov, C. Talbot, Initiation of salt diapirs with frictional overburdens: numerical experiments, *Tectonophysics, 228*, 199-210, 1993.

Poliakov A.N.B., Podladchikov Y., Dawson E.C., Talbot, C., Salt diapirism with simultaneous brittle faulting and viscous flow, Salt Tectonics, *Geol.Soc.Spec.Publ. 100*, 291-302, 1996.

Price, N. J., and J. W. Cosgrove, Analysis of Geological Structures, Cambridge University Press, Cambridge., 502 pp., 1990.

Rudnicki, J.W., and Rice J.R., Conditions of the localization of the deformation in pressure-sensitive materials. *J.Mech.Phys. Solids, 23*, 371-394, 1975.

Rubin, A.M., and D.D. Pollard, Origins of Blake-Like Dikes in Volcanic Rift Zones, U.S. Geological Survey Professional Paper 1350, 1449-1470, 1987.

Shaw, W. J. and J. Lin, Model of ocean ridge lithospheric deformation: Dependence on crustal thickness, spreading rate, and segmentation, *J. Geophys. Res.*, (submitted) 1996.

Shaw P.R. and Lin J., Causes and consequences of variations in faulting style at the Mid-Atlantic Ridge, *J. Geophys. Res.*, 98, pp.21839-21851, 1993.

Shelton, G. and J. A. Tullis, Experimental flow laws for crustal rocks, *Trans. Amer. Geophys. Union*, 62, 396, 1981.

Sleep N.H., Sensitivity of heat flow and gravity to the mechanism of sea-floor spreading, *J. Geophys. Res.*, 74, 542-549, 1969.

Small, C., A global analysis of mid-ocean ridge axial topography, *Geophysical Journal Intl., 116*, 64-84, 1994.

Solomon, S. C., P.Y. Huang and L. Meinke, The seismic moment budget of slowly spreading ridges, *Nature, 334*, 58-60, 1988.

Tapponier, P., and J. Francheteau, Necking of the lithosphere and the mechanics of slowly accreting plate boundaries, *J. Geophys. Res., 83*, 3955-3970, 1978.

Thatcher, W. and D. P. Hill, A simple model for fault-generated morphology of slow spreading ridges, *J. Geophys. Res., 100*, 561-570, 1995.

Tolstoy, M. A., A. Harding, and J. Orcutt, Crustal thickness on the Mid-Atlantic Ridge: Bull's eye gravity anomalies and focused accretion, *Science, 262*, 726-729, 1993.

Toomey, D.R., S.C. Solomon, and G.M. Purdy, Microearthquakes beneath the median valley of the Mid-Atlantic Ridge near 23°N: Tomography and Tectonics, *J. Geophys. Res., 93*, 9093-9112, 1988.

Tryggvason, E., Widening of the Krafla Fissure Swarm During the 1975-1981 Volcano-tectonic Episode, *Bull. Volcanol., 47, 1*, 47-69, 1984.

Tucholke, B. E. , J. Lin , and M.C. Kleinrock, Megamullions and mullion structure defining oceanic metamorphic core complexes on the Mid-Atlantic Ridge, *J. Geophys. Res., 103*, 9857, 1998.

Vermeer, P.A. and R. de Borst, Non-Associated Plasticity for Soils, Concrete and Rocks, *Heron*, v. 29:1, 1984.

Vermeer P.A., The orientation of shear bands in biaxial tests, *Geotechnique* 40, 2: 223-232, 1990.

Vening-Meinesz, F.A., Les graben africains resultant de compression ou de tension dans la croute terrestre?, *Kol. Inst. Bull., 21*, 539-552, 1950.

Weissel J.K. and Karner G.D., Flexural uplift of rift flanks due to mechanical unloading of the lithosphere during extension, *J. Geophys. Res., 94*, 13919-13950, 1989.

Zuber. M. T., and E.M. Parmentier, Lithospheric necking: A dynamic model for rift morphology, *Earth Planet. Sci. Lett., 77*, 373-383, 1986.

A. Poliakov, Laboratoire de Geophysique et Tectonique, CNRS UMR 5573, Universite Montpellier II, Montpellier Cedex 05, France; email:aljosha@indri.dstu.univ-montp2.fr.

W.R. Buck, Lamont-Doherty Earth Observatory, Palisades, NY 10960, email: buck@ldeo.columbia.edu

Normal Fault Growth in Three-Dimensions in Continental and Oceanic Crust

Patience A. Cowie

Department of Geology and Geophysics, The University of Edinburgh, Edinburgh, U. K.

In the past, structural geology and mid-ocean ridge research have been divided disciplines; the former relying on field observations from the continents, the latter being based on remote sensing data from the deep-oceans. However, faulting is a ubiquitous deformation process with no regard for academic boundaries. The aim of this article is to review recent advances in our understanding of normal fault development in continental extensional provinces and to discuss how these ideas might be applied to fault formation at mid-ocean ridges. The main topic to be covered here is that of *fault scaling relationships*. These relationships describe the spatial clustering, the displacement patterns and the size-frequency statistics which characterize fault populations formed in the brittle crust. There are obvious practical applications in crustal deformation studies if fault population systematics can be characterized by specific scaling exponents. However, the aim of this paper is to focus on what these relationships may be indicating about the underlying physical mechanisms of crustal deformation. The scaling properties of continental fault data sets are compared and contrasted with those from Iceland and sections of the mid-ocean ridge system having different spreading rates. The implications for fault formation in a sea floor spreading setting are discussed, in particular the initiation and growth of faults near the axis, the distance at which faulting becomes inactive, and the influence of magmatic processes and oceanic crustal rheology on fault development.

INTRODUCTION

Large strains in the upper brittle part of the Earth's crust are primarily achieved by the accumulation of displacement on faults. Observations show that deformation is not accommodated by a single fault but by a population of faults and fractures that range in size from millimeters to 1000s of kilometers. Regardless of tectonic setting - extensional, compressional, or strike-slip - the faults which form during progressive crustal deformation exhibit key similarities. In particular, it is widely observed that larger (longer) faults in a population have more displacement on them than smaller (shorter) faults. This is consistent with fault growth, i.e. the length of a fault increasing as displacement accumulates. Furthermore, a systematic distribution of fault sizes occurs with many more small faults than large faults so that the deformation is accommodated on a wide range of scales. Finally, the faults tend to cluster so that the resulting strain distribution is heterogeneous with relatively undeformed blocks surrounded by zones of more intense deformation. Such systematic patterns of faulting have been widely observed even in areas of complex crustal composition and/or prolonged tectonic history. Thus fault pattern evolution has been interpreted as an example of self-organisation in natural systems. The aim of this paper is to demonstrate how fault population systematics may be related to the physics of fault growth and the mechanism(s) of self-organisation in both continental and mid-ocean ridge settings.

Fault scaling relationships are a statistical description of the size-frequency distributions of fault attributes (length, displacement, etc.), the correlation between displacement on a fault and its length, and the spatial pattern of faulting. Considerable emphasis has been placed on the practical applications of these relationships, such as estimating the relative numbers and the strain contribution of faults too small to be resolved by a particular imaging/mapping technique. Furthermore, the relationships have been combined to build more realistic structural models for the

Faulting and Magmatism at Mid-Ocean Ridges
Geophysical Monograph 106
Copyright 1998 by the American Geophysical Union

crust [e.g., *Gauthier and Lake*, 1993]. Such models provide only probabilistic predictions of the fault distribution in a given area, but they reflect the level of structural complexity which is important for addressing problems directly associated with faulting and fracturing, such as seismicity and rock permeability. The impetus for fault population analysis originated from the hydrocarbon industry, initially coal mining but more recently oil and gas production as well. Thus, a great deal of what we know about fault scaling is based on empirical relationships derived from normal fault systems in sedimentary rocks subject to moderate amounts of strain. However, increasing amounts of data from a wider range of tectonic settings has shown that systematic relationships are ubiquitous. This has led to theoretical modelling and analogue experiments aimed at trying to understand the physical mechanisms which control fault propagation and fault array evolution. A Special Publication of the Journal of Structural Geology [vol. 18, nos. 2/3, 1996] is devoted to recent work on fault and fracture scaling properties.

In 1976, *D. Elliott* published a figure showing the maximum displacement, *d,* on thrust faults as a function of their along-strike length, *L* (Figure 1). These faults occur in the foothills and front ranges of the Canadian Rockies (Figure 1a). The estimates of *d* and *L* were obtained by determining stratigraphic separation along thrusts varying in length from about 10 km to 500 km. *Elliott* [1976] found a strong correlation between *d* and *L,* described by the relationship $d = 0.07L$ (Figure 1b). To my knowledge, this is the first published example definitively showing a displacement-length scaling relationship for faults. A 1975 study by *MacMillan* argued for such a relationship but it was based on data derived from a literature search which included several oceanic transforms and plate boundary faults. A systematic relationship between *d* and *L* is only expected for "bounded faults", i.e., those faults which do not correspond to edges of tectonic plates. The San Andreas Fault in California and the Alpine Fault in New Zealand are examples of unbounded faults; the displacements on these structures are determined by global plate kinematics.

Ten years after Elliott's classic paper on thrust fault evolution, interest in displacement-length scaling was renewed by the work of *Watterson* [1986] and then *Walsh and Watterson* [1987, 1988]. These workers compiled a data set of *d* and *L* measurements for faults in a variety of different tectonic settings and rock types, including their own data from normal faults in British coalfields. In addition to highlighting the persistence of a *d-L* correlation over several orders of magnitude, Watterson and Walsh presented a fault growth model in which $d = PL^2$, where *P* is a function of rock properties. A non-linear correlation appeared to be consistent with fault length data available at that time in the scale range of kilometers to 100's of kilometers.

The non-linear *d-L* relationship proposed by Walsh and Watterson has been the source of disagreement ever since. One school of thought [*Cowie and Scholz* 1992*a,b*] maintains that fault data sets should be separated according to tectonic setting and rock type and concludes that each individual data set actually shows a linear relationship between *d* and *L,* as found, for example, by *Elliott* [1976]

(a)
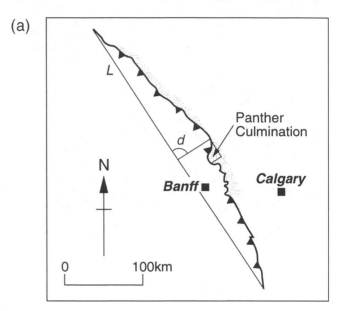

(b)
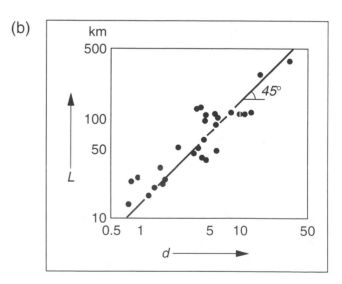

Figure 1. (a) Elliott's [1976] "bow-and-arrow" model for thrust faults in the Canadian Rockies. The displacement varies from zero at the fault tips to a maximum near the centre and is reflected by the curvature of the thrust in map view. (b) Displacement-length relationship for thrusts derived by Elliott [1976]. Line fit given by $d = 0.07L$. Reprinted from Elliott [1976] with permission from The Royal Society of London.

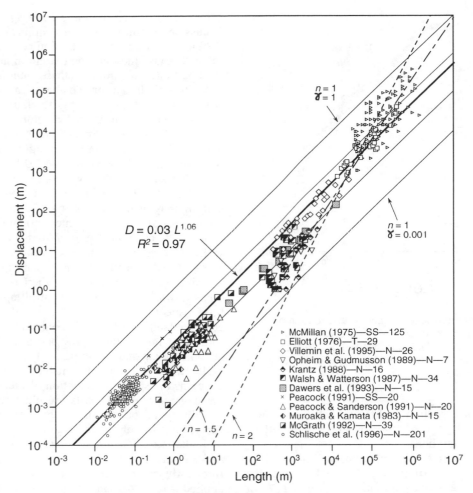

Figure 2. "Global" displacement-length scaling relationship for faults from a variety of tectonic settings and rock types compiled by *Schlische et al.* [1996]. Over nearly nine orders of magnitude the ratio of displacement to length is in the range 1×10^{-3} to 1×10^{-1} with an average ratio $\gamma = 0.01$. Two other lines are shown for $n = 1.5$ and $n = 2$, which are the non-linear correlation for faults in the large scale range proposed by *Walsh and Watterson* [1987], *Marrett and Allmendinger* [1991], *Gillespie et al.* [1992]. See text for discussion. Reprinted from Schlische et al. [1996] with permission from The Geological Society of America.

and more recently demonstrated by *Dawers et al.* [1993] [see also *Clark and Cox*, 1996]. The opposing school of thought [e.g., *Marrett and Allmendinger*, 1991; *Gillespie et al.*, 1992] argues that such restricted data sets, when taken alone, can not provide a reliable correlation and that data sets must be combined to provide meaningful statistics. The combined data sets studied by *Gillespie et al.* [1992] and *Marrett and Allmendinger* [1991] yielded a non-linear relationship of the form $d = PL^{1.5}$. However, this relationship, when extrapolated to smaller scales predicts that faults with lengths in the range 1 m to 10 m should have displacements of less than a millimeter, whereas several workers have found that this is not in fact the case [e.g., *Muraoka and Kamata*, 1983; *Peacock and Sanderson*, 1991; *Schlische et al.*, 1996]. The most recent

compilation of a global fault data set, which includes smaller scale faults, indicates that a linear relationship between d and L is a good approximation over more than eight orders of magnitude [*Schlische et al.*, 1996] (Figure 2). Although there is significant variability in the d/L ratio, *Schlische et al.* [1996], find that the average value is 0.01, i.e., the maximum displacement on a fault is on the order of about 1% of its length (see Figure 2). In spite of this work there is still some disagreement over linear versus non-linear scaling for some fault data sets simply because of the ambiguities and limitations inherent in these types of data, especially the problem of defining fault length in areas that are geologically complex. The disagreement has fueled several detailed fault studies, with the result that the underlying mechanisms of fault growth and the

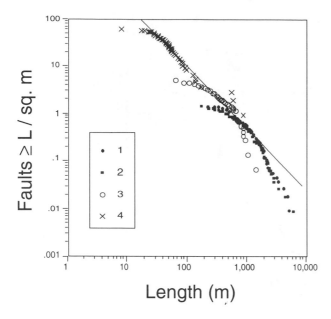

Figure 3. (a) Cumulative size-frequency distribution of fault lengths in Volcanic Tablelands ares of Owen's Valley, California, USA [*Scholz et al.*, 1993]. The four data sets are from the same geographic area but obtained from field mapping and aerial photo analysis at different scales.

consequences for *d-L* scaling in different settings are now becoming clearer. It is this latter aspect which will be discussed in more detail below.

The second of the fault scaling relationships which I discuss here is the size-frequency distribution, i.e., the number of faults in a given population with length, *L*, or displacement, *d*. In continental regions it has been found that the distribution, in most cases, is best described by a power law relationship, e.g., $N_{\geq L} \propto L^{-c}$ (Figure 3) [*Kakimi*, 1980; *Villemin and Sunwoo*, 1987; *Childs et al.*, 1990; *Scholz et al.*, 1993]. The advent of fractal concepts in earth science research, led to the proposal that fault populations may be scale invariant and thus that the size-frequency distribution defined at one scale may be extrapolated over several orders of magnitude [e.g., *Main et al.*, 1990; *Turcotte*, 1992; *Yielding et al.*, 1992]. The assumption of scale invariance permitted the estimation of relative numbers and strain contribution of faults below the limit of resolution [*Scholz and Cowie*, 1990; *Walsh et al.*, 1991, *Marrett and Allmendinger*, 1992]. The size of the exponent *c* determines the relative contributions of faults of different sizes to the total fault strain. There is a debate as to whether *c* is a universal exponent [e.g. *Sornette and Davy*, 1991], or whether it varies in some systematic way and if so over what range of values. Part of the reason for this uncertainty is, once again, due to the difficulty of fully characterizing a fault population over a large enough range of scales without introducing sampling artifacts. (See *Marrett and*

Allmendinger [1991] and *Pickering et al.* [1995] for discussion of fault population sampling). *Scholz and Cowie* [1990] argued that *c* must be limited to a certain range of values because the total fault strain in a deforming region must be finite. According to these authors, for a population of faults with dimensions smaller than the thickness of the brittle crust *c* < 3.0, whereas for faults greater than this thickness, i.e., 2-dimensional structures, *c* < 2.0. If *c* << 3.0 (or *c* << 2.0 for the 2-dimensional case) then the smallest scale faults do not contribute significantly to the total strain, whereas if *c* ≈ 3.0 (or *c* ≈ 2.0 for the 2-dimensional case) then small faults may account for as much as 40-50% [*Walsh et al.*, 1991]. While the contribution of small-scale faulting may or may not be significant, the assumption of scale invariance allows some estimate of the "missing" strain to be calculated [e.g., *Marrett and Allmendinger*, 1992]. Consequently, there is considerable emphasis now being placed on trying to understand the origin of power law size-scaling, the instances where scale invariance may break down, and the factors controlling the value of *c*. In this respect, it has proved particularly instructive to compare fault populations in continental regions with those formed at mid-ocean ridges because of the contrasts in tectonic setting and rock type.

The third scaling relationship which I will discuss concerns the spatial distribution of faults in a region, i.e., the degree to which faults are clustered and the lateral variability of brittle strain. Fractal analysis has been applied extensively to characterize the spatial distribution of faults [e.g., *Hirata*, 1989; *Davy et al.*, 1990; *Vignes-Adler et al.*, 1991; *A. Sornette et al.*, 1990, 1993]. Most frequently a technique called box-counting analysis is applied (Figure 4). This method involves superimposing a grid over a fault map, and calculating how many of the grid elements or boxes are occupied by faults (Figure 4a) (see *Walsh and Watterson* [1993] and *Yielding et al.* [1996] for more details). By repeating the calculation for different grid sizes it has been shown that, for faults in continental regions at least, the number of occupied boxes, *N*, is given by the relationship $N = r^D$ where *r* is the size of the box and *D* is the fractal dimension (Figure 4b). If *D* = 2.0 then the pattern of faulting is homogeneous over the map area, whereas *D* < 2.0 means that the deformation is more localized. For example, if the faults are concentrated along linear zones then *D* ≈ 1.0, whereas diffuse patches of faulting would give *D* < 1.0. A similar approach can be used for a 3-dimensional fault distribution, using cubes of size *r*, or for faults mapped along transects or cross sections, using linear intervals of length *r* [*Turcotte*, 1992; *Korvin*, 1992; *Gillespie et al.*, 1993; *Yielding et al.*, 1996].

Another fault scaling relationship which has been discussed extensively in the literature is the observed correlation between the displacement on a fault and the thickness of the gouge zone developed along the fault plane (see *Scholz* [1990, chapter 3] for a summary of this topic).

While this observation is relevant to the issue of fault development it will not be discussed in this paper.

INTERPRETATION OF FAULT SCALING RELATIONSHIPS

Various deformation processes and regional tectonic constraints have been shown to play an important role in fault pattern evolution. These include the propagation of fault tips through rock with varying mechanical properties, interaction and linkage between neighboring faults, kinematic constraints imposed by fault geometry, the rheology and mechanical heterogeneity of the lithosphere, and the boundary conditions causing the deformation. Each influences fault development in a different way. In this section I review the concepts and models presented in the literature to explain the impact of these processes on fault scaling relationships. Methods for identifying the importance of different mechanisms, via fault population analysis, will also be discussed.

Growth of Individual Faults

There are two mechanisms by which a fault can grow in size: by propagation of the fault tips through intact rock, or by linkage with adjacent structures. Both mechanisms are likely to be operating in any given tectonic setting although one may dominate over the other in certain situations. Tip propagation will be affected by rock properties in the vicinity of the fault, whereas linkage will be a function of the preferred growth direction, local fault density and orientation. In this section I will focus mainly on lateral growth of faults and introduce the basic concepts. Down-dip growth and 3-dimensional aspects of fault kinematics will then be discussed.

The first models which aimed to explain the correlation between maximum displacement, d, and length, L, focused on tip propagation by considering isolated single faults and the rheological properties of rock in the tip region. For example, *Elliott* [1976] introduced the idea of a "ductile bead" which is a zone of folding and fracturing formed ahead of a propagating thrust fault. According to *Elliott* [1976], lengthening of the fault is achieved by progressive intensification of the deformation in the bead until a new shear surface is formed. The first time rock properties were explicitly related to the d/L ratio was in the geometrical growth model proposed by *Watterson* [1986] and developed by *Walsh and Watterson* [1988]. These workers assume that the total fault displacement is the sum of slip events produced by earthquakes which rupture the whole fault. In this case, the slip increment each time the fault ruptures is given by $u = 16R\Delta\sigma/(7\pi\mu)$, where $\Delta\sigma$ is earthquake stress drop, μ is the elastic shear modulus of the rock in which the fault grows, and R is the fault radius [see *Kanamori and Anderson*, 1975]. Thus the parameter P in

the expression for growth, $d = PL^2$ [*Walsh and Watterson*, 1988], is given by $8a(7\pi\mu/(16\Delta\sigma))^2$, where a is the size of successive slip increments which is assumed constant. The key feature of Walsh and Watterson's [1988] model is that

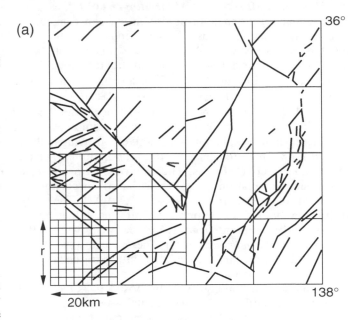

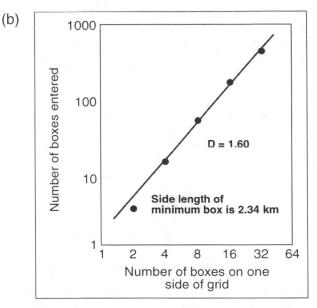

Figure 4. Example of fractal analysis of the spatial distribution of faults in Japan using the box-counting method [*Hirata*, 1989]. (a) Shows the system of successively smaller grid cells used in the analysis; (b) indicates how the fractal dimension, D, of the distribution is determined from the slope of the line relating the number of boxes, N, occupied by faults to the size of the boxes, r. Reprinted from Hirata [1989] with permission from Birkhauser Verlag AG.

incremental tip propagation does not depend on the stress concentration at the fault tip. In fact, they assume that stresses at the tip are not sustained between slip events. Modified versions of this model, described by *Gillespie et al.* [1992] and *Marrett and Allmendinger* [1992], assume instead that tip propagation is driven by, and completely relaxes, the instantaneous stresses imposed by only the most recent slip event. According to this model, fault growth in a material with constant fracture toughness follows the relationship $d = PL^{1.5}$. In summary then, a non-linear relationship between d and L (i.e., $d = PL^n$, $n > 1$) is predicted if complete stress dissipation occurs in the tip region between, or during, fault propagation episodes. Such a model does not explain how an actively growing fault is loaded between earthquakes if stress dissipation is such a rapid and effective process. The empirical evidence for a non-linear correlation ($n > 1$) was based on combining different data sets [*Gillespie et al.,* 1992] The only individual data set which exhibits a value for n significantly different from 1.0 actually shows $n = 0.5$ [*Fossen and Hesthammer,* 1997]. These data are specific to small-scale shear bands associated with pore collapse in high-porosity sandstones.

The model for fault growth proposed by *Cowie and Scholz* [1992b] differs in a fundamental way from the those just described in that the stresses at the fault tip depend on the total accumulated displacement along the fault not just the last slip event. *Cowie and Scholz* [1992b] proposed a model to explain the *d-L* correlation based on the theory of post-yield fracture mechanics. The fault plane is represented by a discrete crack embedded in a material with finite elastic strength, such that inelastic yielding occurs in the tip region. The yield zone is related to a "process zone" around the perimeter of the fault plane, similar to the ductile bead of *Elliott* [1976], and the peak stress at the tip is constrained to be finite. According to their model, an actively growing fault maintains a critical displacement profile such that the stress level at the tip is always just equal to the shear strength of the surrounding rock. The predicted relationship between displacement and length for a fault growing in rock with uniform mechanical properties is given by $d = \gamma L$. The parameter γ depends mainly on the ratio of two material properties: the shear strength of the surrounding rock and the elastic rigidity. Thus fault populations in different tectonic settings and rock types are predicted to exhibit different *d/L* ratios. The theory developed by *Cowie and Scholz* [1992b] may be applied to all types of brittle faults - normal, strike-slip, and reverse - because the model does not depend on the orientation of principal stresses nor the details of the deformation processes occurring at the fault tip. The work of *Dawers et al.* [1993] set out to test the predictions of the *Cowie and Scholz* [1992b] model by analysing a population of faults formed in one tectonic episode in a mechanically homogeneous rock type. These workers found a strong

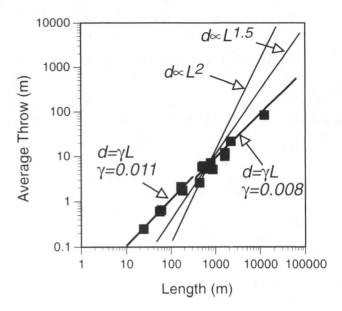

Figure 5. Displacement-length correlation obtained by *Dawers et al.* [1993] for isolated single faults in the Volcanic Tableland in Owen's Valley, California, USA. The data are fit by a line $d = \gamma L$. The shift in the data at 300-400m corresponds to a change in γ from 0.011 to 0.008; the lower value relates to faults with lengths greater than the thickness of the brittle layer, which is the Bishop tuff in this case. The other lines indicate the non-linear fits which have been proposed by other authors (see text for discussion). Reprinted from Dawers et al. [1993] with permission from The Geological Society of America.

linear relationship between the average displacement and the length of isolated single segment faults over three orders of magnitude with $\gamma \sim 0.01$ (Figure 5).

Growth of Simple Fault Arrays

So far the discussion has focused on tip propagation of single isolated faults and has shown how the *d/L* ratio may be related to rock properties. In natural fault populations, linked fault arrays are very common and plots of *d* versus *L* typically show a large degree of scatter about a mean value. Some of this variability may be explained in terms of data collection methods and local variations in rock properties [*Walsh and Watterson,* 1987; *Peacock and Sanderson,* 1991; *Cowie and Scholz,* 1992a; *Burgmann et al.,* 1994]. However, it has been proposed that fault linkage plays an important role in fault growth and leads naturally to large deviations from the mean *d/L* ratio. *Cartwright et al.* [1995] analyzed normal fault arrays in the Canyonlands area of east-central Utah and found that lengthening of faults occurs predominantly by linkage. These authors suggest that linking of two faults is followed by a period of displacement accumulation until the critical displacement

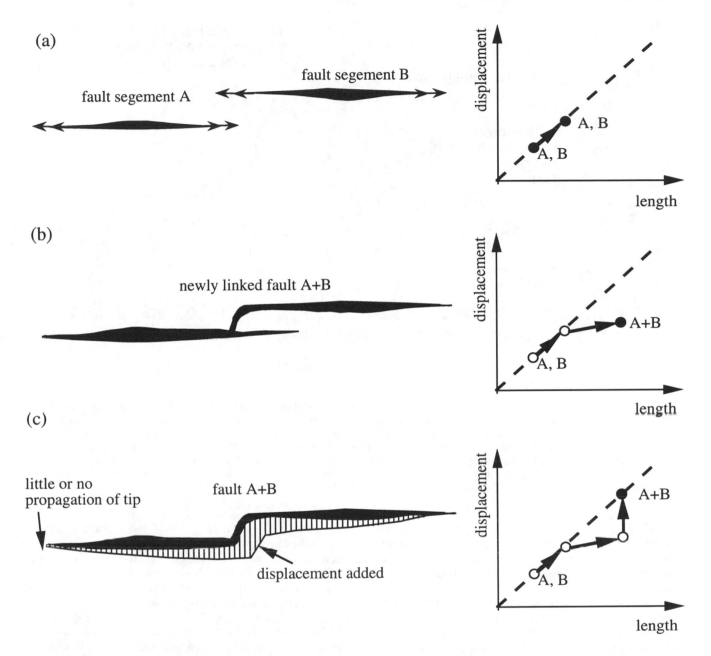

Figure 6. Model for fault growth proposed by *Cartwright et al.* [1995]. (a) Bilateral tip propagation of a single fault and growth according to the ideal displacement-length relationship (dashed line, right-hand panel). (b) Deviation from a simple displacement-length correlation when linkage occurs. The linked structure is initially "under-displaced" and will accumulate displacement until the critical profile for tip propagation is re-established (c). Reprinted from *Cartwright et al.* [1995]. with permission from Elsevier Science Ltd.

profile is re-established and tip propagation resumed (Figure 6). According to *Cartwright et al.* [1995], faults that plot below the growth curve for tip propagation must have linked relatively recently and are "under-displaced" (see Figure 6b). However, in a parallel study, *Dawers and Anders* [1995] found that the overall *d/L* ratio for the fault

array they mapped is the same as that for a single segment fault (Figure 7 and 8). Although one could argue that the array mapped by *Dawers and Anders* [1995] is in the early stages of linkage, the overall profile shape is not, in fact, significantly "under-displaced" relative to the single segment faults (Figure 8). One mechanism that reconciles

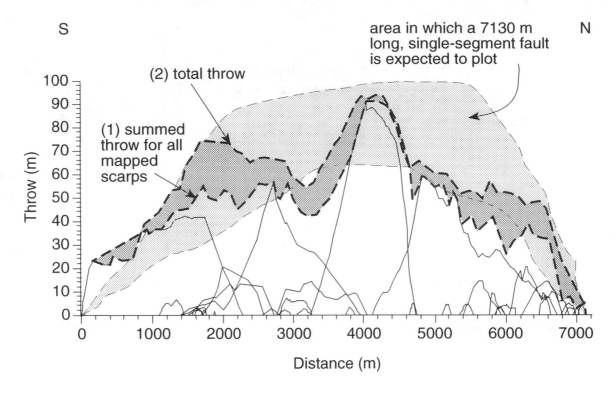

Figure 7. Fault throw profile mapped by *Dawers and Anders* [1995] along the length of a linked fault array in the same area as that of Figure 5. Four main segments are clear from the large throws (>40 m). Also shown in (1) a summed profile, constructed by summing the throws measured on all of the mapped scarps; and (2) a total throw profile, which includes the elevation change related to tilting between fault segments. The stippled region represents the area in which we expect a throw profile for a single-segment fault of the same length (7130 m) to plot, taking into account the natural variability in profile shape [see Figure 5; *Dawers et al.*, 1993]. Note that the total throw profile (2) is more irregular than that for the single-segment faults. Reprinted from *Dawers and Anders* [1995] with permission from Elsevier Science Ltd.

the study of *Dawers and Anders* [1995] with *Cartwright et al.*'s [1995] model is elastic interaction between growing faults which modifies their displacement profiles even before linkage has been established [*Willemse*, 1997; *Cowie*, 1998].

The displacement profiles of linked arrays show some interesting features in detail (Figure 7). For example, the individual segments of the fault array typically show a systematic change in *d/L* ratio with large values near the center of the array and decreasing values towards either end (Figure 7). Secondly, the overall profile of the linked system is more irregular than that for single segment faults. Displacement minima and subsidiary faulting usually occur in the zones of segment linkage [*Walsh and Watterson* [1991].

From the discussions above it is clear that detailed mapping of fault displacement profiles can yield information about the rheological properties of the crust. For example, lateral variations in fault displacement gradients may be interpreted in terms of local changes in rock properties through which the fault is growing. Such changes can often be related directly to compositional

variations [*Burgmann et al.*, 1994], but interaction between faults in close proximity results in regions of enhanced compression or dilation which may, respectively, increase or lower the effective shear strength of rock in the tip region. *Peacock and Sanderson* [1991] have documented systematic variations in displacement gradients in regions of fault interaction, and the effects have been modelled by *Willemse et al.* [1996] and *Willemse* [1997]. Displacement profiles may be used as an indicator of how linkage has evolved and, moreover, which faults are behaving as kinematically coherent arrays in areas where linkage zones are not well resolved [*Walsh and Watterson*, 1991]. The relative importance of linkage versus tip propagation also may be evaluated by analysing the variability in *d/L* ratios for a given fault population. For example, a large scatter in *d/L* ratio with faults exhibiting irregular displacement profiles may be indicative of a linkage-dominated environment, whereas simpler profiles and a tighter *d/L* distribution may indicate areas where linkage is inhibited. Identifying changes in scaling behavior is an important part of this type of detailed structural analysis. For example, the shape of fault displacement profiles has been shown to

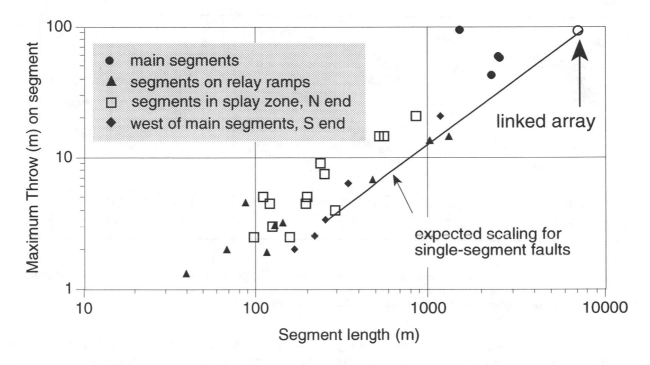

Figure 8. Maximum throw versus segment length for the data shown Figure 7. The line represents the *d* versus *L* scaling shown in Figure 5. Note that the data point for the entire linked array falls along this line, whereas the data for fault segments plotted individually generally lie above the line. Reprinted from *Dawers and Anders* [1995] with permission from Elsevier Science Ltd.

change when the fault plane dimensions exceed the thickness of the deforming brittle layer [*Dawers et al.*, 1993; *Shaw and Lin*, 1996].

Growth of Fault Populations

For scaling relationships to be established and preserved, fault development at one scale can not be independent of faults and fractures forming at smaller or larger scales of deformation. In other words, fault scaling relationships imply that faults of all sizes are "genetically related" and thus organized in some way. One of the first models to describe this *organization* of fault populations was a fractal hierarchy structure presented by *King* [1983]. According to this model, relative movement on faults intersecting at a triple point will generate sets of smaller scale faults to accommodate strain around the intersection (Figure 9a). Regions of overlap between adjacent fault segments are also sites of subsidiary faulting and fracturing associated with the linkage process (Figure 9a) [*Cartwright et al.*, 1996]. A tip process zone is another example of a genetic association between faults and fractures on different scales. In this case an aureole of brittle deformation forms ahead of a propagating fault tip and is integral to the growth process [*Anders and Wiltschko*, 1994; *Scholz et al.*, 1993] (Figure 9b). The process zone has been defined in terms of

microcracking at the grainscale [*Vermilye*, 1996]. However, field observations show that, at the outcrop scale minor faults and fractures develop on either side of a fault surface forming damage zones, in some cases out to distances of 10s to 100s of meters [*Knott et al.*, 1996]. This type of damage may be a relict of fault linkage that is further intensified by frictional slip once the fault plane is established. Complex cross-cutting relationships indicate that the damage evolves as displacement accumulates [*Little*, 1996]. Figure 9c shows a third common example, that of hanging wall deformation above a non-planar normal fault. The minor faults which form in this case are usually antithetic to the master fault and they develop gradually while the master fault accumulates displacement [*White et al.*, 1986].

In contrast to these processes for producing additional small-scale faults, we have already discussed how larger structures form by linking smaller faults together. *Wu and Bruhn* [1994] related these two end-member processes via a model for normal fault development in which small faults form preferentially near the tips of a larger structure, due to high stresses in this region, and growth occurs by hierarchical linkage. On a larger scale, the evolution of an entire fault population depends on the relative rates of nucleating new faults, growing existing faults and coalescing structures on many different scales. The

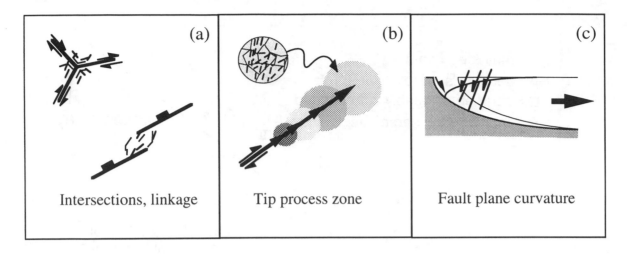

Figure 9. Geometrical associations between faults on different scales (a) Deformation around fault intersections and zones of linkage. (b) Predicted pattern of microfractures formed in the process zone of a propagating fault tip. (c) Small faults formed in the hanging wall of a normal fault, antithetic to the main structure.

combination of these processes will control the resulting size-frequency statistics and the spatial clustering characteristics.

A general framework for understanding the origin and evolution of fractal fault patterns was proposed by *A. Sornette et al.* [1990] and developed theoretically by *D. Sornette et al.* [1990]. These workers proposed that fractal fault patterns form in the upper crust as a consequence of the threshold nature of brittle failure and the redistribution of the resulting strain perturbations according to the equations of elasticity. Because the elastic strain field obeys Laplace's equation, stress perturbations decay algebraically with distance producing interactions, and thus correlation, at long range. At the same time, strong stress enhancements occur at fault tips and intersections. A combination of screening and enhancement effects thus arise spontaneously over a range of scales as a fault population evolves. The effect is to concentrate the deformation in some areas while leaving intervening regions undeformed. In other words the deformation is *self-organizing*.

Davy et al. [1990] and *A. Sornette et al.* [1990; 1993] showed experimentally how a fractal fault pattern evolves with increasing amounts of deformation using a case study of strike-slip faults formed during the collision between India and Asia. A number of numerical models and other analogue experiments have now been developed to simulate the evolution of fault patterns in space and time for a range of boundary conditions [e.g., *Cowie et al., 1993a, Sornette et al.,* 1994; *Poliakov et al.,* 1994; *Davy et al.,* 1995; *Ishikawa and Otsuki,* 1995; *Fossen and Gabrielsen, 1996; Ackermann et al., 1997*]. The motivation for these studies is that the interplay between different

deformation processes and the temporal sequence of events can be observed in detail, in contrast to a single snap-shot of the deformation as preserved in natural examples. The key feature which distinguishes this body of work is the spontaneous evolution of many faults at a distance from applied boundary conditions, unlike more traditional elastic dislocation or crack models which provide static solutions for one or a few pre-defined faults. Recent results from rock mechanics experiments are also relevant to this area of research through studies of damage accumulation and shear localization in brittle rocks [e.g., *Moore and Lockner,* 1995; *Lockner et al.,* 1991].

The origin of self-organisation in fault pattern evolution can be traced back to the heterogeneity as well as the overall rheology of the lithosphere. In all the models and experiments which exhibit self-organizing behavior, the upper part of the lithosphere is represented by a heterogeneous brittle layer which can support elastic stresses below a critical threshold. Once the failure threshold is reached rupture results in an abrupt stress drop on the fault and a perturbed stress field in the surrounding crust. Lower layers in the lithosphere, represented by ductile rheologies, play a modifying role in the evolution of faulting by dissipating or redistributing stresses [e.g., *Davy et al.,* 1995; *Heimpel and Olsen,* 1996]. The presence of heterogeneity in the brittle layer has been shown to be a key element in fault development. In some cases the heterogeneity is proscribed [e.g., *Cowie et al.,* 1993a; *Sornette et al.,* 1994], while in other models it enters via numerical "noise" [*Poliakov et al.,* 1994] or, in the case of analogue models, via microscopic variations in the sand grains used to represent the brittle crust [see *A. Sornette et al.,* 1990]. Grain-scale heterogeneities also play a role in fracture pattern development in rock mechanics experiments [*Moore and Lockner*, 1995].

A heterogeneous crustal layer means that, when deformation begins, faults and fractures nucleate at any points of low failure strength and strain accumulation is accommodated initially by an increase in the number of faults. Continued nucleation and propagation eventually leads to a situation in which the stress fields of individual structures begin to interfere. Growth and coalescence of existing faults then gradually dominates over nucleation of new faults. This is the regime in which power law size-scaling and a fractal spatial pattern develop in these models [*Cowie et al.*, 1995]. As larger faults gradually form, many faults which formed earlier are shielded by these structures and become inactive. Eventually a stage is reached in which additional strain is accommodated by repeated slip on a few major faults or fault zones and the intervening areas are tectonically quiescent. A fractal fault structure is achieved after only a few percent strain.

An example of a simulated fault pattern using a thin-plate model for lithospheric deformation is shown in Figure 10 [*Cowie et al., 1993a*]. The model consists of a lattice of elements which break when their strength is exceeded. Stress is redistributed onto neighboring elements according to the equations of elasticity, each time a rupture occurs. The elements reheal after rupture and can therefore rerupture many times, each time accumulating more displacement. A constant velocity of 10 cm/yr is applied along one edge of the lattice ($y = 0$) which imposes a uniform anti-plane shear strain rate across the lattice. The elements thus fail as shear fractures oriented orthogonal to the plane of the model, qualitatively comparable to steeply dipping normal faults viewed from above. In Figure 10a the broken elements are shown in shades of grey indicating the amount of accumulated displacement from zero (white areas) to the maximum displacement (black). Note that larger displacements occur on longer faults, consistent with the *d-L* correlation discussed above. Figure 10b shows the evolution of fault activity through time for the same simulation by plotting the y co-ordinate of each rupture event. Note the gradual transition from uncorrelated ruptures at times $t < 20000$ years, the appearance first of rupture localization in time (earthquakes), and then in space (faults). The localization in time occurs first, indicated by neighboring elements breaking at the same instant. This is followed by concentration of the rupture activity along zones while adjacent areas become inactive.

Cowie et al. [1993] demonstrated that power law fault scaling relationships develop even when the undeformed crustal structure is randomly heterogeneous. Thus, although pre-existing geological structure may influence the details of fault development, fractal fault patterns are an example of self-organisation rather than inherited organization. The combination of material heterogeneity and elastic interactions at both short-range and long-range are crucial

in producing a wide range of characteristics documented by field observations, for example, the wide spacing of major faults, variations in fault clustering and fault zone complexity, and also intermittent rupture activity on different structures [see also *Sornette et al.*, 1994].

Quantitative analysis has shown how the scaling properties of a fractal fault pattern evolve with increasing strain [e.g., *A. Sornette et al.*, 1993]. The results for analyzes in 2-dimensions show a systematic decrease in the value of exponent c, which defines the size-frequency distribution, from approximately 2.0 to a value close to 1.0 once a major fault has formed. The fractal dimensions which characterize the spatial pattern of faulting follow a similar systematic trend indicating that the strain becomes strongly localized into zones where the faulting is most intense and the displacements are largest [*Cowie et al.*, 1995]. The decrease in c is reflecting the development of large faults at the expense of small faults formed during the initial phases of the deformation. The fault strain becomes more strongly convergent as the total amount of deformation increases [*Scholz and Cowie*, 1990; *Cowie et al.*, 1995]. In other words, the largest faults gradually account for a greater percentage of the strain and the smaller faults in the population become less significant.

Sornette and Davy [1991] argued theoretically that $c = 1.0$ is the "attracting" value for a fault population in a self-organized critical state in which deformation proceeds only by fault growth and not by any new fault nucleation. They ignored the role of coalescence and assumed that tip growth is controlled by conservation of energy during crack propagation in an elastic-brittle crust. *Cladouhos and Marrett* [1996] reached a different conclusion using a mathematical model to explore explicitly the competing effects of growth by tip propagation versus growth by linkage. *Cladouhos and Marrett* [1996] found that, if the deformation occurs at a constant strain rate (i.e., constant moment release rate), power law size-scaling can only develop and be maintained if fault linkage plays an important role in the growth process. Furthermore, they found that, when linkage is permitted, the slope of the size-frequency distribution decreases systematically as strain accumulates, similar to that observed in the numerical and analogue models described above.

Scholz and Cowie [1990] estimated a value for $c \sim 1.0$ using data from faults in Japan and found that faults smaller than the thickness of the brittle crust contribute less than 10% to the total fault strain budget. This Japanese data set consists primarily of major (crustal-scale) strike-slip faults. In contrast, *Walsh et al.* [1991] and *Marrett and Allmendinger* [1992] analyzed faults in extensional basins in the North Sea and most of these faults have dimensions much less than the thickness of the crust. For these smaller-scale, less developed, faults they obtained $c \sim 1.5$ (in 2 dimensions) so that as much as 40-50% of the total strain in this case may be accounted for by faults with

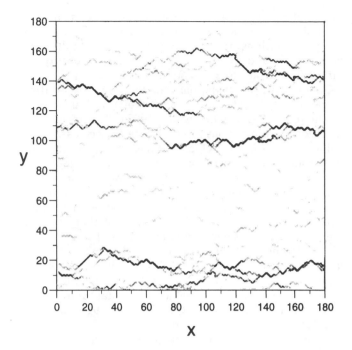

displacements of less than 100 m. These two examples are consistent with model predictions that c decreases as total strain increases. However, a systematic study of many data sets by *Cladouhos and Marrett* [1996] found that the correlation is actually difficult to define.

The main limitation of the existing theoretical explanations for power law scaling concerns the question of fault nucleation as deformation proceeds. Thus far propagation and linkage of existing faults are considered to be the most important mechanisms operating after an initial

Figure 10. (a) Fault patterns produced using the lattice rupture model described in *Cowie et al.* [1993a]. The fault traces are shown in shades of grey indicating the amount of accumulated displacement. (b) Rupture activity as a function of model time for the fault pattern shown in (a). In (b) each dot represents the rupture, or re-rupture, of a lattice element at any x-position in the lattice, projected onto the y-axis. Simultaneous rupture of adjacent lattice elements plots as a vertical line of ruptures. Note that localisation in time is then followed by localisation into different zones of the lattice.

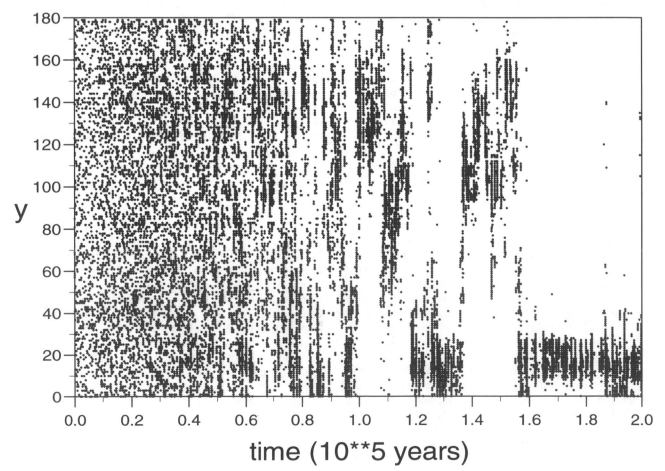

transient phase of nucleation. For a small amount of total strain (i.e., a few percent) this is probably a reasonable assumption. However, Figures 9a and 9c show examples of minor faults forming preferentially around, and subsequent to, larger structures in order to accommodate local strain variations. Small faults and fractures are continuously forming and developing in these regions as the larger structures accumulate displacement. Furthermore, large tectonic rotations may inhibit movement on existing faults so that an entirely new set of faults must develop to accommodate the strain [Scotti et al., 1991; see also Jackson and McKenzie, 1983]. These factors have not yet been explicitly considered in fault population modelling.

Fault Plane Geometry and Kinematics

The discussion so far has focused on fault growth in 2-dimensions. Information about the 3-dimensional geometry of faults has been derived from coal mine plans, seismic reflection surveys, earthquake focal mechanisms and aftershock distributions. Gravity modelling and geodetic inversions are also used to infer the dip and depth extent of normal faults [Stein et al., 1988; Escartin and Lin, 1995]. Low-angle extensional detachment faults and their interpretation are a special topic in their own right, particularly in the context of mid-ocean ridges as a consequence of the observations documented by Cann et al. [1997]. It is beyond the scope of this article to discuss the topic in detail. Instead the reader is referred to another paper in this volume by J. Karson.

Using detailed mine plans, Rippon [1985] showed that normal faults in sedimentary rocks have an elliptical tip line. The tip line defines the perimeter of the fault plane in three-dimensions where the displacement dies to zero. The displacement contours on the fault plane also have an elliptical shape with the maximum displacement usually near the center of the fault plane [Walsh and Watterson, 1987]. Nicol et al. [1996] have subsequently synthesized a wide variety of faults in order to assess the factors controlling the ellipticity. These authors found that in the majority of cases fault planes are elongated along a sub-horizontal axis with an aspect ratio greater than 2:1. Their data indicate that shape is independent of the slip vector orientation and fault dip, whereas rheological layering and interaction with neighboring faults play a significant role in controlling tip line shape. According to Nicol et al. [1996], the differences between Mode II (edge dislocation) and Mode III (screw dislocation) fracture energies seems to have less impact than bulk rheology on determining fault plane shape. Temporal evolution of fault plane shape and displacement distributions in three-dimensions have been described by Childs et al. [1995], using a technique called structural back-stripping applied to high-resolution seismic images of normal faults formed in sedimentary strata.

The geometrical evolution through time of normal fault systems has been studied in great detail via analogue experiments. Layered rheologies are used and the focus is usually on the development of faults in cross-section. For example, Fossen and Gabrielsen [1996] show fault sets forming in experiments using plaster overlying a layer of barite. Large degrees of extension are simulated in these experiments (>10% up to 50% usually) and thus rotation and internal deformation of crustal blocks is significant. Complex conjugate fault sets, with different orientations and different generations of faults, are typically observed. So far the scaling properties of fault orientations has received little attention compared to the other scaling properties discussed here. Ouillon et al. [1996] present a wavelet technique for analysing structural trends in fault patterns at different scales, but scaling exponents are not derived.

We have already discussed how bends in fault traces can lead to the formation of secondary structures (Figure 9a, c). The presence of antithetic structures in the hanging wall of a normal fault may be used to infer down-dip curvature of the master fault when direct observations are not available [White et al., 1986]. Jackson and McKenzie [1983] show how listric faults form due to the brittle-ductile layering of the crust. In this case a synthetic-antithetic fault pair develop and form a graben, where the width of the graben depends on the depth to the ductile layer or basal decollement. This geometrical relationship has been used, by Plescia [1991] for example, to interpret fault patterns and rheological layering of Martian lithosphere [see also lunar studies by Golombek and McGill, 1983].

Depth of Fault Nucleation

The depth at which faults nucleate and the direction in which they propagate are also relevant to our understanding of how deformation progresses. Lin and Parmentier [1988] presented a fracture mechanics model for a normal fault in which the threshold for propagation and instantaneous tip extension is determined by the shear stress intensity factor, K_{II}. They find that faulting initiates at the free surface and slip on the fault migrates down-dip. Down-dip propagation occurs and a listric fault plane is produced in their models if K_{II} is small compared to the tensile stress intensity factor, K_I. One of the underlying assumptions of the *Lin and Parmentier* [1988] model is that the pre-deformation stress regime in the crust is lithostatic at all depths, i.e., $\sigma_{xx} = \sigma_{yy} = \sigma_{zz}$. In this case the normal stress on any plane equals the mean stress and the differential stress is zero. Inevitably, the application of a far field tensional stress will lead to failure initiating at the surface, in a material which obeys the Mohr-Coulomb failure criterion, because the normal stress is lowest in this region. An alternative assumption is that of zero horizontal strain under a vertical lithostatic load [Price, 1966; p. 69]. In this case, assuming a Poisson's ratio for the crust of 0.25, $\sigma_{xx} = \sigma_{yy} = \sigma_{zz}/3$. A ratio of $\sigma_{zz}/\sigma_{xx} = 3$ is compatible with incipient normal faulting at all depths on optimal planes with a coefficient of friction of approximately 0.6 [*Zoback and Healy*, 1984]. This is consistent with the concept of self-organized criticality in crustal deformation which suggests that the state of stress is everywhere close to the failure threshold [*Bak and Tang*, 1989; *Sornette, D. et al.*, 1990]. Thus, in this ideal theoretical scenario, there should be no preferential depth of nucleation.

At mid-ocean ridges the stress regime is more complex because both dyke injection and faulting occur. The stress field allowing dyking (i.e. tensile fracture) is significantly lower than that for required for shear failure. Rather than invoking alternating periods of magmatic accretion and faulting, Poliakov and Buck (this volume) show how the two processes can co-exist if the faults initiate in the mantle lithosphere.

FAULT SCALING RELATIONSHIPS AT MID-OCEAN RIDGES

The aim of this section is to review the available data on fault populations formed at mid-ocean ridges, including Iceland, and to compare these data sets with those compiled for continental faults described above. Assessing the similarities and differences yields new information about the underlying physical processes controlling fault formation in general. The discussion will focus on the normal faults which develop at the ridge axis, subparallel to the plate boundary. This brittle deformation is an integral part of the sea floor spreading process at all spreading rates and leads to the formation of significant sea floor topography [*Macdonald*, 1982]. At fast-spreading ridges the maximum fault scarp height is of the order of a few hundred meters whereas at slow-spreading ridges scarp heights of >1 km are common. Detailed interpretations of these faults have been derived from sonar backscatter images and bathymetric maps of the ocean floor and the parameters which have been systematically measured include the following: fault length, L, fault scarp height, h (which is a measure of fault throw, and thus displacement, along dip slip faults), and fault spacing, S.

There are a number of complications in the interpretation of oceanic fault distributions. There is the problem of limited resolution that affects all remote sensing data sets to some extent. Furthermore, near the ridge axis fault scarps may be partially or totally obscured by volcanic flows, particularly at fast-spreading ridges [see for example *Macdonald et al.*, 1996]. The volcanic flows may exploit zones of linkage in the fault scarps facing away from the axis (outward-facing faults), thus obscuring the continuity and true length of the structures (Figure 11). Accumulation of lavas in the hanging-walls of inward-facing faults means that scarp heights observed at the surface may not directly reflect tectonic displacements. This is a problem at fast-spreading ridges in particular where volcanic flows are more extensive (Figure 11). Further off axis, sediment

deposition obscures the scarps of smaller faults at all spreading rates. Finally, talus ramps develop along the base of active fault scarps as they degrade, making the dip of the faults difficult to constrain (Figure 12). These processes will influence measurements of spacing, displacement and length.

In spite of these problems with interpretation, several detailed fault population studies have been completed [e.g., *Carbotte and Macdonald*, 1994; *Searle*, 1984; *Cowie et al.*, 1994 and references therein]. Such studies have been possible with the advent of high-resolution sidescan sonar imaging. The fault scarps show up as bright acoustic reflections and a great deal of detail about fault scarp morphology and linkage geometry may be observed [*Allerton et al.*, 1995; *McAllister and Cann*, 1996] (Figure. 12). The most significant difference between these oceanic fault populations and those observed on the continents is that the size frequency distributions of h, S, and L are best described by exponential functions (Figure 13). Independent workers have obtained similar exponential distributions from different parts of the mid-ocean ridge system using different types of data. The significance of exponential scaling is that a characteristic length-scale, for example average fault length $<L>$, can be defined whereas a fractal distribution has no intrinsic scale. Taken at face value, this evidence suggests that fault formation at ridges is occurring in a fundamentally different way from that in continental rift zones. The only example, to my knowledge, of a power law length distribution for faults within the oceanic rift system was found by *Gudmundsson* [1987a,b] working in Iceland, although the range of lengths sampled in this study was limited to between several tens of meters and a few kilometers.

Width of the Fault Generation Zone

One explanation for the exponential distributions comes from the general discussion of the *Growth of Fault Populations* presented above. In that section, fractal fault patterns are explained in terms of short-range and long-range elastic interactions between many faults forming simultaneously. These interactions allow spatial correlations in the fault pattern to develop. If the width of the fault generation zone at mid-ocean ridges is relatively narrow, then only a small number of faults are active for a short period of time, perhaps even just one fault. In this case each new fault, or "package" of contemporaneous faults evolves independently and is unrelated to preceding fault formation episodes. According to such a model, sequential fault formation episodes would produce an exponential population of faults on the ridge flanks [*Malinverno and Cowie*, 1993]. Fractal statistics therefore might be limited to scales smaller than the width of the active rift zone over which simultaneously growing faults will interact. Examples at this scale might include the fault trace itself, antithetic faults, fault splays, and en echelon

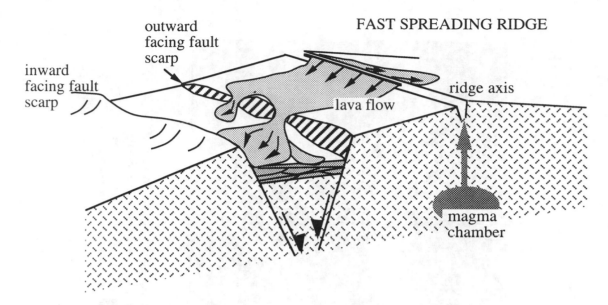

Figure 11. Sketch illustrating the growth fault model of Macdonald et al. [1996] for the East Pacific Rise near 9°N showing the influence of active faulting on volcanic flows from the axis.

fractures emanating from a fault tip. This interpretation could explain the power law distributions observed in the Icelandic rifts by *Gudmundsson* [1987a, b].

There is, however, no consensus in the literature over the width of the active zone of faulting at mid-ocean ridges. At slow-spreading ridges, the first normal faults form within 2-3 km of the neovolcanic zone and active faulting clearly continues out to distances of 8-10km, indicated by fresh talus fans and earthquake activity [*Macdonald and Luyendyk*, 1977; *Kong et al.*, 1988; *Lin and Bergman*, 1990, *McAllister and Cann*, 1996]. The crestal mountains of the Mid-Atlantic Ridge continue to evolve morphologically with the formation of tilted and/or faulted slopes dipping away from the rift axis at distance of 10-20km [e.g., *Macdonald and Luyendyk*, 1977]. In contrast, evidence from the slow spreading Asal rift in Djibouti shows that only faults within ±2km of the rift axis were active simultaneously during the 1978 rifting event [*Stein et al.*, 1991]. Holocene activity in Asal affected only those faults located ≤4 km to the southwest and ≤2 km to the northwest of the axis [see also *De Chabalier and Avouac*, 1994]. These width estimates are considerably less than that inferred for the Mid-Atlantic Ridge.

The evidence from fast-spreading ridges is also inconclusive. *Lee and Solomon* [1995] presented an analysis of fault development along the East Pacific Rise using SeaBEAM data, and estimated a width of several tens of kilometers [see also *Macdonald et al.*, 1996]. Thermal models for fast spreading ridges predict relatively uniform plate thicknesses across the axis, supporting the argument that active faulting may not be confined to a narrow zone of young lithosphere [*Shaw and Lin*, 1996]. However,

using SeaMARC II data also from the East Pacific Rise, *Edwards et al.* [1991] argued that the maximum width was only of the order of a few kilometers. Detailed observations from deep sea dives presented by *Macdonald et al.* [1996] suggest that faults are active up to at least 6 km from the East Pacific Rise axis. The evidence for this comes from the distribution of lava flows which are seen to drape over outward-facing fault scarps and pond against inward-facing faults (those facing towards the axis). Whether the ponding of flows against inward facing faults is evidence for them being active is not clear, as the flows may be infilling previously formed topography.

A narrow zone of active faulting is not the only explanation for exponential fault statistics. An alternative is that the thickness of the brittle crust rather than the width of the active zone is the controlling length-scale. Analogue experiments of extensional deformation, using a thin layer of wet clay over a latex sheet, initially develop a power law fault population but the distribution evolves into an exponential once the largest faults have propagated through the clay layer [*Ackermann et al.*, 1997]. The finite layer thickness limits the maximum fault size that can develop and further deformation is taken up by the formation of new small faults. The fault spacing gradually decreases through time and becomes more uniform during these experiments.

Either explanation of exponential size-scaling depends on fault interaction being suppressed but for different reasons determined by a physical length scale in each case: the narrow width of the fault generation zone, or the brittle layer thickness. The first explanation may be more applicable to slow-spreading ridges where there are

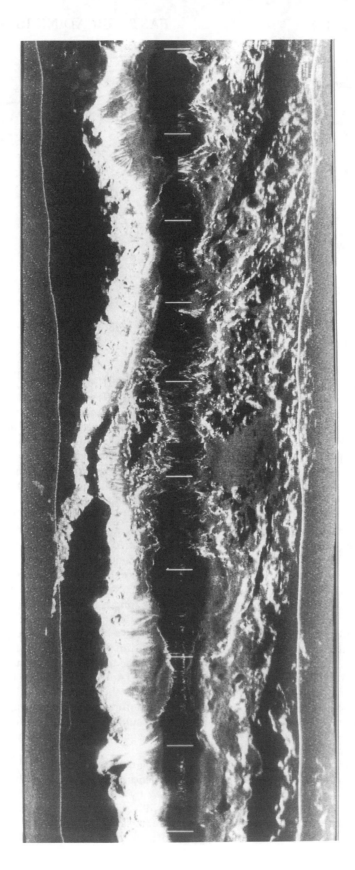

Figure 12. (a) Side scan sonar image and (b) line drawing interpretation of a fault linkage structure on the east flank of the Mid-Atlantic Ridge near 29°N at a distance of 15 km from the neovolcanic ridge. Note the accumulations of talus against the fault scarps *A* and *B*, in some places covered in sediment indicating that the fault has not recently been active. *R* indicates the relay ramp formed in the overlap of the two fault segments. Figure modified from *Searle et al.* [1998].

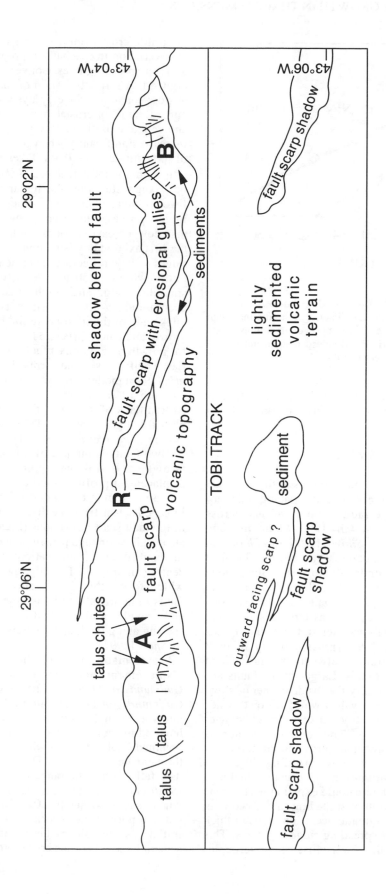

Figure 12. Continued.

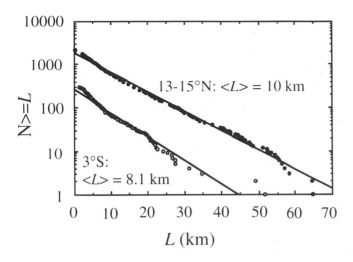

Figure 13. Fault length-frequency distributions of fault populations in two areas of the East Pacific Rise and the mean fault lengths calculated for each of these distributions. Data derived from Gloria and SeaMARC II sidescan sonar images the East Pacific Rise [from *Cowie et al.*, 1994].

significant temperature gradients across the axis. At fast-spreading ridges, it may be the brittle layer thickness that is more important.

Fault Development and d-L Scaling

Thermal structure, composition, and crustal thickness are the main factors thought to influence the spacing and scarp heights of mid-ocean ridge faults both along and across axis [*Shaw*, 1992; *Shaw and Lin*, 1993; *Shaw and Lin*, 1996; *Escartin et al.*, 1997]. *Shaw* [1992] and *Shaw and Lin* [1993] noted the strong contrast in fault development between the center and ends of slow-spreading ridge segments which they attributed to lateral changes in thermal structure. Fault scarps are smaller and more closely spaced at the segment centre, whereas large displacement faults with a wider spacing characterize the segment ends (Figure 14; *Searle et al.* [1998]). Asymmetry in fault development is also common between the inside and outside corners of segment ends. Larger throw faults and more complex geometries typify the inside corner of slow-spreading ridges, with the first valley wall fault on this side having a scarp of height typically in excess of a kilometer [Shaw, 1992] (Figure 14). The outside corner is characterized by many more small throw faults. *Shaw and Lin* [1996] compared fast and slow spreading ridges and concluded that the combination of crustal thickness variations and thermal structure could explain the change in style of faulting as a function of spreading rate. According to their model, focused magmatic accretion leads to thick crust at the centre of slow-spreading ridge segments. This results in a weak zone in the lower crust which decouples

the brittle crust from the upper mantle, causing an along-axis variation in mechanical properties that does not arise at faster spreading rates. Conversely, *Escartin et al.* [1997] argued that serpentinisation of the oceanic crust, common at slow-spreading ridges, plays a more important role than either thermal or crustal thickness variations. These authors showed that both fault spacing and fault scarp height increase significantly when the low-coefficient of friction of serpentinite ($\mu_f \sim 0.3$) is taken into consideration.

In terms of fault lengths, *McAllister and Cann* [1996] have suggested that the linear geometry of ridge axes favours linkage between co-linear small faults and fissures in young weak lithosphere. Thus, long ridge-parallel faults may develop easily. For example, the abrupt appearance of the first axial valley bounding fault seen at slow-spreading ridges may be reflecting the critical threshold for linkage of faults which nucleated at different points along the spreading segment. Such linked structures can then accommodate all the brittle strain until cooling and thickening of the lithosphere inhibits its lateral and vertical growth and a new structure forms nearer the axis. Evidence for growth by along-axis fault linkage also comes from the East Pacific Rise, and forms the basis for *Macdonald et al.*'s [1996] model for abyssal hill formation.

Figure 12 shows an example of fault linkage from the eastern flank of the Mid-Atlantic Ridge near 29°N, imaged with the high-resolution TOBI sidescan sonar surveying system [*Searle et al.*, 1998]. The zone of overlap between the two faults is marked by a relay ramp, *R* (Figure. 12b). Growth by linkage typically results in irregular displacement profiles (Figure 7), and in some cases lower *d/L* ratios (Figure 6). Evidence for along-strike (along-axis) linkage is clear from the discontinuous and variable widths of the fault scarp reflections (Figure 14). The inside corner also exhibits linkage perpendicular to the axis, indicated by the hooked lateral terminations of faults near the segment center (Figure 14). The ridge-perpendicular fault segment which forms the "hook" allows displacement on these faults to be transfered to an adjacent fault further off-axis and provides a mechanism to explain the change in fault spacing along the length of slow-spreading segments noted by *Shaw* [1992].

A systematic study of the correlation between fault displacement and length has been carried out by *Gudmundsson* [1992] for the Holocene rift zone of Iceland. *Gudmundsson* [1992] found a linear correlation between throw and length for normal faults ranging between several hundred meters and about 10 km in length. The γ value he obtained is about 0.003 which is significantly lower than the estimates obtained by *Dawers et al.* [1993]. The Iceland data falls within the range of γ values observed globally [*Schlische et al.*, 1996], but is lower than the average value, which is approximately 0.01. *Gudmundsson* [1987b] points out that normal faults forming in the rift zones of Iceland initiate as tension fractures and can achieve lengths of several hundreds of meters before accumulating any throw.

The formation of tension fractures is due to the high cohesive strength of the basaltic lava flows. *Hatton et al.* [1994] showed that the apertures of the tension fractures also scale with fracture length. As a result of this early growth history the Iceland faults typically "gape" at the surface, although as fault throw accumulates the surface opening gradually diminishes. Linkage also plays a dominant role in the development of these fault arrays *Gudmundsson* [1987b].

Cowie et al. [1994] used a combination of SeaMARC I and GLORIA sidescan data, plus high-resolution bathymetry profiles, to construct a fault scarp height-length correlation plot for the East Pacific Rise near 3°S and 12°N (Figure 15). Although large uncertainties are inherent in their analysis, *Cowie et al.* [1994] concluded that d and L are correlated, and the average d/L ratio they inferred was about 0.007. *Carbotte and Macdonald* [1994] found a similar value (0.008) in their analysis of fault populations on the East Pacific Rise using SeaMARC II and high resolution bathymetry. All of these data sets show at least a factor of two variability about the average d/L value, but they are on the lower side of the range exhibited by the global fault data set [*Schlische et al.*, 1996].

Subsurface Constraints on Fault Geometry

Iceland has yielded some information on three-dimensional fault plane geometry and orientations [*Forslund and Gudmundsson*, 1992]. These workers carried out a detailed study of normal faults exposed in the Tertiary and Pleistocene lava series of western Iceland. In this area, faults are exposed at about 1km below the original land surface, due to erosion. In contrast to the presently active zones of rifting where the faults planes are vertical at the earth's surface, the faults mapped by *Forslund and Gudmundsson* [1992] dip at between 70° and 75° on average, but nowhere were fault dips of less than 50° found. Faults with larger throws tend to show shallower fault dips but the variation observed is only about 5°. Normal faults mapped in ophiolites indicate that much shallower dips may be possible. For example, *Varga* [1991] mapped listric detachment faults in the Troodos ophiolite which flatten out at the contact between the sheeted dykes and the intrusive sequence. It is not clear whether these formed *in situ* at the ridge axis, or later as the ophiolite was obducted and emplaced. Evidence from slow-spreading centers indicates that dips of <45° may be quite common, particularly where the crust is serpentinised [*Escartin and Lin*, 1995; *Mutter and Karson*, 1992, see also paper by J. Karson this issue].

Seismic Versus Aseismic Faulting

Another aspect of faulting at mid-ocean ridges is the evidence for significant aseismic slip, especially at fast-spreading ridges [*Riedesel et al.*, 1982; *Cowie et al.*,

1993b]. The seismicity that has been detected along the East Pacific Rise, using ocean bottom seismometers, is of very low magnitude and occurs mostly in zones of complexity such as overlapping spreading centres [*Wilcock et al.*, 1992]. The seismic fault strain (i.e. the seismic moment release on faults) in these areas is only a few percent of the total fault strain [*Cowie et al.*, 1993b]. Larger events, up to magnitude 5.0-6.0, have been observed teleseismically from slow-spreading ridges [*Huang and Solomon*, 1988] and on-site studies have also shown significant levels of microseismicity near the axis [e.g, *Toomey et al.*, 1985; *Wolfe et al.* 1995]. Available data from the Mid-Atlantic ridge indicates that the seismic moment release is comparable to the observed fault strain, e.g., *Macdonald and Luyendyk* [1977], *Solomon et al.* [1988], *Cowie et al.* [1993b]. However, earthquake catalogues represent the activity for a short, possibly unrepresentative, period of geological time and on-shore analogues suggest that aseismic faulting may be significant even at slow spreading ridges. *Stein et al.* [1991] calculated the ratios of seismic to aseismic deformation during the 1978 rifting event in the Asal Rift, Djibouti. These workers estimated that only 8-28% of the geodetically measured extension was released seismically. *Stein et al.*'s [1991] inversion of the geodetic data included contributions from both dykes and faults. If the slip on normal faults is considered separately, then the seismic moment released during this event was still much less (~30%) of the total fault related extension, i.e., many faults must have moved aseismically.

Experimental work carried out on oceanic rock types has shown that common serpentine minerals such as antigorite and lizardite tend to exhibit stable, as opposed to stick-slip, sliding in frictional sliding experiments. *Reinen et al.* [1994] have shown that unstable slip can only occur in these mineral phases at high sliding velocities. Thus seismic slip is unlikely to initiate in serpentinized zones although ruptures initiated within unserpentinized portions could continue to propagate. Serpentinization, plus pore-pressure pressure fluctuations, caused by flux of fluids through the oceanic crust will influence fault development significantly, particularly in the chemically reactive environment of hydrothermal circulation [*Escartin et al.*, 1997].

SUMMARY AND CONCLUSIONS

Fault populations in a wide range of tectonic settings exhibit systematic scaling properties which can be characterized statistically. These properties include the spatial pattern of faulting, the displacement and length size-frequency distributions, and the ratio of displacement to length of individual structures. Theoretical and analogue models which reproduce these observations rely on a lithospheric rheology which is elastic-brittle at shallow depths and is furthermore heterogeneous. Deeper

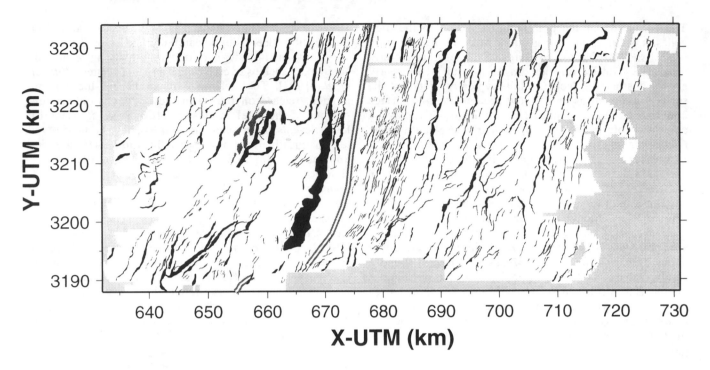

Figure 14. Fault interpretation of the axial region of the Mid-Atlantic Rudge near 29°N based on side scan sonar images obtained using the deep tow system TOBI. The width of the fault scarps is indicated by black areas and is a measure of the displacement on the faults. Figure modified from *Searle et al.*, [1998].

lithospheric layers which behave in a ductile manner may play a significant modifying role, and certainly influence the three-dimensional geometry of crustal scale faults. Modelling of fault displacement profiles has been used to estimate rock properties and the distribution of stress in the shallow crust. The size-frequency distribution and spatial clustering of fault populations are determined by the rates of fault nucleation, growth and linkage which in turn depend on the total amount of strain and the kinematics of the deformation. The systematics of fault patterns have been interpreted as an example of self-organisation in natural systems, reflecting the complex interplay between brittle rupture and the long-range nature of the elastic strain field. Comparing oceanic with continental fault populations reveals important differences in terms of: (a) size-frequency distributions which are exponential rather than power law, (b) fault scarp spacing which is more uniform (less clustered), (c) d/L ratios that tend to be slightly lower than the global average, and (d) significant fault movement which occurs without producing large earthquakes, especially on fast-spreading ridges. These differences may be attributed to the thermal structure of the ridge axis (both along- and across-axis), the elongate geometry of spreading centres which enhances linkage, and the petrological thickness of oceanic crust. The mafic crustal composition, and hydrous alteration of mafic minerals to serpentinite,

play a significant role in determining both the initation of faults and the subsequent localisation of brittle strain.

Acknowledgements. I would like to thank the organizing committee of the 1995 Ridge Theoretical Institute for inviting me

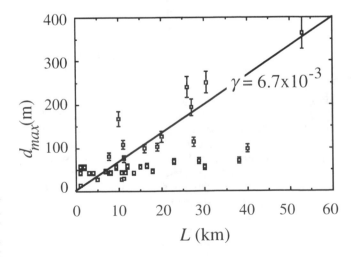

Figure 15. Correlation between maximum displacement and length derived from SeaMARC I sidescan sonar data near 12°N on the East Pacific Rise [*Cowie et al.*, 1994].

to attend a very stimulating meeting and for giving me the opportunity to expand my ideas on fault evolution via this lengthy article. Jian Lin and Suzanne Carbotte provided useful reviews of the manuscript. P. Cowie is supported by a University Research Fellowship from The Royal Society of London.

REFERENCES

Ackermann, R. V., R. W. Schlische, and M. O. Withjack, Systematics of an evolving population of normal faults in scaled physical models, *Geol. Soc. Amer. Abstracts with Programs*, *29*, no. 6, 198, 1997.

Allerton, S., B. J. Murton, R. C. Searle, and M. Jones, Extensional faulting and segmentation of the Mid-Atlantic Ridge North of the Kane Fracture Zone (24°N to 24°40'N), *Mar. Geophys. Res.*, *17*, 37-61, 1995.

Anders, M. H., and D. V. Wiltschko, Microfracturing, paleostress and the growth of faults, *J. Struct. Geol.*, *16*, 795-816, 1994.

Bak, P., and C. Tang, Earthquakes as a self-organized critical phenomenon, *J. Geophys. Res.*, *94*, 15635-15637, 1989.

Burgmann, R., D. D. Pollard, and S. J. Martel, Slip distributions on faults: effects of stress gradients, inelastic deformation, heterogeneous host-rock stiffness, and fault interaction, *J. Struct. Geol., 16*, 1675-1690, 1994.

Cann, J. R., D. K. Blackman, D. K. Smith, E. McAllister, B. Janssen, S. Mello, E. Avgerinos, A. R. Pascoe, and J. Escartin, Corrugated slip surfaces at ridge-transform intersections on the Mid-Atlantic Ridge, *Nature*, *385*, 329-332, 1997.

Carbotte, S. M., and K. C. Macdonald, Comparison of seafloor tectonic fabric at intermediate, fast and superfast spreading ridges: Influence of spreading rate, plate motions, and ridge segmentation on fault patterns, *J. Geophys. Res.*, *99*, 13609-13631, 1994.

Cartwright, J. A., B. D. Trudgill, and C. S. Mansfield, The growth of faults by segment linkage: Evidence from the Canyonlands grabens of S. E. Utah, *J. Struct. Geol.*, *17*, 1319-1326, 1995.

Cartwright, J. A., C. S. Mansfield, and B. D. Trudgill, Fault growth by segment linkage, in *Modern Developments in Structural Interpretation*, Geol. Soc. Lond. Spec. Pub., vol. 99, edited by P. C. Buchanan and D. A. Nieuwland, pp. 163-177, 1996.

Childs, C., J. J. Walsh, and J. Watterson, A method for estimation of the density of fault displacements below the limits of seismic resolution in reservoir formations, in: *North Sea Oil and Gas Reservoirs - II*, edited by A. T. Buller, E. Berg, O. Hjelmeland, O. Torsaeter and J. O. Aasen, pp. 309-318, Graham and Trotman, London, 1990.

Childs, C., J. Watterson, and J. J. Walsh, Fault overlap zones within developing normal fault systems, *J. Geol. Soc. Lond.*, *152*, 535-549, 1995.

Cladouhos, T. T., and R. Marrett, R., Are fault growth and linkage models consistent with powerlaw distributions of fault length? *J. Struct. Geol.*, *18*, 281-194, 1996.

Clark, R. M., and S. J. D. Cox, A modern regression approach to determining fault displacement-length scaling relationships, *J. Struct. Geol.*, *18*, 147-154, 1996.

Cowie, P. A., and C. H. Scholz, Displacement-length scaling relationship for faults: Data synthesis and discussion, *J. Struct. Geol.*, *14*, 1149-1156, 1992a.

Cowie, P. A., and C. H. Scholz, Physical explanation for displacement-length relationship for faults using a post-yield fracture mechanics model, *J. Struct. Geol.*, *14*, 1133-1148, 1992b.

Cowie, P. A., C. Vanneste, and D. Sornette, Statistical Physics Model for the Spatio-temporal Evolution of Faults, *J. Geophys. Res.*, *98*, 21809-21822, 1993a.

Cowie, P. A., C. H. Scholz, A. Malinverno, and M. H. Edwards, Fault Strain and Seismic Coupling on the East Pacific Rise, *J. Geophys. Res.*, *98*, 17911-17920, 1993b.

Cowie, P. A., A. Malinverno, W. B. F. Ryan, and M. H. Edwards, Quantitative Fault Studies on the East Pacific Rise: A Comparison of Sonar Imaging Techniques, *J. Geophys. Res.*, *99*, 15205-15218, 1994.

Cowie, P. A., D. Sornette, and C. Vanneste, Multifractal scaling properties of a growing fault population, *Geophys. J. Int.*, *122*, 457-469, 1995.

Cowie, P. A., A healing-reloading feedback control on the growth of seismogenic faults, *J. Struct. Geol.*, in press, 1998.

Davy, P., A. Sornette, and D. Sornette, Some consequences of a proposed fractal nature of continental faulting, *Nature*, *348*, 56-58, 1990.

Davy, P., A. Hansen, E. Bonnet, and S-Z. Zhang, Localisation and fault growth in layered brittle-ductile systems: Implications for deformation of the continental lithosphere, *J. Geophys. Res.*, *100*, 6281-6289, 1995.

Dawers, N. H., M. H. Anders, and C. H. Scholz, Growth of normal faults: displacement-length scaling, *Geology*, *21*, 1107-1110, 1993.

Dawers, N. H., and M. H. Anders, Displacement-length scaling and fault linkage, *J. Struct. Geol., 17,* 607-614, 1995.

DeChabalier, J-B., and J-P. Avouac, Kinematics of the Asal Rift (Djibouti) determined from the deformation of Fieale Volcano, *Science, 265*, 1677-1681, 1994.

Edwards, M. H., D. J. Fornari, A. Malinverno, W. B. F. Ryan, and J. Madsen, The regional tectonic fabric of the East Pacific Rise from 12°50'N to 15°10'N, *J. Geophys. Res.*, *96*, 7995-8018, 1991.

Elliott, D., The energy balance and deformation mechanisms of thrust sheets, *Phil. Trans. Roy. Soc. Lond.*, *A283*, 289-312, 1976.

Escartin, J. and J. Lin, Ridge offsets, normal faulting, and gravity anomalies of slow spreading ridges, *J. Geophys. Res.*, *100*, 6163-6177, 1995.

Escartin, J., G. Hirth, and B. Evans, Effects of serpentinization on the lithospheric strength and the style of normal faulting at slow-spreading ridges, *Earth and Planet. Sci. Letts.*, *151*, 181-189, 1997.

Forslund, T., and A. Gudmundsson, Structure of Tertiary and Pleistocene normal faults in Iceland, *Tectonics*, *11*, 57-68, 1992.

Fossen, H., and R. H. Gabrielsen, Experimental modeling of extensional fault systems by use of plaster, *J. Struct. Geol.*, *18*, 673-687, 1996.

Fossen, H., and J. Hesthammer, Geometric analysis and scaling relations of deformation bands in porous sandstone, *J. Struct. Geol.*, *19*, 1479-1494, 1997.

Gauthier, B. D. M., and S. D. Lake, Probabilistic modeling of faults below the limit of seismic resolution in Pelican Field, North Sea, offshore United Kingdom, *Bull. Am. Assoc. Pet. Geol.*, *77*, 761-777, 1993.

Gillespie, P. A., J. J. Walsh, and J. Watterson, Limitations of displacement and dimension data for single faults and the consequences for data analysis and interpretation, *J. Struct. Geol., 14*, 1157-1172, 1992.

Gillespie, P. A., C. B. Howard, J. J. Walsh, and J. Watterson, Measurement and characterisation of spatial distributions of fractures, *Tectonophysics, 226,* 113-141, 1993.

Golombek, M. P., and G. E. McGill, Graben, basin tectonics and the maximum total expansion of the Moon, *J. Geophys. Res., 88*, 3563-3578, 1983.

Gudmundsson, A., Tectonics of the Thingvellir fissure swarm, SW Iceland, *J. Struct. Geol., 9*, 61-69, 1987a.

Gudmundsson, A., Geometry, formation and development of tectonic fractures on the Reykjanes Peninsula, SW Iceland, *Tectonophysics, 139*, 295-308, 1987b.

Gudmundsson, A., Formation and growth of normal faults at the divergent plate boundary in Iceland, *Terra Nova, 4,* 464-471, 1992.

Hatton, C. G., I. G. Main, and P. G. Meredith, Non-universal scaling of fracture length and opening displacement, *Nature, 367*, 160-162, 1994.

Heimpel, M., and P. Olsen, A seismodynamical model of lithospheric deformation: Development of continental and oceanic rift networks, *J. Geophys. Res., 101*, 16155-16176, 1996.

Hirata, T., Fractal dimension of fault systems in Japan: fractal structure in rock fracture geometry at various scales, *Pageoph., 131*, 157-170, 1989.

Huang, P. Y., and S. C. Solomon, Centroid depths of mid-ocean ridge earthquakes: Dependence on spreading rate, *J. Geophys. Res., 93*, 13445-13477, 1988.

Ishikawa, M., and K. Otsuki, Effects of strain gradients on asymmetry of experimental normal fault systems, *J. Struct. Geol., 17*, 1047-1053, 1995.

Jackson, J., and D. Mckenzie, The geometrical evolution of normal fault systems, *J. Struct. Geol., 5*, 471-482, 1983.

Kakimi, T., Magnitude-frequency relation for the displacement of minor faults and its significance in crustal deformation, *Bull. geol. Surv. Japan, 31*, 467-487, 1980.

Kanamori, H., and D. L. Anderson, Theoretical basis of some empirical relations in seismology, *Bull. seism. Soc. Am., 65*, 1075-1095, 1975.

King, G. C. P., The accommodation of large strains in the upper lithosphere and other solids: The geometrical origin of seismic b-value, *Pageoph., 121*, 761-816, 1983.

Knott, S. D., A. Beach, P. J. Brockbank, J. L. Brown, J. E. McCallum, and A. I. Welbon, Spatial and mechanical controls on normal fault populations, *J. Struct. Geol., 18*, 359-372, 1996.

Krantz, R. W., Multiple fault sets and three-dimensional strain, *J. Struct. Geol., 10,* 225-237, 1988.

Kong, S. L., R. S. Detrick, P. J. Fox, L. F. Maya, and W. B. F. Ryan, Morphology and tectonics of the Mark area from Sea Beam and Sea MARC I observations (Mid-Atlantic Ridge 23°N), *Mar. Geophys. Res., 10*, 59-90, 1988.

Korvin, G., *Fractal Models in the Earth Sciences*, Elsevier, 1992.

Lee, S-M, and S. C. Solomon, Constraints from Sea Beam bathymetry on the development of normal faults on the East Pacific Rise, *Geophys. Res. Lett., 22*, 3135-3138, 1995.

Lin, J., and E. A. Bergman, Rift grabens, seismicity, and the volcanic segmentation of the Mid-Atlantic Ridge: Kane to Atlantis Fractures Zones (abstract), *Eos Trans. AGU, 71*, 1572, 1990.

Lin, J., and E. M. Parmentier, Quasistatic propagation of a normal fault: a fracture mechanics model, *J. Struct. Geol., 10*, 249-262, 1988.

Little, T., Fault related displacement gradients and strain adjacent to the Awatere strike-slip fault in New Zealand, *J. Struct. Geol., 18*, 321-342, 1996.

Lockner, D. A., J. D. Byerlee, V. Kuksenko, A. Ponomarev, and A. Sidorin, Quasi-static fault growth and shear fracture energy in granite, *Nature, 350*, 39-42, 1991.

Macdonald, K. C., and B. P. Luyendyk, Deep tow studies of the structure of the mid-Atlantic Ridge crest near lat 37°N, *Bull. Geol. Soc. Amer., 88*, 621-636, 1977.

Macdonald, K. C., Mid-Ocean Ridges: Fine scale tectonic, volcanic and hydrothermal processes within the plate boundary zone, *Ann. Rev. Earth. Planet. Sci., 10*, 155-190, 1982.

Macdonald, K. C., P. J. Fox, R. T. Alexander, R. Pokalny, and P. Gente, Volcanic growth faults and the origin of Pacific abyssal hills, *Nature, 380*, 125-129, 1996.

MacMillan, R.A., *The orientation and sense of displacement of strike-slip faults in continental crust,* BSc Thesis, Carlton University, Ottawa, 1975.

Main, I. G., P. G. Meredith, P. R. Sammonds, and C. Jones, Influence of fractal flaw distributions on rock deformation in the brittle field, in *Deformation Mechanisms, Rheology and Tectonics, Geol. Soc. Lond. Spec. Pub.*, Vol. 54, edited by R. J. Knipe and E. H. Rutter, pp. 71-79, 1990.

Malinverno, A., and P. A. Cowie, Normal faulting and topographic roughness of mid-ocean ridge flanks, *J. Geophys. Res., 98,* 17921-17939, 1993.

Marrett, R., and R. W. Allmendinger, Estimates of strain due to brittle faulting: sampling of fault populations, *J. Struct. Geol., 13*, 735-738, 1991.

Marrett, R., and R. W. Allmendinger, Amount of extension on "small" faults: An example from the Viking Graben, *Geology,.20*, 47-50, 1992.

McAllister, E., and J. R. Cann, Initiation and evolution of boundary wall faults along the Mid-Atlantic Ridge, 25-29°N, in *Tectonic, Magmatic, Hydrothermal and Biological Segmentation at Mid-Ocean Ridges, Geol. Soc. Lond. Spec. Pub.*, Vol. 118, edited by C. J. MacLeod, P. A. Tyler and C. L. Walker, pp. 29-48, 1996.

McGrath, A., *Fault propagation and growth: a study of the Triassic and Jurassic from Watchet and Kilve, North Somerset,* MSc Thesis, Royal Holloway and Bedford New College, The University of London, 1992.

Moore, D. E., and D. A. Lockner, The role of microcracking in shear fracture propagation in granite, *J. Struct. Geol., 17*, 95-114, 1995.

Muraoka, H., and H. Kamata, Displacement distribution along minor fault traces, *J. Struct. Geol., 5*, 483-495, 1983.

Mutter, J. C., and J. A. Karson, Structural processes at slow-spreading ridges, *Science, 257*, 627-634, 1992.

Nicol, A., J. J. Walsh, J. Watterson, and C. Childs, The shapes, major axes orientations and displacement patterns of fault surfaces, *J. Struct. Geol., 18,* 235-248, 1996.

Opheim, J. A., and A. Gudmundsson, Formation and geometry of fractures and related volcanism of the Krafla fissure swarm, northeast Iceland, *Bull. Geol. Soc. Amer., 101*, 1608-1622, 1989.

Ouillon, G., C. Castaing, and D. Sornette, Hierarchical geometry of faulting, *J. Geophys. Res.*, *101*, 5477-5487, 1996.

Peacock, D. C. P., Displacements and segment linkage in strike-slip fault zones, *J. Struct. Geol.*, *13*, 1025-1035, 1991.

Peacock, D. C. P., and D. J. Sanderson, Displacements, segment linkage and relay ramps in normal fault zones, *J. Struct. Geol.*, *13*, 721-733, 1991.

Pickering, G., J. M. Bull, and D. J. Sanderson, Sampling powerlaw distributions, *Tectonophysics*, *248*, 1-20, 1995.

Poliakov, A. N. B., H. J. Herrmann, Y. Y. Podladchikov, and S. Roux, Fractal plastic shear bands, *Fractals*, *2*, 567-581, 1994.

Plescia, J. B., Graben and extension in northern Tharsis, Mars, *J. Geophys. Res.*, *96*, 18883-18895, 1991.

Price, N. J., *Fault and Joint Development in Brittle and Semi-Brittle Rocks*, Pergamon Press, Oxford, 1966.

Reinen, L. A., J. D. Weeks, and T. E. Tullis, The frictional behaviour of lizardite and antigorite serpentines: Experiments, constitutive models, and implications for natural faults, *Pageoph*, *143*, 317-358, 1994.

Riedesel, M., J. A. Orcutt, K. C. MacDonald, and J. S. McClain, Microearthquakes in the Black Smoker Hydrothermal Field, East Pacific Rise at 21°N, *J. Geophys. Res.*, *87*, 10613-10623, 1982.

Rippon, J. H., Contoured patterns of the throw and hade of normal faults in the Coal Measures (Westphalian) of northwest Derbyshire, *Proc. Yorkshire Geol. Soc.*, *45*, 147-161, 1985.

Schlische, R. W., S. S. Young, R. V. Ackerman, and A. Gupta, Geometry and scaling relationships of a population of very small rift-related normal faults, *Geology*, *24*, 683-686, 1996.

Scholz, C. H., and P. A. Cowie, Determination of total strain from faulting using slip measurements, *Nature*, *346*, 837-839, 1990.

Scholz, C. H., *The Mechanics of Earthquakes and Faulting*, Cambridge University Press, New York, 1990.

Scholz, C. H., N. H. Dawers, J.-Z. Yu, M. H. Anders, and P. A. Cowie, Fault growth and fault scaling laws: Preliminary Results, *J. Geophys. Res.*, *98*, 21951-21962, 1993.

Scotti, O., A. Nur, and R. Estevez, Distributed deformation and block rotation in three dimensions, *J. Geophys. Res.*, *96*, 12225-12243, 1991.

Searle, R., GLORIA survey of the East Pacific Rise near 3.5°S: Tectonic and volcanic characteristics of a fast-spreading mid-ocean ridge, *Tectonophysics*, *101*, 319-344, 1984.

Searle, R. C., P. Cowie, N. C. Mitchell, S. Allerton, J. Escartin, C. J. MacLeod, S. M. Russell, P. A. Slootweg, and T. Tanaka, Detailed evolution of a slow-spreading ridge segment: The Mid-Atlantic Ridge at 29°N, *Earth and Planet Sci. Lett.*, *154*, 167-183, *1998*.

Shaw, P. R., Ridge segmentation, faulting and crustal thickness in the Atlantic Ocean, *Nature*, *358*, 490-493, 1992.

Shaw, P. R., and J. Lin, Causes and consequences of variations in faulting style at the Mid-Atlantic Ridge, *J. Geophys. Res.*, *98*, 21839-21851, 1993.

Shaw, W. J., and J. Lin, Models of ocean ridge lithospheric deformation: dependence on crustal thickness, spreading rate, and segmentation, *J. Geophys. Res.*, *101*, 17977-17993, 1996.

Solomon, S. C., P. Y. Huang, and L. Meinke, The seismic moment budget of slowly spreading ridges, *Nature*, *334*, 58-60, 1988.

Sornette, A., P. Davy, and D. Sornette, Growth of fractal fault patterns, *Phys. Rev. Lett*, *65*, 2266-2269, 1990.

Sornette, A., P. Davy, and D. Sornette, Fault growth in brittle-ductile experiments and the mechanics of continental collisions, *J. Geophys. Res.*, *98*, 12111-12139, 1993.

Sornette, D., P. Davy, and A. Sornette, Structuration of the lithosphere as a self-organized critical phenomenon, *J. Geophys. Res.*, *95*, 17353-17361, 1990.

Sornette, D., and P. Davy, Fault growth model and universal fault length distribution, *Geophys. Res. Letts.*, *18*, 1097-1081, 1991.

Sornette, D., P. Miltenberger, and C. Vanneste, Statistical physics of fault patterns self-organized by repeated earthquakes, *Pageoph*, *142*, 491-527, 1994.

Stein, R. S., G. C. P. King, and J. B. Rundle, The growth of geological structures by repeated earthquakes 2: Field examples of continental dip-slip faults, *J. Geophys. Res.*, *93*, 13319-13331, 1988.

Stein, R. S., P. Briole, J-C. Ruegg, P. Tapponnier, and F. Gasse, Contemporary, Holocene, and Quaternary deformation of the Asal rift, Djibouti: Implications for the mechanics of slow-spreading ridges, *J. Geophys. Res.*, *96*, 21789-21806, 1991.

Toomey, D. R., S. C. Solomon, G. M. Purdy, and M. H. Murray, Microearthquakes beneath the median valley of the Mid-Atlantic Ridge near 23N: Hypocentres and focal mechanisms, *J. Geophys. Res.*, *90*, 5443-5458, 1985.

Turcotte, D. L., *Fractals and Chaos in Geology and Geophysics*, Cambridge University Press, 1992.

Varga, R. J., Modes of extension at oceanic spreading centres: evidence from the Solea graben, Troodos ophiolite, Cyprus, *J. Struct. Geol.*, *13*, 517-537, 1991.

Vignes-Adler, M., A. Le Page, and P. M. Adler, Fractal analysis of fracturing in two African regions, from satellite imagery to ground scale, *Tectonophysics*, *196*, 69-86, 1991.

Villemin, T., and C. Sunwoo, Distribution logarithmique self-similaire des rejets et longuers de failles: Examples du Bassin Houiller, Lorraine, *C. R. Acad. Sci. Paris.*, *305*, 1309-1312, 1987.

Villemin, T., J. Angelier, and C. Sunwoo, Fractal distribution of fault lengths and offsets: Implications of brittle deformation evaluation-Lorraine Coal Basin, in *Fractals and the Earth Sciences*, edited by C. C. Barton, P. R. LaPointe, Plenum Press, New York, pp. 205-226, 1995

Walsh, J. J. and J. Watterson, Distribution of cumulative displacement and of seismic slip on a single normal fault surface, *J. Struct. Geol.*, *9*, 1039-1046, 1987.

Walsh, J. J., and J. Watterson, Analysis of the relationship between displacements and dimensions of faults, *J. Struct. Geol.*, *10*, 239-247, 1988.

Walsh, J. J., and J. Watterson, Geometric and kinematic coherence and scale effects in normal fault systems, in: *The Geometry of Normal Faults*, Geol. Soc. Lond. Spec. Pub., Vol. 56, edited by A. M. Roberts, G. Yielding and B. Freeman, pp. 193-203, 1991.

Walsh, J. J., and J. Watterson, Fractal analysis of fracture patterns using the standard box-counting technique: valid and invalid methodologies, *J. Struct. Geol.*, *15*, 1509-1512, 1993.

Walsh, J. J., J. Watterson, and G. Yielding, The importance of small-scale faulting in regional extension, *Nature*, *351*, 391-393, 1991.

Watterson, J., Fault dimensions, displacement and growth, *Pageoph*, *124*, 365-373, 1986.

White, N. J., J. A. Jackson, and D. P. McKenzie, The relationship between the geometry of normal faults and that of the sedimentary layers in their hanging walls, *J. Struct. Geol., 8*, 897-909, 1986.

Wilcock, W. S. D., G. M. Purdy, S. C. Solomon, D. L. DuBois, and D. R. Toomey, Microearthquakes on and near the East Pacific Rise, 9-10°N, *Geophys. Res. Letts.*, *19*, 2131-2134, 1992.

Willemse, E. J. M., D. D. Pollard, and A. Aydin, Three-dimensional analysis of slip distributions on normal fault arrys with consequences for fault scaling, *J. Struct. Geol.*, *18,* 295-310, 1996.

Willemse, E. J. M., Segmented normal faults: Correspondence between three-dimensional models and field data, *J. Geophys. Res., 102,* 675-692, 1997.

Wolfe, C., G. M. Purdy, D. R. Toomey, and S. C. Solomon, Microearthquake characteristics and crustal velocity structure at 29N on the Mid-Atlantic Ridge: The architecture of a slow-spreading segment, *J. Geophys. Res.*, *100*, 24449-24472, 1995.

Wu, D., and R. L. Bruhn, Geometry and kinematics of active normal faults, South Oquirrh Mountains, Utah: Implications for fault growth, *J. Struct. Geol., 16,* 1061-1075, 1994.

Yielding, G., J. J. Walsh, and J. Watterson, The prediction of small-scale faulting in reservoirs, *First Break*, *10*, 449-460, 1992.

Yielding, G., Needham, and H. Jones, Sampling of fault populations using subsurface data: a review, *J. Struct. Geol.*, *18*, 135-146, 1996.

Zoback, M. D., and J. H. Healy, Friction, faulting and *in situ* stress, *Annales Geophysicae*, *2/6*, 689-698, 1984.

Patience A. Cowie, Department of Geology and Geophysics, The University of Edinburgh, West Mains Road, Edinburgh, EH9 3JW, UK; e-mail: cowie@glg.ed.ac.uk